Martinique Revisited

Gulf of Mexico
Florida
Bahama Islands
Cuba
ATLANTIC OCEAN
Yucatan
Cayman Islands
Dominican Republic
Haiti
Puerto Rico
Anguilla
Barbuda
Antigua
St.Croix
St.Kitts
Montserrat
Guadeloupe
Dominica
Jamaica
Bay Islands
Honduras
Martinique
Caribbean Sea
St.Lucia
St.Vincent
Barbados
Nicaragua
Aruba
Bonaire
Grenada
Curaçao
Tortuga
Margarita
Tobago
Trinidad
Costa Rica
Panama
Colombia
Venezuela
0 1000 km
78°
70°
62°
26°
18°
10°
Cartographics 1986

Clarissa Thérèse Kimber

Martinique Revisited

The Changing Plant Geographies of a West Indian Island

Texas A&M University Press

College Station

The paper used in this book meets the minimum requirements
of the American National Standard for Permanence
of Paper for Printed Library Materials, Z39. 48-1984.
Binding materials have been chosen for durability.

Library of Congress Cataloging-in-Publication Data
Kimber, C. T. (Clarissa Thérèse), 1929–
Martinique revisited.
Bibliography: p.
Includes index.
1. Phytogeography—Martinique—History. 2. Botany—Martinique—Ecology—History. 3. Vegetation dynamics—Martinique—History. 4. Man—Influence on nature—Martinique—History. 5. Martinique—History. I. Title.
QK231.M35K56 1987 581.97298'2 86-30203
ISBN 0-89096-297-9

To the memory of
my mother who taught me to look
at the landscape and
my father who insisted that it be
explained

Contents

Plates

Figures

Maps

Tables

Preface

A taxi ride from Fort-de-France to Basse-Pointe in early June of 1975 turned out to be an encounter that phenomenologists would cheer. The landscapes of my mind's eye, created in 1962–63 and cherished for almost fifteen years, were being shunted aside by new constructions which only partially resembled the old ones.

Traveling the freeway with five Martiniquais bound for the *paie plat* of the northeast coast, I am struck by the fact that the sugarcane fields I remembered are now second-growth logwood and acacia savanna. These lands are being left idle as speculators wait for urban and industrial expansion within the metropolitan area to raise the real value of the land. We bypass the old traffic bottleneck of Lamentin and take the broad, curved road recently built to modern standards for the eastern, windward side of the island. The Lamentin Plain still has sugarcane fields, but it is the giant sprinkler systems watering the large-scale banana plantings that I notice. As we climb the series of low hills connecting the ancient volcanic mornes of the south with the younger volcanic mountains of the north, I notice that provisions grounds and complex dooryard gardens are fewer than I had remembered. There is more of what my friends in Jamaica call "ruinate," second-growth shrublands. Through the harbor town of Trinité, with its waterfront now neatly bounded by masonry, we pass on to Ste. Marie. The sugar mill looks closed down, abandoned. Out into the countryside we go again. Very occasionally, row crops show up on the hillsides. There are provisions grounds, yes, but a new species has become important, eggplant or melongine, which I later learn was bound for the metropolitan market. The hills are covered with dry woodland containing lots of poirier or pink pouie and other sun-loving volunteering plant species. Only the valley bottoms along the small stream courses are planted in crops.

It is beginning to grow dark; the driver switches the car radio to a quiz program. Apparently everyone listens to quiz programs. Like those at home, the program is interrupted by the real business, advertisements. Soap, yogurt, and other consumer goods are advertised by supermarkets or by some local producer of beverages such as Bière Lorraine or Fanta. It is hard to see outside but, as we turn into the side road leading to the Plantation Leyritz, it is not sugarcane fields I am looking at but more bananas and large fields of pineapples on either side. I had expected the refurbished old plantation house but not the sight of banana leaves shredded by the Trade Winds.

It was clear on the first day of my return that the island revisited was a different place from that I had learned thirteen years before. I had come to witness the changes in plant geography. The changes were substantial. I wrote a paper on my findings after the 1975 trip. Now, after my most recent field session in 1985, I have written a book.

This book examines closely what had happened to the vegetated part of the rural landscapes of a West Indian island over the past 350 years. I have speculated on gross changes in the plant cover for 1,500 years before that. I have been more concerned with changes in boundaries of vegetation types than with phytosociological changes within the vegetation. However, the large number of introduced exotics have been traced as well as possible. The places these migrant species have taken in the vegetation are documented. Plant taxonomy and ecology, historical accounts, diaries, journals, and official records all contributed to this reconstruction of the past plant geographies of Martinique. I have been eclectic in my methodology.

The presentation of the plant geographical records is by periods which correspond to those critical to French Caribbean history. The first chapters set the stage in terms of the physical geography of the island. Then follow the chapters outlining, first, the human occupation with the settlement activities, the developing technologies, and the land use following from local and international forces. The second half of each chapter describes the results of these processes in the plant geography of the island for the period. Process and result thus are tied together for each segment of the island history. The last chapter integrates the results of processes by focusing on the impact of each process on different vegetation types. Afterward there is a discussion of the geography of change in vegetation for the island throughout the periods of modern Caribbean history.

I have attempted to be selective in illustration but detailed enough to witness the effects of the various agents and processes on the evolving plant geography of Martinique.

Acknowledgments

This work began more than twenty-five years ago when I started my studies in tropical plant geography. Many kind people helped me along the way, some of them far more than they ever realized. I should first thank those who gave me my early training in plant geography: Dr. Herbert Mason, Dr. John Curtis, and especially Dr. Jonathon Sauer. To the study of historical plant geography I carried an appreciation of the historical method inculcated by Dr. Carl O. Sauer and Dr. Andrew Clark. Dr. Clarence Olmsted's thoughtful questions about agricultural systems focused my attention on planted as well as on wild vegetation. Dr. Jonathon Sauer first proposed the study of Martinique as an example of a tropical setting where I might examine whether the plant ecological theories derived from the mid-latitudes really do work for the tropics. His help in the formation stages was enormous and I owe a great deal to him. Investigations proceeded through the years, honed by the observation and questionings of my colleagues Dr. Homer Aschmann, Dr. Carl Hansen, Dr. Christopher Salter, Dr. Robert Stanton, Dr. Rezneat Darnell, and Dr. Joseph Sonnenfeld. I could not have worked without the help of systematists Dr. Hugh Ilts and Dr. Richard Howard in identifying my specimens, and Dr. Jacques Fournet's impressive work on the flora of the French West Indies greatly facilitated the sorting of thorny questions of synonymy. Over the years the lessons of my mentors and colleagues have stood me in good stead.

Working in a small West Indies island is an expensive proposition and it was the generosity of funding sources which has made it all possible. A grant from the Office of Naval Research in 1962–63 permitted the first two seasons of field work in the Caribbean, most of which were spent in Martinique. Another grant, from the Texas A&M University's Dean of Geosciences Discretionary Fund in 1975, made possible an extended three months' study on the island during which time I was able to resurvey the sample areas originally surveyed in 1963. An academic study leave awarded by Texas A&M University permitted an extended stay in Martinique during the spring of 1985. These periods of research were supplemented by shorter stays in 1972 and 1979 funded as grants to myself. I owe much to those funding my field research.

In my own community many people helped me during my research and during the preparation of the manuscript. The helpful Inter-library-Loan staff of the Sterling C. Evans Library provided me with materials not available in Martinique. The staff of the Tracy Herbarium have curated my West Indian collections and helped in identifying specimens and resolving questions of synonymy. Christopher Müller-Wille contributed his sophisicated understanding of cartography to turn my maps into reasonable images. I thank him and his able staff for executing the maps and figures in the text. Dr. Ann Marie Elmquist helped me catch the elusive diacritical mark and put it on the correct letter. I am indebted to all for their help. For the errors that remain I accept full responsibility and hope my readers will point them out to me. Mrs. Reba Ragsdale and her staff supported me during the many versions of the text.

In Martinique many field sessions have been aided by local scholars and scientists. I thank Dr. and Mrs. Jacques Petitjean-Roget, Mr. Emile Hayot, Mr. Jacques Daudin, Dr. Henri Petitjean-Roget, and most especially Drs. Henri and Madelaine Stehlé. Great field companions sharpened my observational skills, and I thank Miss Lucienne Maurice, Mr. Jean-Pierre Fiard, and Dr. Jacques Portecôp. The latter graciously offered me space in his laboratory and facilitated field work in many ways during my most recent trip in 1985. Readers will find that I have used much of his ecological research. My archival and library studies were facilitated by the kind help of Alexandre Sénet, Liliane Chauleau, and Jean-Paul Hervieu, who made the departmental archives in Martinique and Guadeloupe open to me. Bernard Petitjean-Roget and Emile Hayot graciously shared many photocopies of texts made from the archives in Paris. Mr. V. Henry and Miss G. Leger gave me space in the Bibliothèque Schoelcher for my research. Henri Petitjean-Roget opened to me the library of the Musée Edgar Clerc and that of the Ecomuseum at Hon. Murat. Marie Galante. Mr. Jean-Pierre Fiard gave me several of his superb photographs to supplement those of my own. To all these I give my sincere thanks.

The hospitality of Mr. and Mrs. André Asselin, Mr. and Mrs. Charles de Lucy de Fossarieu, Dr. and Mrs. Jacques Petitjean-Roget, and Miss Lucienne Maurice made my early stays pleasant cultural experiences. The departmental officials from Martinique and the *métropole* were invariably courteous and helpful when I needed data, information, or that elusive reference. Most important, however, has been my experience over the years of the land and people of Martinique, *"Le pays des revenants,"* "The land of those who return."

Martinique Revisited

1 Introduction: Plants and People in Place through Time

The vegetation of the tropical island of Martinique has been subjected to human exploitation for more than two thousand years. Interference in the physical and biological processes has been so effective that the term *domesticated* may properly be applied to most of the island landscapes. This domestication has involved changes in the constitution of vegetation by the introduction of exotic species and the selective culling of native species. It has caused changes in the distribution of native vegetation by disturbance, clearing, and burning, and the creation of entirely alien vegetation types. The result is a plant cover composed of a few traditional plantation crops with a cosmopolitan tropical weed population inserted among remnants and modifications of wild New World tropical vegetation.

The major part of the island surface today is in either planted or spontaneous vegetation (see map 1). A smaller part is covered with planted vegetation in which there are many spontaneous elements. The present recorded pattern of planted, spontaneous, and mixed vegetation is, in part, a transitory one, an accident of current land use. Yet parts of the vegetation pattern are old, originating in the colonial periods or before. The relative persistence of some vegetation types in parts of the island and the turnover in vegetation types in other parts suggest that environmental characteristics leading toward stability are present in some areas and not in others. The geography of recent vegetation change in Martinique is the result of biological evolution, plant ecology, and human history. This book recounts the recovered history of the plant geography of Martinique.

A Reconstructed History

This account is basically chronological. It begins with a reconstruction of vegetation change based on paleogeology and ecology, and on island biogeographical theories. The following précis of that reconstruction sketches the main points that will be made in this book.

Raised by back-arc volcanic action far from the nearest continental land mass in the early Tertiary period, the island was partly or completely planed off by erosion and covered with marine sediments, only to be rebuilt as a high volcanic oceanic island in later geological time. During the Pleistocene epoch, changing sea levels, volcanism, and tec-

Map 1. Spontaneous and planted vegetation, Martinique, F.W.I.
Source: Compiled by author.

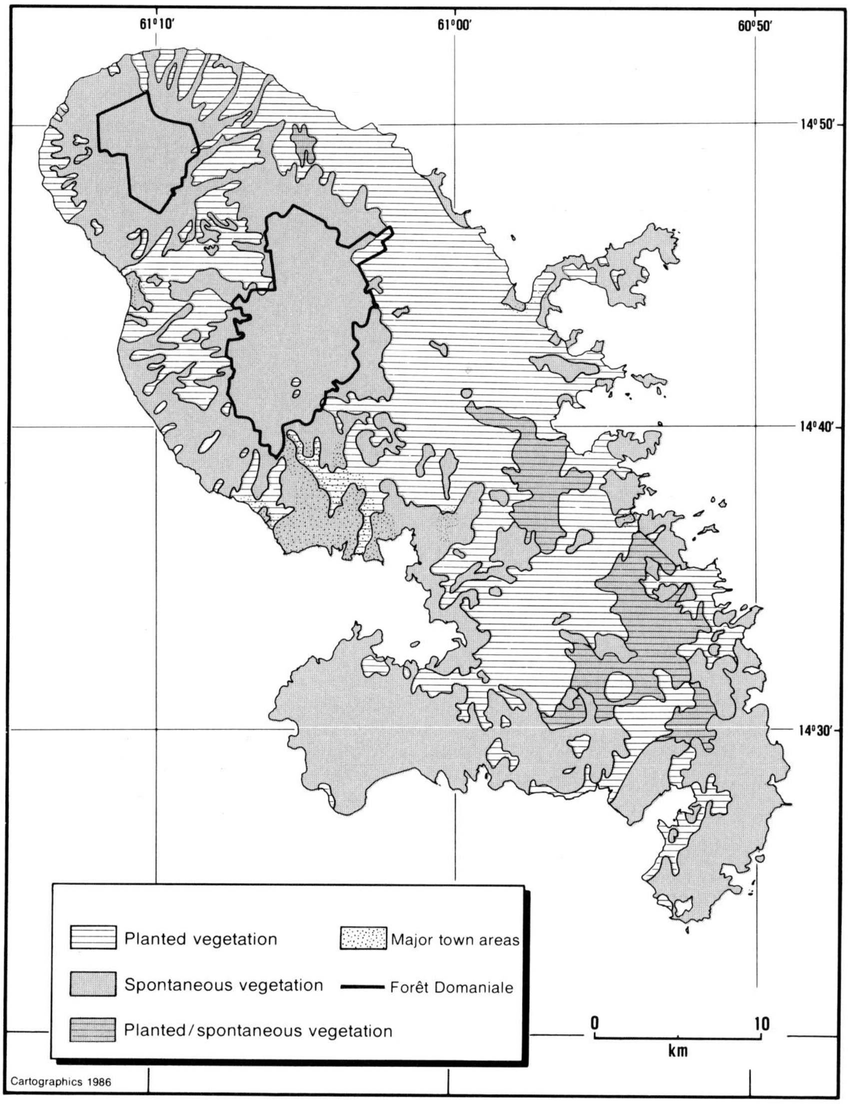

tonic uplift modified the island's size and coastal configuration. The tropical climate has continued throughout geological history, but during the periods of high sea stands the precipitation values may have been lower instead of higher, as the current evidence indicates cooler and more arid conditions in tropical regions during the oscillations of the Ice Age. Volcanic activity has continued into the twentieth century. Degradation under tropical climatic conditions began almost simultaneously with the earliest mountain building. Soil formation and colonization by waif flora and fauna began shortly thereafter, to continue to the present day.

Martinique, like new volcanic islands elsewhere, received an inoculation of numerous species by long-range dispersal as soon as a stable land mass was established. These included marine plants and animals (especially the reef corals) as well as terrestrial species. After the pioneer period, when only sun-loving species tolerant of raw mineral soil colonized, plant communities diversified and became more complex, and the number of species in the flora increased markedly.[1] As younger islands filled in the spaces between Martinque and the South American continent and the larger islands of the Greater Antilles, the rate of immigration of plant species increased.[2] The successful invaders came from all continents save one, long-distance dispersal being as effective as it can when there are so many agents and agencies of transport and a long period of time available for colonization to take place. However, some source areas have contributed more species than others. The colonization was a dynamic process likely accompanied by some turnover in species. As the build-up of species increased to a theoretical saturation level, elimination of new colonists was more common. Changes in immigration rates or creation of niches within the island vegetation modified the number of species in the potential equilibrium.

Because the visitation rates of new plant species—resulting from natural dispersal by ocean currents, birds, wind systems and storms, and waif animals—fluctuated during geologic periods, particularly during the Pleistocene, establishment rates varied. Changing coastal configurations with changes in sea stands during the Pleistocene caused modifications in coastal environments. Low-lying areas expanded and contracted, putting stress on the plant populations and perhaps accelerating biological evolution.[3] Steep marine slopes mean that the central islands of the Lesser Antilles were never connected, nor were the islands at either end of the chain ever connected to the North or South American continental land masses.

We may presume that after the early period of island formation an increase in habitats was not common. Some sort of dynamic equilibrium was set up by the late Holocene.[4] Tree fall gaps, hurricane swaths, and natural primary colonization on riverbank washouts and on volcanic ash and lava flows provided niches in which pioneer and early successional species persisted in the flora.[5] Exposure of marginal seafloors during low sea stands expanded island area that was later covered again and reduced in size. With time, adaptive radiation provided taxa, the endemics, for "unfilled," newly created niches.[6] Considered as an ecosystem, including the animals and physical components, Martinique approached a condition in which the greatest possible proportion of available ecological niches was filled and the fullest possible use of the resources made.[7] Stress and disturbance factors changed in relative importance as the incidence of impact changed through time.[8] The island was a self-contained, functioning system with few introductions, and of those introductions, more were rejected than were established.

With the advent of seafaring Amerindians, the relatively stable conditions of the island ecosystem were upset.[9] The first aboriginal populations were South Americans, bringing their livelihood activities of hunting, fishing, gathering, and agriculture. These migrants modified habitats, created new ones, and introduced new species. Whatever species equilibrium had existed was thrown into some imbalance. Humans set fires that introduced a newly important selection pressure on native species. Their domesticated species and associated weeds proved hardy competitors for the native species in the newly created environments. The rate of immigration of new species was accelerated. There is no reason to suppose that the elimination of species proceeded at the same rate as immigration, or indeed that it increased at all in the early periods of human occupation. Some wide-amplitude plants were circumscribed in habitat by more competitive, or rather, more specially adapted plants.

A dynamic equilibrium might or might not have become established during the aboriginal period. Certainly, regeneration was a characteristic part of the plant cover after colonization by the first islanders. The importance of the effect of the human immigration on the island's vegetation was a function of numbers and technology. Old agricultural plots were abandoned and new plots were cleared and burned, so plant succession and regression took place. But in terms of the island as a complete, open system, the new and old components, including humans, may have arrived at a self-replicating dynamic equilibrium with a virtual steady state of energy exchange (Margalef 1981). Assaults on the plant cover were balanced by regeneration processes.

However, one major change in the island had taken place. The island no longer could be thought of as a single unit neatly circumscribed by its own boundaries with sea and air. The seafaring people who became resident islanders were in regular, if not continuous, contact with other islands and the southern continent. Hence, part of the regular interactions of members of the system involved extraterritorial interactions. These functional relations were significantly different from the occasional waif inoculations of the preceding geologic periods, no matter how important those dispersals had been in constituting the pristine flora. The isolated ecosystem was no more.

The advent of the Europeans and Africans introduced further complications. There were even more extreme modifications of habitat and major invasions of plants and animals. Large grazing herbivores introduced an entirely new process for change in the island. Large areas were taken out of plant production entirely and converted to housing and industry. Fully as important for change as grazing and permanent buildings was the beginning of the export orientation of the agricultural system. Although there was some contribution from abroad in the form of advanced technology, capital, and people, considerable sums of energy in terms of calories were extracted and exported to Europe. As a *colonie d'exploitation,* the system felt a net loss for many years. The island was sustained by infusions of capital and by a hypothesized degree of "play" in the ecosystem, but symptoms such as epidemic disease in crop plants, the abandonment of estates after the uprising of the slaves in the southern peninsulas, and the enlargement of areas in secondary bushlands and desert cactus shrub all point to imbalances at different times.

The island continues in imbalance. The ecosystem today tends toward simplification. Only significant infusions of capital from metropolitan France in the form of direct subsidies, guaranteed quotas, and credit facilities shield the human population from great malaise and the plant cover from destruction.

Interpreting the Changes in the Plant Geography

Change occurs in response to assaults on the system from without and from within. The mechanisms or strategies for coping with changes are part of the structural and functional relationships among plant species that compose the vegetation and of the relationships between the vegetation and the environment. There are stresses on the vegetation arriving from outside the local site.[10] There are disturbances from within such as those caused by species population build-ups. There are the changes that occur over relatively long periods of time such as evolution, migration, land-sea oscillations, climatic variability, or volcanism associated with plate boundary motions. There are the transformations of natural systems under domestication. There are differences in scale that make real differences in vegetation response. The fall of a canopy tree opens a small area for colonization. A landslide carries away the whole slope, exposing a large area. The processes set in train are different for each scale.

It is a truism that in tropical islands the last great wave of colonists to occupy the oceanic islands was of human beings and that they have had the most important impact on vegetation in the short term.[11] Aboriginal peoples with their social organization, technologies, and values have had considerable effect on vegetation, and changes in the plant cover have been the consequence. The advent of Europeans with their social organization, technologies, and values had other effects. Much attention will be paid to the institutional contexts of land-use decisions, subliminal or conscious as the case may have been, that affected the way the plant cover was treated and how the vegetation responded.

Determining causes for changes in the plant cover of any island is difficult on several counts. One is a consequence of the nature of available empirical evidence and the difficulties in interpreting the historical record. Another is the difficulty of distinguishing between mere correlations in time and place and real causal connections between agent and process. There is little experimental evidence of the effects on vegetation of causes for change under controlled conditions except in regeneration studies. As a result, the processes must be inferred from the implications of the activities reported in the historical evidence. So far, these have not proved susceptible to testing.

Further complicating any analysis of causation is the fact that several controls or processes may have similar effects on the vegetation. Several processes did change in importance in Martinique at the same time, and assigning the relative importance of each has not been easy. It may be fruitless to try. Last, it should be recognized that the organization of natural processes and their interaction may not be analogous to causalities as we know them. We have proceeded as if they were, but the possibility that this sort of analysis is improper cannot be ignored.

If vegetation can be considered as a part of an ecosystem composed of living and nonliving components interacting directionally to produce a stable system, then one may consider modifications of the components or processes within the system as probable causes for change in the vegetation as well as for changes in the dynamic status of the system. Vegetation is a function of available habitats and existing flora, so modifications of either flora or habitats will lead directly to changes in the vegetation. These changes include retrogression, succession, or cyclic replacement.

In the succeeding chapters I report the effects of modifying the existing floral composition by selective utilization, plant introductions, and the elimination of suitable habitats. I also report the effects of processes instigated by humans that modified habitat conditions (clearing and burning of the plant cover, deforestation, grazing of the vegetation by introduced livestock, changes in crops grown, changes in technologies associated with agriculture, and industrialization).[12] For the same time periods, I recount the modifications of habitats made by natural catastrophes such as volcanic explosions and hurricanes.

Throughout the period during which these interferences with the stability of the system were being made, evolutionary processes continued to operate, although at rates modified by human interference. Degradation and aggradation processes in the physical landscape, continued long-range dispersal of propagules by agents other than humans, and the regenerative sequences characteristic of wild vegetation went on. Successional trends within the vegetation following the fallowing and abandonment of agricultural lands took place, as did regressive trends.

To find evidence for causes of change, I investigated the plant geography of Martinique by proceeding along several lines. The first investigation was an extended examination of the modern vegetation of Martinique, begun with field investigations in 1962–63 and continued intermittently until 1975, when another field season in the summer was possible. Since then, short periods in the field have been useful to clarify particular points. Climatological, soil, and ecological data collected on the island by French scientists have supplemented these plant investigations.

The second investigation was an examination of the historical record from primary and secondary sources. The local departmental archives have some documents from the colonization period, as well as others from the eighteenth, nineteenth, and twentieth centuries. Contemporary accounts of the vegetation, of plant introductions, and of landscapes provided the direct evidence for the landscape reconstructions at different periods and, in some cases, the identity of those responsible for the changes. The identification of periods was made in terms of island and Caribbean historical events rather than in terms of traditional European periods (Adelaide-Merland 1977,54).

The island is particularly well served by religious and economic historians who have carefully used materials from archival collections located in Paris, especially the Archives Nationales and the Archives de la Marine. Their works have been made available to me and greatly aided me in the reading of archival materials as well as the secondary sources.

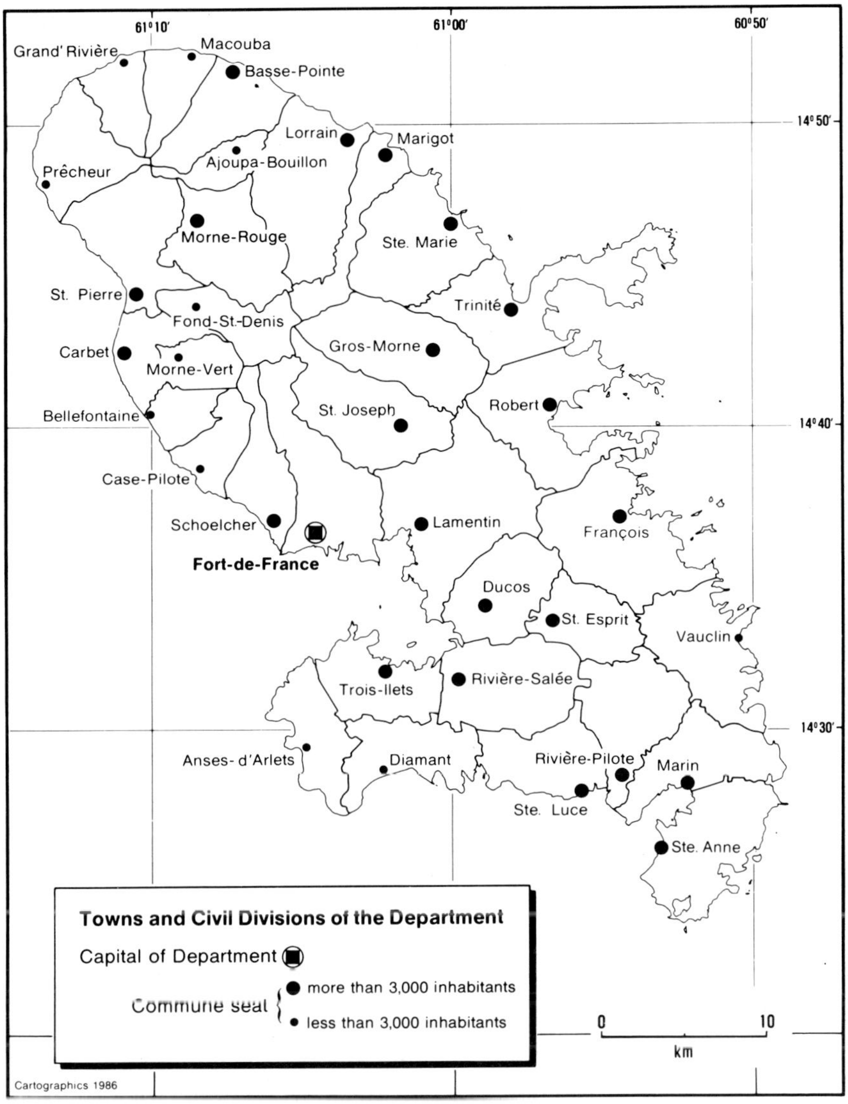

Map 2. Towns and civil divisions of the department. Source: After France, Centre National de la Recherche Scientique 1977, Distribution of the Population, plate 12.

Additionally, historical materials on population size, density, and distribution and on settlement patterns and road networks provide evidence for the specific location of culturally induced processes. Varying agricultural systems and crop-livestock combinations through time were used as indirect evidence to piece together the activities affecting the environment and the vegetation directly. The political organization of the department will serve as a reference grid for much of the discussion (see map 2).

Third, I examined reports of the archeological record in order to reconstruct the prehistoric vegetation and the processes responsible for its presence. Archeological excavations conducted over a thirty-year period provided useful information on livelihood activities, stone and ceramic industries, and the location of settlements so that both the geography of land use and the particular affecting processes might be reconstructed.

Last, in these investigations my interpretation of evidence was guided by the increasingly important literature on island ecosystems, tropical vegetation, and human influences on the earth.

A Typology for Considering Change in Plant Geography

There are four ways by which the plant cover can be modified in an established island ecosystem. Changes may occur in species membership of the flora and fauna, in the interaction processes, in habitats or environments, and in the dimensions of the system. Each of these, in turn, affects the whole island system. These changes are witness to the effects of stress and disturbance beginning at the species level, but as these effects accumulate they are expressed at the vegetation level.

There can be change in the membership of the system by addition or subtraction. An increase in the number of species increases the competition for niches. This competition may eliminate the less well adapted types or narrow the niches of different species. Some new members may bring constellations of species peculiarly adapted to themselves, as did humans with their weeds and inquilines. This change would increase the complexity of the system and, presumably, tie up more energy, as indicated by greater mass. Loss of members would have a reverse effect, and although the plant cover might remain continuous, its complexity would be reduced. In simplified communities loss from epidemic disease would be an increased threat.

There can be change in the processes characteristic of the system and new causal relationships established with the system, increasing the potential for complexity. Carefully managed burning and grazing create new species compositions that are in potential equilibrium with burning and grazing. Under this condition the processes may be said to have created additional subsystems, thereby increasing the complexity of the entire system. Industrialization and urban growth create wastes that have not been absorbed by the natural systems, and pollution becomes a new environmental determinant of vegetation types. In actual fact, change in process more often reduces the complexity. For example, when overgrazing and agricultural activities increase erosion and reduce soil fertility and when pollution introduces toxic substances to air, water, and soil, the diversity of the plant cover is consequently reduced.

There may be changes in habitats in which new environmental settings are created. The most dramatic of these are associated with volcanic explosions and the deposits of raw mineral rock that bury already established vegetation. More important in the modification of habitats is clearing and burning for agriculture, which do not immediately modify the soil but change all other micro-environmental characteristics and, ultimately, the soil as well. Only highly specialized or pioneer communities can exist in these new habitats. With urbanization the vegetation is stripped off, runoff is accelerated, and waste is accumulated. The larger towns create heat islands that modify the microclimates. Industrialization introduces toxic substances, chemical and bacteriological, and increases the weight of transported goods, leading to compaction and, consequently, interference with natural ground and surface drainage systems. Other habitats are created more subtly by the continuous attrition of such disturbances as traffic, grazing, selective culling, and so on.

A fourth way change can be introduced is by expanding the dimensions of the system. Examples are the modifications encountered during the trading activities of the Amerindians and the modifications resulting from the inclusion of the island into the mercantilist colonial system of France. The ecosystem is no longer coincident with the island borders but has dimensions overseas, to include parts of the southern continent, or the *métropole,* or trading areas now serving and being served by the island.

Each of these four kinds of changes has had a real effect on the plant cover of this island. The effects have never been restricted to any one period in history, but they have varied in importance. Habitats changed with the results of degradation and aggradation processes as well as with the evolutionary development of more complex plant communities. During the Miocene, planation was most important in modifying the island's cover, for it reduced the number of habitats available to plants, perhaps eliminating all. Later the island was rebuilt by volcanism to the present elevation.

Higher and lower sea stands modified the size of low-lying rim lands. Plant migrations and the evolution of local taxa reconstituted the flora. Both American Indians and Old World peoples modified plant migration rates, since they were effective dispersal agents, introducing many new plant species into old and newly developed habitats. Clearing and burning, and the planting associated with the agricultural activities of humans, modified habitats and changed the plant membership profoundly. But for the last three hundred years colonial plantation agriculture with its export-market orientation maintained by duties, subsidies, and quotas has had the dominant, most pervasive influence on the plant cover of the island. However, the thirty-year downward trend in population and the persistent, increasingly important migration from country to city may significantly reverse the results. It is also possible that, with the growing sense of nationalism, the islanders will be separated from France. Should that happen, substantial changes in plant cover will take place.

Notes

1. There is a substantial literature on theoretical island biogeography. The modern treatments were codified in MacArthur and Wilson's now classic *Theory of Island Biogeography* (1967). Many scientists have used their models to good effect (Heathwole and Levins 1972, 1973; Simberloff 1974); however, distinguished investigators have pointed out the severe difficulties with some of the assumptions, especially J. D. Sauer (1969) and Gilbert (1980). I find it a useful device for keeping track of membership changes.

2. Simpson (1952) described the concepts of stepping-stones and sweepstakes routes in the dispersal of species to islands.

3. It has been hypothesized that fluctuating environments stress species populations and thus trigger speciation (Elson 1969; Parsons 1969; Bousfield and Thomas 1975; Haffer 1969; Vuilleumier 1971; Bradshaw 1971).

4. Dynamic equilibrium or steady-state concepts are replacing traditional monoclimax and polyclimax concepts. See Tansley 1935, 1939 and then Stoddart 1965 and D. Watts 1970.

5. Odum (1981) suggests that severe stress can set back a plant community to an earlier succession state. Van Valen (1965) related the morphological variation as witness to sorting out the ecological niche. A. R. Hill (1975) focused on stability and stress caused by human activities. First colonizers into open ground are referred to as *pioneers.* J. D. Sauer (1961, 1965) and Byrne (1980) used this concept in their island studies.

6. For a discussion of niche overlap see Pianka 1975 and Shure and Hunt 1981. Gause's (1934) statement is still valuable, but the concept of niche is variously used. A good introduction to the different concepts can be found in Levins 1968 and Whittaker and Levin 1975. Whittaker (1970) used the concept of competitive exclusion in his gradient studies.

7. As the biotic community develops, there will be an increase in organization (biomass also) and in the efficiency of the functioning of the system. See Margalef 1957 and Odum 1981.

8. Stress effects on natural ecosystem have excited much recent attention. Barrett and Rosenberg's (1981) compilation is a good introduction. Geographers have taken another approach (Sauer 1952; Butzer 1972).

9. Glacken (1967) reviewed the divergent views associated with the man-nature relation. Goudie (1982) has a systematic review of differential impacts of humans on the earth they inhabit. However, the negative effects of human impacts on natural systems is most commonly brought to mind, as in Evans et al. 1975, Simmons 1979, and Iltis 1981. The positive position is well represented by René Dubos's *The Wooing of the Earth.*

10. Stress may be thought of as assaults or constraints that limit the production of all or part of the plant mass, whereas disturbance can be thought of as mechanisms that limit biomass, causing its partial or total destruction (Grime 1979,21, 39).

11. Fosberg (1983) has a scathing denunciation of such conditions. Several geographers have examined closely the story of such impacts. A. H. Clark (1959) examined the New Zealand situation; Merrill (1958), Saint Kitts and Nevis; D. Harris (1965), Antigua and Barbuda; D. Watts (1966), Barbados; and Byrne (1980), the Bahamas.

12. This account is largely empirical and descriptive. It deals with vegetation as a landscape unit rather than as a phytosociological unit. However, changes in floristic membership have been monitored.

2 The Pearl of the Antilles: The Natural Setting

Martinique is one of the Lesser Antilles, which mark the eastern border of the Caribbean Sea. It is situated near the center of the archipelago, lying astride lat. 14°40′ N and diagonally across long. 61° W. A true oceanic island, it was built up from a submarine bank by volcanic activity and was never connected with either Saint Lucia 30 km to the south or Dominica 40 km to the north. The bank was not connected to either the South American continent or to the Greater Antillean land mass. The east-west submarine profile is asymmetrical; the southeastern shelf slopes are gentle, with scattered keys and banks extending 15 km beyond the island's coastline; the western shelf falls directly into the Grenada Trough, which is more than 4,000 m deep and separates the island from the Aves Swell. Fort-de-France Bay is shallow, and during the lowest sea stands the island area may have been one-tenth to one-third larger than at the present.

The marine temperatures surrounding the island vary between 26 and 28°C, while average water salinities are 36 ppt. On the north side of the island a strong current, derived from the North Equatorial Current, removes sediments eroded by wave action along the coast between the mouth of Capot River and Cape Saint Martin (see map 3). High cliffs along this coast are pierced by deeply incised rivers. This coast has some of the wildest and most romantic scenery in Martinique, with waterfalls tumbling to beaches 30 to 60 m below. Sections of the south coast are scoured by an analogous current flowing through the Saint Lucia Channel, but the southeasternmost peninsula protects the coast immediately to the west of it from this erosion. As with the rest of the Caribbean islands, the tidal range is 0.5 m at the greatest, and there is generally one tidal cycle per day. Almost equidistant from the Greater Antillean world and South America, Martinique was subjected to physical and biological processes derived from each direction. This chapter introduces the consequences of those processes as a setting for the plants whose geographies are the burden of this book.

Topographic Sketch

The island extends 64 km from north northwest to south southeast. It is 26 km in its largest east-west dimension, and 10 km through its narrowest. It is just under 1,100 sq km in area, with a generally smooth, regular coastline on the west, north, and northeast but an extremely in-

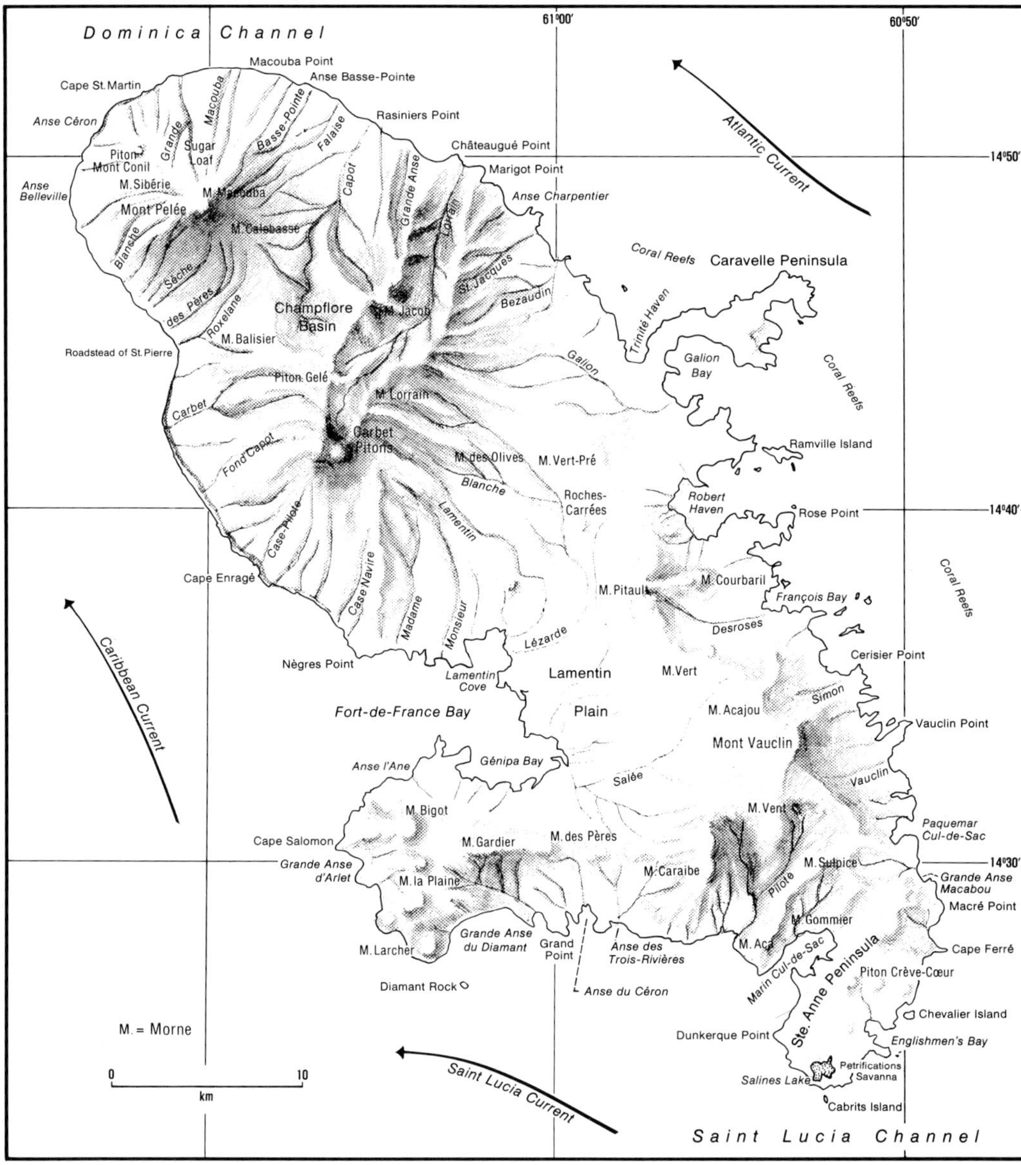

Map 3.
Physical features.
Source: After France, Centre National de la Recherche Scientifique 1977, *La Martinique,* plate 2, and Oro-Hydrography, plate 3.

dented coastline on the south and east. Three bays have been important harbors: the Fort-de-France Bay, the Marin Cul-de-Sac, and the Robert Haven. The Caravelle Peninsula, which projects some 9 km into the Atlantic Ocean, gives the island its distinctive outline. The shoreline extends 350 km, with considerable variation from the cliffy northern rim to the drowned riverine estuaries on the southern coasts.

Because of the island's shape, no interior point is more than 15 km from the coast. Yet, such is the orientation of the people with regard to the sea that they speak of going to the "interior" or of living in the "interior." Folk vocabularies generally reflect local realities, and they do here, for as soon as one is out of sight of the sea, especially in the mountainous north, the character of the climate and landscape differs distinctly from that of the coast. The distinction is not so marked in the lower-lying, southern regions.

Martinique is a "high island," a complex volcanic pile dominated by Mont Pelée and the Carbet Pitons, which provide steep environmental gradients, creating landscapes with much diversity, which effectively isolated parts of the population throughout most of the settlement history. Most of the area is hilly or mountainous. Some of the landforms have distinctive folk labels such as *piton,* a high, steep-sided mountain; or *morne,* a low, more gently outlined mountain or hill, although some high peaks have this term applied to them; or *fond,* a steep-sided valley. Plains areas are few and most are very small. The island has many rivers, but generally their courses are incised and short, their drainage basins small, and their regimes highly variable. Most perennial streams that reach the sea come out of the northern and central mountains.

The island is dominated by two northern volcanic massifs. Mont Pelée rises with steep, straight slopes to 1,397 m. Together with its foothills, it constitutes most of the northern part of the island, about one-eighth of the total surface. This still-active volcano abuts an older massif, Piton Mont Conil of the extreme northwest (Westercamp and Tazieff 1980,27). More than twenty-five streams drain the massif. They have short (6 to 7 km) courses and flow directly to the sea. Only on the eastern side are there tributaries to a major stream, the Capot River. That river's catchment area includes not only the eastern side of Mont Pelée but also the Champflore Basin and the northern flanks of both Morne Jacob and the high Carbet Pitons. Despite their short length, the rivers can be wildly destructive during heavy rainstorms. None have been dammed, although several have stoneworks built during the colonial years to divert water into irrigation canals. The Peléean interfluves on the west are long and narrow, with steep slopes to the river bottoms that fall rapidly in the upper reaches, and have relatively level courses as they approach the Caribbean Sea (see figure 1). This topography contrasts conspicuously with the gentle landscape of the northeastern flanks of Mont Pelée, which was very early called the *paīe plat.* In this region wide spurs cut by deeply incised streams are virtually all divided between gentle slopes of the interfluves and very steep slopes of the valley walls.

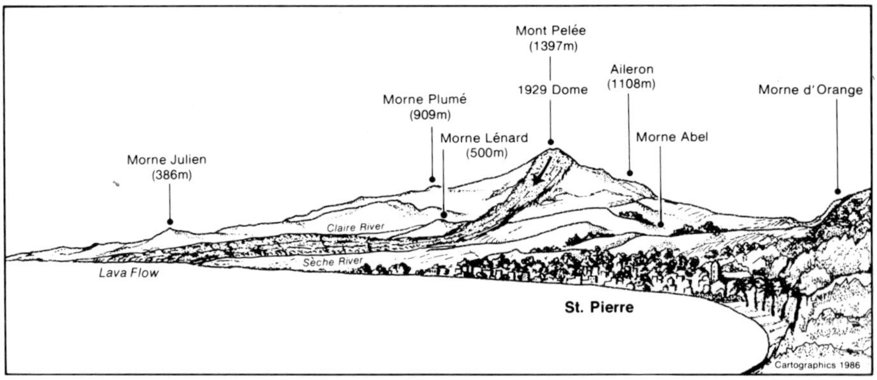

1. *View of the Mont Pelée massif from the Caribbean side. Source: After Westercamp and Tazieff 1980,46.*

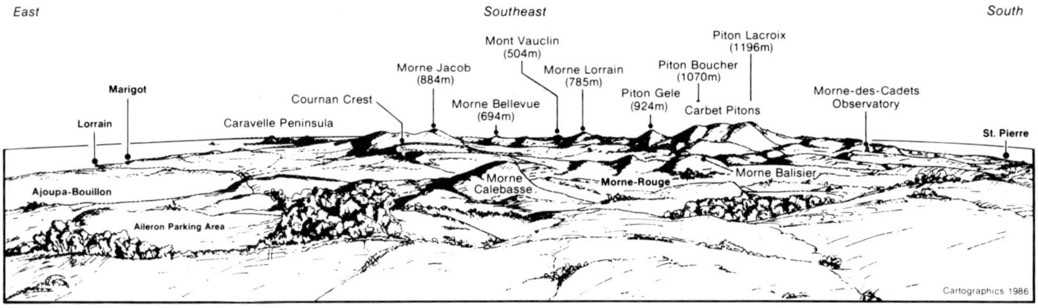

2. Panorama from the Aileron Parking Area of the Carbet Pitons and Morne Jacob.
Source: After Westercamp and Tazieff 1980,49.

The two areas of contrasting slopes have markedly different water regimes, only the Atlantic side having dependable, permanent flow.

The Carbet massif, connected with Mont Pelée by a low ridge, has a string of conical summits running north-northeast–south-southwest, of which Piton Gelé (924 m) and the Carbet Pitons (1,196 m) are the highest. To the east are Morne Jacob (884 m) and Morne Lorrain (785 m). Due west of the Carbet Pitons is an ancient caldera, the floor of which lies some 800 m below the summits (see figure 2). Landslides frequently disturb the unstable upper slopes of all these mountains.

The narrow western valleys penetrate a relatively short distance (6 to 7 km) into the Carbet mountains. The Lorrain River, which flows into the Atlantic Ocean just south of the town by that name, has cut a sizable gorge through Morne Lorrain deep into the Carbet massif, well over halfway across the island. The Lézarde River and its tributary, the Blanche River, flow south into the Lamentin Plain. These last two have the largest valleys on the island.

Connecting the mountains of the northern two-thirds of the island with those of the south is a line of hills known as the Barre-de-l'Ile, of which Morne Vert-Pré, the Roches-Carrées, and Morne Pitault are higher elevations (near 300 m). To the southeast is Mont Vauclin (502 m), which has the highest summit in the southern part of the island and is the only peak besides Mont Pelée commonly referred to as a "mountain."

South and westward across the width of the southern portion of the island is a series of well-dissected volcanic deposits of which Morne Vent and the massif of Morne Caraïbe are the major eastern landmarks. Separated from these are a series of hills collectively known as the Mornes des Pères, higher than those to the east and separated from them by a narrow lowland. West of the Morne des Pères is a series of isolated peaks rising above a fine-grained, well-dissected surface, with generally reduced and gentle slopes.

The southeastern Sainte Anne Peninsula has a karst landscape with a series of small hills, composed of both mogotes and intrusive volcanic necks (100 to 140 m), which are high enough to have significantly different microclimates from the plain below. Areas lying at 25 to 40 m above sea level, including Salines Lake, are part of a low-lying wave-cut plateau which is dissected by faults. Parts of stream courses follow these fault lines, and occasionally an upthrust is visible as a low ridge (Butterlin 1977,157).

Only five substantial rivers originate in the southern mountains. Two are large. La Manche River drains the southern end of the Morne Pitault region, and the Salée River drains Mont Vauclin and the Morne Acajou uplands. Both rivers carry detritus from these old volcanic mountains during flood stage. In all southern rivers, water sometimes reaches the sea until July, but, more often, these channels are nothing but a series of artificially dammed pools, except after the onset of heavy rains. However, ground water taken from the streams beds generally is adequate for domestic and livestock use and is tapped by the larger trees throughout the year. Water is generally available from wells at short distances from the coast, even in the south. Still, droughts of three and four years may cause water deficits even in humid areas (Bonnot 1974).

The largest area of level land on Martinique is the Lamentin Plain (75 sq km), which is a result of slight uplift and infilling of a shallow Tertiary sea that separated the northern and southern parts of the present-day island (Butterlin 1956;294). It is drained by the Lézarde River and its tributary the Blanche River, which rise in the Carbet Massif, and by the Salée River, which runs from the low hills and mountains east of the plain. These rivers cut down into the floor when the shallow sea was exposed during low sea stands of the Pleistocene. A lowering of the sea level of only 20 m would completely expose the floors of the Lamentin Cove and Génipa Bay. Alluvial infilling is evident on all margins of the plain but is most marked on the border of the Génipa Bay (Revert 1949a,111; France, Centre National de la Recherche Scientifique 1977, plate 4). The plain surrounding the Salines Lake in the extreme southeast of the island is likewise an emergent old bay filled with material eroded from the surrounding hills (Butterlin 1956;294).

Some streams have created small floodplains, the most important of which are along the eastward trending Galion and Desroses river valleys, which enter the Robert Haven and François Bay, and along the lower stream course of the Lézarde River, which meanders over the Lamentin Plain. The gradient is here so low that much of the plain was originally ill-drained marshland. Before the establishment of drainage canals, there were widespread flooding of the middle and lower stream courses during the rainy season of the year.

In summary, plains areas are few and not large. Collectively, "flat" land constitutes about 112 sq km, or 10 percent of the island surface. Four-fifths, or 80 percent, of the island is covered in mornes averaging 100 to 250 m high, with very steep slopes. South of the line from Génipa Bay to Robert Haven a dense network of streams bound these low hills. North of the line the large massifs control the drainage and, although the valley walls are steep, they are straight and the crests between valleys are uninterrupted. It is the northern part that the 75 sq km of land above 500 m is located. Although less than 7 percent by area, these high mountains physically dominate the landscape in a pervasive manner (France, Centre National de la Recherche Scientifique 1977, plate 3). However, very steep slopes can be found in the southern half of the island, especially in the southwestern portion around the higher mornes.

Climate

Climate is the primary control on vegetation. Rainfall is the most variable element on this island and controls the distribution of the major plant formations. The primary controls over the weather and climates of Martinique are the seasonal changes in the prevailing pressure systems and the insular topography. The two sets of controls interact to produce markedly different climatic conditions in different parts of the island, ranging from dry-subhumid, near hot desert conditions, to perhumid, cool montane rainforest conditions. Throughout the existence of the island mass, climates have reflected changes in those controls. During the Pleistocene not only did surface area change, especially the amount at low elevations, but rainfall amounts and patterns were different. The more recent climatological interpretations suggest the island was drier but still regionally humid. Mont Pelée was building during this period and would have created a large mountainous mass between Piton Mont Conil and the Carbet Pitons, increasing the mid- and high elevation area. There is no evidence that elevation changes caused by sea level change resulted in any stress to plants as a result of freezing or very low temperatures at summits. It is assumed that all present-day climatic types existed and that there was more coastal and lowland area. The climatic conditions described in this section are those of modern times.

The yearly pressure regimes are affected by the changing positions of the Bermuda and Azores Highs, the intertropical convergence zone (ITCZ), and the passage of easterly waves. The centers of the Bermuda and Azores Highs regularly move between 20 and 25° N, and the air masses passing over the island, set in motion by the highs, vary in constituent properties according to their source regions, whether the Atlantic, or lands immediately to the east of the Atlantic. The ITCZ moves north during the summer, bringing cloudy days, weak winds, and fre-

Table 1. Average temperatures in Martinique

Station	Elevation	Jan.	Feb.	Mar.	Apr.	May	June	July	Aug.	Sept.	Oct.	Nov.	Dec.	Year
Lamentin (central plain)	4 m	24.7	24.6	25.1	25.2	26.2	26.8	26.7	26.8	26.5	26.3	25.8	25.1	25.8
Paquemar (south windward)	10 m	25.9	25.9	26.1	26.6	27.3	27.5	26.6	27.9	27.6	27.5	27.1	26.4	26.9
St. Pierre (north leeward)	40 m	24.7	24.8	25.5	26.5	27.0	27.1	26.9	27.1	27.1	27.0	27.1	26.4	26.4
Fort-de-France (low-midelevation)	144 m	24.3	24.2	25.0	25.6	26.1	26.2	26.0	26.5	26.6	26.3	25.9	25.0	25.6
Morne-des-Cadets (upper elevation)	510 m	21.4	21.4	22.1	23.0	23.7	23.8	23.5	24.1	24.3	24.1	23.3	22.2	23.0

SOURCES: France, Ministère de la France d'Outre-Mer 1958, 2, 24; France, Direction de la Météorologie Nationale, Service Météorologie de la Martinique 1985.

quent storms. The generally weak temperature inversion in the lower atmosphere can be temporarily destroyed, and thunderclouds of considerable height and intensity bring stormy weather that often lasts five or six hours at a time. Hurricanes, called tropical cyclones in Martinque, develop any month in the year except April, but are most common in July, August, September, and October. Those originating near the Cape Verde Islands are the most damaging to Martinique; those originating in the western Atlantic generally are not so destructive (Bonnot 1974,7; Garnier 1976).

The climate of Martinique and the ocean surrounding it is humid tropical. The mean annual temperature about Fort-de-France (sea level) is 25.6°C. At Mornes-des-Cadets which is at 510 m elevation, the mean annual temperature is 2.5°C lower (see table 1). The average annual precipitation at sea level ranges from 1,000 to 2,000 mm, depending upon exposure, and is highly variable from year to year. The length of day is nearly constant, varying little over an hour and a half per day from December to June. Relative humidities are generally high, averaging 78 percent at sea-level stations and ranging throughout the year from 72 percent in March to 82 percent in October (France, Centre National de la Recherche Scientifique 1977, Climatology). Those stations with relatively low humidity values, such as Prêcheur, have average annual relative humidities of 74 percent. Lying as it does in the path of the Trade Winds, the island characteristically has steady movements of air in exposed positions, averaging 12 knots and blowing northeasterly and easterly 85 percent of the year. However, the wind may become capricious, veering rapidly during the frequent storms that form within the region or pass through it. When severe storms pass to the south of the island, high seas may invade the low-lying lands of the Lamentin Plain, drowning coastal roads and homes, and bringing salt and brackish waters onto land ordinarily well above marine intrusions. In interior sheltered valleys, where the wind dies down completely, people may comment on how "oppressive" the weather is. Relative humidity is consistently high.

Folk climatology distinguishes two main seasons: a dry season called *la carème* and a rainy season called *l'hivernage.* The dry season generally begins in December and lasts until May, but it is most marked during the period February through April, which is the *carème* proper. During this season the Bermuda High dominates, and the dry, descending air brings little rain. Precipitation during the previous two months is primarily orographic in origin. The Trade Winds are regular and strong during this period. The rainy season generally begins in May or June but may be delayed until July, and it lasts through November, during which time the Azores High dominates. Rains occur most consistently during July, August, and September. On the long-term average (see Table 2), highest values are recorded in September, October, and November for most stations. These high precipitation values are the result of strong rains derived from tropical cyclones and the depressions associated with the easterly waves moving along the ITCZ.

Table 2. Average precipitation in Martinique, 1951–70 (in millimeters)

Station	Elevation in meters	Jan.	Feb.	Mar.	Apr.	May	June	July	Aug.	Sept.	Oct.	Nov.	Dec.	Year
Acajou L	102	122	93	79	113	167	202	274	265	290	241	227	163	2243
Anses-d'Arlets L	15	57	47	48	50	70	108	175	125	192	178	115	85	1242
Basse-Pointe W	50	124	93	75	128	147	200	250	258	260	258	261	198	2302
Carbet L	20	43	28	30	41	47	113	190	176	160	138	120	60	1153
Diamant L	5	55	41	38	60	79	135	176	144	159	181	146	97	1311
Deux-Choux UE (W)	609	370	297	282	363	367	427	654	483	462	452	433	417	5033
Fort-de-France C	144	111	81	68	96	142	194	265	249	267	228	198	150	2049
François C	8	86	58	50	91	116	146	167	167	218	191	158	138	1586
Galion W	45	99	81	62	100	135	67	236	202	273	249	240	182	2026
Ilet Cabrits L	8	69	39	50	39	66	121	132	119	146	164	119	110	1174
Lamentin C	4	108	82	72	91	147	199	251	238	291	241	201	164	2085
Lorrain W	23	123	108	96	101	166	181	255	327	252	318	330	264	2511
Marin L	30	83	68	44	68	73	113	156	138	170	185	147	122	1365
Morne des Cadets UE (L)	510	216	175	157	163	192	299	481	384	327	311	271	238	3214
Petit-Bourg C	4	94	78	71	89	136	178	214	220	241	222	179	145	1867
Prefontain UE	230	165	108	109	147	180	197	258	248	278	256	244	189	2550
Rabouchon UE (W)	431	216	172	166	226	234	346	477	397	393	365	335	316	2590
Paquemar W	10	65	62	43	77	108	113	166	149	173	175	149	118	1390
Prêcheur L	8	58	47	66	42	47	112	226	229	184	169	139	87	1396
Ste. Marie W	10	84	59	55	103	129	155	206	213	229	242	228	163	1863
St. Pierre L	5	105	87	78	85	94	177	320	269	240	224	175	124	2078
Trois-Ilets L	3	88	55	48	63	95	157	186	184	182	204	157	118	1526

SOURCES: France, Centre National de la Recherche Scientifique 1977, Climatology, plate 1, 5, 7.

NOTE: L = leeward; W = windward; C = central plain; UE = upper elevation.

The annual rainfall patterns are functions of position relative to the path of the Trade Winds. Three rainfall regimes can be distinguished: windward, leeward, and upper elevation regimes. The windward regime has drier months but none with evapotranspiration values limiting plant growth. The leeward regime has from one to six months in which evapotranspiration exceeds precipitation and moisture stress is an environmental condition. The upper elevation regime has no effective drier season. Averages conceal the fact that precipitation is erratic from year to year and long droughts may occur (Campan 1959;3), the most recent of which began in 1972 and increased in length and intensity each year to 1975 (Bonnot 1974).

Squalls, thunderstorms, and hurricanes are the three kinds of storms experienced in Martinique. In Martinique the term *grain* is used for local squalls of minor intensity. These are characterized by a slight lowering of the barometric pressure followed by an abrupt rise. Temperature falls and humidity rises. There are several minutes of strong wind and rain. Occasionally, more violent squalls are accompanied by thunder and torrential rains, often causing tree-fall gaps.

Thunderstorms lasting several hours occur mostly during the rainy season. Fort-de-France has an average of thirty-six of these storms per year (Bonnot, 1974,6). Fifty such storms have been recorded in one year on the flanks of Mont Pelée. In the mountains they may have important consequences, for after a succession of these storms landslides may occur.

The hurricanes bring the heaviest rains and winds and are potentially the most damaging storms. These large-scale (150 to 250 kilometers across) weather systems have winds equal to or greater than 65 knots. Combining these with other tropical cyclones yields the plot in map 4. They occur most often from August through November. In the period 1886–1984 no hurricanes passed near Martinique during the *carème*. In the month of July for the period, one passed over Martinique and nine others had storm tracks which passed within three degrees of the island. The months of August and September are the most important, with twenty-one and nineteen hurricane strikes in the area, respectively. October had four hurricanes. For the same period there were eighty severe tropical storms passing over or by Martinique (U.S. Department of Commerce 1985). Yet there were 35 years when no major storm track passed within three degrees of the island. For the 112 years of record there were 1.2 storms per year or 0.4 hurricanes per year with possible respites of 6 years with no such storms. High winds and torrential downpours are part of the normal environment of the island system. During geologic time when different conditions existed, the frequency of hurricane passes would have been greater or fewer. When they were fewer and further apart, processes of plant succession would have a longer time within which to operate.

The most pronounced climatic differences from place to place are local responses to elevation and to orientation to Trade Winds. Tempera-

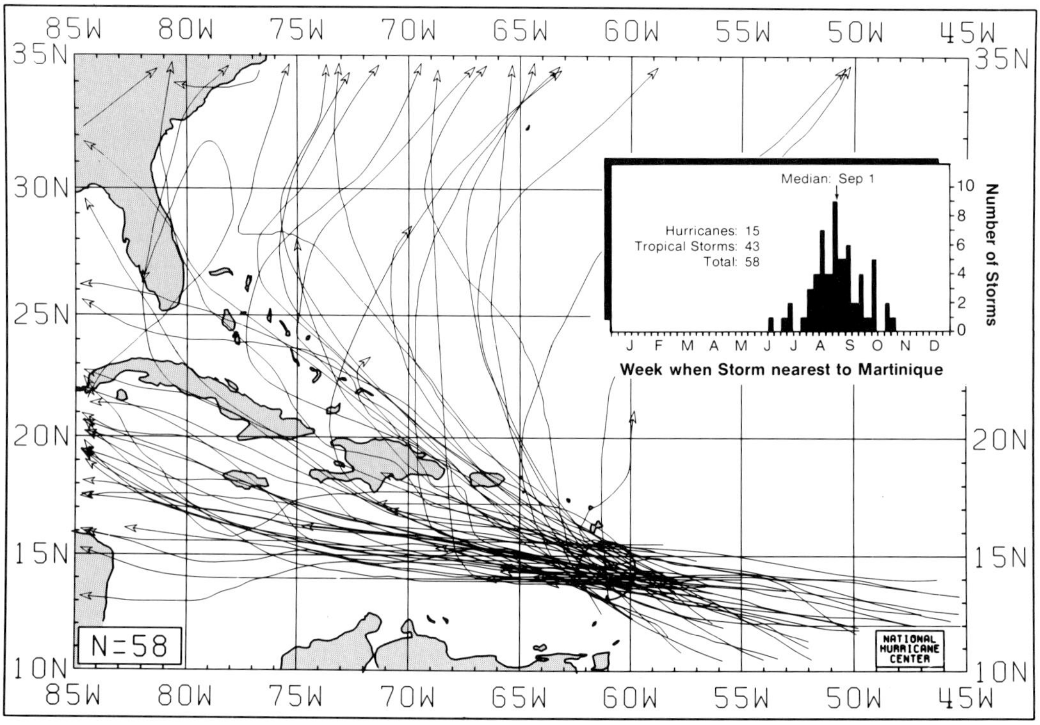

Map 4.
Tracks of hurricanes and severe storms passing within 75 nautical miles of Martinique, 1886–1984, and the dates of storm arrival.
Source: U.S. Department of Commerce, courtesy of NOAA, 1984.

ture decreases with increasing altitude by about 1°C per 100 m to level of condensation, usually 700 m on the windward coast and 800 to 900 m on the leeward side. Seasonal temperature differences exist at higher elevations, but the higher the altitude, the less pronounced the difference. Cloudiness is important at higher elevations. Mont Pelée is exposed for eleven hours per day only fifty-four days out of the year. Orographic lifting of warm oceanic air masses causes much rainfall on the northern windward Atlantic side, and precipitation values are displaced upward on the Caribbean or lee, side (see map 5). The south coast is relatively protected and is in the lee of low hills that create droughty conditions somewhat similar to those on the Caribbean slope. Even the windward slopes have more sunshine than in the north because of the lack of important orographic lifting immediately onshore.

Land-sea breezes are noticeable in all parts of the island that are not blocked from the coast by hills or major mountain spurs. The particular orientations of the valleys varies the importance of the land-sea breeze. There is a slight rotation of the Trade Winds at the end of the day toward the east on the Atlantic coast and from the north on the Carib-

Map 5.
Mean annual rainfall.
Source: After France, Centre National de la Recherche Scientifique 1977, Climatology, plates 1,7.

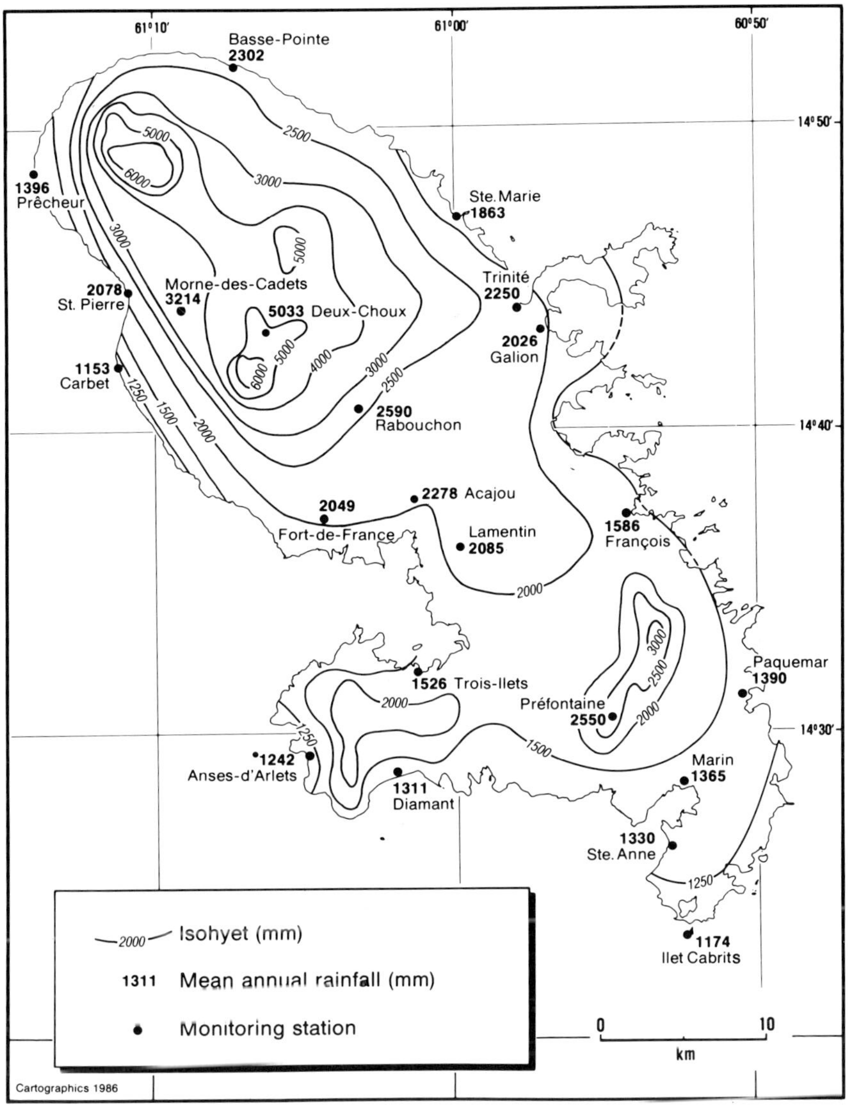

bean coast (Perruset 1974,6). The regular diurnal movement of air may be accentuated by the chinook effect on the leeward side in valleys with the proper orientation on both the Caribbean and Saint Lucia Channel coasts. Morne-Vert Commune is famous for its winds, which blow all the time, except at midday. The orientation of this valley system with regard to the high plateau between Mont Pelée and the Carbet Pitons and the coastal lowland provides a significant flow of air to be added to the land breeze. The regular drying effect of this descending air explains why Morne-Vert Commune is able to support plants not tolerant of heavy rainfall, whereas Fond-Saint-Denis immediately to the north with similar rainfall totals and temperatures cannot.

Further analysis of the climate of Martinique according to the

Thornthwaite system modified for tropical conditions by Malik illustrates more precisely the local regions of deficit and excess moisture. Four zones of actual minimum evapotranspiration exist. The southern two, in the extreme southwest about Anses-d'Arlets and in the southeast about Salines Lake, are associated with low annual precipitation. The other two in the interior at high elevations, on Mont Pelée and in the Morne-Vert Commune of the Carbet Pitons region, are associated with high relative humidities. Two regions of maximum evapotranspiration are the Lamentin Plain and the northeast coast about the town of Ste. Marie and from Lorrain to Basse-Pointe. Excess moisture conditions over 4 m are found in the mountainous interior from Mont Pelée to the Carbet Pitons. There is a water surplus in the region of the Vauclin massif caused by orographic lifting, but the amount and its importance can hardly compare with that of the northern massifs (Cherbonnier 1963).

The zonation of climatic conditions with altitude is so strongly differentiated that climatic belts or bioclimatic zones are virtually concentric. Following a terminology first introduced by Thornthwaite and Holdridge and Associates, the climatic characterizations by Cherbonnier (1963), and Portecôp (1979,14–16) distinguish seven bioclimatic zones based on temperature, rainfall, index of aridity, and number of dry months. Combining Portecôp's two drier zones yields the bioclimatic map for Martinique (map 6). The six zones are characterized in Table 3.

Nowhere is temperature generally limiting for plant growth. At high elevations cloudiness with its corresponding low incidence of light is probably limiting, and on summits the exposure to mechanical stress by constant wind is limiting. The critical problem for plant growth is moisture. Variability of precipitation from year to year stresses plants and produces uneven biomass and uneven reproduction of species. Extended drought puts severe pressure on vegetation that is regularly subjected to and, therefore, adapted to annual drought stress. Long-term records permit the identification of areas with normal annual surplus and annual deficits. Affecting much smaller areas are the unpredictable, but relatively common, severe winds that cause openings in the most dense forest to be colonized by a number of small trees until one gains dominance over the others. Occasionally, high winds associated with hurricanes destroy large areas of vegetation, which are then colonized by spontaneous vegetation. The native vegetation evolved so that it has many wide-amplitude plants that possess strategies for coping with a variety of stressful conditions as well as those plants that depend on more benign tropical conditions.

Geology, Geomorphology, and Soils

The island's geological history is fundamental to its present geomorphology and soils as well as its biogeography. The relative age and area of different parts of the island give clues as to the time available for the evolution of the island surface, the soils developed on that surface, and the potential colonization by plants and animals. The geologic history of the

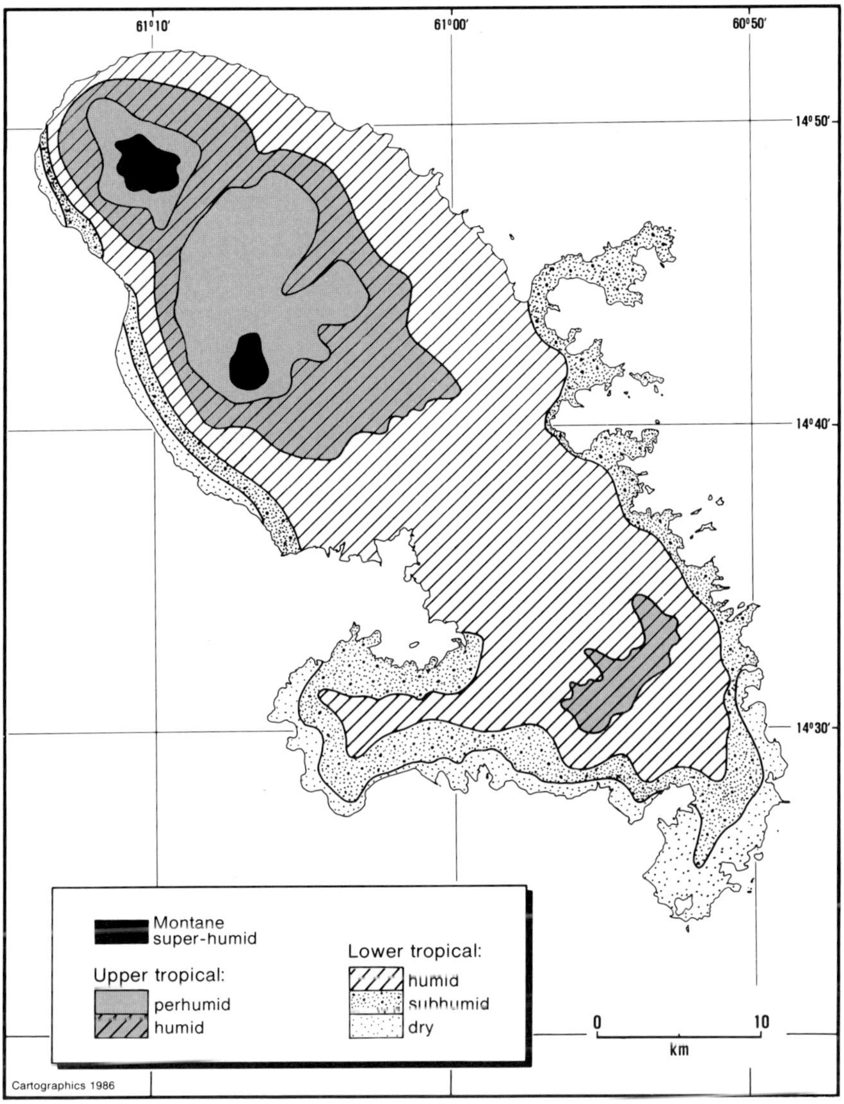

Map 6. Bioclimatic zones. Source: After Portecôp 1979,15.

Caribbean region has been obscure, but confusion has begun to be resolved with the knowledge gained from plate tectonics (Grünewald 1965; Butterlin 1956; Westercamp 1974; France, Centre National de la Recherche Scientifique 1977, plate 1; Case and Holcombe map 1980). Martinique as an island dates from the Paleogene, judged on the basis of the age of rocks in the southeast portion of the island (Andreieff et al. 1974; Stipp et al. 1974). The Lesser Antilles Island Arc Ridge was created by the subduction of the Atlantic Plate under the Caribbean Plate (Mann and Burke 1984,316). Earthquake epicenters seem to derive from crust-crust contact and thus do not produce major earthquakes locally, although there are many faults throughout the island and evidence of

Table 3. The bioclimatic zones of Martinique

Bioclimatic Zone	Distribution	Temp. (°C)	Rainfall mm	Index of Global Humidity[a]	Number Dry Months[b]	Index Aridity (IA)[c]
Lower montane, tropical, superhumid	Highest mountain regions of Mont Pelée and Carbet Pitons	18–20	over 5,000	A_{3-13}	0	329
Upper tropical, perhumid	Humid upper slopes of the interior, Deux-Choux, Morne des Cadets	20–25	2,700–6,200	A_{1-2}	0	72–164
Upper tropical, humid	Lower windward and interior of northern massifs and summit of Mont Vauclin	20–25	1,900–2,700	B_{2-4}	0	70–80
Lower tropical, relatively humid	Largest area Atlantic windward coast to central lowlands higher leeward slopes	25–26	1,500–2,200	B_1	0	53–68
Lower tropical, subhumid	Coastal regions from Caravelle Peninsula to Petit-Bourg, inside leeward belt	26	1,300–1,700	C_2	1–2	39–48
Lower tropical, dry	Coastal regions of leeward coasts	26 27	1,100–1,300 900–1,100 est.	C_1 C_1	2–4 5–6	32–39 30

SOURCES: After Campan 1959, 5; Cherbonnier 1963, 1–6; Portecôp 1979, 14–16.
[a]After Thornthwaite 1948, modified by Malik 1960.
[b]Months with 50 mm or less of precipitation.
[c]$IA = \frac{P + T}{10}$.

much surface disturbance (Grünewald 1965). Subareal volcanic activity arose from this submarine platform. Considerable work has been done to date this mass.

A recent revision of stratigraphic correlations suggests three conditions: (1) This platform from which the volcanoes rise may be at least mid-Tertiary in age (Lewis and Robinson 1976,339); (2) volcanism was renewed in the late Tertiary continuing into historic times; and (3) the central islands, including parts of Martinique, are the oldest land masses of the chain.

The existing islands of the Lesser Antilles have different characteristics north and south. Traditionally these have been called the limestone arc islands and the volcanic arc islands. The northern islands do lie along two arcs: an outer, older arc of islands now largely covered by calcareous deposits of older materials and of the accretionary prism, and an inner,

Map 7.
Two island arcs join in Martinique.
Source: After Tomblin 1975,468.

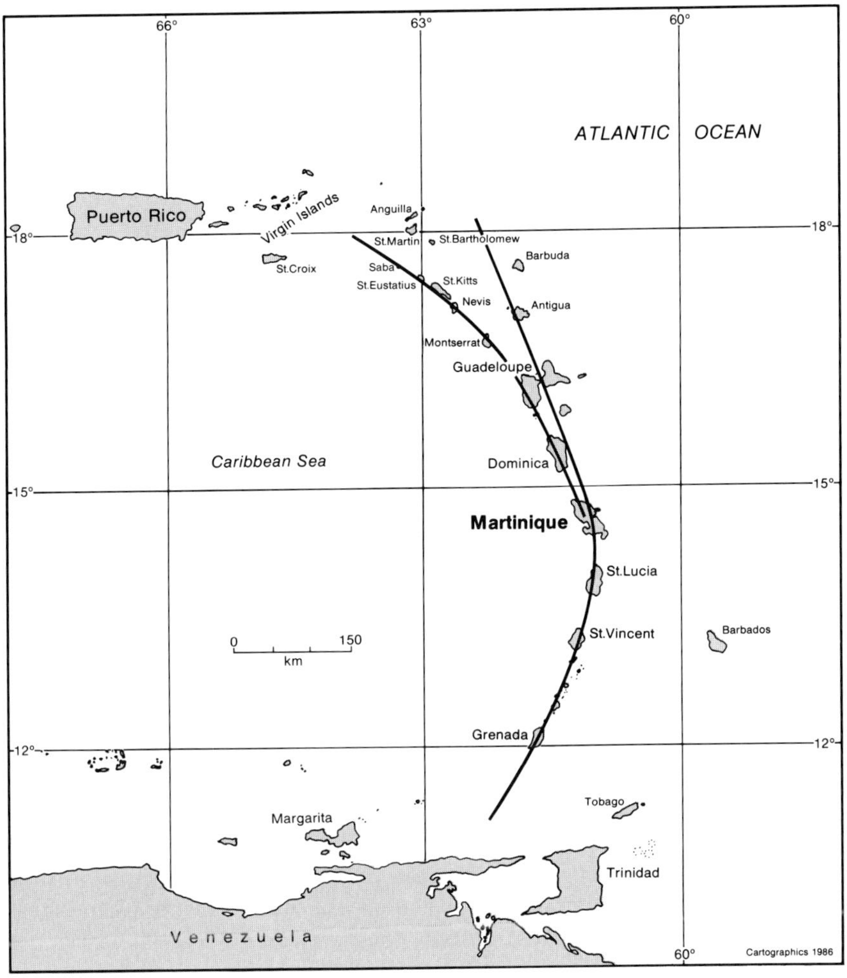

younger arc consisting essentially of volcanic deposits (see map 7). The southern islands, beginning with Martinique, are in a single arc and contain rocks spanning the age range of the two northern arcs (Bowin 1976,22). On Martinique, the older rocks, corresponding in age with the northern outer arc, are in the eastern part of the island from Sainte Anne Peninsula to the Caravelle Peninsula (see table 4). The island arc ridge, passing through from the towns of Marin to Ste. Marie, roughly divides the rocks of the outer and inner arc sectors of the island, thus separating old tholeiites from the calc-alkaline rocks of the western side (Mervoyer and Westercamp 1974,7). Tectonic movements have resulted in major and minor fault systems. Only the major faults are mapped (see map 8).

Eleven lithologic units have been distinguished to subsume the island's complex series of volcanic and sedimentary deposits. These units have been used on map 9, simplified and adapted from the twenty-seven mapping units of the geologic map of the island compiled by Wester-

Table 4. Martiniquan geologic formations and their ages in millions of years

Recent arc materials—1 to 10
- Quaternary sedimentaries—3.5 or later
- Peléean volcanics—0.5 or later
- Recent calc-alkaline vulcanism—0.8 to 2
- Tuffs of Bassignac—3.5 to 2.2
- Morne Jacob submarine and terrestrial volcanics—2 to 6
- Fort-de-France submarine volcanic complex—6.2 to 6.4
- Earlier calc-alkaline vulcanism: Sainte-Luce, Rivière-Salée, and Trois-Ilets Peninsula—2 to 8
- Volcanic flows of Morne Pavillon and Morne Caraïbe—6.4 to 10

Ancient arc flows—10 to 36
- Mont Vauclin vulcanisms—10.1 to 11.5
- Tuffs and other vulcanism of Vauclin and Rivière-Pilote—11.5 to 14.5
- Marin limestone and tuffs—11.0 to 14.2
- Vauclin tuffs—11.5 to 15.5 million years
- Acajou-Bois Soldat intrusives—15.0 to 15.5
- François conglomerates—15.5 to 20.5
- Caravelle, Robert, and Sainte-Anne basalts—17 to 18
- Reef corals of Lower Miocene—20.5 to 22
- Fond Moustique tuffs and limestones—17.8 to 22
- Basal complex—20 to 36

Source: France, Centre National de la Recherche Scientifique 1977, Geology, plate 4.

camp (1975) and from plates in the *Atlas des Départements d'Outre-Mer* (France, Centre National de la Recherche Scientifique 1977, Geology, plate 4). Six groupings of the volcanics fall for the most part west of the tholeiitic demarcation, and four from the earlier arc activity lie east of that demarcation. The breakdown of minerals to clays is well advanced in all the old deposits. These formations now exist in a landscape of well-dissected, isolated mornes. The recent volcanics have the higher summits with the steeper slopes and the soil formations not so advanced.

The earliest sedimentary rocks are the "basal conglomerates" that are composed of altered tuffs and limestones. Potassium/argon measurements place this complex as upper Miocene-Pliocene (Stipp et al. 1974); however, other data suggest an age of lower Miocene-Oligocene (France, Centre National de la Recherche Scientifique 1977, Geology, plate 4). Traces of these conglomerates have been found scattered throughout more than half of the island, always below other sedimentary rocks, except where penetrated and overlain by volcanic rocks. Only a fraction of the outcroppings is shown on the geological map because of their small size. Other limestones, differing in the proportions of corals within the limestone and in the amount of intercalated beds of tuffs, are concentrated in the drier parts of the southeastern portion of the island. Recent sediments include alluvial deposits, volcanic ejecta, and reef debris collected along the shore. The largest area of stream alluvium is in the Lamentin Plain, but small ribbon-like deposits exist along the lower courses of the principal rivers.

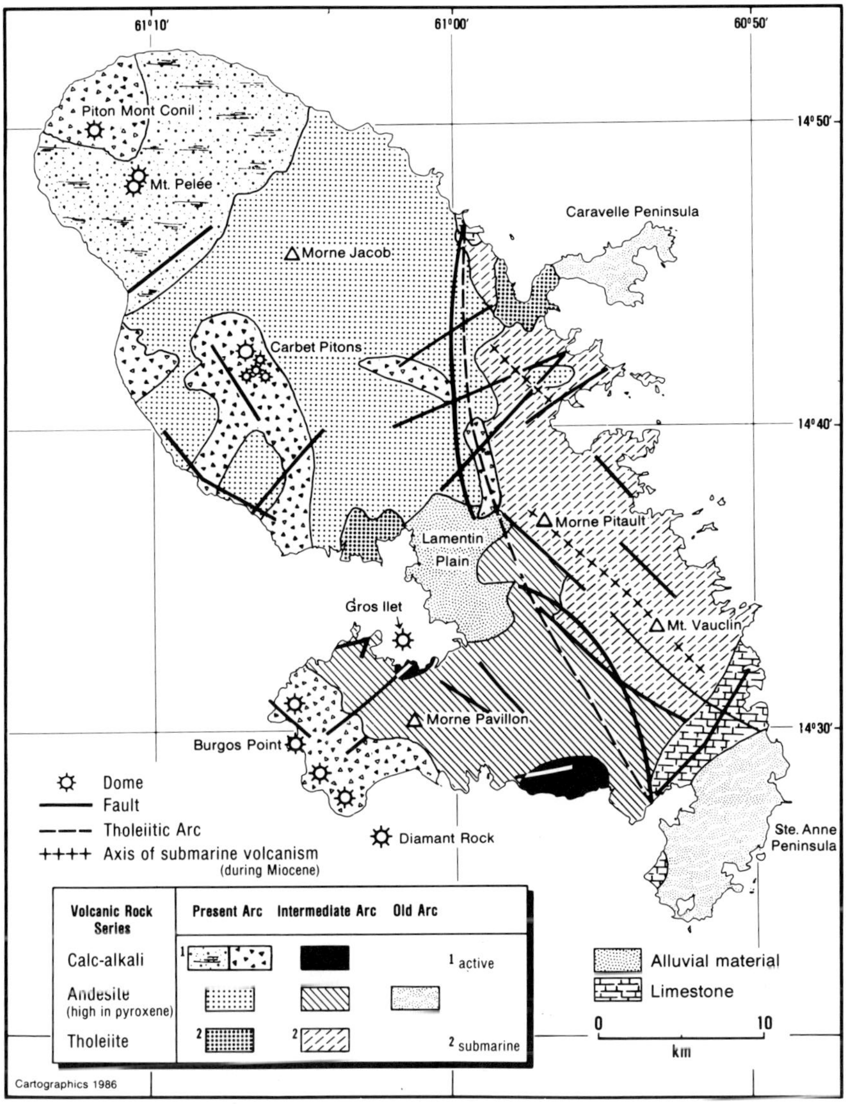

Map 8. Volcanism and associated structural relationships. Source: After Westercamp and Tazieff 1980,26, 28.

Sand beaches appear to be in dynamic equilibrium in Martinique. The sands are constantly being deposited, but the removal by longshore currents (the North Equatorial, Caribbean, and Saint Lucia currents) has prevented any large accumulations of sands. These sands are two types: those derived from coral reef materials, found only in the southeast, and those built of black sands derived from volcanic materials found elsewhere. Small Pleistocene raised beaches and elevated terraces, particularly abundant in the northwest coast and southeastern peninsula areas, provide specialized sites and habitats.

According to carbon-14 dating methods, extensive eruptions of Mont Pelée occurred six thousand years ago and again in A.D. 135, 295, and 1000. During the period of recorded human history, there have been

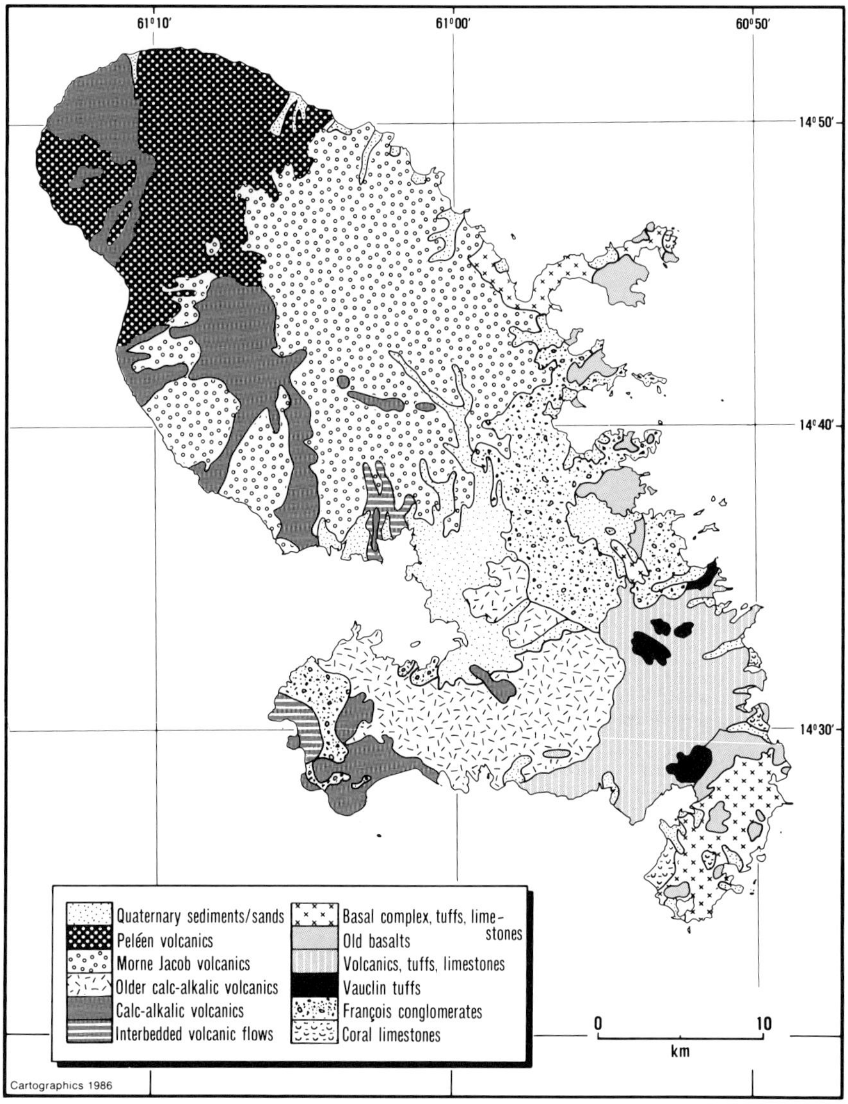

Map 9.
Major lithologic units.
Source: After France, Centre National de la Recherche Scientifique 1977, Geology, plate 4.

several eruptions of Mont Pelée. A series of historic pumice flows and eruptions were deposited in the eastern Basse-Pointe to Lorraine area. The earliest historically documented of these was the mild explosion of 1792. The second, in August of 1851, consisted of emissions of cinders and mud and the formation of more active fumaroles. Similar activity is recorded for 1876. The most violent of all recorded eruptions was that of 1902, which buried much of the upper slopes of the windward side and destroyed St. Pierre (see map 10). The last series of eruptions was in the years 1929–33 (Lacroix 1904,29ff.; Dupuget 1976).

These eruptions have been substantial, even locally catastrophic, removing or burying all surface materials, destroying vegetation, and

Map 10. Summit of Mont Pelée and view from the southeast. Source: Westercamp and Tazieff 1980,50.

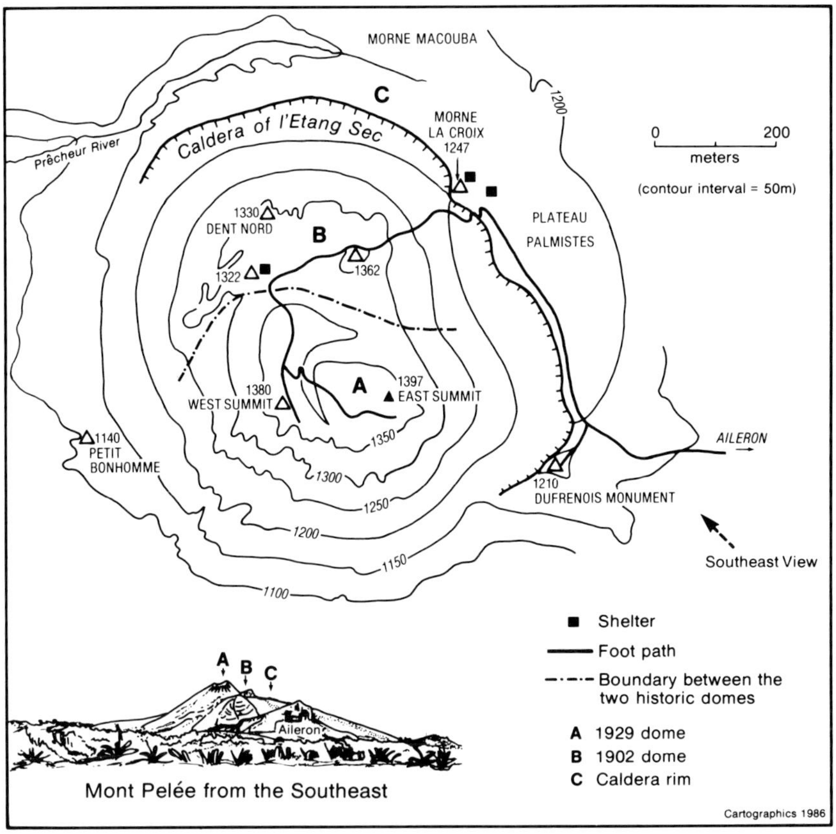

killing animals and human inhabitants. Roobol and Smith (1976,402, 405) think, on the basis of particle size analysis of the low elevation erosion debris, that there has been no substantial reduction in elevation of the major peaks in the north. If that is so, there has been no loss of high elevation habitat.

Soils derived from these rocks are the second great control on vegetation. Since soils are formed on geologic structures through time under the control of climate, both soils and climate show spatial variations generally coincident with vegetation, which is a function of both. The soils of Martinique, which are extremely diverse, are fundamentally a consequence of the relative age, topographic situation, chemical composition of the rock, zonal climatic conditions, and the vegetation cover. These soil-forming processes combine and recombine in different ways. Because of the very steep environmental gradients, differences in soils may occur over short distances. Soils developed on geological formations of similar age have evolved under different local climates. Formations of the same chemical and mineralogical composition vary in age and have had different erosion histories. Parent material of similar or close chemical composition may develop into soils with different characteristics, including textures. According to Colmet-Daage (France, Centre National de la Recherche Scientifique 1977, Pedology, plate 10) the soil types may be grouped into those formed on hard rock or old volcanic tuffs (including the ferrisols, red and brown montmorillonitic soils, and the vertisols derived from volcanics or coralline limestones), those formed on recent permeable ash falls (including the allophanic brown earths of the humid regions, sometimes including gibbsite), and those formed on alluvium (see map 11).

Colmet-Daage has identified some one hundred different soil types for Martinique (Colmet-Daage, 1964; France, Centre National de la Recherche Scientifique 1977, Pedology, plate 10). These have been gener-

Map 11. Major soil groups. Source: After France, Centre National de la Recherche Scientifique 1977, Pedology, plate 10.

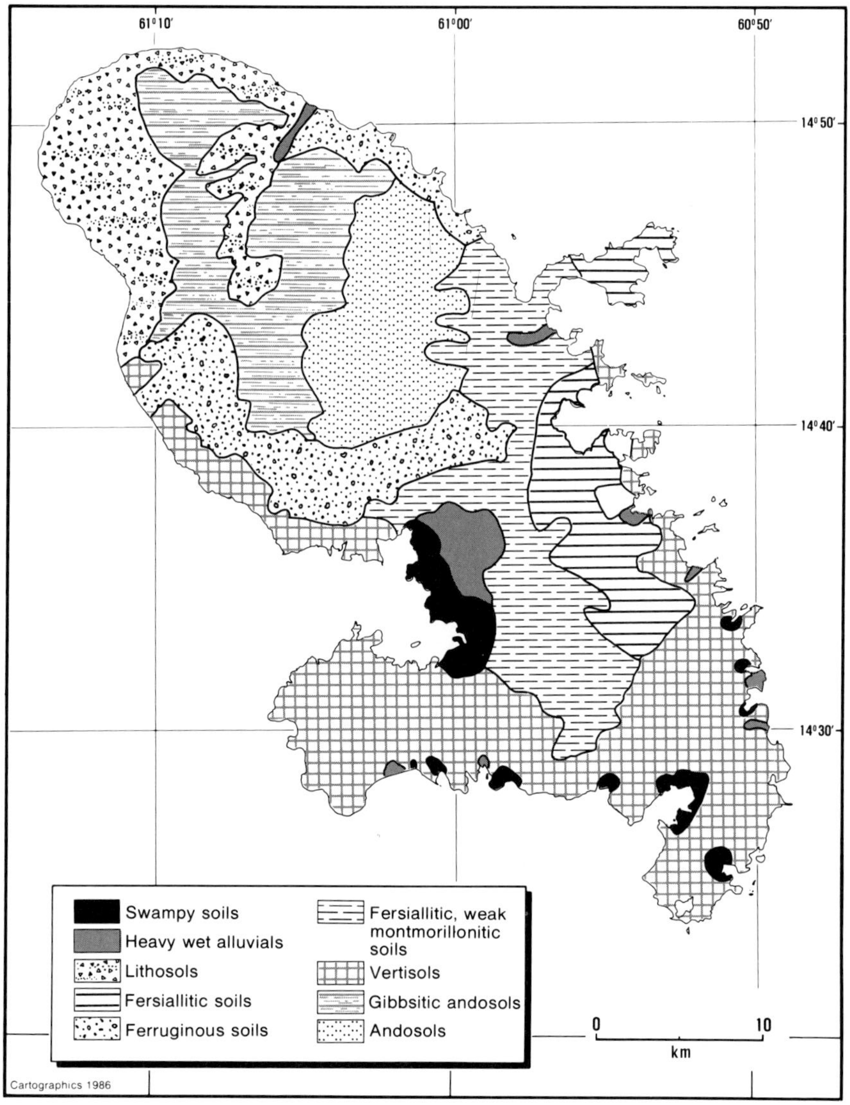

Table 5. The soil groups of Martinique

1. *Marine swamp soils.* Fluvisols, Entisols, and Inceptisols. Wet phases of coastal alluvium where detritus mixes with saline ooze, periodically flooded. Salinity values vary with quantity of river discharge. Substrate varies—limestones, lateritic debris, coral sands, and marine-deposited tuffs and breccias. Extremely localized except in the Lamentin Plain.

2. *Heavy wet alluvial soils.* Derived under moderately low rainfall from a variety of old parent material, including lavas, breccias, especially andesitic and labradoritic or dacitic tuffs. Vary directly with moisture availability. Catenas formed from morne top, flanks, and foot. Largely calcimorphes, soils are basic (pH 7.8) except in poorly drained areas whose soils are acidic (pH 4 to 4.5). Base exchange capacity 25 to 70 me/100.

3. *Clayey alluvial soils.* Riverine, derived from various aged parent materials with kaolinitic and montmorillonitic clay fractions but without a vertic horizon. Fertile, low permeability or with good structure. Base exchange capacity between 20 to 25 me/100. Phosphorus availability determined by proportion of kaolinite clays. Other river alluviums with more sand. Limited distribution. These soils of the middle valley floor are more friable and more easily worked and were exploited early agriculturally. Blanche River and Lamentin River valleys have largest amount.

4. *Colluvium.* Soils lacking in horizonation mixed by slumping, sheet erosion. Mapped with 3.

5. *Lithosols.* Inceptisols and Entisols on volcanic ash. Poorly developed soils derived from ash on andesitic pumice. Sandy, lacking in significant clay fraction and that halloysitic. Highly variable in base exchange capacity and pH, but mostly low in organic matter. Found on the most recent Peléean ejecta and on the very steep slopes of older Peléean materials.

6. *Ferrisols.* Oxisols and Ultisols, red earths. Zonal soils formed on lavas and breccias of andesite and labradorite composition of Pliocene and pre-Miocene ages. Includes some lateritic soils. Formed under humid 1,500 to 3,000 mm precipitation well distributed throughout the year. Some 40 different soil types have been described by Colmet-Daage, but he has reduced them to 10 groups. Base exchange capacity is relatively high, 15 to 20 me/100. These soils dominate the central portion of the island from Trinite to Sainte-Luce communes. Clay fraction is kaolinite or a mixture of kaolinite metahalloysite. Red and yellow colors predominate in the B horizon.

7. *Fersiallitic soils.* Oxisols. Zonal soils formed of old andesitic and labradoritic parent material. Clay fraction is largely montmorillonitic, giving the B horizon a brownish yellow color. Highly variable soil group, with one set of soils having highly unstable clay fraction with base exchange capacity 50 to 70 me/100. Another set of soils, less well-developed, montmorillonite clay fraction and a base exchange capacity of less than 10 me/100. Soils found in the central portion of the island from Robert to Rivière-Pilote and Sainte-Luce communes.

8. *Rust-brown ferruginous soils.* Nitosols. Reddish brown lateritic soils with well-developed soil horizons derived from volcanic ejecta under a humid climate with a dry period. Lustrous sheen throughout the profile, clay formation well started. On the Caribbean side the B horizon may have considerable amounts of halloysitic clays. May be formed on top of ferrisols. Many facies varying in degree of clay fraction development and steepness of slope. Found in a ring about the northern massifs, the major groupings are in the Sainte-Marie, Lorrain, Carbet, and Schoelcher communes.

9. *Vertisols.* Vertisols and Alfisols. Black earths and grumisols. Black soils with a marked color contrast in the B horizon to olive beige or yellow beige. These soils contract severely during drought, developing deep cracks. When wet, very sticky. Base exchange

Table 5. *continued*

capacity varies from 25 to 70 me/100; pH is close to 4 to 5.5; and organic matter relatively poor (2 to 4%). Found on the dry littoral, Caravelle Peninsula, south coast, and Caribbean slope. Soils on slopes and under more humid conditions are less deep and may have more unaltered parent material incorporated in the C horizon.

10. *Andosols.* Ultisols, yellow earths. Red-yellow latosols formed on the Pliocene and Quaternary ejecta of cinders and pumice, the Jacob andesites, Carbet dacites, and Pelée andesites. Formed under high precipitation and moderate temperatures. Presence of allophanes in large quantities gives a soapy feel. Yellow color at depths in spite of 2 to 3% organic matter. Base exchange capacity variable 1 to 10 me/100. Structure and high soil moisture make these vulnerable to soil slumping and on steep slopes to severe sheet erosion. The pH is 6 or higher. Even among older yellow earths and the very humid yellow earths the iron and aluminum weathering has not progressed to bauxite.

    Three different soil groupings may be distinguished: (1) young soils on the upper elevations of Basse-Pointe and Lorrain communes and Ajoupa-Bouillon with a lot of sand and gravel fractions coated with clays; (2) dark-colored soils with organic matter, over 30% on flats or in areas of impeded drainage, swamp phases of the upper montane forest; and (3) very humid soils in regions with over 5,000 mm of rainfall where the allophanes may form a gel.

SOURCES: France, Centre National de la Recherche Scientifique 1977, Pedology, plate 10; Hardy 1973.

alized into ten soil groupings and reconciled insofar as possible with the British West Indian soils classification system for ease of comparison with the other islands of the Lesser Antilles (Hardy and Rodriguez 1941, 47–51). Table 5 summarizes the characteristics of each soil group. When feasible, the corresponding FAO/UNESCO classification term has been added. The order of the table follows that of the soil map legend (map 11).

Certain general points can be made about the relationships between soil type and vegetation. First, there is a close correlation between continuously wetted soils and vegetation. Soil type regions may correspond closely to vegetation borders. For example, the mangrove swamps and coastal meadows are determined by the saturated soils. Second, lithosols and vertisols correlate well with dry vegetation types. Third, for other soil types and topographic situations, climate, and its derivative, soil moisture, has had the overarching importance and soil type has apparently little control over vegetation.

In summary, on the island of Martinique the major geosystems have combined to produce many different habitats within a very small geographical area. The geological systems have been expressed differently at different times in the geologic record. Aggradational and degradational processes on rocks of different chemistries have contributed toward much diversity in soils. The climatological systems have affected the island surface directly as well as indirectly, as mitigated by oceanic influences. In this study of the changing geography of the plant cover of Martinique, the physical environmental conditions considered are those of the Holocene and Recent, with particular attention to the Recent. Occasional references to the Pleistocene epoch will be made, especially in chapter 3, but that period is beyond the scope of this work. However, it should be recognized that Martinique has been a geologically unstable platform subjected to swings in climatic conditions, especially during the periods of changing sea levels. Temporal and spatial variations in the expression of the climatological, geological, and pedological systems have created many different habitats.

Following from the richness in habitats, Martinique has a rich and varied vegetation. This complexity would make the island of great interest to biogeographers whether or not human beings had entered into the ecosystem defined by the land-sea-air interface. Their presence, however, has added a particular spice to the story. This work focuses on what happened to plant geographies during the time Martinique has been occupied by people.

3 Flora and Vegetation of the Island

The first recorded descriptions of the island's vegetation came after at least 2,000 years of human occupation (Du Tertre 1667–71). Judging by the number of species now in the relatively undisturbed remnant communities, the island had a rich native flora. The variety of physical conditions found in different parts of the island provided numerous habitats for plants. Plants having tolerance limits sufficiently similar coexisted in space, constituting what can be called vegetation types, each type having a particular pattern of species abundance (Whittaker 1975). Hence, the island ecosystem was differentiated into different plant landscapes. Martinique with its comparatively rich set of habitats and comparatively small flora tended toward a stability of vegetation composition for each habitat, but no stand is exactly like any other.

The vegetation types have changed through time, even during that period of completely wild vegetation before humans entered the island boundaries. These changes were initiated by disturbances in the environments of these plant groupings and by evolution and secular change. The disturbance factors may have been initiated by geophysical events of large scale such as a hurricane, landslide, or volcanic eruption; or they may have been larger in scale—a rise in sea stand, the crowding of old crust materials against new, or the regional chilling of the climate. They may have been small in scale, as with the toppling of a large canopy tree. Evolution was a species event initiated by in-migration of exotic species and consequent niche partitioning and niche overlap (Whittaker and Levin 1975, 231–69), or species replacement (MacArthur and Wilson 1967). The species composition changes from one assemblage to another as biological response to changing physical conditions and the historical interactions of plants with those factors.

In this chapter the cast of characters or the plant membership is discussed along with some identification of the sources of the native species. The animal species are not discussed, although it is recognized that animals act as dispersal agents and do modify habitats. The vegetation types of long standing and significant areal extent are described as to geometry and species composition. More attention will be devoted to the apparently self-replicating types, often called climax, than to the seral or successional types. It is against this flora and vegetation that the cultural plant geography has been played out.

The Plant Membership and Sources of the Flora

Although the whole flora has not been completely characterized, Jacques Fournet (1978) has compiled the scattered literature for the phanerogams. Estimates vary from 1,958 to 2,700 species of vascular plants (Proctor 1977; Fournet 1978; and Stehlé 1944). Only 58 of these are endemic to Martinique, but 21 more are found only on Guadeloupe and Martinique. Even more are endemic to the Lesser Antilles (Stehlé 1937a,b; Stehlé, Stehlé, and Quentin 1937–49; Fournet 1978; Howard 1979).

When one takes the French West Indies as a whole, the listed native vascular flora of Martinique and Guadeloupe consists of 797 genera, with 1,958 species. Of these, 320 species in 57 genera are cryptogams. More important for this study are the native phanerogams, which number 1,668 species in 730 genera (Fournet 1978). The phanerogamous species-to-genus ratio of 2.3 to 1 is low and is consistent with oceanic islands like Mauritius and the Canary Islands (Carlquist 1974,112).[1] The floristic richness as defined by Lebrun (1967) is high. The figure is 568 and is derived by applying the formula, total number of species (1,958) divided by log of the square of the total area (log 2,800 km^2).

There are some 186 plant families in the angiosperms, gymnosperms, and cycads of the islands (Fournet 1978,11–16). The largest native plant families are the cosmopolitan Poaceae, Orchidaceae, Asteraceae, Leguminosae, and Cyperaceae. The tropical families Rubiaceae, Euphorbiaceae, Myrtaceae, and Melastomaceae, similarly well represented, each have over forty native species. Some ten other families have more than twenty natives each. Thirty-six families are monotypic, a situation characteristic of oceanic islands (Ridley 1930,690; Carlquist 1974,111). The native genera of the West Indies are small; most average less than five species per genus. Those genera having ten or more native species are *Eugenia* in the Myrtaceae, *Croton* and *Phyllanthus* in the Euphorbiaceae, *Panicum* and *Paspalum* in the Poaceae, and *Piper* and *Peperomia* in the Piperaceae.

The affinities of the whole flora have not been worked out, but all workers in the Lesser Antilles have come to much the same conclusion (Stehlé 1944,1945–46,1949; Questel 1951; Fournet 1978; Howard 1952,1973, 1974). As might be expected, there is a pan-Caribbean element as well as greater Antillean, Central American, and Southern Continental elements that extend into the Lesser Antilles. However, the proportions of each element differ from vegetation belt to vegetation belt (Howard 1973,19).

When Beard tabulated the affinities of 243 native forest tree[2] species for the Lesser Antilles, he found that nearly half of the montane and rainforest species, 52 and 42 percent, respectively, were endemic to the island chain. The rest came in fairly equal proportions from the Greater Antilles and South America. Of the tree species in Beard's dry zone forests (including the littoral and dry woodlands of this work), 20 percent of the flora consisted of wide-ranging tropical species, 12 percent were endemic, and of the rest, twice as many came from the Greater Antilles as came from South America. On the other hand, all the swamp trees are wide ranging. The freshwater *Annona glabra* is found throughout the neotropics, and the mangrove species are found on the tropical coasts of both America and Africa (Beard 1949b,29,39).

When the island flora of Martinique is compared with the continental flora of Costa Rica at the same latitude, few families in either flora have a significant number of species (4 percent or more of the native flora). Those that do are all cosmopolitan families: the Poaceae (8.8 and 4.7 percent, respectively), the Orchidaceae (5.1 and 14.3 percent), the Leguminosae (8.8 and 4.7 percent), and the Asteraceae (5.3 and 4.3 percent). For other families consult table 6. The flora of Martinique, like that of oceanic islands elsewhere, has a relatively small number of species compared with continental areas. Whether this is because of rare colonization events from source areas or the ecological poverty of the island itself is unknown (MacArthur and Wilson 1967,3; Lack 1976,

Table 6. The modern phanerogamic flora of Costa Rica and Martinique

	Costa Rica[a]		Martinique[b]			
Illustrative Families	Species	% of Flora	Total Species	% of Flora	Natives	% of Native Flora
Orchidaceae	ca. 1,000	14.3	144	5.1	100	4.0
Poaceae (Graminae)	ca. 300	4.3	125	4.5	125	7.5
Bromeliaceae	ca. 160	2.3	31	1.1	28	1.7
Arecaceae (Palmae)	ca. 100	1.4	58	2.1	8	0.5
Araceae	121	1.8	47	1.7	21	1.3
Cyperaceae	ca. 125	1.8	85	3.1	82	4.9
Piperaceae	540	7.7	31	1.1	24	1.4
Moraceae	62	0.8	20	0.7	6	0.3
Loranthaceae	39	0.6	20	0.7	17	1.0
Ranunculaceae	8	0.1	1	—	1	—
Lauraceae	48	0.7	30	1.1	26	1.6
Brassicaceae (Cruciferae)	ca. 20	0.3	13	0.5	3	0.1
Leguminosae (Fabaceae, Caesalpineaceae, Mimosaceae)	ca. 330	4.7	249	8.8	122	7.3
Euphorbiaceae	112	1.6	110	4.0	78	4.7
Sapindaceae	52	0.7	19	0.6	15	0.8
Melastomaceae	213	3.1	46	1.7	40	2.4
Myrsinaceae	42	0.6	16	0.5	13	0.8
Ericaceae	50	0.7	3	—	3	0.1
Apocynaceae	50	0.7	25	0.9	7	0.4
Lamiaceae (Labiatae)	59	0.8	23	0.8	11	0.7
Scrophulariaceae	45	0.7	17	0.6	11	0.7
Solanaceae	135	1.9	52	1.9	33	2.0
Gesneriaceae	100	1.4	23	0.8	10	0.6
Rubiaceae	260	3.7	103	3.7	65	3.9
Asteraceae (Compositae)	300	4.3	146	5.3	93	5.6

[a]Total number of species about 7,000 according to Standley 1937–39.
[b]Total number of species 2,779, with 1,668 natives, according to Fournet 1978.

47). This lack suggests a loss of potential variability in disturbance responses (Doyle 1981) and fewer vegetation types within plant associations than in continental areas (Beard 1949a).

Plants probably colonized the Lesser Antilles when the first islands of the limestone arc (including the accretionary prism) were formed. The mineralized wood of the Petrifications Savanna date from the Middle Tertiary (Martin-Kaye 1977,182). However, since limestone islands were subsequently submerged, the modern flora was initiated when the volcanic islands were formed in the late Tertiary. As an oceanic island that was never connected with the South American or Greater Antillean land masses nor connected with islands to the north or south, Martinique could have received these plants or their ancestral forms only by overseas dispersal, a process operating through time, but presumably less effective later since there would be duplications of previously established species (MacArthur 1972; Carlquist 1974). On the other hand, new introductions, which were previously unable to become established because of the lack of more evolved communities, could have found or created suitable niches as the vegetation became more complex.

Pleistocene fluctuations in sea level associated with the advance and retreat of continental glacierrs in the mid- and high latitudes would have repeatedly exposed and covered the coastal margins of the island, thus varying its size. Because of the present submarine profile of the island, it can be postulated that the Caribbean littoral, as well as the littoral about the southeastern peninsulas, would have been broader, whereas the north Atlantic and Dominica Channel coasts would have not increased substantially. Low sea stands would have changed altitudinal zonation only slightly, and only the mountain summit vegetation and the low elevation flat lands would have been significantly enlarged in that area. The additional elevation would not have reduced ambient temperatures to within frost possibilities. The mangrove swamps of the Fort-de-France Bay may well have been less extensive in low sea stands. According to the changes in sea level of 20 m, the area of the island would have increased or decreased by no more than 150–300 sq km, assuming no major change in configuration of the submarine surface. I postulate no significant changes in habitat types, although the area of the dry zone may have been much increased during low sea stands. More important for the development of the flora would have been the cyclic migrations of plants as new land was exposed or older land was inundated. This kind of stress may have been responsible for some local evolution of endemic species.

The agencies and agents of natural plant dispersal in the Lesser Antilles are ocean flotation, wind, birds, bats, and perhaps turtles.[3] Ocean currents today bring plant material to the West Indies from the Canary Islands, the Mediterranean basin, and the coast of Africa, as well as from the American mainland. On walks along Martiniquan beaches one sees seeds and plant parts at the berm line. During the flood stage much plant material is discharged from the Amazon, Guiana, and Orinoco rivers into the Caribbean and Antillean currents (Guppy 1917,19).[4] Contributions are being made to West Indian drift materials by the Magdalena and Cauca rivers of Colombia and by small Central American streams flowing into the Caribbean (Simpson 1956,12). Velocities of the Caribbean current are about 17 meters per day (m/day), and those of the Antillean current are 34 to 50 m/day. This would bring drifts from Brazil in 24 days, from Venezuela in 10 days, and from Central America in 140 to 188 days. Drift of African origin could arrive in 118 days (Guppy 1917,71).

Mangrove, riverine, and many species of the dry littoral formations have seeds capable of effective, long-range dispersal by ocean currents. The major mangrove species *Rhizophora mangle, Avicennia germinans* and *Laguncularia racemosa* are all capable of transoceanic dispersal. According to Adams (1972,349,377), plants found in riverine habitats of West Africa and the West Indies include the common *Geoffroea inermis* and the scandent vine *Entada gigas.* According to Guppy (1917,91) *G. inermis* has seeds incapable of floating across and remaining viable, whereas *E. gigas* has seeds that are capable of so doing. Unfortunately, much of the literature is anecdotal, without controlled experiments to support the explanation thought apparent. Pods of certain leguminous plants float and may be found accumulated with fruits and seeds of other plants at the high-water mark on most beaches. Whole plants and fragments of the well-represented Asteraceae and Poaceae may float, as do some species of families less well represented (Gunn and Dennis 1976). It is probable that some of the endemics of the interior vegetation types were derived from these oceanic introductions (Beard 1949a,48; Guppy 1917, 130; Bond 1948,207–29).

Several families with large numbers of species in the flora of Martinique have seeds adapted to wind dispersal, notably the Orchidaceae, Bromeliaceae, Bignoniaceae, and Asteraceae (Ridley 1930; van der Pijl 1972). The first two families furnish many species and the conspicuous tank epiphytes to the forest vegetation types, the montane woodland and bushland, as well as to the seasonal and dry woodlands. Lianas of the Asteraceae and Malpighiaceae are found in the forest types of the island, and shrubby components of upper elevations are frequently plants adapted to wind dispersal. Although few in species, the wind- or water-dispersed Bombacaceae are conspicuous members of seasonal forests and dry woodlands as well as pioneers in rainforest vegetation. Canopy species and pioneers more often tend to be wind dispersed than do the understory plant members, which are likely to be transported by animals and insects.

Martinique is regularly visited by birds migrating between North and South America, and by birds following only sections of these long migration routes (Bond 1960,iv–vi; Lincoln 1952,54–55; Lack 1976, 149–50). During the rainy season numerous migrating species settle down on the swamps of the Salée River (Sylvestre and Lotaut 1947,2: 288). No plants with strictly North American affinities show up in the flora, presumably because they have not been able to find or develop a proper niche; the differences in day length may be the most critical factor. Species of Poaceae, Cyperaceae, Leguminosae, and Malvaceae are commonly carried by birds. A third of the grasses belong to four cosmopolitan genera, and more than half of the sedges of the meadow vegetation types belong to another four cosmopolitan genera known to be transported by birds. The upland forests and woodlands have plants with fleshy, pulpy fruits and seeds attractive to birds and bats, and few large, heavy-seeded plants, although a number of such species occur in the lowland flora. A considerable number of lowland plants also have sticky fruits carried by birds. The terrestrial fauna of Martinique is clearly of waif origin and thus probably accounted for the arrival of few plants (Pinchon 1967,186–204).

Shorter range dispersal by inter-island movements was possible and probably increasingly important as neighboring islands of the chain were covered with plants and began the differentiation of local species (Simpson 1967; MacArthur and Wilson 1967; Grisebach 1864). The system of currents within the Caribbean facilitates this exchange. Wind patterns within the Caribbean suggest a regular transfer from the Greater Antilles to the Lesser Antilles, although exceptionally high winds capable of transporting heavier disseminules may reverse the movement. There is a regular movement of birds between islands of the Greater Antilles and islands of the Lesser Antilles, although the strong territoriality of most island land birds (pigeons to passerines) and all bats requires that many plants would have had to be taken to Martinique by a sequence of journeys.

Vegetation Types and Their Distribution

However the species constituting the native flora got to Martinique, they formed a complex natural vegetative cover, as indicated by surviving relics and accounts left by early travelers. This evidence has been used to postulate the types, composition, and geographic patterns of the immediately precultural vegetation of the island. In the interest of simplicity and brevity, the naming of types has followed popular usage except where doing so would cause confusion. The growth form terms of *forest, woodland, bushland, desert,* and *meadows* have been used for the primary categories. *Meadows* is used in preference to *grasslands* since species other than grasses may dominate. Habitat and even floristic terms have been used in the terminology to be concise.

The natural spontaneous vegetation has been classified into five major categories by structural criteria that make possible the construction of a relatively simple key to the vegetation (see table 7). These are further subdivided by structure, leaf habit, growth form, and habitat into twenty-two types.[5] The four forest types differ in complexity of structure and in the degree of evergreen and deciduous habit. The rainforest has three stories, and the montane and seasonal forests have two. The rainforest and the montane forests are evergreen; the seasonal and semi-

Table 7. Structural organization of the wild vegetation of Martinique

Forests: continuous tree cover with at least two stories of trees over 4 m in height

Rainforest: three tree stories; evergreen habit

Montane: two tree stories; evergreen habit; upland sites

Seasonal: two tree stories; canopy level has deciduous elements

Semideciduous: two tree stories; canopy level has more deciduous elements; southern sites; longer period of leaflessness for individual trees

Woodlands: continuous tree cover of a single tree story at least 4 m or more in height

Littoral: coastal sites; evergreen habit; moss scarce; epiphytes and climbers rare

Dry woodlands: mostly closed canopy woodland of evergreen and deciduous trees; dry areas

Swamp woodlands: inundated coastal and riverine sites; evergreen habit

Mangrove: tidal brackish water and salt water; rarely over 5 m; stilt roots and/or pneumatophores

Annona swamp: fresh-water riverine sites; occasional trees 10 m or more; clean branches

Pterocarpus swamp: palustine swamp; trees to 10 m

Laurel: montane sites on exposed slopes; dense canopy; clearly dominated by laurel-leafed species

Elfin: high mountains; evergreen; stunted and gnarled trees thickly covered with mosses, epiphytes, and climbers; rarely over 7 m

Silverleaf-balsa stands: within rainforest areas, dominantly fast-growing pioneer trees with pale bark and large leaves; light-colored boles

Tree fern colonies: high to midelevation community composed largely of tree ferns with caudices 4 to 10 m tall; frequently covered with epiphytic ferns

Palm brake: mountain slope community composed mainly of palms; abundant mosses; epiphytes; complete herbaceous ground cover

Bushlands: generally continuous woody cover at a level below 4 m in height

Montane bushland: stunted, gnarled bushes covered with mosses and lichens

Littoral hedge: wind-pruned, evergreen coastal community; bare branches and sparse ground cover under tight canopy

Leguminous thorn scrub: more or less closed canopy of microphyllous, thorny bushes, and small trees

Mixed cactus scrub: open vegetation with abundant columnar cactus and prickly pear; amount of thorny trees varies; relatively abundant herbaceous ground cover

Desertic Types: sparse cover with large patches of bare soil

Cactus desert: exceedingly sparse growth; *Opuntia* species common

Desert rock pavement: windswept rocky promontories with scattered woody growth and ephemeral annuals; frequently encrusted with salt

Sandy beaches: littoral community dominated by vines and rhizomatous grasses

Rocky cliffs: steep slopes colonized by herbs and terrestrial bromeliads

Table 7. *continued*

Meadows: continuous herbaceous cover with no important woody growth

Littoral meadow: wet meadows in coastal areas; largely spontaneous herbaceous growth dominated by sedges (and locally) ferns

Lower montane meadow: montane community associated with springs; dominated by grasses and sedges

deciduous are mixed evergreen and deciduous with different seasonal characteristics.

The woodland types are the most numerous and are distinguished mainly on the basis of habitat: the littoral occupies the dry coastal locations; the dry woodland, the interior in droughty situations; the swamp, on tidal and wet riverine sites; the laurel, on middle mountain slopes; and the elfin, at high elevations. The silverleaf-balsa stands are distinguished by the large, parasol-like leaves. The tree fern colonies are distinguished by the predominance of erect trunk-like caudices reaching well over a person's head. The palm break is distinguished by its peculiar growth form. All woodlands are primarily evergreen, although the littoral and dry woodland do include many deciduous elements.

The other woody types differ by height and habit. Among the bushlands, the evergreen montane and littoral types are distinguished by habitat. The mixed cactus scrubs tend to be more open, lower vegetation with columnar cacti much in evidence. The leguminous thorn scrub was probably relatively smaller in total area in precultural times than was the mixed cactus scrub. The four desertic types differ in the sites occupied and by the degree of barrenness, the cactus desert having the most dense coverage.

The herbaceous communities are distinguished by differences in relative proportion of ferns, grasses, and sedges as well as by location. Both of these meadows are presumed to have been much smaller in extent than they are today.

The complexity of vegetation stands matched by that of environmental variables, which can be generalized into environmental zones. Five environmental zones or belts are recognized for the volcanic islands of the Lesser Antilles (Beard 1949a,55). These are littoral belt, dry belt, middle belt, mountain belt, and upper mountain belt. Subdivisions in the dry belt and middle belt bring the number of bioclimatic units to six, as described in the preceding chapter. These belts are arranged in more or less concentric rings about the interior mountain massifs, with the exception of the dry belt, which is found only on the leeward sides and on the Caravelle Peninsula (see figure 3). The middle belt may reach the coast, where rainfall criteria are met. The leeward sides of the mountains tend to have higher elevation boundaries than the windward sides because of the warming of descending air of the Trade Winds. This arrangement of belts provides a guide, together with the location and ecology of par-

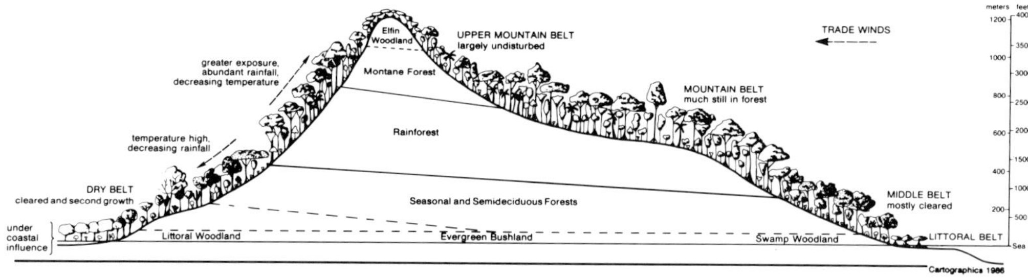

3. Schematic arrangement of vegetation belts in Martinique. Sources: After Beard 1949a and Stehlé 1945–46.

ticular vegetation stands in the present century, to the distribution of original wild vegetation.

The mean annual rainfall is a reasonable guide to the location of climatic belts, which correlate well with the belts of spontaneous forest vegetation. The littoral belt with swamps and littoral forests has in general the regional climatic characteristics previously described for the island and summarized on the profile diagram in the lower tropical dry and subhumid zone. It circumscribes the island at levels below 50 m, locally somewhat higher on cliffs. On the windward side it seldom extends more than 100 m inland and usually less. On the leeward coasts it may extend 2 km into the interior on low ground, as in the Lamentin Plain. The number of days with rain varies from four to nine per month in the dry season, and from eleven to sixteen in the rainy season. Salt spray and wind may be significant microclimatic factors.

The dry zone with dry woodlands and dry forest exists only on the leeward sides of the island, extending from the littoral belt up to 150 m. Within this zone the drought season is more severe than in any other belt and may be as long as four to six months. Rainfall averages only 1,300 to 1,700 mm, and in dry years it may fall below 1,000 mm. Rainfall estimates to local meteorologists are for lows of 300 mm in some parts of the dry belt for years of drought, although there are no stations recording such low values. The mean annual temperature at three recording stations is 26° C, and the relative humidity is 76 percent.

The middle belt with seasonal and rain forests has a drought season that is less severe, rarely longer than three months, including upper and lower tropical humid bioclimatic zones. The average annual precipitation varies from 1,500 to 2,700 mm. Annual temperature averages 25°C. At La Tracée, just north of Fort-de-France, a record of three years showed a mean daily range of about 10°C. The humidity varies seasonally, as in the dry belt and by altitude. Short-term records kept at forest guard stations suggest humidity values of 70 to 90 percent as being characteristic of this climatic belt. Practically no day is cloudless, and much of the time there is a complete cloud cover. Winds are locally strong, especially toward the upper border on the eastern slopes.

The mountain belt with montane forest circumscribes the mountain core of the island extending from 250 to 900 m elevation, depending on exposure of the upper tropical perhumid bioclimatic zone. In this zone the precipitation varies from 2,500 to 6,200 mm. Above 600 m there are 250 to 300 days of rain per year, and the dry and wet seasons may be detected in the different number of days of rain per month, 16 to 17 and 22 to 28, respectively. The air is often saturated and always high in humidity. Mean annual temperatures are 23°C. Morne-des-Cadets (Fond-St.-Denis, 510 m), has recorded an absolute low of 16.2°C. According to Campan (1959) this value is commonly approached but not exceeded. The sky is typically overcast and constant winds are characteristic.

The upper mountain belt with elfin woodlands and montane bushlands is restricted to cloud-capped peaks, the tropical, lower-montane superhumid bioclimate of Portecôp. Here the climate is always moist; over 6,000 mm of precipation is estimated to fall every year. Calculated averages place temperatures generally about 16° to 20°C. Winds are constant; the air is usually saturated and frequently cloud filled. The sun rarely shines directly on the vegetation, and when it does, usually only in the early morning. All these vegetation belts contain a number of vegetation types.

It is presumed that in prehuman time natural disturbances were highly localized and that the island was therefore dominated by tropical forests with patches and narrow zones of woodland, bushland, and meadows. These, for the most part, would have been dynamic, self-replicating types recognized in the ecological literature as climax vegetation. Referring to them as climax does not mean that the forests were changeless but rather that the perturbations in the controls operating on the vegetation would not have made the community unrecognizable and that the communities have relatively stable species complements and recognizable structure. Although secondary vegetation of pioneer and seral types consequent to natural disasters and biological evolution would have been always present and provided niches for plants peculiar to such communities, it is impossible to estimate the locations and areas involved for most of these types. For instance, the regions of andosols on steep slopes where landslides are frequent must have had large areas in pioneer vegetation. While these are assignable regionally on a probability basis, they were locationally ephemeral through time and are not represented on the map of reconstruced native vegetation (see map 12). Montane bushland, strand, and rocky cliffs occupied such a small area that they are not shown at this scale.

In order to trace the impact of human exploitation of the plant cover throughout the estimated 2,000-year history, I briefly describe vegetation type, as identified in table 7, in terms of structure and especially in terms of species composition. Species are identified by their scientific names, which are listed by family in appendix A. For those who are acquainted with the creole and French names and would like to use them, appendix B is an alphabetical listing of vernacular names and the scien-

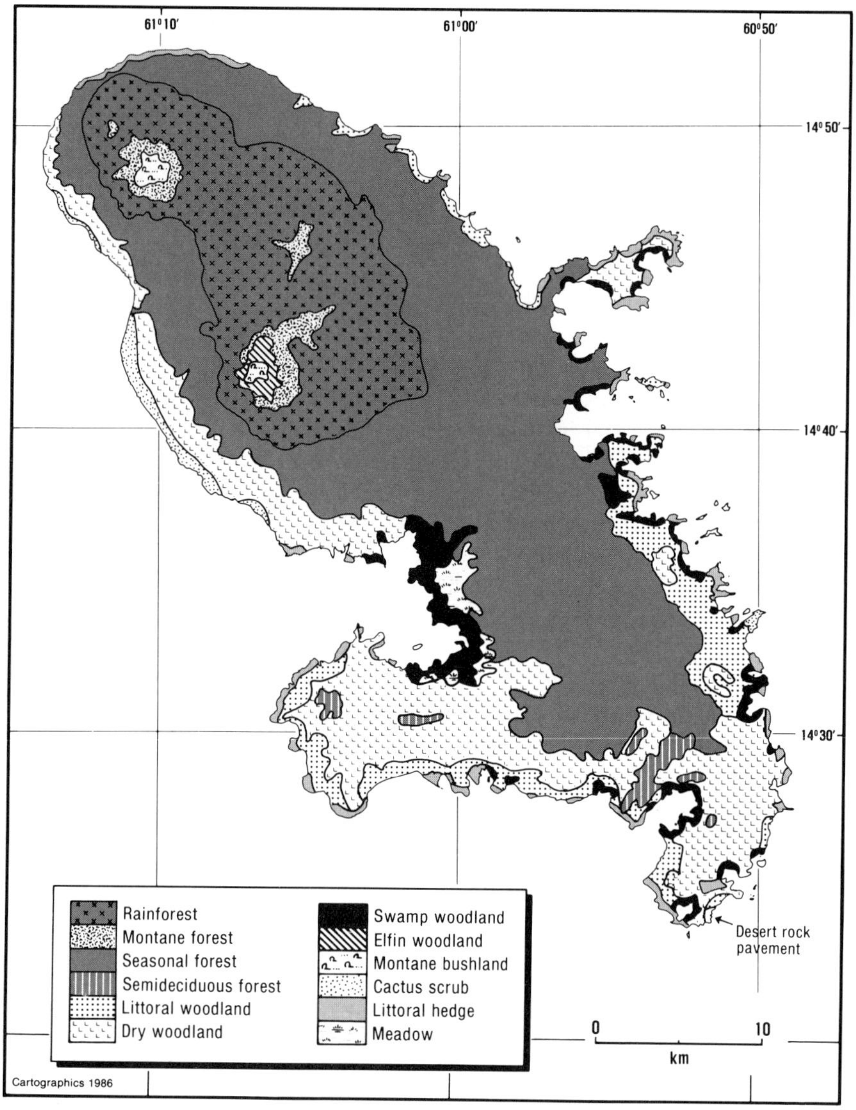

Map 12. Reconstructed distribution of native vegetation. Source: Compiled by author.

tific names associated with them. Usage follows Proctor 1977 for the ferns and fern allies and Fournet 1978 for the others, except as nomenclature has been revised by Howard (1979).

Forest Types

Rainforests are the most luxuriant of the Martiniquan forests and would have been the tall, broadleaf, evergreen forests covering much of the interior between 300 and 900 m on the windward side and 500 and 1,000 m on the leeward slopes. They developed with allophanes and other andosols. There are enough relict sites to demonstrate that the aboriginal rainforests of two millennia ago were of the classic type, replete with many strata and growth forms.

In the present relict rainforest, dominant trees vary from medium to tall, those of the canopy reaching heights of 35 m or more (see figure 4). Diameters vary, with large boles up to 3 m around. Trunks are occasionally buttressed. They are long-stemmed, the first branches well above 15 m from the ground. Crowns of the dominant trees are round or flat; those of the understory trees are more pyramidal and elongated. Stratification is characteristic and often distinguishable to the casual observer.

The dominant stratum forming the canopy today is surmounted by occasional emergents with umbrella-shaped canopies and underlain by a discontinuous middle stratum averaging 15 to 25 m in height. Below these is a fairly continuous stratum of trees and saplings 5 to 12 m high. The shrub layers are also stratified. In some forests the herbaceous layer is complex, although typically the floor is quite open. Rope lianas,

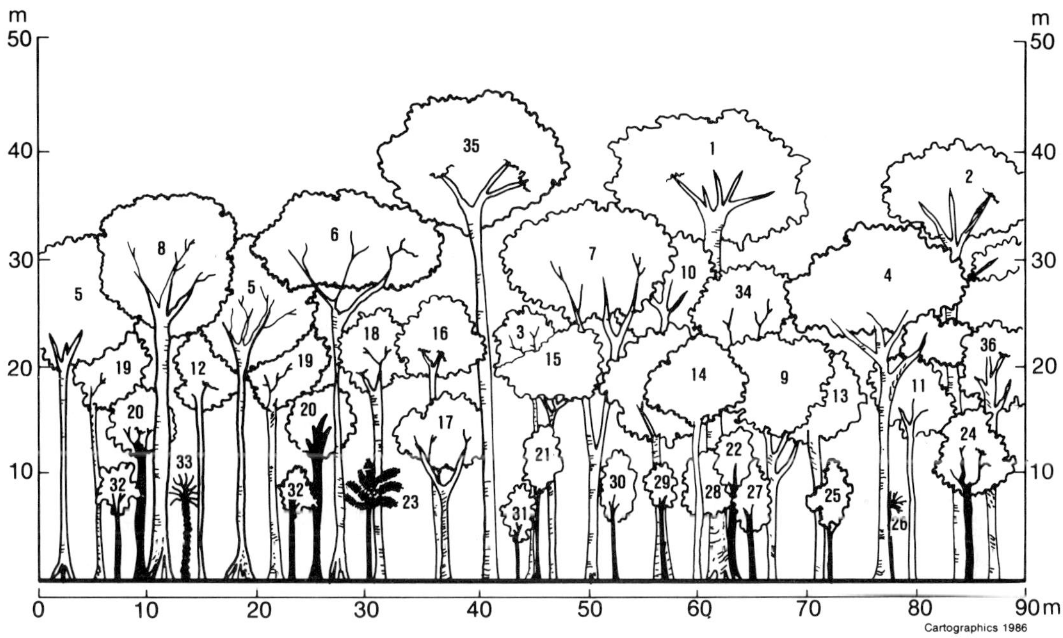

1. *Sloanea dentata*
2. *Dacryodes excelsa*
3. *Cordia collococca*
4. *Chimarrhis cymosa*
5. *Tovomita plumieri*
6. *Sloanea truncata*
7. *Carapa guianensis*
8. *Tapura latifolia*
9. *Pouteria hahniana*
10. *Licania ternatensis*
11. *Nectandra dominicana*
12. *Pouteria chrysophylloides*
13. *Manilkara bidentata*
14. *Guatteria caribaea*
15. *Ternstroemia obovalis*
16. *Pouteria dussiana*
17. *Geoffroea inermis*
18. *Eugenia octopleura*
19. *Nectandra coriacea*
20. *Ocotea leucoxylon*
21. *Phoebe elongata*
22. *Eugenia gregii*
23. *Cyathea arborea*
24. *Aniba bracteata*
25. *Charianthus corymbosus*
26. *Prestoea montana*
27. *Endlicheria sericea*
28. *Marila racemosa*
29. *Cordia laevigata*
30. *Hedyosmum arborescens*
31. *Tetrazygia discolor*
32. *Miconia tricotoma*
33. *Aiphanes minima*
34. *Sterculia caribaea*
35. *Talauma dodecapetala*
36. *Ocotea cernua*

4. Rainforest profile diagram.
Source: After Portecôp 1979,43.

Table 8. The floristic composition and ecological grouping of rainforest species

Tree Strata	Shrub Strata	Herb Strata
Sloanea massoni	*Piper dilatatum*	*Thelypteris reticulata*
Dacryodes excelsa	*Piper aduncum*	*Heliconia caribaea*
Cordia collococca	*Psychotria caribaea*	*Selaginella flabellata*
Chimarrhis cymosa	*Hemitelia muricata*	*Thelypteris balbisii*
Tovomita plumieri	*Psychotria tenuifolia*	*Polybotrya cervina*
Carapa guianensis	*Eugenia pseudopsidium*	*Heliconia bihai*
Tapura latifolia	*Eugenia albicans*	*Scleria secans*
Pouteria chrysophylloides	*Conostegia calyptrata*	*Stigmatopteris rotundata*
Talauma dodecapetala	*Conostegia icosandra*	*Besleria lutea*
Cyathea arborea	*Cestrum megalophyllum*	*Adiantum fragile*
Nectandra antillana	*Clidemia umbrosa*	*Grammitis serrulata*
Pouteria hahniana	*Miconia furfuracea*	*Panicum laxum*
Ternstroemia obovalis	*Piper aequale*	*Isachne disperma*
Ocotea leucoxylon	*Heisteria coccinea*	*Ichnanthus pallens*
Geoffroea inermis	*Palicourea crocea*	*Psychotria antillana*
Manilkara bidentata	*Gonzalagunia hirsuta*	*Scleria latifolia*
Sloanea truncata	*Faramea occidentalis*	*Columnea scandans*
Sterculia caribaea	*Clibadium erosum*	*Borreria ocymoides*
Pouteria dussiana	*Piper dussii*	*Begonia odorata*
Licania ternatensis	*Cephaëlis axillaris*	*Nautilocalyx melittifolius*
Guatteria caribaea	*Psychotria uliginosa*	*Boehmeria ramiflora*
Cordia elliptica	*Miconia laevigata*	*Teliostachya alopecuroidea*
Ocotea cernua	*Miconia prasina*	*Nepsera aquatica*
Aniba bracteata	*Miconia striata*	*Pilea inequalis*
Tetrazygia discolor		*Lindernia rotundifolia*
Phoebe elongata	Species of the higher transition zone	*Ophioglossum reticulatum*
Endlicheria sericea		*Pilea nummulariaefolia*
Eugenia octopleura	*Baccharis pedunculata*	*Blechnum occidentale*
Nectandra coriacea	*Symphysia racemosa*	*Salpichloena volubilis*
Aiphanes minima	*Miconia globuliflora*	*Tectaria trifoliata*
Rondeletia parviflora	*Ilex macfadyenii*	*Gleichenia bifida*
Sloanea dussii	*Rapanea coriacea*	*Borreria verticillata*
Marila racemosa	*Conostegia montana*	*Sauvagesia erecta*
Hedyosmum arborescens		*Pityrogramma calomelanos*
Charianthus corymbosus		*Asplenium formosum*
Sapium caribaeum		*Carludovica insignis*
Eugenia gregii		
Cordia sulcata		
Miconia mirabilis		
Miconia trichotoma		
Psychotria spp.		
Ilex sideroxyloides		
Cecropia peltata		
Ochroma pyramidale		
Oreopanax capitatum		
Syzygium jambos		
More mesophytic species		
Simaruba amara		
Inga laurina		
Inga ingoides		
Species of the higher transition zone		
Podocarpus coriaceus		
Clusia plukenetii		
Prestoea montana		
Inga coruscans		
Myrcia martinicensis		
Calyptranthes fasciculata		
Freziera undulata		
Byrsonima martinicensis		

Sources: Modified from Stehlé 1937, 1945–46, 215–28; Portecôp 1979, 44–46; Kimber 1969, 69–72.

Table 8. *continued*

Epiphytes	Lianas	Parasites
Asplundia rigida	*Mikania hookeriana*	*Dendrophthora martinicensis*
Guzmania lingulata	*Mikania cordifolia*	*Phoradendron hexastichum*
Philodendron giganteum	*Mikania micrantha*	*Phoradendron chrysocarpum*
Nephrolepis rivularis	*Cissus sicyoides*	
Jacquiniella globosa	*Cissampelos pareira*	
Hymenodium crinitum	*Alloplectus cristatus*	
Epidendrum nocturnum	*Philodendron oxycardium*	
Epidendrum ibaguense	*Philodendron lingulatum*	
Elleanthus dussii	*Sarchorhachis incurva*	
Dichaea muricata	*Anthurium palmatum*	
Peperomia nigropunctata	*Monstera pertusa*	
Ophioglossum reticulatum	*Coccoloba ascendens*	
Peperomia rotundifolia	*Passiflora suberosa*	
Marcgravia umbellata	*Paullinia vespertilio*	
Peperomia hernandifolia	*Anthurium cordatum*	
Hymenophyllum hirtellum	*Hillia parasitica*	
Peperomia trifolia	*Smilax guianensis*	
Asplenium salicifolium	*Mikania latifolia*	
Elaphoglossum dussii	*Dioscorea polygonoides*	
Hymenophyllum polyanthos	*Centropogon berterianus*	
Pleurothallis wilsoni	*Clusia rosea*	
Grammitis asplenifolia	*Asplundia rigida*	
Grammitis serrulata		
Elaphoglossum feei		
Peperomia trifolia		
Trichomanes alatum		
Pleurothallis ruscifolia		
Elaphoglossum lingua		
Polypodium glaucophyllum		
Epidendrum dussii		
Epidendrum pallidiflorum		
Peperomia hirtella		
Isochilus linearis		
Stelis scabrida		
Pleurothallis imraei		
Cochilidium linearifolium		
Elaphoglossum glabellum		
Epidendrum nocturnum		
Eleanthus cephalotus		
Ionopsis satyrioides		
Epidendrum ciliare		
Epidendrum difforme		
Leiphaimos aphylla		
Octomeria graminifolia		
Peperomia questeliana		
Phyllogonium sp.		
Rhynchelytrum repens		
Plagiophila sp.		

climbing vines, and some epiphytes add to the diversity of growth forms in the rainforests, as do the rarely observed constrictors of the *Clusia* and *Ficus*.

Floristic diversity in the rainforest is today, and must have been earlier, the richest of all vegetation types on the island. It is similar to the *Sloanea-Dacryodes* forests described for neighboring islands (Beard 1949a, 1955). Henri Stehlé in the 1930s and 1940s, Jacques Portecôp in the mid-1970s, and Kimber in the early 1960s made collections and relevées in these forests (Stehlé 1937, 1945–46; Portecôp, 1979; Kimber 1969). Table 8 summarizes within growth-form categories the modern species and gives some indication of the relative richness of rainforests.[6] A large number of tree species having a low number per hectare dominated the forests. Nevertheless, there were some trees that had wide distribution throughout the forest. The three chataigniers—especially *Sloanea massoni* together with *Pouteria hahniana, Talauma dodecapetala,* and *Licania ternatensis*—must have been abundant in aboriginal times. *Dacryodes excelsa* would also have been frequent in the canopy.

By present evidence, the middle story of trees was dominated by the members of the Lauraceae, especially species of the genera *Nectandra* and *Ocotea*. In addition, *Tovomita plumieri, Tapura latifolia, Sterculia caribaea,* and *Byrsonima martinicensis* would have been common. *Podocarpus coriaceus,* the only gymnosperm of the island, was more often in this stratum of the forest then confined to montane forests, as it is today. The

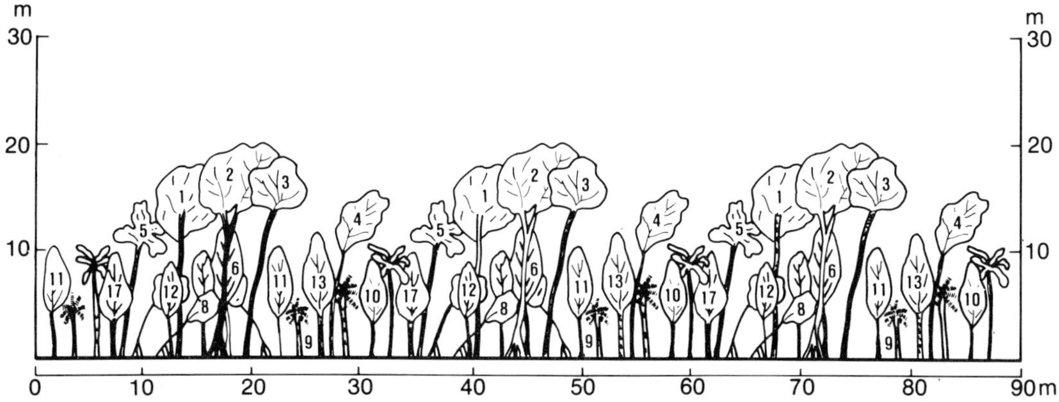

(original transect 35 m)

1. *Cordia elliptica*
2. *Tovomita plumieri*
3. *Pouteria chrysophylloides*
4. *Nectandra dominicana*
5. *Cecropia peltata*
6. *Pouteria dussiana*
7. *Podocarpus coriaceus*
8. *Clusia plukenetii*
9. *Prestoea montana*
10. *Inga coruscans*
11. *Calyptranthes fasciculata*
12. *Myrcia martinicensis*
13. *Frezeria elegans*
14. *Cyathea arborea*

5. *Upper montane forest profile diagram.*
Source: After Portecôp 1979,48.

lower story of trees must have been filled with members of the Aquifoliaceae such as *Ilex sideroxyloides,* Lauraceae such as *Endlicheria sericea,* Myrtaceae such as *Eugenia* spp., and especially those called "café grand bois" of the Rubiaceae, various *Psychotria* species.

Montane forests located on the northern mountain massifs occupied a much smaller area than the rainforests. On the windward side they were found at elevations between 900 and 1,000 m; on the leeward side, between 1,000 and 1,100 m. They typically occupied entisols, and the yellow earths or andosols developed on volcanic rock. In their physiognomy they reflected the cooler temperature, more moisture, steeper slopes, and increased wind speeds of higher elevations. They were similar in morphology to rainforests but lacked emergents and the upper canopy level; only two tree strata were represented at about 20 and 10 m levels (see figure 5). An interrupted shrub layer reached 3 m.

The thinner canopy passed more light to the floor than in rainforests, so that mosses on trunks, buttresses, and stilt roots were more prevalent than in rainforests. Occasionally the trees were burdened with tank epiphytes beyond their capacity to support them, and the trees collapsed under the burden. This kind of opening may have been more often the site of gap regeneration than in wind throws of the rainforests. There is no reason to doubt that, then as now, the montane forest canopy became lower with elevation and the forests thinned abruptly to pole thickets in exposed sites. Table 9 lists species commonly present in montane forests.

Tree species were fewer in montane forests than in the rainforests, but it is still difficult to name characteristic dominants. Frequent canopy trees would have been *Chimarrhis cymosa* and the endemic *Pouteria hahniana* of the rainforest, as well as *Sterculia caribaea.* Lower-story trees were *Byrsonima martinicensis, Hedyosmum arborescens, Genipa americana,* and various species of *Rondeletia, Myrcia,* and *Ilex.* Locally conspicuous would have been the *Prestoea* (formerly *Euterpe*) palms, *Podocarpus coriaceus,* and the tree ferns *Cyathea arborea* and *C. muricata.* The epiphytic *Clusia plukenetii* started in a tree axil in the usual manner, but later it became earth-bound with large boles.

The scandent *Dioscorea polygonoides, Manettia dominicensis,* and *Scleria* spp. may have been draped over the shrub layer. Pure stands of colorful anthuriums, ferns, and a profusion of orchids were probably abundant in prehuman times, as they are today.

Seasonal forests occupied the largest areas of the vegetation types for the prehuman period. They occupied very diverse habitats in terms of soils—lithosols, andosols, fersiallitics, ferralitics, and allophanes—but generally were confined to localities drier and warmer than the rainforest. They were found at elevations ranging from 10 to 500 m depending upon exposure. These forests were less tall, lacking the topmost layer, although emergents reached as high as 40 m (see figure 6). Judging by remnants in the present landscape, these forests had two tree stories, without sharp stratification, and possessed locally elaborate shrub and field layers.

Table 9. The floristic composition of montane forests

Tree Strata	Shrub Strata	Herb Strata	Epiphytes	Lianas
Cordia elliptica	*Conostegia calyptrata*	*Erythrodes plantaginea*	*Asplenium salicifolium*	*Hillia parasitica*
Chimarrhis cymosa	*Conostegia icosandra*	*Blechnum occidentale*	*Elaphoglossum petiolatum*	*Manettia dominicensis*
Tovomita plumieri	*Miconia trichotoma*	*Apteria aphylla*	*Hymenophyllum hirtellum*	*Anthurium cordatum*
Pouteria chrysophylloides	*Conostegia montana*	*Ischaemum latifolium*	*Pleurothallis wilsonii*	*Dioscorea polygonoides*
Nectandra dominicana	*Miconia globuliflora*	*Besleria lanceolata*	*Stelis scabrida*	*Centropogon berterianus*
Sterculia caribaea	*Rapanea coriacea*	*Begonia macrophylla*	*Pleurothallis rucifolia*	
Cecropia peltata	*Ilex macfadyenii*	*Scleria secans*	*Peperomia trifolia*	
Pouteria dussiana	*Symphysia racemosa*		*Peperomia hernandifolia*	
Podocarpus coriaceus	*Baccharis pedunculata*		*Peperomia hirtella*	
Clusia plukenetii				
Prestoea montana				
Inga coruscans				
Calyptranthes fasciculata				
Myrcia martinicensis				
Freziera undulata var. *elegans*				
Cyathea arborea				
Myrcia deflexa				
Byrsonima martinicensis				
Hedyosmum arborescens				
Genipa americana				
Ilex spp.				
Pouteria hahniana				
Rondeletia spp.				

Sources: Modified from Stehlé 1937; Portecôp 1979, 48–49; Kimber 1969, 72–73.

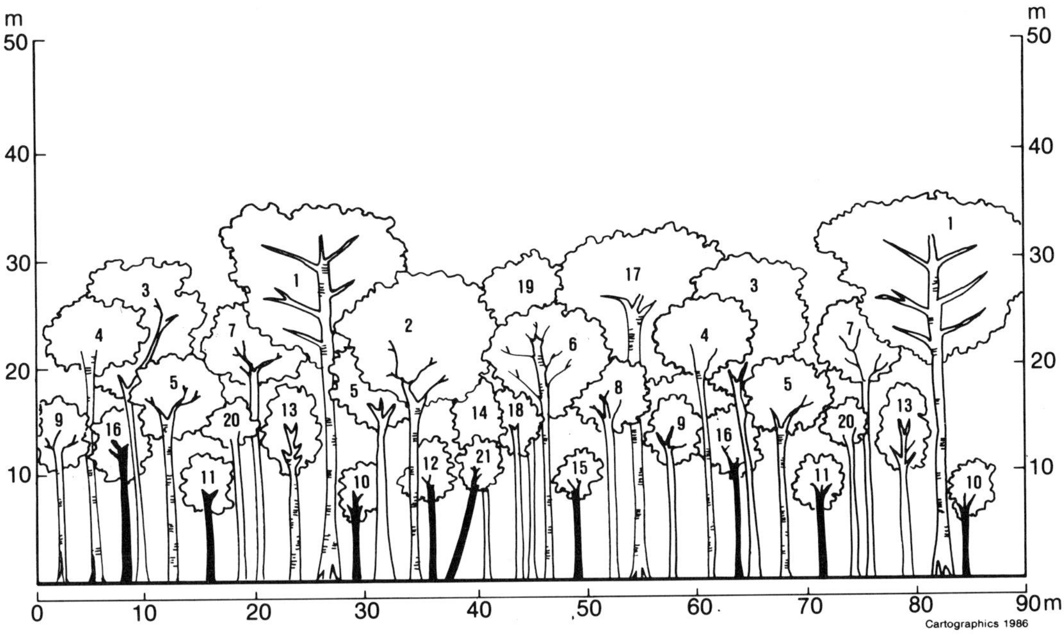

(original transect 65 m)

1. *Ceiba pentandra*
2. *Hymenaea courbaril*
3. *Cedrela mexicana*
4. *Inga ingoides*
5. *Inga laurina*
6. *Simaruba amara*
7. *Lonchocarpus pentaphyllus*
8. *Sapium caribaeum*
9. *Nectandra patens*
10. *Ficus crassinervia*
11. *Myrciaria floribunda*
12. *Daphnopsis caribaea*
13. *Ocotea leucoxylon*
14. *Pimenta racemosa*
15. *Byrsonima coriacea*
16. *Phoebe elongata*
17. *Chimarrhis cymosa*
18. *Geoffroea inermis*
19. *Manilkara bidentata*
20. *Cecropia peltata*
21. *Calophyllum calaba*
22. *Cordia alliodora*

6. Seasonal forest profile diagram.
Source: After Portecôp 1979,37.

A few important seasonal forest trees had buttresses, but these were less common than in rainforests. The most conspicuous of these was the very large kapok or fromager, *Ceiba pentandra*,[7] with its high plank buttresses. Basal swelling was rare in these forests, although superficial roots were found in soils with rapid drainage, and aerial and stilt roots where the soil was saturated. Lianas and epiphytes would have been less abundant and thorny plants more common. Although the largest proportion of the trees remained evergreen, a distinct seasonal variation was characteristic of the canopy layer of all stands. The deciduous habit ranges from total leaf-fall in *Cedrela mexicana* through facultatively deciduous species such as *Hymenaea courbaril* to those species that lose only a few of their leaves at any one time. Generally, leaves are replaced at the same time they are shed, so there was frequently little witness to the deciduous habit for the forest as a whole. It is possible that the present proportion of trees possessing the deciduous habit is greater than that in earlier times because of selective exploitation of evergreens.

Table 10. The floristic composition and ecological grouping in seasonal forests

Tree Strata	Shrub Strata	Herb Strata	Other
Characteristic Mesophytic Species	*Tabernaemontana citrifolia*	*Olyra latifolia*	EPIPHYTES
Cordia sulcata	*Piper dilatatum*	*Oplismenus hirtellus*	*Epidendrum ciliare*
Inga laurina	*Miconia virescens*	*Blechum brownei*	*Epidendrum difforme*
Byrsonima coriacea	*Miconia prasina*	*Pharus latifolius*	*Polystachya concreta*
Simaruba amara	*Miconia striata*	*Scleria latifolia*	*Polypodium polypodioides*
Inga ingoides	*Psychotria tenuifolia*	*Alternanthera sessilis*	*Polypodium aureum*
Lonchocarpus pentaphyllus	*Piper dussii*	*Desmodium triflorum*	*Polypodium piloselloides*
Ceiba pentandra	*Odontonema nitida*	*Herissantia crispa*	*Polypodium astrolepis*
Eugenia monticola	*Palicourea crocea*	*Sida acuta*	LIANAS
Ocotea martinicensis	*Mimosa ceratonia*	*Scutellaria havanensis*	*Mucuna urens*
Eugenia lambertiana	*Piper amalago*	*Adiantum tetraphyllum*	*Bauhinia excisa*
Myrciaria floribunda	*Solanum torvum*	*Spiranthes torta*	*Securidaca diversifolia*
Daphnopsis americana	*Clidemia hirta*	*Ponthieva petiolata*	*Lasiacis divaricata*
Sapindus saponaria	*Lepianthes peltata*	*Borreria parviflora*	*Macfadyena unguis-cati*
Drypetes dussii	*Myrcia splendens*	*Emilia sonchifolia*	*Clusia rosea*
Guazuma ulmifolia	*Gonzalagunia hirsuta*	*Scoparia dulcis*	*Asplundia rigida*
Hymenaea courbaril	*Lantana camara*	*Cyperus rotundus*	PARASITES
Cedrela mexicana	*Solanum triste*	*Commelina diffusa*	*Phoradendron trinervium*
Homalium racemosum	*Chiococca alba*	*Laportea aestuans*	
Beilschmiedia pendula	*Urena lobata*	*Blechnum occidentale*	
Chionanthus compactus	*Psidium guajava*	*Asplenium formosum*	
Myrcia fallax	*Eugenia lambertiana*	*Adiantum tenerum*	
Margaritaria nobilis			
Ormosia monosperma			
Ficus trigonata			
Cecropia peltata			
Ochroma pyramidale			
Hura crepitans			
Swietenia macrophylla			
Goeffroea inermis			
Sapium caribaeum			
Zygia latifolia			
Species More Xerophytic			
Tabebuia pallida			
Calophyllum calaba			
Zanthoxylum martinicense			
Citharexylum fruticosum			
Bucida buceras			
Cordia alliodora			
Pimenta racemosa			
Species More Hygrophytic			
Nectandra coriacea			
Cordia elliptica			
Chimarrhis cymosa			
Ocotea leucoxylon			
Sapium caribaeum			
Manilkara bidentata			
Aniba bracteata			
Phoebe elongata			
Spondias mombin			
Mammea americana			
Meliosma herbertii			

SOURCES: Portecôp 1979, 38–39; Kimber 1969, 73–76.

Table 10 lists the species composition recovered in seventeen *relevées* by Portecôp in the mid 1970s (Portecôp 1979,38–39).

Although there were no species generally dominant throughout stands, judging from studies on the island, there were regional dominants. Locally important in the northern part of the island were the canopy trees *Simaruba amara, Geoffroea inermis, Manilkara bidentata, Hymenaea courbaril, Calophyllum calaba, Sapium caribaeum, Ocotea* spp., *Inga ingoides, Eugenia* spp. *Chimarrhis cymosa,* and *Meliosma herbertii.* Understory trees included *Daphnopsis americana, Pimenta racemosa, Nectandra coriacea, Ocotea leucoxylon, Inga laurina,* and *Cordia sulcata.*

Conspicuous by their sheer size, but probably not abundant in any of the forests, were *Mammea americana, Cedrela mexicana,* and *Ceiba pentandra.* The last two may grow to reach 2 m in diameter. *Hura crepitans* was associated with these forests.

In the southern seasonal forests leguminous trees, especially savonettes of the genus *Lonchocarpus, Geoffroea inermis,* and *Zygia latifolia,* were probably as conspicuously dominant trees in the riverine or gallery forest facies of the aboriginal seasonal forests as they are today. Though these are useful species and may have been concentrated culturally, woody legumes are important members of rich tropical forests. Other canopy species are *Cedrela mexicana, Hymenaea courbaril,* and *Byrsonima coriaceae.* The understory trees are much the same as in the northern part of the island.

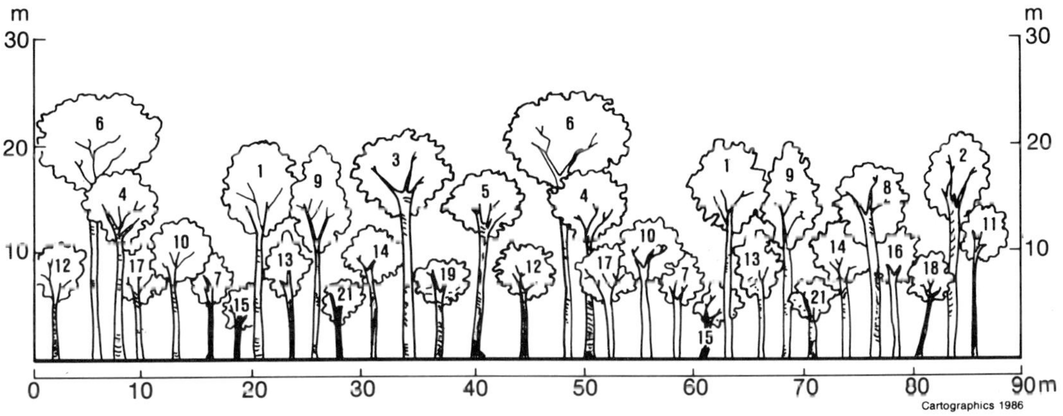

(original transect 60 m)

1. *Tabebuia pallida*
2. *Coccoloba diversifolia*
3. *Bursera simaruba*
4. *Bucida buceras*
5. *Inga laurina*
6. *Simaruba amara*
7. *Cassipourea elliptica*
8. *Lonchocarpus benthamianus*
9. *Pimenta racemosa*
10. *Citharexylum fruticosum*
11. *Pisonia fragrans*
12. *Guettarda scabra*
13. *Erythrina corallodendron*
14. *Zanthoxylum martinincense*
15. *Anacardium occidentale*
16. *Beurreria succulenta*
17. *Picramnia pentandra*
18. *Myrcia citrifolia*
19. *Casearia decandra*
20. *Haematoxylon campechianum*

7. *Semideciduous forest profile diagram.*
Source: After Portecôp 1979,33.

Table 11. The floristic composition of some semideciduous forests

Tree Strata	Shrub Strata	Herb Strata	Other
Tabebuia pallida	*Randia aculeata*	*Sporobolus indicus*	LIANAS
Coccoloba swartzii	*Zanthoxylum spinifex*	*Chloris barbata*	*Guzmania lingulata*
Bucida buceras	*Croton bixoides*	*Scleria lithosperma*	*Wittmackia lingulata*
Bursera simaruba	*Croton hircinus*	*Dichromena nervosa*	*Clusia rosea*
Inga laurina	*Ardisia obovata*	*Kalanchoe pinnata*	EPIPHYTES
Simaruba amara	*Coccoloba pubescens*		*Tillandsia utriculata*
Cassipourea guianensis			*Polypodium aureum*
Lonchocarpus benthamianus			PARASITES
Pimenta racemosa			*Phoradendron chrysocarpum*
Citharexylum fruticosum			*Phoradendron trinervium*
Pisonia fragrans			
Guettarda scabra			
Erythrina corallodendron			
Zanthoxylum martinicense			
Anacardium occidentale			
Beurreria succulenta			
Picramnia pentandra			
Myrcia citrifolia			
Casearia decandra			
Haematoxylon brasiletto			

Sources: Portecôp 1979, 33.

On the dry margins of the seasonal forest along the south coast is the semideciduous forest, which rarely exceeds 20 m and which represents a vegetation type once much more extensive (see figure 7). It occupies, and did occupy, the lower tropical subhumid bioclimate on gently sloping land covered by vertisols high in montmorillonitic clays. The canopy consisted of *Bursera simaruba* and *Tabebuia pallida*, two deciduous trees, from complete leaf-fall in the height of the *carème* from February to March, as well as the evergreens *Coccoloba swartzii, Bucida buceras, Inga laurina, Simaruba amara, Pimenta racemosa, Citharexylum fruticosum, Lonchocarpus benthamianus*. The understory included entirely evergreen species *Guettarda scabra, Picramnia pentandra, Zanthoxylum martinicense*, and *Myrcia citrifolia*. Shrubs, grasses, and sedges as well as four lianas typically found in this forest type are found in table 11.

These forests occupied an estimated 77 percent of the vegetation cover just prior to the advent of South American Indians to Martinique.[8] The rainforests would have accounted for 22 percent (roughly 25,000 ha), the montane forest for less than 5 percent (some 3,000 ha), the seasonal forests for 52 percent (some 55,000 ha). These are the vegetation types that best fit the category often mentioned in the first accounts of the island, *les bois de haute futaie* or *les grands bois*. Typically, the rainforests are brought to mind by these phrases, but both the seasonal and the montane forests were called thus by the European visitors and early colonists (Coppier 1645).

Woodland Types

The woodlands, though collectively occupying much less area than the forests, extended over a wider ecological range. They were physiognomically and geographically quite disparate.

Littoral woodland had a windward coastal distribution varying in width from less than 1 km in the north to possibly 3 km in the south. It was bounded on the seaward side by littoral hedge and mangrove swamp woodland. On the interior it bordered dry woodland. Some strips of woodland too narrow to represent on the map probably bordered the coastal margins of the dry woodland in protected, low, but well-drained areas of the southern peninsulas, and the seasonal forest of the Lamentin Plain.

In terms of physiognomy and structure, the littoral woodland probably resembled the present examples to be found in the Caravelle Peninsula and Marin Commune. These relict woodlands illustrate a continuous single-story canopy that in moist and protected places, rose to 20 m, but that more often was windshorn, with occasional tufts of dead branch extending from the canopy. Where protected, the tree story graded imperceptibly into the top of the shrub and sapling layer. The canopy let through flecks of sunlight to the floor. Even so, the ground was mostly bare, with a number of coarse herbs and grasses growing in the thin soils. There were few lianas, but epiphytes were conspicuous and abundant. Salt spray, wind, and brackish water all exerted some controlling influence on this vegetation (see figure 8).

Tabebuia pallida and *Calophyllum calaba* are characteristically the tall trees in the modern woodland. These would have been joined by some tall trees of the present seasonal forest such as *Bursera simaruba* and *Lonchocarpus* spp. Reaching less high, but far more extensive throughout the woodland, were *Coccoloba swartzii* and other species that

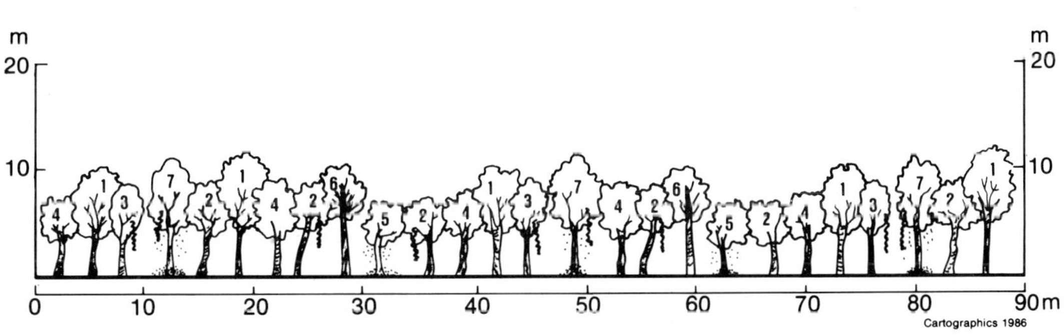

(original transect 45 m)

1. *Hippomane mancinella*
2. *Thespesia populnea*
3. *Tabebuia pallida.*
4. *Coccoloba uvifera*
5. *Capparis indica*
6. *Pisonia fragrans*
7. *Erythrina berteroana*

8. Littoral woodland profile diagram.
Source: Compiled by author.

Table 12. The floristic composition of littoral and dry woodlands

Tree Strata	Shrub Strata	Herb Strata	Other
Bursera simaruba	*Croton bixoides*	*Sporobolus indicus*	EPIPHYTES
Lonchocarpus benthamianus	*Eugenia ligustrina*	*Oxalis corniculata*	*Tillandsia utriculata*
Tabebuia pallida	*Randia aculeata*	*Kalanchoe pinnata*	*Polypodium aureum*
Pisonia fragans	*Myrcia citrifolia*	*Scleria filiformis*	*Clusia rosea*
Citharexylum fruticosum	*Tabernaemontana citrifolia*	*Ruellia coccinea*	
Zanthoxylum martinicense	*Lantana camara*	*Desmodium canum*	LIANAS
Zanthoxylum monophyllum	*Coccoloba pubescens*	*Stachytarpheta jamaicensis*	*Abrus precatorius*
Coccoloba swartzii	*Capparis flexuosa*	*Dichromena nervosa*	*Stigmaphyllon cordifolium*
Erythrina berteroana	*Cornutia pyramidata*	*Dichanthium isachaemum*	*Passiflora suberosa*
Calophyllum calaba	*Calliandra tergemina*	*Furcraea tuberosa*	*Petrea kohautiana*
Erythrina corallodendron	*Plumieria alba*	*Trimeza martinicensis*	
Beurreria succulenta	*Eugenia lambertiana*	*Chloris barbata*	PARASITES
Bunchosia nitida	*Rauvolfia viridis*	*Jatropha gossypifolia*	*Cuscuta americana*
Ximenia americana	*Zanthoxylum spinifex*	*Abilgaardia monostachya*	*Cassytha filiformis*
Casearia decandra	*Erythroxylon brevipes*	*Wedelia calycina*	*Phoradendron trinervium*
Pimenta racemosa	*Cassia chamaecrista*	*Crotalaria retusa*	*Phoradendron chrysocarpum*
Cordia alliodora	*Zygia latifolia*	*Scleria secans*	
Ceiba pentandra	*Psychotria nervosa*	*Pectis martinicensis*	
Myrcia citrifolia	*Bernardia corensis*	*Vernonia arborescens*	
Myrcia splendens	*Jacquinia arborea*	*Cassia obtusifolia*	
	Cephalocereus nobilis	*Pilea microphylla*	
	Chiococca alba	*Mimosa pudica*	
	Pithecellobium unguis-cati		
	Allophyllus racemosus		
	Acacia macrantha		
	Acacia parvifolia		
	Croton hircinus		
	Croton flavens		
	Ardisia obovata		
	Leucaena leucocephala		
	Daphnopsis caribaea		
	Parkinsonia aculeata		

SOURCES: Portecôp 1979, 30–31; Kimber 1969, 76–78.

can take the form of a shrub or a tree, depending on local environmental factors. The species characteristics of this vegetation are in table 12.

Dry woodland, it is assumed, was an interior, lowland vegetation type present in areas of severe annual drought with regular drying winds and shallow soils with impeded drainage. These conditions obtain in the southern peninsulas and on the lower midslopes of the Caribbean north coast. There woodlands had, it is supposed, an uneven, occasionally interrupted canopy to 15 m high and a dense, well-developed shrub layer (see figure 9). Terrestrial bromeliads, lianas, scandent cactus, and platyopuntias also constituted the floor stratum. The proportion of deciduous trees varied, as it does today, from being almost 50 percent of the individuals to being virtually unrepresented. In the more evergreen subtypes the tree leaves were predominantly hard and thickly cutinized.

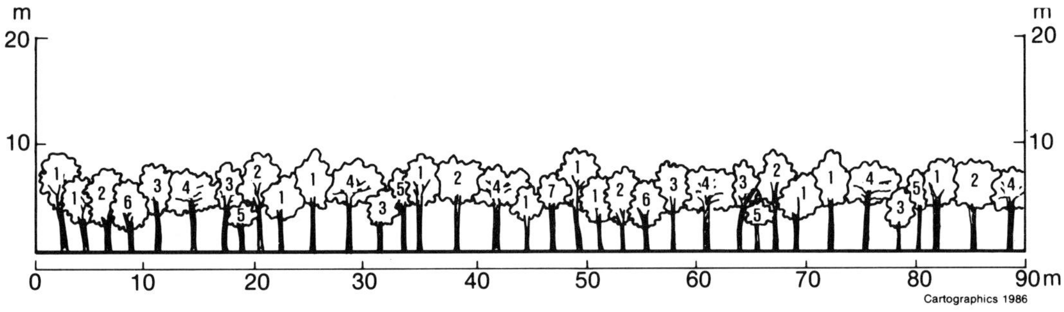

(original transect 40 m)

1. *Tabebuia pallida*
2. *Bursera simaruba*
3. *Lonchocarpus benthamianus*
4. *Pisonia fragrans*
5. *Casearia decandra*
6. *Zanthoxylum martinicense*
7. *Beurreria succulenta*

9. Dry woodland profile diagram.
Sources: After Portecôp, 1979; and author.

Other shrubs were thorny, an occasional tree buttressed, and palms would have been present in varying amounts from place to place.

Tall trees up to 23–25 m that are common today in the dry woodland are *Pisonia fragans, Bursera simaruba, Tabebuia pallida,* as well as the giant *Ceiba pentandra.* The shorter bay trees *Pimenta racemosa* and *Cordia alliodora,* as well as the *Myrcia* species, may have occurred in large numbers but not in the high proportions they do now (see table 12). Shrubs commonly found in the dry woodland were *Pithecellobium unguis-cati, Parkinsonia aculeata, Randia aculeata, Coccoloba swartzii, Rauvolfia viridis, Daphnopsis caribaea,* and *Psychotria nervosa,* to judge from the present-day floristic composition. Shrubby species related characteristically to the upper story, such as *Zanthoxylum punctatum* and *Myrcia citrifolia,* were not infrequent. Numerous trees and shrubs were parasitized by different mistletoes of the genus *Phoradendron.*

Swamp woodlands were dominated by low trees with wide crowns of leathery, broad, evergreen leaves. They were restricted to coastal or tidal waters. It is estimated that in precultural times over 4,000 ha were covered by these edaphic, wet woodlands. In contrast to most vegetation types, these were very simple, being composed of very few species. In Martinique there were three kinds of swamp woodlands: the mangrove swamp referred to as a "palétuvier" or "manglier," the *Annona* swamp woodland called "mamain" swamp, and the *Pterocarpus officinalis* swamp called "mangle medaille."

Mangrove swamp woodlands grow in the quiet reaches of inlets and in the lee of promontories, where sheets of tidal salt water wash across muddy and rocky bottoms. They are and were the commonest of the three swamp woodlands. There were readily recognized by stilt roots or pneumatophores on the seaward sides and by tangled branching root

Table 13. The floristic composition of mangrove and swamp woodlands

Tree Strata	Shrub Strata	Herb Strata	Lianas and Epiphytes
Saline and Brackish	*Machaerium lunatum*	Wet Areas	*Cydista aequinoctialis*
Rhizophora mangle	*Parkinsonia aculeata*	*Acrostichum aureum*	*Rhabdadenia biflora*
Avicennia germinans	*Pavonia spicata*	*Eleocharis geniculata*	*Tillandsia polystachya*
Laguncularia racemosa		*Cyperus alopecuroides*	*Polypodium aureum*
Conocarpus erecta	Dry Areas	*Cyperus elegans*	
Pterocarpus officinalis	*Morinda citrifolia*	*Mariscus planifolius*	
	Bontia dapnoides	*Torulinium odoratum*	
Palustrine	*Acacia farnesiana*	*Mariscus lingularis*	
Pterocarpus officinalis		*Pycraeus polytachyos*	
		Fimbristylis dichotoma	
Dry Areas		*Cyperus luzulae*	
Annona glabra		*Fimbristylis complanata*	
		Paspalum distichum	
		Ludwigia octovalvis	
		Ludwigia hissopifolia	
		Dry Areas	
		Batis maritima	
		Sporobolus virginicus	
		Philoxerus vermicularis	
		Heliotropium curassavicum	
		Stenotaphrum secundatum	
		Sesuvium portulacastrum	

SOURCES: Modified from Portecôp 1979, 19; Kimber 1969, 87.

patterns. In contrast to regions farther south, even in the undisturbed, original mangrove swamp, the trees were probably not much more than 10 m tall and, more commonly, they were 5 m to 8 m tall. Of this floristically simple vegetation type, particular examples consisted of pure strands of the stilt-root *Rhizophora mangle*[9] behind which was a zone of combined *Avicennia germinans* and *R. mangle* followed by a zone of *Laguncularia racemosa* and, occasionally, *Conocarpus erecta*. These trees covered a shrub layer containing *Machaerium lunatum, Parkinsonia aculeata,* and *Pavonia spicata* (see table 13). In other locations, drier and sandier, the sequence may have always been less clear, although there is a tendency for *Rhizophora* and *Avicennia* to be seaweed of the other mangrove species.

In the inner margins, especially where the sand becomes evident at the surface and the water is practically fresh, a small tree or large shrub, *Annona glabra,* is dominant and the four mangrove tree species are absent.[10] These fresh-water swamp areas are locally referred to as "mamain" swamps. Along river flank depressions where standing fresh water forms swampy, marshy ground, *Pterocarpus officinalis* or mangle medaille grew to 20 m in height. The other tall tree in the palustrine vegetation type is *Calophyllum calaba,* a member of the littoral woodland.

The understory in the two fresh-water swamps was on slightly acid muds covered with herbaceous growth. Whether there were more lianas, epiphytic ferns, and arboreal bromeliads in the precultural period than now is quite uncertain, judging from sites on other Lesser Antillean islands. On seaward muddy bottoms the aquatic *Thalassia testudinum* and *Cymodocea filliformis* formed colonies. In muddy lagoons not colonized by mangrove, the seagrass *Ruppia maritima* also formed colonies on the shallow bottom.

Annona swamp woodland probably always occupied much less territory than mangrove and may well have been confined to widely scattered regions behind sandy beach ridges. The largest areas were in the flats behind Anses d'Arlets, Lamentin, at Larienty, about Marin and the region surrounding the Marigot de Bellevue in Sainte-Marie Commune, and by the inner southern shore of Robert Haven. The *Pterocarpus* woodland is known today only in one location north of the Galion River and may have always been highly restricted and limitied to that particular environment.

Laural woodlands were found on the crests of exposed spurs on slopes of the Pitons, on the Morne Lorrain, and in the Champflore Basin region, where exposure to drying winds results in lower physiological humidities. Patches of woodland were also to be found on the high spurs above Prêcheur, Grand' Rivière, Macouba, and Lorrain (Stehlé 1937, 251–52). The soils on these ridges are andosols developed on the older Pliocene and Pleistocene volcanics. This woodland had a moderately high (15–20 m) dense canopy composed of hard, mesophyll evergreen leaves. There were few lianas and virtually no shrubs, but the ground flora was rich, including the *Heliconia* spp., which grew to 3 m, and many species of ferns (see table 14). This vegetation was easily penetrated except for the poor footing. The dominant trees are members of the Lauraceae, especially species of *Nectandra* and *Ocotea*, which were second-story trees of the rainforest. The only exception is *Styrax glaber* of the Styracaceae.

Table 14. Selected species from the laural woodlands

Tree Strata	Shrub Strata	Herb Strata
Nectandra antillana	*Endlicheria sericea*	*Eupatorium magdalenae*
Nectandra dominicana	*Chrysobalanus cuspidatus*	*Heliconia bihai*
Nectandra coriacea	*Dahlbergia monetaria*	*Heliconia caribaea*
Ocotea cernua	*Hedyosmum arborescens*	*Blechnum ryanii*
Styrax glaber	*Clibadium erosum*	*Pteris biaurita*
Phoebe elongata	*Charianthus corymbosus*	*Pityrogramma chrysophylla*
Persea urbaniana	*Cestrum alternifolium*	
Beilschmiedia pendula	*Piper reticulatum*	
Aniba ramageana	*Piper dussii*	
	Piper glabrescens	
	Piper aequale	

SOURCES: Modified from Stehlé 1937; Kimber 1969, 79–80.

Laural woodlands formed islands in the rainforest, where wind stress was severe or where erosion of the soil caused poor water retention. These woodlands were confined to middle elevations below the cooler montane forest zone. This vegetation may occupy a particular site for a long time, so it can be considered an arrested seral stage in the rainforest.

Just as in descending to the coast the forest yielded to woodlands, so in ascending the higher mountain slopes the complex, many-layered vegetation again became reduced in stature and simplified in structure. Strata were eliminated and forests yielded to woodland. Trees were low, often less than 10 m tall. The tops of hard, evergreen leaves were windblown from the almost constant buffeting of the winds off the Atlantic Ocean. Seen from within, the trunks and branches formed a tight and narrow canopy. Mosses and lichens blanket branches, trunks, and the ground. Virtually undisturbed by humans, the present aspect probably resembles that of precultural times.

Both the northern mountain massifs had a ring of elfin woodland. The woodland must have had a checkered history on Mont Pelée, for it was directly vulnerable to volcanic eruptions. In all probability, elfin woodland has always occurred on the mountain after the type was constituted, but at times it must have been attenuated and fragmented. On the Carbet Pitons and on Piton Gelé this type had, presumably, a more stable recent history (France, Office des Forêts Nationales 1981).

The characteristic dominant was, and is today, *Clusia plukenetii*, called "palétuvier du montagne" or "manglier alpine" (Pinchon 1971, 54). *Charianthus nodosus, Rondeletia stereocarpa,* and *Didymopanax* tree melastomes and myrtles were common members in the canopy (see table 15). The shrub *Cephaëlis axillaris* was conspicuous by its blue berries, as were the epiphytic orchids by their lovely white or purple flowers. *Charianthus corymbosus* and *Rapanea coriacea* add their own colors in the openings of the woodland. *Peperomia, Psychotria,* and *Elaphoglossum* were common epiphytes.

There are other woodlands whose ancient presence can be considered certain but that cannot be shown on the map. They are spatially ephemeral, for they colonize in the wind throws, ash falls, landslides, and areas with excessive erosion. Three deserve mention here because they must have constituted a significant portion of the plant cover: silverleaf-balsa stands, tree fern colonies, and palm brakes.

Silverleaf-balsa *Cecropia-Ochroma* stands can be seen from great distances because of the white undersides of the characteristically fluttering, large leaves of the dominant trees. Among the first of the weedy tree species to invade openings in rainforest are *Cecropia peltata* and *Ochroma pyramidale.* These two aggressive pioneer trees would have formed extensive populations when hurricanes passed, leaving flattened vegetation behind, or where the landslides common in allophanes destroyed the forest. These 15-m trees mix with other weedy trees such as *Sterculia caribaea, Sapium caribaeum,* and *Chimarrhis cymosa,* more permanent members of the vegetation.

Table 15. The floristic composition of elfin woodlands and associated bushlands and meadows

Tree Strata	Shrub Strata	Herb Strata	Lianas and Epiphytes
Clusia plukenetii	*Cephaëlis axillaris*	*Tibouchina chamaecistus*	*Manettia dominicensis*
Charianthus nodosus	*Rapanea coriacea*	*Gaultheria swartzii*	*Mikania hookeriana*
Didymopanax attenuatum	*Charianthus corymbosus*	*Pitcairnia spicata*	*Grammitis asplenifolia*
Rondeletia stereocarpa	*Gesneria ventricosa* var. *montana*	*Blechnum ryanii*	*Peperomia nigropunctata*
Prestoea montana	*Besleria lutea* var. *alpestris*	*Lobelia flavescens*	*Psychotria guadelupensis*
Podocarpus coriaceus	*Urena lobata*	*Isachne rigidifolia*	*Elleanthus dussii*
Cyathea arborea	*Rondeletia martinicensis*	*Machaerina restioides*	*Elaphoglossum plumieri*
Dunalia arborescens	*Miconia martinicensis*	*Conyza bonariensis*	*Epidendrum miserrimum*
		Viola stipularis	*Peperomia hirtella*
		Rhynchospora longiflora	
		Lycopodium reflexum	
		Ischaemum latifolium	
		Hemidiodia ocymifolia	
		Nepsera aquatica	
		Polygala paniculata	
		Symphysia racemosa	
		Thelypteris normalis	
		Anatherum bicorne	
		Epidendrum ramosum	
		Nautilocalyx melittifolius	
		Begonia hirtella	
		Philodendron giganteum	
		Pilea parietaria	
		Dichanopteris flexuosa	
		Lycopodium clavatum	
		Lycopodium cernuum	

Sources: Modified from Portecôp 1979, 53–54; Stehlé 1937, 228–32; Kimber 1969, 86–90.

Tree fern colonies form a pioneer woodland type, scattered in patches, in the rainforest and montane forest matrix, pioneering on landslips, especially in ferrisols and recent ash deposits in protected areas. Two ferns, *Cyathea muricata* and *C. arborea,* form arborescent communities 5 to 10 m tall. This simple community contains the aroids *Oreopanax capitatum* and *Didymopanax urbanianum* and the bromeliads *Guzmania plumierii* and *Pitcairnia* sp., along with numerous lycopods, mosses, and some grasses. The fern *Blechnum binervatum* is epiphytic and climbs on the trunks of tree ferns.

Palm brakes on very exposed slopes, vulnerable to landslide, are common but never extensive throughout the windward and interior mountain slopes between 250 and 800 m. Palms of the rainforest and montane forest dominate. The dominant palm known for many years as *Euterpe globosa* but renamed *Prestoea montana* grows in a patchy, uneven-aged distribution pattern varying in height from 3 to 10 m. Palm brakes of *Rhyticocos amara* can be found on north-facing slopes of midelevations of the northern mornes. Growing among the palms are saplings of *Hedyosmum arborescens, Byrsonima martinicensis,* and *Ilex sideroxyloides.* Most trees have small boles and a twisted, depressed growth habit. Shrubs are characteristically absent, but the ground is covered with large herbs such as *Heliconia bihai, Anthurium* spp., and *Asplundia rigida.* Smaller herbs such as *Selaginella flabellata* and lycopods are everywhere abundant, occasionally forming pure populations. The trunks and branches are covered with mosses and ferns. Lianas may be present.

Bushland Types

Bushlands, woody plant communities with a canopy less than 4 m in height, occupied comparatively small areas during the prehuman period. Dry bushlands can be clearly differentiated into several types today, but probably not so for the period before colonization. They were certainly along the Caribbean coast in the dry belt on steep slopes and especially on heavy montmorillonitic clay soils. These bushlands were dense or open, depending on the dominants of a particular community. They were composed of varying proportions of leguminous shrubs, including *Pithecellobium unguis-cati, Cordia* spp., *Croton* spp., and *Lantana* spp. Evergreen species probably present in the dry bushland included *Crossopetalum rhacoma, Pimenta racemosa, Erythroxylon brevipes, E. ovatum,* and several species of *Capparis.* Mosses and ferns were absent. There probably were few lianas or epiphytes, as they are rare today. Some shrubs have thorns; others have water-conserving adaptations, as in *Bursera* and *Pisonia.*

Montane bushland, composed of mounding shrubs with stiff branches and evergreen leathery leaves, was found above elfin woodland at the summits of the Carbet Pitons and in scattered patches on Mont Pelée. As in the elfin woodland, mosses, bromeliads, and orchids blanketed the stems of woody vegetation and the ground. Dominant shrubs are, and were, in this early time the two endemics *Rondeletia martinicensis* and *Miconia martinicensis,* as well as *Didymopanax attenuatum, Charianthus* spp., and *Ilex macfadyenii.* Melastomes and ericads were abundant. Leaves in this vegetation are typically rolled, hairy, or heavily cutinized. Whether montane bushland was ever important on Mont Pelée is not known. Volcanic eruptions and intense fumerol activity shift position and, as new areas are affected, the montane bushland suffers temporarily, only to reestablish itself at a later time. This vegetation type is well represented today on the Carbet Pitons and Morne Jacob and was so earlier.

Littoral hedge is a dense thicket at the seaward margin of the littoral woodland that shares its flora. In late prehuman times this hedge was, presumably, a virtually continuous narrow band on the north Atlantic coast and occupied the coastal margins of the peninsulas of the south Atlantic coast except in the most exposed promontories, where it was replaced by desert rock pavement and cactus shrub. The tight canopy of hard evergreen leaves is pruned by wind and salt spray to an even surface. The interior is open except for the leggy, bare branches over a scanty ground flora. Characteristically, heavily cutinized leaves form a dry, loose, undecomposed litter on the floor.

The dominant shrub was the common sea grape, *Coccoloba uvifera,*

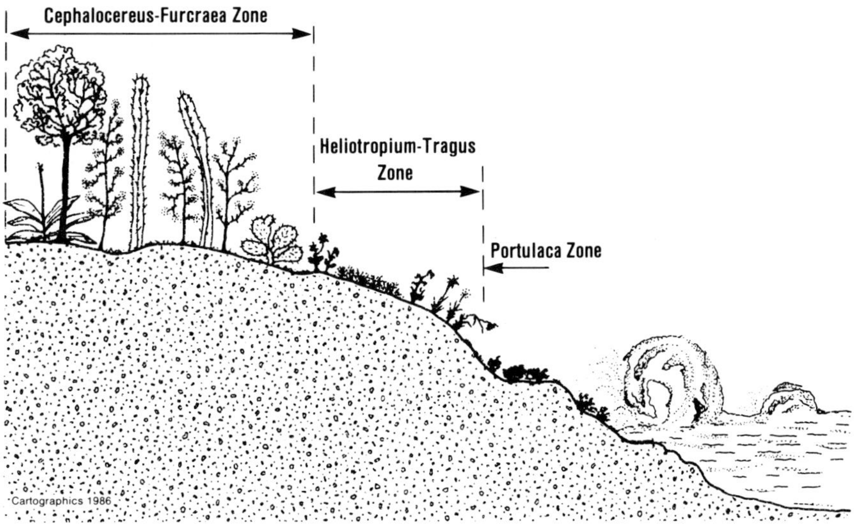

10. Mixed cactus scrub and transect to the sea. Source: After Portecôp 1979,28.

but a half-dozen other kinds of shrubs were frequently important, as they are today. *Hura crepitans, Zanthoxylum spinifex,* and *Capparis frondosa* were often dwarfed in the littoral hedge. The ecologically broad spectrum *Tabebuia pallida* joins *Eugenia axillaris, Erithalis fruticosa,* and *Maytenus elliptica.* The infamous poisonous *Hippomane mancinella* is important on sandy beach ridges, sometimes growing straight and tall enough to be classed as woodland.

The mixed cactus scrub in prehuman time probably did not occupy more than a fraction of its present range. It was found in leeward cliffy sites and those steep slopes with droughty soils of the southeastern peninsula. It grades to littoral, herbaceous vegetation (see figure 10). Related in physiognomy and floristics to the dry bushland, it differs from it mainly in the high proportion of cacti and century plant forms. The columnar and branched *Cephalocereus nobilis* were numerous and rose above the tops of a thin scrub of thorny legumes, *Randia aculeata* and *Croton* spp. On the ground were *Opuntia* and *Furcraea tuberosa,* blooming conspicuously during the long dry season when most woody plants were leafless. Herbs and grasses are today present during the rainy season, and dry grasses remain visible throughout the rest of the year. The native grasses included *Sporobolus indicus, Cenchrus echinatus,* and *Dactyloctenium aegyptium.*

Desert Vegetation

In prehuman times "deserts" in Martinique were probably confined to the very small areas of desertic rock pavement on exposed, calcareous headlands of the windward side, from the Caravelle Peninsula southward and the rocky cliffs of the Caribbean headlands. The rock pavement is dry because of low precipitation and continuous salt spray off the Atlantic Ocean. This is a sere vegetation composed of low, windswept shrubs and cacti, widely spaced or in scattered drifts, that do not cover the ground surface. Even immediately after rains the ground is a relatively barren patchwork of limestone rock with colonies of grasses and sedges, and occasional mats of prostrate plants. The scattered shrubs of this formation include *Sophora tomentosa,* which is found in the bushlands farther inland, and *Conocarpus erecta* and an occasional opuntia. Color is contributed by the generally inland herb *Evolvulus sericeus,* a bright blue color, and the coastal herbs *Pectis humifusa,* a bright yellow, and *Portulaca,* white to yellow.

Seaward among the rocks wet by spindrift are *Portulaca halimoides, Heliotropium curassavicum, H. ternatum,* and the grasses *Chloris inflata, Tragus berteronianus, Stenotaphrum secundatum.* Landward are the trailing acacia-like *Desmanthus depressus,* the opuntia *D. dussii, Alysicarpus vaginalis,* and the shrubby euphorbs *Croton flavens* and *C. bixoides.* This vegetation type grades into the mixed cactus scrub with the conspicuous *Cephalocereus nobilis,* the rare *Agave americana,* and *Furcraea tuberosa* among the woody acacias.

Rocky cliffs are desertic twice over because of the excessive drainage on the steep slopes of tuffs and their location along the dry Caribbean coast. The winds of the Dominica Channel also keep the high tuff cliffs of the north coast droughty. The dry southwestern cliffy coast about Morne Larcher is composed of older, more weathered rocks. The bases of the cliffs are colonized by *Ipomoea pes-caprea* on sandy shores and by *Pectis humifusa, Portulaca halimoides* on the rocky margins (see figure 11). The grasses *Dactyloctenium aegyptium, Cynodon dactylon,* and *Sporobolus virginicus* are generally present. Above the base but still in reach of salt spray during storms are the bromeliad *Pitcairnia angustifolia,* two composites *Wedelia trilobata* and *Egletes prostrata* in among the woody legumes, *Randia aculeata,* and an occasional *Coccoloba uvifera.*

Herbaceous Types

There was no natural savanna in Martinique, and herbaceous communities were few, occupying very specialized local habitats associated with saturated ground: littoral meadows and montane meadows. These were dominated by plants special to the types but also shared many herbaceous species with surrounding vegetation.

Littoral meadows, bearing a mosaic of low halophytic plants, were situated in brackish coastal flats. The largest areas were in the Lamentin Plain and in Sainte-Anne Commune. Patches of sedges were dominated by species of *Eleocharis, Rynchospora,* and *Scleria.* The creeping beach grass *Sporobolus virginicus* was also common. Mat plants included *Salicornia ambigua, Sesuvium portulacastrum,* and *Batis maritima.* Clumps of small ferns such as *Dryopteris gongyloides, D. excelsa,* and *Nephrolepis biserrata* were locally important.

Swamp fern (*Acrostichum aureum*) fringed the large Lamentin Plain grassland with a wide band behind the formerly extensive *Rhizophora-Avicennia* mangrove swamp and covered enough area to be considered a subtype of the littoral meadow. It was also present in large areas behind the mangrove swamp at Point Marin. This fern is found today in the

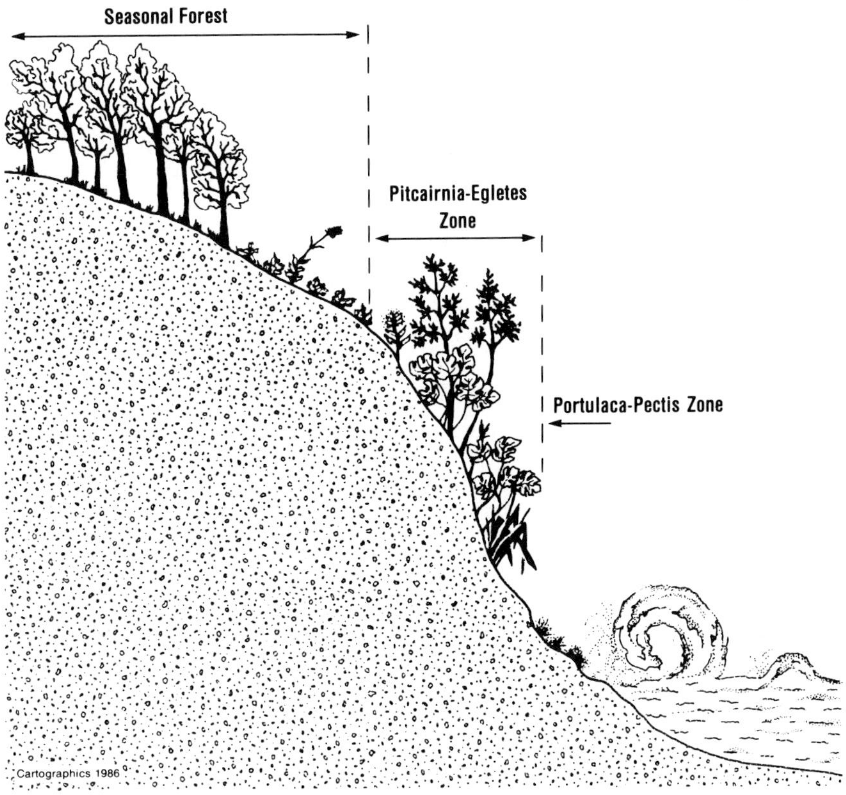

11. Rocky cliffs–seasonal forest transition profile diagram and transect to the sea coast.
Source: After Portecôp 1979,27.

landward margins of the mangrove swamp and in the wet margins of littoral woodland.

Montane meadow areas were too small to appear on the small-scale map of vegetation. Where the ground is continuously saturated with slowly moving water, as from a spring in the mountain region, local areas were occupied by a soggy sward of moderately tall sedges with clumps of ferns. These are, and probably were, found in the Carbet Pitons and on the western high slopes of Mont Pelée. Dominant plants in this vegetation type were the sedges *Eleocharis* spp. and *Kyllinga squamata* and the grass *Isachne rigidifolia.* The common fern was *Dicranopteris bifida.* Richer, drier communities at lower elevations included *Hyptis atrorubens, Tradescantia* sp., *Commelina diffusa,* and *Sauvagesia erecta.*

The Present Classification System in Relation to Others Used in the Caribbean

The vegetation types recognized as having been important in the plant geography of Martinique have been distinguished by several kinds of criteria to permit description and analysis of vegetation change. The criteria for separating types are pragmatic and mainly physiognomic rather than floristic-systematic, so they could be easily recognized in the field and in the historical record. The types are not of equal ecological rank; some are climax, others are essentially pioneer associations, and still others are distinctly successional communities.

Eighteen vegetation types of the spontaneous class are interpreted as stable, self-replicating types. The pattern of distribution corresponds to the local climatic and specialized edaphic conditions. The other three are pioneer or successional vegetation, and thus indicate some disturbance. The silverleaf-balsa (*Cecropia-Ochroma*) stands are pioneer woodlands leading to rainforest, whereas palm brakes lead to montane forest; both are characteristically found in continuously disturbed habitats affected by hurricanes or on steep slopes subject to slumping, landslides, or severe sheet erosion. Tree fern colonies extend from rainforest to elfin woodland areas in wet situations. The mixed cactus scrub, though always present, has increased in extent because of a historical deterioration of the edaphic environment through erosion.

At the Pan-Caribbean Forestry Meeting in 1946, Beard's classification system for the vegetation of tropical America was adopted until such time when further work should deny its utility (Beard, 1944a). The present work has been much influenced by Beard's ecological studies, although the final form of the classification used is somewhat different from his. According to Beard, the plant formation is primarily a physiogomic unit that contains a number of floristic associations united by common structure and life form. The formations themselves are arranged into series according to similar habitat, and for convenience in discussing ecological relationships, the series are divided into "climatic" and "edaphic" groups (Beard 1942). A similar dichotomy of climatic and edaphic formations is used here to identify basic categories of the spontaneous vegetation. However, presumed climax, seral or successional, and pioneer vegetation types have been all given consideration and rank as landscape features.

Stoffers's excellent study of the vegetation of the Netherlands Antilles was especially useful in selecting the distinguishing criteria for the several dry communities of spontaneous vegetation as was that of Asprey and Loveless (Stoffers 1956; Asprey and Loveless 1958). Similarly, Asprey and Robbins's (1953) study of Jamaican vegetation was useful in developing the criteria for distinguishing the montane vegetation types.

The most careful published work to date on the composition of the plant communities of Martinique was done by Jacques Portecôp (1979, 1–78), after the original work by Kimber (1969), who built on the work of the earlier works of the Stehlés. Conversations with the Stehlés and Portecôp and joint field excursions were occasions for lively discussions on the vegetation and its ecological status. These discussions refined my understanding of the plant ecology of the island, and the present treatment is the result of these refinements.

All communities defined floristically by Stehlé have been identified in the field as in Martinique, except the *Eugenia* woodlands. Where differences exist between two systems, they are the result of combining some of Stehlé's communities that were distinguished on the basis of lithic substrate correlations and are not particularly useful for this work. A comparison of the classification systems used by Beard, Stehlé, Stof-

Table 16. Correlations among classification systems used for woody vegetation in Caribbean basin studies

Kimber, this work	Portecôp 1978	Stoffers 1956	Stehlé 1945–46	Dansereau 1966	Beard 1944a	Holdridge 1971
Montane bushland	Prairie altimontaine montaine à Ananas	Elfin woodland	Forêt altitudinale rabougrie (*a*) alluvial facies (*b*) volcanic facies (*c*) culminating	Montane scrub	Elfin woodland	Bosque Pluvial Premontano
Elfin woodland	Foure semi-arbore à Aralie-Fuschia					
Montane forest	Forêt hygrophile de transition à Laurier rose–Clusia	Montane thicket	Forêt hygrophytic de transition	Montane forest	Montane rainforest	Bosque Muy humido Tropical
Rainforest	Forêt hygrophile à Chataignier-Gommier	Rainforest	Forêt hygrophytique	Lower montane forest	Lower montane rainforest	
Seasonal forest	Forêt mesophile à Pois doux-Mahot	Seasonal evergreen forest	Forêt mesophytique (*a*) calcareous facies (*b*) volcanic facies	Seasonal evergreen forest	Evergreen seasonal forest	Bosque humido Tropical
	Forêt meso-xerophile à Poirier-Galba	Semi-evergreen seasonal forest			Semi-evergreen seasonal forest	
Semi-deciduous forest	Forêt-xerophile à Poirier-Bois rouge	Dry seasonal evergreen	Forêt xerophytic (*a*) sandy facies (*b*) calcareous facies (*c*) volcanic facies	Semi-deciduous forest	Xerophytic rainforest Deciduous seasonal forest	Bosque Seco Tropical
Dry woodland	Forêt xerophile à Poirier-Gommier rouge	Deciduous seasonal forest		Littoral forest		
Littoral woodland	Type sableux	Littoral woodland	Facies sableux	Tropical beach	Littoral woodland	
Mangrove swamp Mamain swamp Woodlands	Mangrove	Mangrove woodland	Mangrove (*a*) marine facies (*b*) riparian facies	Mangrove complex	Mangrove woodland	
Rock pavement	Type rocheux	Rock pavement				
Strand/rocky cliffs	Type sableux	Littoral woodland	Facies sableux	Strand complex	Rock pavement	

fers, Portecôp, Dansereau, and Holdridge in this work can be found in table 16.

Close comparisons of different systems show that although the names of the associations differ and there has been much splitting and lumping of the vegetation types, most vegetation types used here can be recognized in all other systems. This is a result of the small size of the island, its relatively few habitats, and its relative remoteness from continental source areas so that although the number of formations is large, there is generally only one plant association within each, using *association* in the broad sense.

It should be here reemphasized that these vegetation types are essentially artifacts of an attempt to describe changes in the plant geography and to account for those changes. No vegetation type is homogeneous across all its spatial extent, and with the passage of time the exact species complement changes.

Changes in the Plant Geography

The only available information about severe impacts on the vegetation in the period before human occupation comes from the geological record. Some of that information can be read directly, as with volcanic eruptions, whose extent and time can be mapped and dated. The information concerning changes in sea level can be read only from the geomorphic record combined with the information from the analysis of deep sea cores. The changes in regional climate are similarly arrived at indirectly from cores from the different parts of the Atlantic bottom and from the floor of the Caribbean. Changes in the hydrological regimes and in groundwater can be recovered from landform analysis and from stalagmites in caves and sinkholes, as has been shown by Wilson and his associates.

The early building of Mont Pelée was by lava and pyroclastic flows. Later volcanism consisted exclusively of pyroclastic flows. Traineau (1982) prepared a series of maps illustrating the spatial extent, chemical composition, and depth of the deposits associated with these eruptions. For any one explosion, the deposits extended as far south as Case-Pilote, as far east as Basse-Pointe and Macouba, or as far west as Prêcheur. Their depth varied considerably. Maps (map 13) selectively illustrate some of these events. Traineau distinguished three major building periods associated with (1) the ancient structure age 400,000–40,000 Y.B.P.; (2) the intermediate structure age 40,000–19,500 Y.B.P., after which there was a hiatus of 6,000 years during which no activity has been discovered; and (3) the recent structure 13,500 Y.B.P.—to present (Traineau 1982,14). The four eruption events associated with intermediate structure were compiled on maps that have been redrawn at a smaller scale (see map 13a).

The eruptions building the recent structure were all pyroclastics with pumice explosions alternating with *nuées ardentes* eruptions (Traineau, Westercamp, and Coulon 1983, 241). The first event after the rest period was a *nuée ardente* event of huge volume. It covered the whole of

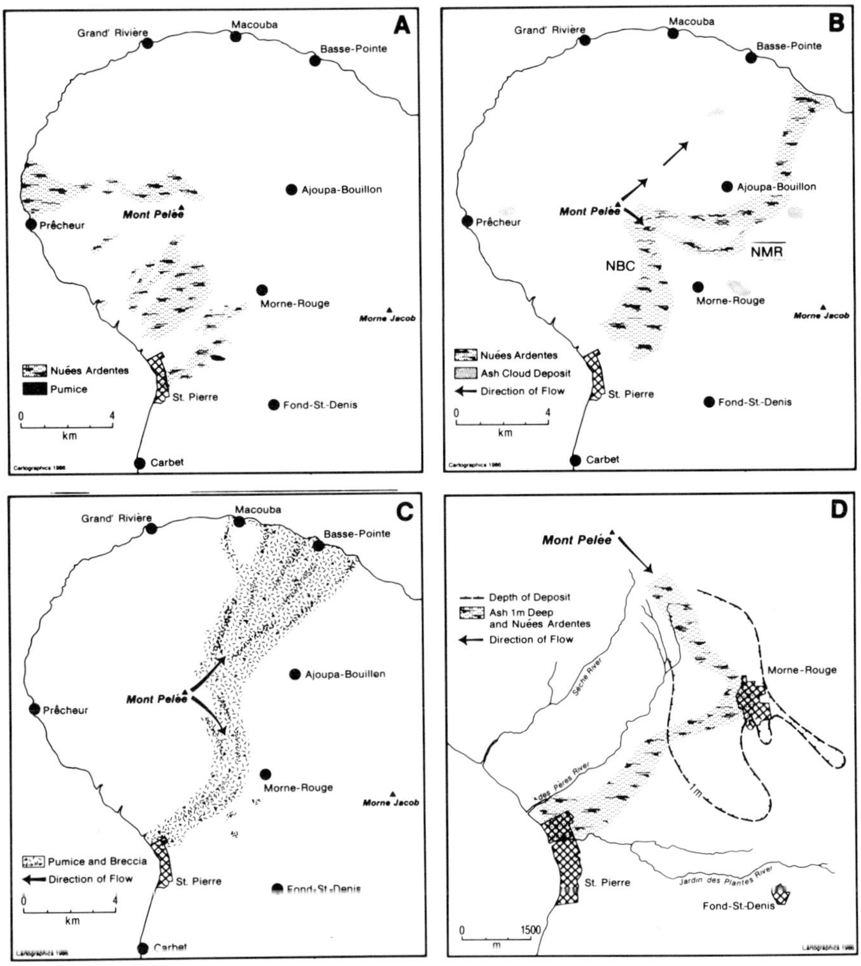

Map 13. Selections from volcanic eruptions prior to 5,000 Y.B.P. Source: After Traineau 1982, 73, 83, 84, 89.

the southern flanks of the volanco centering on Balisier Calave (NBC) and Morne-Rouge (NMR) (see map 13b). What is especially interesting about the NBC deposits is that they lie on top of a paleosol that had formed during the hiatus (Traineau 1982.65). Dating by carbon-14 methods yields an average age of 13,500 ± 300 Y.B.P. After two more events was a pumice episode that varied in depth from 0.5 to more than 25 m thick, which has been dated at 7,800 Y.B.P. (see map 13c). Again, after two more recovered eruptions, the next was a *nuée ardente* event that exploded directly toward Morne-Rouge and then followed the Roxelane River to St. Pierre. The ash cloud associated with the eruption extended into the Champflore Basin (see map 13d). Several more eruptions have been recovered, dated at 4,610 Y.B.P., 4,060 Y.B.P., 3,980 ± 30 Y.B.P., 3,710 ± 30 Y.B.P., and 2,440 Y.B.P., which are the last explosions before the known arrival of colonists from South America.

The average of 450 years between eruptions for the intermediate period is well under the length of time most investigators consider necessary to reconstitute a "mature" rainforest. It is probably too short to reassemble completely the species and geometry of the seasonal forest. The other vegetation types may have had time to achieve a floristic and physiognomic character similar to the self-replicating types. What this means is that for the Mont Pelée massif as a whole, the rainforest and seasonal forests would have species complements in areas different one from another. The differences noted in remote stands of the present-day forest may be more like that of the late geological period than was previously thought, and the variability may be caused as much by disturbance pattern and biological events such as species colonization histories than any habitat differences such as substrate or microclimate.

From the plotting of the eruption flows it is clear that the plant cover on Mont Pelée has been dynamic and that it may be more accurate to call some of the forest formation late succession, recognizing that there may be multiple pathways in succession rather than the static picture presented in the discussions associated with map 12.

Indeed, unpublished *relevées* made in crest locations by Fiard and Kimber (1985) and Rolet and Fiard (1985) suggest that the dominant trees in that vegetation may be mid-succession-sequence species rather than mature vegetation dominants. What is clear is that the vegetation fabric on Mont Pelée was probably a mosaic of plant assemblages and vegetation types rather than continuous belts, as existed on the Carbet Pitons, Morne Jacob, and Morne Lorraine massifs.

During the recent geologic past, sea levels changed vertically to lower and higher apparent sea stands. Globally, sea levels in 125,000 Y.B.P. (134,000–118,000 Y.B.P) were probably higher than sea levels in 103,000 Y.B.P., 82,000 Y.B.P., and 60,000 Y.B.P. by amounts close to 8 m, 6 m, and 21 m (Stearns 1976,449). About the perimeter of Martinique there are a number of marine terraces indicating that sea levels were higher than today at several times in the past. Numerous small features can be found on the northern perimeter about the Piton Mont Conil massif. The largest number are found from Anse Couleuvre to Fond Moulin just to the west of Grand' Rivière. In the south at elevations of 1 to 12 m are surfaces consolidated by calcareous algae. There are certain small islands covered with sands 4 m deep. In the Marin Cul-de-Sac and Fort-de-France Bay there are a number of such features. Diamant Rock has a terrace at 8 m elevation that is 5 m deep.

The submarine valleys of the Lamentin River and Salée River and the drowned mouths of southern rivers indicate that sea levels have been lower than they are today. Studies in Barbados (Pittman 1974,1811–20) and Bonaire (Bandoian and Murray 1974,1243–52) have demonstrated that several changes in sea level have been complicated by tectonic movement. These are certainly different physical processes in Barbados than obtained in the mid-Lesser Antillean arc, but in Martinique certain other tectonic activities took place. Arguing from this kind of evidence, one can conclude that sea levels may have stood as much as 70 m higher and 20 m lower than the position of the sea on the island land mass today. Besides the obvious gains and losses in area, there has been some

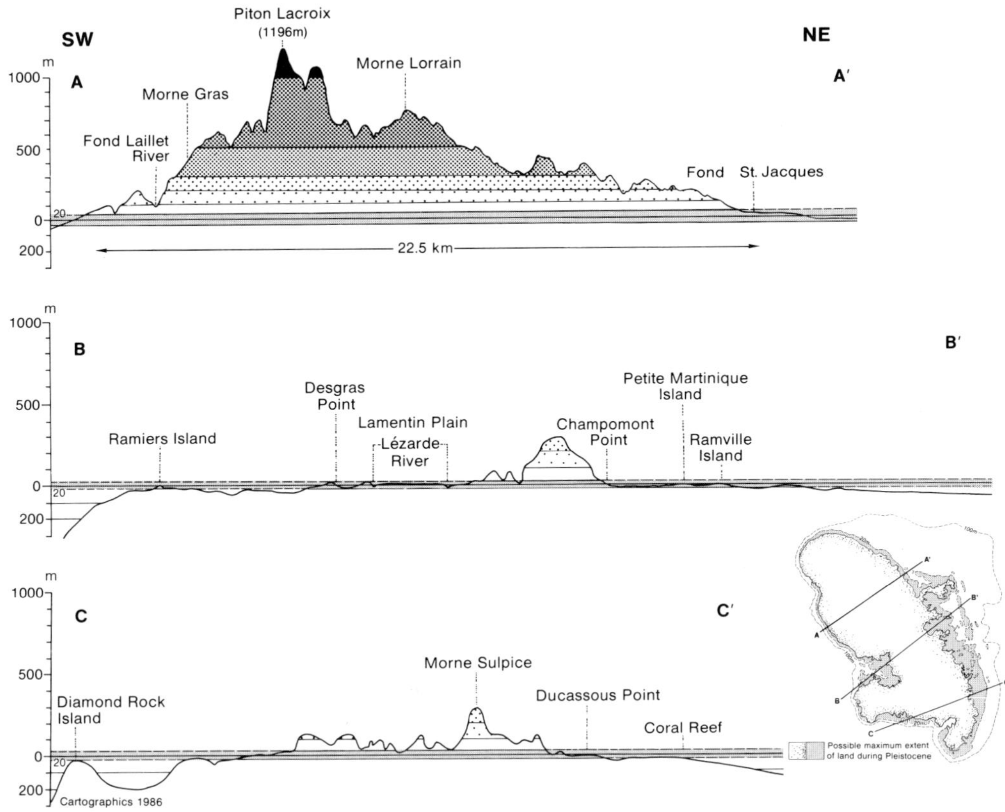

12. Profile diagrams illustrating the possible expansion of land area at Pleistocene low sea stand of 20 m and 100 m. Sources: France, Centre National de la Recherche Scientifique 1977, Geomorphology; U.S. Defense Mapping Agency 1985.

change in the kinds of areas available to lowland vegetation types (see figure 12). If the island has remained relatively stable across its extent, the exposed land areas for low sea stands would resemble the areas diagrammed. Probably the transect B–B' across Fort-de-France Bay shows more land exposed than would have been present during the Pleistocene. All vulcanism in the south had ceased, and the erosion process would not have filled the bay to that extent. The sum of 90 m in elevation change would have changed the summit elevations not enough to have made frost conditions at the mountain tops. Mountain meadows may have expanded in area, but more likely they are edaphically controlled, and slight movements of the altitudinal zone boundaries would have merely enlarged the montane thicket and forest rather than the meadowland.

Changes in regional climate would have lowered the rainfall values from those of the present. This would have caused an expansion of the dry vegetation types of the southern peninsula and the Caribbean coastlands. It would seem that seasonal forest would have moved up in elevation if one uses the Holdridge model, which plots mean annual temperature against average total precipitation and potential evapotranspiration (Holdridge 1967,16). However, such speculation needs to be critically examined. There is a real need for palynological work to provide some hard evidence for changes in vegetation or species composition over geologic time.

Notes

1. The problem of mathematical ecology is that one species is not like any other species; however, biologists have found it useful to compare species lists in the aggregate. For instance, Mauritius has a species-to-genus ratio of 2 to 1 and the Canary Islands have a similar ratio of 2.1 to 1.

2. A tree was defined as "a woody plant capable of attaining a height of five meters at maturity"; thus, that life form in the flora was underestimated rather than overestimated (Beard 1949b,39).

3. The long-range dispersal of the earliest geologic periods of the island's existence would have been supplemented by shorter-range dispersal as the intermediate, younger island masses were built up. See the rich literature search of Alan Graham, "Literature on Vegetational History in Latin America," in Graham (1973), as well as the classic of Ridley (1930).

4. Three trans-Atlantic currents bring drift from Europe and Africa, picking up New World drift as they pass northward into the Caribbean and along the Lesser Antilles Atlantic. The North Equatorial Current orginates from the southward-flowing Canary Current, which passes northwest Africa, turning westward at the Cape Verde Islands and flowing between 10° and 20° N. It enters the Caribbean north of the island of St. Vincent; the rate of movement as recovered from drift bottles can be 13 m/d (Guppy 1917,69).

   The Main Equatorial Current lying between 2° and 10° S originates in the Gulf of Guinea, being the result of the joining of the Guinea Current from the north and west and the Benguela Current from the south that turn westward across the Atlantic joining the Guiana northward-flowing current toward the West Indies, following the trend of the coasts of northern Brazil, the Guianas, and eastern Venezuela into the Caribbean between Trinidad and Grenada. Drift material may originate from the river mouths of tropical wet Africa and from the mouths of the Amazon, Guiana, and Orinoco rivers. The passage across the Atlantic is much more rapid than with the North Equatorial Current, and along the coast of Brazil as it moves north much faster.

   The South Equatorial Current passes across the Atlantic between 10° and 20° S, reaches Brazil, and bifurcates into the Brazil Current flowing southward and the Guiana Current flowing northward, and parallels the waters of the Main Equatorial Current. This current, because of the mixing in the Benguela Current, may carry material from southern Brazil that has crossed the southern Atlantic before turning northward to the Benguela. The trip across the Atlantic as determined by drift bottles takes approximately twelve weeks from Africa to Brazil.

5. Physiognomic criteria differentiate relatively easily the communities found in certain habitats. While *agreeing* with the individualistic concept of plant distribution and *recognizing* continua to be the probable theoretical norm, the distributions of species do aggregate in Martinique because of the steepness of environmental gradients, mosaic of soil types, and erosion factors (see Egler 1968). Plant collections, line transects, and traverses were made to gain familiarity with the more common elements of the flora. Voucher specimens are on file in the University of Wisconsin Herbarium at Madison, Wis. An incomplete set is in the Geography Department Plant Geography Laboratory, Texas A&M University,

College Station, Tex. Jacques Portecôp's specimens are in the Laboratoire Biologie Végétal, Université Antilles-Guyane, Pointe-à-Pitre, Guadeloupe.

6. *Richness* is defined here as species diversity.

7. The *fromager* or kapok is common in Central America, where it is exploited by man and has been ascribed a role in magic. The dispersal mechanisms are adequate for its natural dispersal to the islands. Still, Amerindians may have been responsible for its introduction to Martinique, since virtually all specimens sighted were near places of habitation and on disturbed grounds.

8. These percentages were calculated from areas measured off the map of reconstructed native vegetation at a scale of 1:50,000. My estimates differ from those J. Portecôp reported (France, Centre National de la Recherche Scientifique 1977, Vegetation, plate V). He defines his vegetation boundaries somewhat differently.

9. J. L. de Lanessan 1886a,163–64; de Lanessan noted *Bruguierra gymnorhiza* Lamk., which closely resembles red mangrove or palétuvier in Fort-de-France, Lamentin, the Salée River, and in Vauclin littorals. I have not seen the species, nor is it reported in Fournet 1978. However, a second species of *Rhizophora* has been reported by amateurs and professionals from the islands in Robert Haven. It is possible that in earlier times that species occurred on the west coast as well.

10. The distinction between small tree and tall shrub is often blurred, and the habit of the particular plant is variable depending upon local environmental conditions.

4 The Entry of Humans: The Amerindian Period

It is now generally agreed that the peopling of the Lesser Antilles occured in a series of migrations from the South American continent. Christopher Columbus reported meeting Tainos and Calinas during his voyages in the islands. For many years it was thought that the archeological record revealed two waves of pottery-using peoples, referred to as the Arawaks (Taino) and Caribs (Calina). Using information derived from archeological investigations, ethnolinguistic studies, and reinterpretations of the ethnographic accounts in the early chronicles, researchers have largely agreed upon a new sequence.

The ceramic-processing people were preceded by non-pottery-using cultures beginning at least ± 400 B.C. (Rouse and Allaire 1978).[1] For Martinique, H. Petitjean-Roget (1974,82; 1980,86) reports a surface preceramic site at the coastal Petrifications Savanna in the south, and Allaire and Mattioni (1938,27) report preceramic sites at Boutbois and at Le Godinot in the interior at 250 m elevation in the communes of Morne-Vert and Carbet. The latter sites, found under 50 cm of volcanic debris, have been provisionally dated at 5,000 Y.B.P. (Allaire and Mattioni 1983,31). In the northern Lesser Antilles preceramic sites of Meso-Indians dated by carbon-14 methods of 1600 B.C. have been recovered at Sugar Factory, Saint Kitts (Goodwin 1979), and at Jolly Beach Salt Pond Antigua (Nicholson 1976a). It should be mentioned that during the Archaic the sea level was about 20 m lower than it is today, so most sites probably have not been recorded (Ruppé 1980,334). Thus, the populations may not have been so limited as the small number of recovered sites would indicate.

Colonization by Agricultural Peoples

At the beginning of the Christian era a migration of peoples having ceramic objects passed from the south into the Lesser Antilles and then north. These peoples are thought to have originated in the area of the upper reaches of the Guaporé River and along the coastal and near-coastal areas of western Venezuela. The earliest recovered finds in the Lesser Antilles are at the Morel site in the northeastern part of Grande-Terre, Guadeloupe, dated at A.D. 200 (Clerc 1971,73–81). More recently, Louis Chanlatte-Baik (1983,83) clearly established the reach of these peoples as far as Vieques by the presence of ceramics corresponding in style at the Sorce and La Hueca sites on that island. These Guapoid-Heuca peoples knew manioc, as demonstrated by the presence of platins interpreted as cassava griddles. They had highly stylized modeling traditions that utilized *adorno* representing manatee heads attached to pottery vessels. The pottery was also decorated by punctate and incised designs. The presence of nostril bowls in the sites indicates the use of tobacco and the drinking of juice for ritual and magical purposes. These nostril bowls are still utilized to drink tobacco juice by the Warrau of the upper Orinoco Delta, futher linking this migration with Venezuela.

Fragments of Guapoid-Hueca pottery have been identified at Fond Brulé on the northeastern coast of Martinique. This group of peoples existed in the Antilles for a still undetermined length of time.

About the same time another migration left Venezuela from the lower Orinoco Delta, passing through Trinidad, Grenada, and the rest of the Lesser Antilles. In Martinique evidence of these peoples has been found at several well-studied sites, at Fond Brulé and at Vivé (Mattioni 1973,1984). Archeologists refer to these peoples as Insular Saladoid. They reached Puerto Rico by A.D. 120, as determined by Rouse (Rouse and Allaire 1978) at the Hacienda Grande site, and Hispaniola by A.D. 200.

Judged by successive changes in ceramic decoration and paste, the Modified Saladoid style (A.D. 500 to 700; see table 17) was succeeded by the Terminal Saladoid. Then a new group, originating in the Guyana Plateau, started entering the Lesser Antilles in A.D. 700, supplanting the dominant Terminal Saladoid and then replacing them.[2] Contact between these peoples of different cultures resulted in the newcomers borrowing of Saladoid traits. These post-Saladoid cultures are referred to as Suazoid cultures, and these are now suggested as ancestors of the Insular Caribs which, at the time of European discovery, were called Calina or, in plural, Calinago. The name recorded in Columbus's journal as *Calina* or *Kalinas* was transcribed to *Carib* by English chroniclers and from the English literature *Caraïbe* by the French chroniclers and settlers (J. Petitjean-Roget 1980,115–20). Thus, the usual term for Lesser Antillean Indians has come to be *Caribs*. It is now conventional to refer to the peoples occupying the Lesser Antilles at the time of first European settlement as the Island Caribs.

In the Greater Antilles the evolution of Saladoid led to the Ostonoid styles and then to the Chicoid styles. This last style corresponded to that used by the Taino, referred to as Arawaks, encountered by Columbus during his historical landing on San Salvador in 1492. These people are those referred to as Arawak in many of the historical works and recent anthropological and historical studies.

At the time of Columbus the Island Caribs claimed to have arrived in the Lesser Antilles only a few generations earlier from the Guianas, where a closely related people, the Barima River Caribs, still live. Taylor (1951,71) summarized political geography of the Island Caribs at that time as follows: "[B]y 1493 they had halted in their advance, and except for sporadic feuds of an almost personal character, had established friendly relations with their neighbors to the northwest, the Taino of Pu-

Table 17. The chronology of Amerindian settlement of Martinique

Horizon I: Insular-Saladoid (turn of the Christian era)[a]	
Ia.	Ancient Horizon I[b]
	La Salle, Sainte-Marie Commune, ^{14}C A.D. 180 ± 80
Ib.	Guapoid
	Fond Brulé, ^{14}C A.D. 100
Ir.	Recent Horizon II
	Vivé 1, Lorrain Commune, ^{14}C A.D. 22 ± 100
	[ash layer A.D. 295]
Horizon II: Modified Saladoid (toward the beginning of 3rd C.)[a]	
IIa.	Ancient Horizon II
	Vivé 2, Lorrain Commune, ^{14}C A.D. 420 ± 75
	[ash layer A.D. 500]
IIa/r.	Intermediate Horizon II
	Grande Anse, Lorrain Commune, ^{14}C A.D. 510 ± 80
IIr.	Recent Horizon II
	Diamant 2, Diamant Commune, ^{14}C A.D. 475 ± 60
Horizon II/III/IV: Terminal Saladoid (transition between 6th and 8th C.)[a]	
IIr/IV.	Transition between II and IV horizon
III.	Troumassoid (on Guadeloupe) 6th to 9th C.
	Petroglyphs at Forêt de Montravail, Sainte-Luce Commune
	Paquemar, *pierres à trois points*, Vauclin Commune
Horizon IV: Suazoid 7th C. and by 9th C. to the 16th C.[b]	
IVa.	Ancient or Caliviny Suazoid
IVr.	Suazoid
	Macabou 1, Vauclin Commune, 1100–1200
	Diamant 3, Diamant Commune, no date
Horizon V: Insular Caribs A.D. 1500+	
	Paquemar 3, Vauclin Commune, no date, Macabou 2 about 1500
	Macabou 2, Vauclin Commune, about 1500

SOURCES: H. Petitjean-Roget 1975, vol. 1, and 1982; Allaire 1977, 332; Mattioni 1984.
[a] Called Arawak by earlier investigators.
[b] Called Carib by earlier investigators.

erto Rico, along side of whom they fought as allies in the unavailing attempt of its inhabitants to defend that island from the Spaniards." The Terminal Saladoid (Arawak) had been expelled or overwhelmed in Martinique by the Suazoid (Caribs) at the time of European arrivals, but it is not known when this conquest occurred. It is not unreasonable to assume that the two peoples shared the island for part of the aboriginal period, and that their traditions supplemented one another.

Ecological differences between the early Saladoid and later Suazoid peoples are obscure, and it is uncertain how well the usual distinction between Saladoid series cultures as more dependent upon agricultural products and Suazoid series cultures as more dependent on sea resources applies. However, recent syntheses of archeological investigations indicate some significant differences in technologies and settlement patterns on Martinique (H. Petitjean-Roget 1975,vol.1; Allaire 1977).

These suggest real changes in population numbers, cultures, and exploited environments.

It is supposed that the Saladoid populations reached their peak just before the Suazoid invasions. At most, the Saladoid peoples numbered 10,000 persons on Martinique, a density of 9 persons per square kilometer. The Suazoid populations of Martinique, as indicated by their smaller settlement sites and less intensive agricultural activities, probably never numbered more than 8,000.[3] The ecologies of these peoples have been reconstructed from the remains in the archeological record and from the reports by European observers of Caribbean peoples during the historic contact period.

The peoples called Insular Saladoid by the majority of investigators appear to have colonized the northeast coast of the island about the beginning of the Christian era (see map 14). They cultivated manioc, as witnessed by the presence of microliths that would have been parts of manioc graters. There were axes, suggesting that trees were cut and wood was fashioned. Ceramic spindles indicate the use of cotton or some spinnable vegetable fiber. Because of the absence of fish bones and shell and the relatively small number of bones, one may infer that gathering and agriculture largely supported these early populations.

The peoples called Modified Saladoid differed from earlier peoples by having stone amulets, large manioc beer jars, different painted decorations on the pottery, and *adornos* molded in three dimensions fastened to the pots. Impressions in pottery suggest the Indians had basketry skills. The rare presence of fish bones indicates a limited use of fish and suggests that lines rather than nets were used in fishing.

Animal bones are reported as "possibly more numerous," meaning that hunting for meat may have been more important than in the earlier years. The presence of manioc beer jars, identified by their large size and the characteristic corrosion on the interior of the vessels, suggests a change in the diet. The archeological record indicates that plant food gathering may have become less important as fish and sea turtles became more important in the diet.

The most recent chronology of Indian occupations of Martinique based upon carbon-14 dates and *in situ* archeological sequences has been compiled by Henri Petitjean-Roget (1980,1:75). A modified version of that sequence, which takes into account the work of Louis Allaire (1977,332), H. Petitjean-Roget (1975,20–22,82–85), and Clerc (1968,52), is given in table 17.

Occasionally the Insular Saladoid sites are separated from Modified Saladoid by a layer of ash that has been identified as coming from Mont Pelée and dated at A.D. 295 (dated at University of Florida, Mattioni and Bullen 1973,226). Between A.D. 550 and 600 another violent eruption of Mont Pelée completely destroyed the villages of the northeast coast (Roobol, cited in H. Petitjean-Roget 1975,1:130).

Analysis of recovered sites suggests that these peoples occupied not only the northeast coast but also had begun to settle the calm western

Map 14. Presumed routes of colonization by Indian groups, by name of horizon or industries used in archeological investigations. Sources: After H. Petitjean-Roget 1975, vol. 1; Allaire 1977.

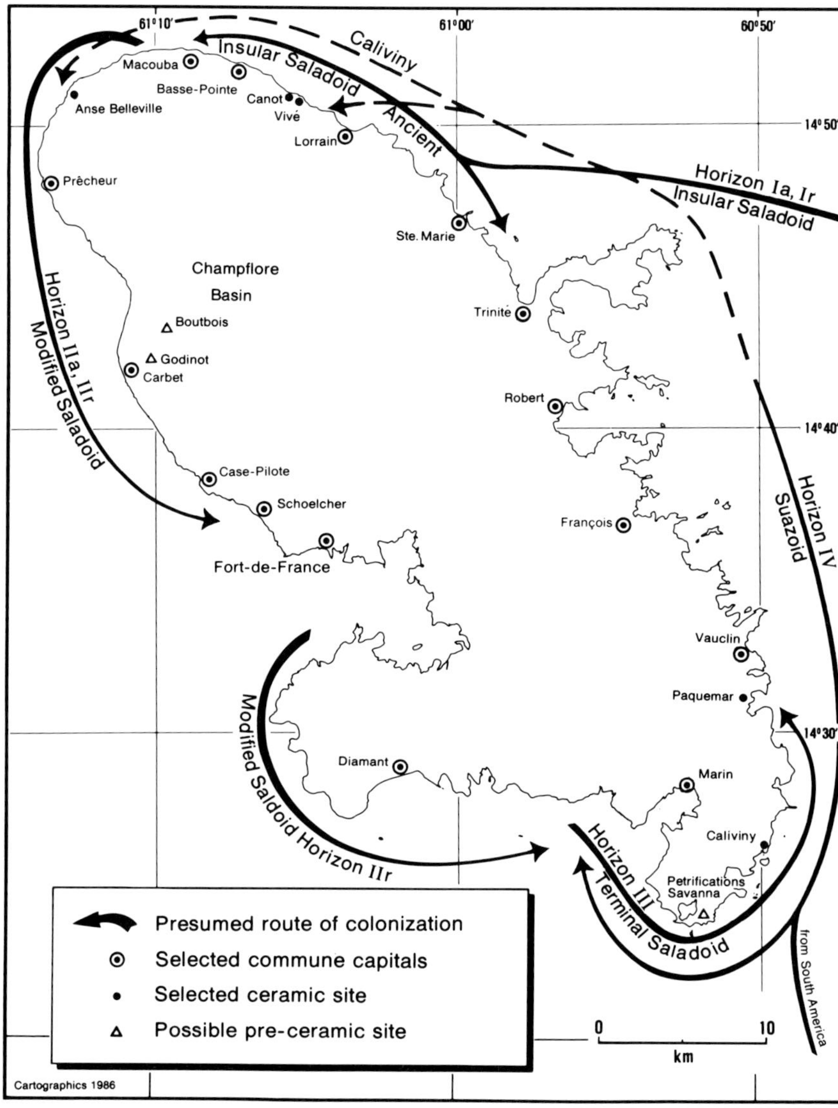

and southern coasts of the island. It is possible that the terrible eruption(s) of the later part of the sixth century drove the people to find village sites in less vulnerable parts of the island coasts. It is also possible that the increased dependency on sea resources, particularly the nearshore resources, meant that these places adjacent to the shallower sea of the Caribbean had become more attractive than the Atlantic Ocean border as settlement sites.

The Terminal Saladoid series peoples occupied the island for about 150 years, roughly from A.D. 450 to 600. They settled on the northwest (Caribbean) coast and southeast coast of Martinique. Shellwork, largely of great conch (*Strombus gigas*) and king helmet (*Cassis tuberosa*), was important. Wood sculpture is inferred from the presence of the little olive

(*Oliva reticularis*), which formed the teeth in wooden effigies. Incised work on pottery had been abandoned and decoration was the white-on-red type, with well-modeled *adornos* representing manatees, turtles, and pelicans. Basketry seems to have been further developed for the construction of fish traps. Cotton fiber use increased. Manioc griddles without feet were common.

By Terminal Saladoid time the inhabitants of Martinique were using a larger number of shellfish species. Seabird and turtle bones became more prevalent in archeological sites. Because the island can be crossed by a person on foot in one day, most of the micro-environments for hunting, gathering, or agriculture were relatively accessible from settlements located on the coasts.

Because of the presence of wooden statues and the presumed importance of bitter manioc, it is proposed that the Indians had developed by this time a tribal cult and that the society was stratified (H. Petitjean-Roget 1975,1:135–36). The estimate of village size for that time—600 to 700 people—is based upon the excavations in Diamant Commune (J. Petitjean-Roget, 1968,67).

The highest densities of Saladoid settlements have been found on the Atlantic coast north of the Caravelle Peninsula. At the time of initial contact between Saladoid, Caliviny, and Suazoid, the reef-protected southeast coast with its drier climate and older clay soils was agriculturally less desirable and may have been an empty area. The Suazoid peoples are reputed to have been better seamen, and they may not have been impeded by reef problems as were the Saladoid, nor restrained by the less agriculturally favorable land. These reefs also may have offered protection against retaliatory raids conducted by the Saladoid from the north and served as an immediate source of the marine conchs that were so important to Suazoid and Island Carib diets.

The peoples called Troummasoid are not well represented on Martinique. However, petroglyphs in the south coast Forêt de Montravail and three-pointed stones found in the upper levels of the Paquemar site lead some archeologists to suppose the people were on some parts of the island. This is because petroglyphs and the stones appear at the very end of Modified Saladoid and are characteristic of those sometimes called Terminal Arawak on other islands in the Lesser Antilles. It is possible that the violent eruptions of Mont Pelée, combined with what has been interpreted as the earliest Insular Carib incursions, caused the removal of some Terminal Saladoid peoples from Martinique to other islands.

There were changes in settlement and livelihood activities during the transition period between the sixth and eighth centuries. The number of objects associated with elaborate ritual behavior increased. Among the recovered objects are wooden statues, three-pointed stones, eared axes, worked-shell amulets, and small mortars and pestles for grinding colors or hallucinogenic seeds.[4] A small carved stone bird head with a long beak (perhaps hummingbird or pelican) having two long grooves in the bird's back has been found.[5] However, the rock engravings are the most important evidence of ritual activity. All of these imply the presence of priests, even a priestly class, and societies supported by surpluses of food staples. Such chiefdoms were reported by the chroniclers of the sixteenth and seventeenth centuries for Hispaniola and Puerto Rico. Hierarchical societies tend to be grouped into large settlements. No large sites have been located on Martinique dating from this period, but one may infer from Terminal Saladoid sites found elsewhere that villages larger than those of the preceding periods existed on the island.

Decorative art pottery declined and largely disappeared. *Adornos* became less and less important. It is possible that baskets and calabash bowls may have replaced pots. The elaborately decorated pottery items, interpreted by some as ritual objects, were replaced by other ritual objects, such as the eared axes, three-pointed stones, and the wooden statues.

Shellfish and manioc were important foodstuffs, as evidenced by the presence of shell and manioc griddles in the sites. There is an abundance of worked shell for tools, amulets, and ornaments. The manatee was used for its oil and meat. The rarity of whole bones is attributed to the habit of fashioning bone tools and other objects (H. Petitjean-Roget 1975,1:141). Gathering may have been less important than in the earlier societies since societies largely dependent on agricultural staples tend to abandon gathering activities.

The early peoples called Suazoid have been interpreted by some as the first of the Carib-speaking invaders from the Guyanas, linked to them by the late Carib mythologies recorded in the chronicles of the sixteenth century. However, most investigators prefer to restrict the Carib speakers to the Calina of the Columbian era (Allaire 1977,337–78).

Traces of early Caliviny are found at Fond-Capot, Basse-Pointe Commune, and Anse Belleville, Prêcheur Commune. These traces are inland, in what would have been seasonal forest vegetation. There was not much work in stone. Shell was primarily for tools, conchs and helmets being used to make axes, adzes, choppers, and scrapers. Pottery found in Caliviny sites (Mattioni and Bullen 1970,103) was substantially different, but it maintained a number of elements found in earlier work, which supports the idea that there was some carryover from the peoples of the intermediate horizons to the last horizon. According to the Island Carib founding myth, the Caribs invaded, killed the Arawak men, and took the Arawak women as wives. The differences in the two traditions that do exist are said to have been perpetuated by the Carib women, who accompanied the raiding men (compare Allaire 1977,41).

The Insular Saladoid sites are all found between the towns of Trinité and Macouba on the northern half of the Atlantic coast of the island (see map 15). All earliest sites are located on low uplands overlooking the Atlantic Ocean. The sites are found in the same area but are also found on the northwestern coast as far south as Schoelcher Commune. The most important of the Terminal Saladoid sites are scattered from Fort-de-France around the southern coasts to Ste. Anne. East coast sites from this horizon are at Simon and Paquemar in Vauclin Commune. Sites labeled transition sites of the sixth through the eighth centuries are very

Map 15. Archeological sites in Martinique. Source: Allaire 1977,100, 314; Mattioni 1973,1984; H. Petitjean-Roget 1982.

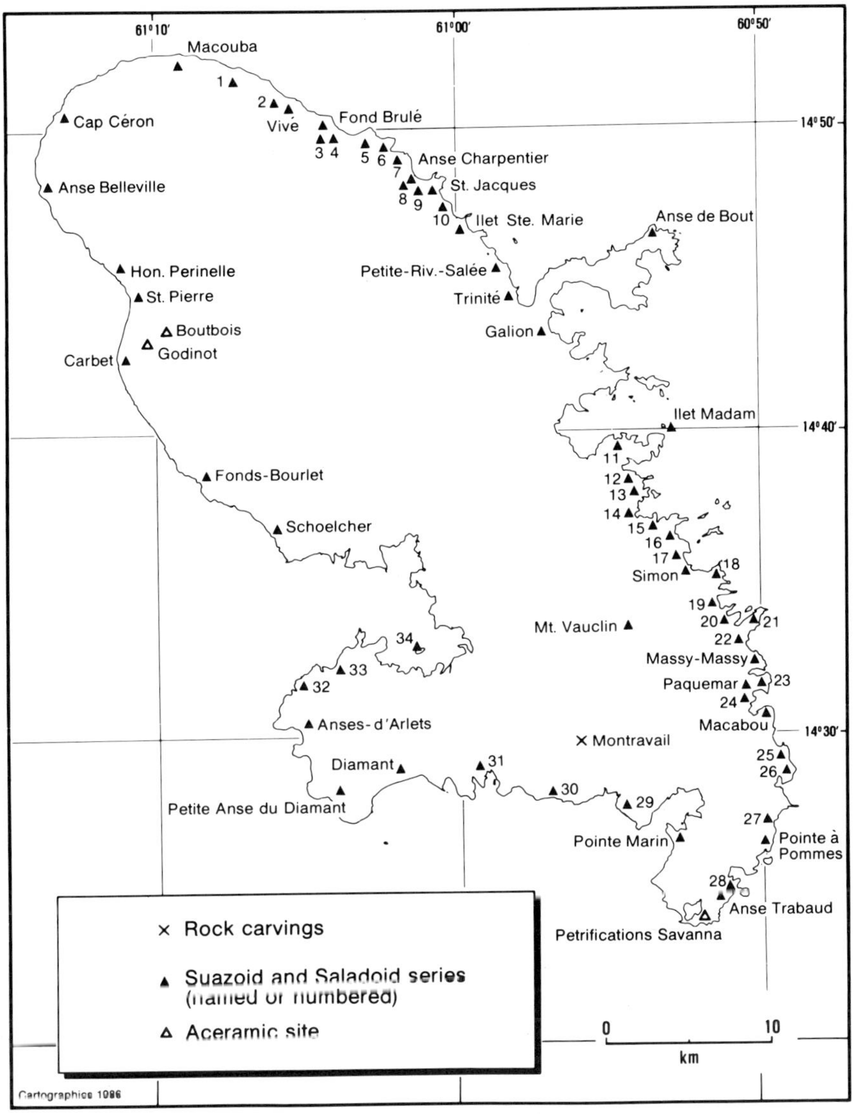

1. Moulin l'Etang
2. Rivière Capot
3. Grande Anse 1
4. Grande Anse 2
5. Galba
6. Adoration
7. Marigot
8. Pain de Sucre
9. St. Jacques W.
10. Lasalle
11. Pte. Royale
12. Morne Courbaril 1
13. Morne Courbaril 2
14. François
15. L'Esperance
16. Dostaly
17. Fregat Cul-de-Sac
18. Pte. Cerisier
19. Sans Souci
20. Unnamed site
21. Point Vauclin
22. Vauclin
23. Pte. Doucassous
24. A-Tout-Risque
25. Gnde. Anse Macabou
26. Pte. Macré
27. Cap Chevalier
28. Pte. Baham
29. Anse Figuier
30. Corps-de-Garde
31. Ilet Sapotille
32. Gnde. Anse d'Arlet
33. Anse à l'Ane
34. Gros Îlet

few. The most recent are on the southeast coast of the island. The recovered sites are found on all coasts, but more are along the southeast coast, an area Amerindians continued to occupy some years after the French took possession of the island. The site and dates of archeological remains indicate that the settlement process began on the northeast Atlantic side of the island and proceeded around the island.[6] Map 15 is a plot of the recovered archeological Amerindian settlement sites.

In summary, the first people on Martinique settled on the northeast coast. Because the seafloor falls away rapidly, seines could not be used to gather food. Shellfish were rare except for *Cittarium pica*. The first settlers were attracted to upland forested sites in a humid, windy coastal area. One of the earliest sites, Vivé 1, extended over an area of 150 m^2 but the deposit is thin (Mattioni 1973,84). Perhaps a relatively large number of people chose to settle briefly in an area not really suitable for the exploitation of sea resources. That is, these were a forest people, not a seacoast people. Local adaptation to the sea environment proceeded slowly but steadily. The importance of shellfish and the number of different species exploited increased with time. The first inhabitants of Martinique arrived as hunting-and-gathering horticulturists. Those who met the first Europeans were fishermen and farmers. If this interpretation of the archeological record is sound, the ecological adjustments suggested by the archeological record must have been profound.

Archeological evidence is not the only source of information about the Indians of the Antilles. The European contact records, especially the early ethnographic accounts, contain information on greater Antillean Arawak life, settlement patterns, and uses of wild plants. For the mainland Caribs, useful sources are Biet 1664, de Goeje 1939, Gillin 1936, and Pelleprat [1665]1857. De Rochefort 1666, Benzoni [1537]1857, Oviedo y Valdes, and Las Casas [1520–61]1875–76 are useful for the Island Arawaks. For the Island Caribs, the most important source is Raymond Breton, who arrived in 1635, 150 years after initial European contact but with the first settlers in the Lesser Antilles (Breton [1665]1892). Bouton 1640, Du Tertre [1667–71]1973, and de Rochefort 1658 provide other published seventeenth-century accounts. Delawarde (1938) reported on a visit to the Carib reserve made in 1936. He reported on the uses the Caribs were making of the forest (hunting and gathering), on their agriculture, housing, and settlement patterns as well as the mélange of ancient lore and European ideas about religion, language, and customs. He was disturbed to find them so *creolisés* (Breton 1938,204). Taylor worked with both the Dominican and British Honduran Caribs of the midtwentieth century and supplied information that, used with caution, provides valuable insight into the remarks of the chroniclers (Taylor 1957). These modern Caribs were until recently only partly acculturated and confined to regions that do not include all the range of microenvironments formerly available to them. I visited the reserve in 1962 and, thanks to Taylor, was able to talk to a number of Caribs about their uses of wild plants.

The Island Carib society at the time of early European voyages was organized into local, self-sufficient, extended matrilocal families. These families were scattered throughout the territory, living in small clearings large enough for the family *carbet* and the associated smaller houses in which the wives lived with their unmarried girls and smaller children (Breton [1665]1892,355; Du Tertre [1667–1671]1973,2:396). Each homestead, also called a *carbet,* contained thirty to one hundred people headed by one man who conducted family rites and extended courtesy to guests but who otherwise exercised little or no authority. More or less stable alliances between *carbets* might be formed by marriage, for these *carbets* were not necessarily friendly with one another. Early reports suggest a larger socio-political unit than the *carbet.* The *carbets* of the leeward side of the island elected a chief whose powers were minimal but whose prestige was sometimes considerable. Those on the windward side had a similar chief.

The Island Carib houses, even the *carbet,* measuring 60 to 100 ft long were generally smaller than Arawak houses. "They were built a little like those of our peasants except the roof and upper parts are made of reeds or palmetto leaves which extend to the ground" (Breton [1665]1892,355;) Du Tertre [1667–71]1973,2:396). Most Caliba settlements were near streams, pointing up the preference for frequent bathing in fresh water rather than in the sea. Although both Caribs and Tainos needed fresh water for drinking, Taino sites are rarely on riverbanks.

Island Carib population size fluctuated. The canoes possessed by the Caribs gave them mobility. In some cases they laid out provisions gardens on uninhabited islands (de Rochefort 1666). Bouton (1640,71), who was in Martinique in 1635, wrote: "It is not possible to state the true number of the Caribs on Martinique because they are continually and actively visiting and passing among those of Dominica and other islands. Sometimes there are many Caribs there and sometimes fewer."[7]

Taino settlements in the Greater Antilles formed loose, unorganized clusters having an open space in the middle onto which fronted the immense house of the cacique. The usual Taino houses during the Columbian era were large multifamily structures (Las Casas 1909, chapter 46). The earlier Saladoid of the Lesser Antilles may not have been so elaborately organized and may have been more similar to the Suazey or Island Caribs, hence the confusion over migration histories mentioned earlier.

The agricultural year had some seasonality of planting, but the harvest, particularly of staple starchy roots, continued year round. Half-wild and wild plants provided some minor foods and many kinds of medicines and useful materials. Some agricultural plants of the Greater Antillean Arawak were grown in fields of hills called *montones,* referred to as *conucos.* Breton ([1665]1892,328) reported that Calina also used *montones,* primarily for manioc. The *conucos* were cleared by burning, and then the earth was piled into hills that varied in height. Benzoni ([1537]1857,85) said that both sweet potatoes and manioc were grown in earth mounds in Trinidad, but Sturtevant suggests that manioc may have also been grown in leveled, loosened soil in the Greater Antilles (Sturtevant 1969). It is true that plants having upright, scandent, or climbing growth habits could be placed in the same mound and that *montones* made possible agricultural utilization of steep slopes with little danger from erosion. However, some plants were not grown in the *conucos* containing *montones,* and the contact accounts are distressingly imprecise in reference to where these plants were grown. At some distance from the houses the Calina had gardens in which staple root crops were grown (Breton [1665]1892,43–80). Tree and bush plants were found about the Carib houses and might well be called dooryard plants (Hodge and Taylor 1957,521). The plants in use were largely root crops for starches and tree fruit crops for sugars.

Both these people spun thread, wove cloth, and knitted hammocks. Both had the mute dog and hunted lizards, parrots, agouti, manatee, and muskrats (Chevillard [1659]1963; Las Casas 1909, Chap. 197, Hodge and Taylor 1957,521; Maurile de Saint-Michel [1652]1963,53). Both had seagoing dugout canoes and were expert swimmers and divers, getting from the open waters and shore much of their food, including several turtles (Coppier 1645,110–12) and marine conchs (Pinchon 1973,264). Some twenty-four marine animals species besides turtles and crabs are reported in the chronicles as having been used for food. Also reported are twenty-four birds, eight useful terrestrial animals, and three venomous snakes. Maurile de Saint-Michel ([1652]1963,53) said the Indian fathers tested the hunting and fishing skills of prospective sons-in-law.

The modern Dominican Caribs have a respectable plant lore involving about 120 species that are native to both Dominica and Martinique. They use species from nearly all of the spontaneous vegetation types that occur in their area, although montane vegetation types are comparatively little used. Of the natural vegetation types, rainforest, seasonal forest, littoral woodland, and littoral hedge contribute most of the useful species. Of the successional types, shrub and orchard savannas and fallow grassland contribute the most. All of this fits the known ecology of aboriginal settlers of Martinique.

The impact of the exploitations by pre-Columbian peoples on the plant geography was locally significant in kind and degree.

Changes in the Plant Geography

Under aboriginal occupation, the vegetation of Martinique underwent many disruptive changes. The composition of natural plant communities was altered by the selective utilization of fire and by the invasion of introduced aliens. The borders of certain natural vegetation types retreated in the face of slash-and-burn agriculture. The vegetation mosaic became more complex as clearings were made. New types of planted and successional vegetation were created from native and introduced species, and there was an expansion of area in successional stages as opposed to the self-replicating types.

The ecosytem model proposed earlier can be used to summarize the effect of the Amerindians on the plant cover, as follows. Modifications resulting from changes in membership include the selective culling of native species in lowland forests, the encouragement of aggressive pioneer species, and the introduction of a large number of exotic food plants, some construction materials, and weeds. Changes resulting from new processes include the consequences of swidden agriculture and burning, including runaway fires. Changes in habitat include the creation of more "edges" in seasonal forests, woodland, and littoral hedges adjacent to agricultural plots and building sites; the opening of clear fields for heliophiles and other pioneers; the deposit near settlements of human and plant wastes that changed the available nitrogen content of the soils. Changes in dimensions of the system, in inter-island and island-continental traffic, increased the dispersal potential of exotic species and the extraction of biomass for use outside the territory of the original system.

Although the plant cover of Martinique at the end of the Indian period had suffered changes, it is probable that sites in the coastal borders less suitable for agriculture were never cleared. If these were disturbed at all, it would have been by runaway fires rather than by clearing and planting activities. The pattern of distribution of both the gardens and the spontaneous successional forms that followed their abandonment included remnant patches of natural vegetation. This pattern was determined by distance from settlements, aboriginally perceived micro-soil distributions, and the fundamental physical environment controls responsible for the original vegetation pattern. A plot showing areas of considerable disturbance reflects the recovered settlement pattern; it is based on a fixed radius representing a normal walk to fields from the location of villages (see map 16).

In the last years before colonization abandoned lands not occupied by Calina reverted to natural vegetation, and secondary growth progressed well along the successional stages toward secondary seasonal forest and littoral woodland. By the time of European contact most of the larger sites abandoned by the earlier settlers would have been reclaimed by developed forest and well-established woodlands and bushlands. The smaller settlements were scattered along the coasts. There were fewer on the northeast and more on the southeast behind the Atlantic barrier reefs. A few settlements fronted the Caribbean Sea.

Probably much of the southern portion of the island south of a line from François to Fort-de-France that was not in mangrove swamp, rock pavement, cactus desert, and littoral grassland was or had been at one time in cultivation. This included land perhaps even to the top of Mont Vauclin, more than 500 m high. North of this line the windward slopes well into the lower margins of rainforest may have been cultivated at one time; certainly much of the seasonal forest, the alluvial valley bottoms, the littoral woodland, and the littoral hedge had been cultivated, except for the steep to cliffy slopes that lace the landscape. On the Caribbean coast it is supposed that fields were used in short rotations and that the

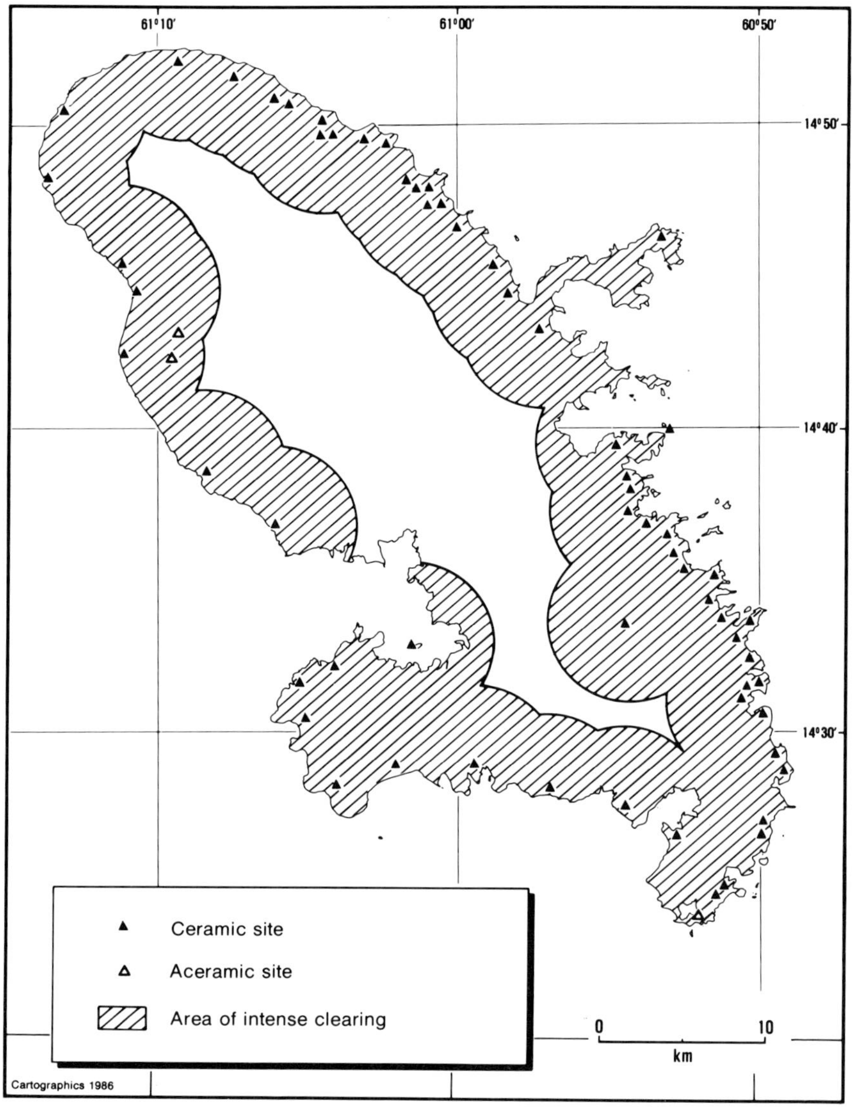

Map 16. Hypothetical perimeter of intense vegetation clearing based on distances from archeological sites. Source: Allaire 1977,314; and compiled by author.

drier strip of coast from Schoelcher to Prêcheur may well have been burned repeatedly by runaway fires during the dry season, when the cleared agricultural plots were burned. The burning of the dry bushlands and seasonal forests throughout the island had probably brought about an increase in thickets composed of woody legumes and others of *Lantana* and *Croton* species. It also seems probable that the burning during the aboriginal times may have aided the expansion of bushlands at the expense of dry woodland, which is the usual vegetation under the prevailing physical environmental controls.

The use of old selective culling of the spontaneous vegetation must have modified significantly the proportional representation of the species in the vegetation or caused changes in the habitat that in turn effected changes in a vegetation type. Many plant uses—for ritual and charm materials, medicines, poison for arrows, barbasco, bird lime, wicks, and minor foods—appear too desultory to have effected much change in the plant cover. Only a few uses likely to have had significant effects and only the species employed in those ways will be discussed here.

For their dugout canoes the Indians needed large trees out of the rainforests with wood that was salt-water resistant. Breton ([1665]1892,253) reported that in Guadeloupe dugouts were made from red cedar (probably *Carapa guianensis*), although many authors suggest *Cedrela mexicana* as the tree (Duss 1897,1:129).[8] The Dominica Caribs use *Dacryodes excelsa,* another large-boled rainforest tree, for their dugouts (Breton [1635]1947,36; [1665]1892,252; Coppier 1645,53). It is quite possible that these trees were significantly reduced in Martinique by Indians in prehistoric times, as the foresters and local rural population believe. Dominica, which has a lower population density, was being culled of these giants for the building of canoes, some of which were bought by Martinique fishermen as recently as 1950. *Mastichodendron foetidissimum,* called *acoma,* was likened to the oak of Europe because it was so large and useful in construction (Breton [1665]1892,245; [1635]1947,35; Bouton 1640,65). *Sloanea massoni,* sometimes regarded as the codominant of *Dacryodes excelsa,* was noted as an excellent wood, but it was far from the coast and difficult to get out of the forest (de Rochefort 1658,59). Other rainforest trees selectively cut out included *Sterculia caribaea,* whose lumber was split into planks to build up the canoe sides, and *Tapura antillana,* whose buttresses were used for paddles.

Species found in the rainforest, but more concentrated in derived second-growth vegetation, were used in house construction. Tree fern trunks were used by the Calina as ridgepoles in the temporary huts called *ajuopas,* as well as in more permanent structures. The Taino at the time of European contact split palm trunks for laths to build their large, bell-shaped houses. The lightweight trunks of *Cecropia peltata* were used in building canoes, and *Ochroma pyramidale* were used for rafts (Pelleprat [1655]1963,30). The leaves of *Heliconia bihai* and *Asplundia rigida* were used as thatch substitutes by both groups and by the Calina for making an impervious layer between the double baskets that were used for carrying and for storage (Coppier 1645,53; Breton 1665,249; de Rochefort 1958,114). Basketry plants included *Ischnosiphon arouma, Asplundia rigida,* and *Stigmaphyllon* spp. (Breton [1635]1947,64 and 1665,246; Du Tertre [1667–71]1973,2:121). These uses and that of gums and latex probably effected little change in the species distributions. However, the disturbances caused by frequent visiting of the same areas may have caused some modifications of the vegetation, creating more open, hence drier avenues or creating a greater abundance of so-called hunter trees by the scattering of fruits and pits.

Because of the settlement pattern, seasonal forest and the littoral and dry woodlands types were the most accessible sources of timber for buildings and for fuel. They share several useful trees that undoubtedly were cut selectively for lumber and boat timber, which reduced their numbers: *Chrysophyllum argentum, Calophyllum calaba, Ceiba pentandra,*[9] and *Cedrela mexicana.* Both *Simaruba amara* and *Tabebuia pallida* were used for firewood and for making household implements. The fiber of the seaside *Hibiscus tiliaceus* was used for making cord and netting. The huge *Hymenaea courbaril* was used to make large war canoes (Bouton 1640,58; Coppier [1664]1963,38). Fruits of the *Rhizophora mangle* were used at times (Oviedo y Valdes [1535]1851,9(9),6), and the wood served for construction and for fuel, but aboriginal use probably did not initiate any substantial changes in borders or composition of the type. These swamp woodlands were good hunting areas for crabs, which were eaten by both Saladoid and Suazoid peoples, but there is no reason to suppose that there was deliberate preservation of this habitat.

Fear of the toxic *Hippomane mancinella* may have caused the aborigines to leave this littoral hedge beach tree alone and, thus, increased its importance with respect to other woody species. *Chrysobalanus icaco* and *Coccoloba uvifera* may have been the deliberately spared near settlements on the windward side because they make good windbreaks and their fruits are tasty. On the other hand, the torch woods—*Erythalis fruticosa, Zanthoxylum spinifex,* and *Amyris elemifera*—may have been reduced (Breton [1665]1892,36; Maurile de Saint-Michel [1652]1963,55). Probably reduced in proportion was *Coccothrinax dussiana,* a fan palm whose wood was used for making bows (de Rochefort 1658,64) and whose broad leaves were used for thatch and broom material (Breton [1665]1892,244). Fronds of the *Rhyticocos amara* were used as thatch (Coppier [1644]1963,42–43; Bouton [1640]1935,18).

Dominica Caliba use the leaves of the agave-like *Furcraea tuberosa* out of the mixed cactus scrub for cordage fiber (Breton 1647,40; Du Tertre [1667–71]1973,2:97), and the inflorescence stalks were formerly used to make rafts. The fiber plant *Ananas comosus,* found near salt ponds and flats, was used to make bowstrings (Coppier [1644]1963,39; Breton [1635]1947,40). Pads and fruits of *Opuntia dillenii* are used for medicine and food. However, since the seeds of both of these species are spread by birds in disturbed sites and they can reproduce vegetatively, they may very well have become more abundant under aboriginal use.

With agricultural activity there would have been a gradual increase in area of scrub and orchard savanna, secondary succession types, which gave a signficantly larger habitat for pioneer and semipioneer plants and exotic weeds. The deliberate harvesting of some grass seeds and roots accentuated the disturbance pattern.

Fiber plants such as the shrubby *Urena lobata* and the suffrutescent *Sida* species were collected for use in twine and broom making. Herbaceous members of these created savannas contributed charm plants and thatching materials. The accidental scattering of seeds and the trampling attendant upon gathering tended to maintain these types.

In Martinique the prehistoric settlement pattern was largely a coastal one, and clearing and burning occurred primarily near coastal areas. A few settlements in the southeast were situated close to the postulated littoral woodland seasonal forest border, and some sites were situated in the riverine facies of seasonal forest, which extend seaward from the main body of the seasonal forest. The leeward settlements were in areas identified as originally dry bushlands and dry woodland and in the valleys occupied by the seasonal forest. Saladoid settlements, with their larger populations, would have produced more compaction of earth, greater enrichment of soils through refuse, larger areas of full sun where there had been only filtered light before. With the Suazoid occupation the degree of such disturbance associated with these clearings in the woodland would have been less (Du Tertre [1667–71]1973,2:395).

It is usually supposed that horticulturists, dependent primarily upon slash-and-burn agriculture, move their settlements periodically as fields become far distant. A divergent opinion holds that the primary reasons for the moving of villages are sanitary considerations. If permanent fields made possible a reasonably fixed settlement, hygiene might well have occasioned the movement of a village. However, with a people tied to a seashore habitat, these moves might be only small shifts in position, and large sites such as that recovered at the Lasalle site in Sainte-Marie could have been built up by accretion, rather than by the contemporaneous presence of a large population.

Agricultural activities required far more extensive clearing than did house sites, but their area and distribution are less easily substantiated. The general pattern of modern swidden agriculturists in the neotropics growing similar starchy root crops involves cultivating a given field for an average of about three years, after which weed invasion and soil depletion will force abandonment. The land is then left fallow for ten to twenty years before it is recleared and cultivated (Leeds 1965,19; Carneiro 1961,48; D. Harris 1970).

The Greater Antillean Taino in the sixteenth century left manioc plants in the ground up to three years, at which time the plant was harvested completely and new cuttings were inserted in the *montones* (Oviedo y Valdes [1535]1851,271–72, chap. 7, para. 2). It is not known how long those gardens were maintained.

The Calibi or Island Carib gardens were apparently short-lived and shifting, although Breton's ([1665]1892,328) reference to *montones* suggests that some low-lying fields may have been more permanent. That modern Carib gardens on Dominica were "cultivated without deep digging or manuring on the primitive system of clearing and burning the forest, gathering one or two crops, and moving farther afield" may reflect a cultural loss through time (Hodge and Taylor 1957,516). According to informants in 1962 the Dominica Caribs left the land fallow some ten to fifteen years before reclearing.

With a good supply of protein food from the sea, the aboriginal people on Martinique probably required no more than one-fifth of a hectare per capita of land in starchy root crops. If the land were cropped three years out of every fifteen to twenty, a total of 1 to 1½ ha would be needed per capita; thus, only 10 percent of the island's surface could have supported nearly 10,000 people. The aborigines may never have been that numerous; even if they were, they could have been highly selective in choosing vegetation types for clearing, since at least 55 percent of the island area is in land on which their staple crops can grow.

The staple root crops did not require constant attention, and we know from contemporary accounts that some gardens were a distance from the settlements (Du Tertre [1667–71]1973,2:373). However, this was not the regular pattern (Breton [1665]1892,123). Agricultural clearings were mostly concentrated in the more accessible coastal woodlands and especially the seasonal forest vegetation. Rather than making new clearings immediately adjacent to old plots, Caribs in Dominica look for certain trees and understory plants, especially *Selaginella flabellata,* as indicators of "fresh soil." This results in a dispersed pattern of agricultural plots. The aboriginal clearing pattern was probably a similar crazy quilt of fields surrounded by old plots in various successional stages of revegetation (Breton [1665]1892,12). Even near villages, patches of natural vegetation may have been left alone, but at one time or another much of littoral woodland, dry woodland, and seasonal forest was cleared or greatly disturbed, thus making those plant formations secondary rather than primary representatives of the vegetation types by the time of European contact.

We can only speculate about the extent and frequency of aboriginal burning outside of agricultural clearings, there being a lack of references to burning in the early literature of the French Antilles. It seems likely that runaway fires did occur in the littoral woodland, dry woodland, and dry bushland types and may have been common enough to modify their composition and physiognomy and to extend the drier seasonal formations at the expense of dry woodland and seasonal forest. Occasionally, a forest fire set by humans may have burned for some time. Lightning-set fires were probably not important even in dry formations because heavy rains accompany or immediately follow lightning storms.

All the leeward coast from Schoelcher, the old Case-Navire, to Saint-Pierre was described when first seen by Europeans as being in *hailler,* a term much used in the seventeenth century to mean scrub or thickets. These would have been derived from the dry woodland by clearing and occasional burning. Certainly fires, when they took place, tended to delay succession on abandoned fields and maintain some fields in grassland and savannas. The lack of savannas on the islands at the time of European arrival does suggest that large-scale burning was not a regular occurrence. The fine-grained, rough terrain may have acted to interrupt the spread of fire.

The migrating Saladoid and Suazoid peoples carried with them many useful plants and, undoubtedly, some weeds. However, sorting out the species they introduced from those that were naturally present in the island is often extremely difficult, and the decision depends mainly on the ecologic behavior of the plants. The case for Indian introduction is clearest with the cultigens that do not occur in natural habitats and lack putative wild progenitors in the West Indies.

Such plants, cultivated by the Antillean Arawaks at the time of discovery, include manioc (*Manihot esculenta*), the staple; maize[10] (*Zea mays*) (Breton [1665]1892,355); yautia or tannia;[11] yampee or kush-kush (*Dioscorea trifida*); peanut or pistache (*Arachis hypogaea*) (Breton [1665] 1892,246); two kinds of beans (probably *Phaselous vulgaris* and *P. lunatus*) (Breton 1664,352; de Rochefort 1658:113); two varieties of sweet potato, the starchy ajes and the sweeter batatas (both *Ipomoea batatas*); two kinds of peppers (probably *Capsicum sinense* and *C. frutescens*); and tobacco (*Nicotiana tabacum* or *N. rustica*). The evidence is less clear for pumpkins or giraumon (*Cucurbita moschata*) (Breton [1665]1892,407; Coppier 1645,91) and bottle gourd (*Lagenaria siceraria*). The latter is spontaneous today but virtually always grows in disturbed habitats. It has been suggested that pumpkins may have come in with corn at the time of the Insular Saladoid, but this is only inferential. All these are ancient crops in tropical South America and can reasonably be assumed to have been brought up through the Lesser Antilles by the early settlers (Du Tertre [1667–71]1973,2:80–140; Coppier [1644]1963,43–44; Breton [1665]1892).

Mackey suggested that the lima bean may have been introduced to the West Indies from Central America, but, as deduced from the modern distribution of subspecific races, it perhaps is a product of postcolonial dispersal. Also, mainly on the basis of modern collections, Hutchinson, Silow, and Stephens (1947) inferred a Central American source for some of the West Indian cotton (*Gossypium hirsutum* var. *punctatum*) (Fryxell 1980). The Greater Antillean Taino not only planted this cotton but also gathered it from the wild. It may actually have reached their area as a sea-dispersed coastal pioneer.

The Saladoid cultures may have domesticated a few other plants in the islands, for example, topitambour (*Calathea allouia*) (Oviedo y Valdes [1535]1959,1:329), yautia or choux caraibe (*Xanthosoma sagittifolium*) (Breton [1647]1929,80), and maraca or toliman (*Canna* sp.) (Hodge and Taylor 1957). The first plant has small tubers clustered on the roots of the plant, prepared by boiling and seasoning. It grows usually in damp places along rivers but may grow in dry habitats as well. As a cultivated plant it may have been brought out of South America, for the contemporary Arawak Campas grow it in Peru.[12] Oviedo mentions roots that probably are species of a *Canna* (Oviedo y Valdes [1535]1851,284–87). If it is *C. edulis* or *C. indica*, then the plant was imported from South America. If it is *C. lamertti*, then it is a local domesticate. Today *Canna* plants are seldom seen in undisturbed places.

The later Suaziod continued to plant these Saladoid crops, whether by a shared horticultural heritage from South America or from the contribution of the early Caribbean neighbors. They drank more of their diet than is recorded for the Saladoid.[13] They evidently introduced two kinds of pineapple (*Ananas comosus*), one selected for fruit to be eaten raw and fermented and the other, *la pité*, for leaf fiber (Bouton 1640,57); Breton [1647]1929,40; Laudonnière [1586]1958,83). They also may have introduced the papaya (*Carica papaya*) from Central America on their trading expeditions. They probably also brought in kidney cotton (*Gossypium barbadense* var. *brasiliensis*), which evidently hybridized with the older Arawak cotton to produce Marie-Galante cotton (*G. hirsutum* var. *mariegalante*) (Pelleprat [1665]1963,31).

Various semicultivated plants (either planted or spontaneous, but cared for) were used by the Amerindians. Several species suggested as brought up through islands by the Saladoid are the guanabana or soursop (*Annona muricata*), sweetsop (*A. squamosa*), indigo (*Indigofera suffruticosa*), hog plum (*Spondias mombin* and *S. lutea*), mamea (*Mammea americana*), annatto or roucou (*Bixa orellana*), guava (*Psidium guajava*), calabash (*Crescentia cujete*), and genipa (*Genipa americana*) (Du Tertre [1667–71]1973,2:17–19,166–77; Maurile de Saint-Michel 1652,73). Other useful plants with ranges extending from South America only as far as the Lesser Antilles have been suggested as Suazey introductions (Breton [1665]1892,356; Loven 1935,363), including the giant arrow grass (*Gynerium sagittatum*); larouman (*Ischnosiphon arouma*); arrowroot (*Maranta arundinacea*), which may have been more important as a poison antidote than as a food plant (Breton [1635]1947,43); locust (*Inga* spp.); cashew (*Anacardium occidentale*); and kunami or nivrage (*Clibadium sylvestre*). The evidence is confused for the castor bean (*Ricinus communis*). It may have been transported to South America and Brazil and, thence, to the Antilles in pre-Columbian times. By the seventeenth century it is noted as an ointment to control vermin and as a vehicle for applying annatto to the body (Chevillard [1659]1963,65).

The list could be extended by including other useful weedy species. No doubt such plants increased in abundance, spread locally by human dispersal and by habitat disturbance. But some are also dispersed by birds and other natural agencies, and these semicultivated species grow spontaneously outside cultivation. Some, such as *Annona, Anacardium,* and *Mammea,* appear in modern times to be members of the native seasonal forest, although they may have entered unoccupied niches in the native forest and are, therefore, indistinguishable from natives. Genipa, for example, is similarly well incorporated into rainforests. Some others, such as roseau, larouman, and arrowroot, are so at home on streambanks that they may have been dispersed by natural agencies after an original human introduction.

The same can be said of all the "useless," weedy species that may be suspected of being prehistoric introductions. A number of weeds may have been introduced for uses no longer suspected. The weedy, leguminous tree *Anadenanthera peregrina,* which produces cohoba, a common drug in South America, has a distribution similar to that of the Arawak-speaking peoples. Safford (1917,6) suggests that it was an introduction by those peoples now referred to as the Saladoid. *Justicia pectoralis,* a common weed of the savannas, may be another drug plant introduced by the Arawak peoples for its hallucinogenic properties, only to have its use lost after the Suazey invasions.

Several new vegetation types were created by the agricultural systems and the settlement patterns of the Amerindians. These are dooryard and provisions gardens, shrub and orchard savannas, leguminous scrub, and the *Randia-Croton-Lantana* thickets. Judging from historic and ethnographic accounts, both the Taino and Island Caribs had plantings about their dwellings, dooryard gardens, and also scattered provisions gardens at varying distances from the villages. Dooryard gardens are classified under planted vegetation, although it is a borderline type because of the number of volunteer elements in it. From what little is known directly of the island Indians, and by extension from better-known groups, the dooryard garden was an anarchistic species mixture about a dwelling with a heterogenous geometry having no distinct line between planted useful weedy invaders and relics of original vegetation left uncleared.

Some plants apparently did not fit into the *conuco* or provisions garden practice as we can reconstruct it. As was noted above, these included such fruit trees as guanabana, sweetsop, hog plum, mamey, guava; calabash trees for gourds; bushes of annatto and indigo for dye and body paint; cannas as ornamentals and for starchy rootstocks; cotton bushes for cordage fibers; peppers for flavorings; and tobacco. The last species needed to be grown in carefully prepared plots devoted solely to that plant and required, in addition, careful attention at various times to its production. It was apparently more important to Saladoid than to Suazoid Amerindians. The native peoples wanted these and other plants conveniently at hand, but the desired quantity did not require extensive planting. Thus, even if the plants were often traded, they were also volunteers or planted around the village.

The first accounts of Island Carib villages mention clean-swept yards between the *carbet* and the smaller houses, but this does not eliminate the idea of surrounding plants (Du Tertre [1667–71]1973,2:396; Breton [1665]1892,85). These people with their interest in fermented drinks and fruit soups may have had more fruit trees in their plantings about their houses. All of the plants inferred for Saladoid dooryards are used by the modern Caribs of Dominica except indigo (*Indigofera tinctoria* and, possibly, *I. suffruticosa),* which were introduced very early after European discovery. Canna, cotton, annatto, and peppers are specifically reported to have been planted by the houses (Hodge and Taylor 1957). These Carib house gardens also contained pineapples, used variously for fresh fruit, a fermented drink, and a fiber source. Some gardens may have included such native medicinal and charm plants as kudjuruk (*Petiveria alliacea*) and *Ambrosia paniculata*. The arrowroot commonly is half-wild on the periphery. *Opuntia dillenii* and physic nut (*Jatropha curcas*) are used in modern Martinique gardens as dividers and fencing and may have been incorporated into these gardens in the drier parts of the island in prehistoric times; certainly they were used medicinally (Du Tertre [1667–71]1973). Where settlements were in littoral woodland and littoral hedge, sea grape certainly must have been in the gardens.

Provisions gardens were the second new vegetation type. Greater Antillean Taino grew their staple crops, all cultigens brought from South

America, in patches cleared by slash-and-burn techniques away from their settlements. The two most important crops, manioc and sweet potatoes, were grown in *montones,* circular earth mounds about 9 feet around and 3 feet high with flat tops. Patterns resembling such earth mounds are reported for the putative Saladoid site at Anse Belleville north of Prêcheur (Reichlen and Barret 1940,248–72).

Maize may have been planted in separate patches also cleared by cutting and burning of the natural cover. An Island Arawak people in Trinidad gang-planted burned-over plots (Benzoni [1537]1857,83). They punched holes at each pace with a dibble for several maize grains. This suggests that those fields were relatively large and flat, and that planting of maize was a communal agricultural activity and may have been a late introduction into the Indian crop inventory. Maize, however, always was of distinctly secondary importance and, by European times in the Greater Antilles, was eaten roasted in the immature state by the Tainos, as it was by the Island Caribs in the Lesser Antilles.

Secondary starchy root crops such as yautia, tuliman, topitambour, kush-kush, and vegetables such as pumpkins, peanuts, and beans were probably also grown in provisions gardens, either by interplanting or in small patches. Tobacco most certainly was grown in special beds.

Suazoid provisions gardens apparently never included the large earth mounds. Otherwise, from what is known, the Suazey gardens did not differ from the Saladoid ones in preparation and differed little in crop content. Today some provisions grounds associated with stream courses are only weakly separated from the surrounding rainforest by the species composition and relative wealth of herbaceous ground cover. Provisions grounds were sometimes high in the interior, and also the Island Caribs laid out provisions grounds on the uninhabited islands.[14] Laudonière (1586,40,288,532–33) reported that his men moved directly from "gardens" to the houses (Hakluyt, 1903–1905,19:123).

The presence of *Erythroxylon coca* and mani or copal derived from *Moronobea coccinea* (Hodge and Taylor 1957,586–87) can be attributed to trade, rather than to local cultivation. There was trade between the Island Caribs and the South American Barima River Caribs well into the nineteenth century, and there is no reason to believe that they were not trafficking with their relatives during the prehistoric period.

The abandonment of provisions gardens provided for small areas of grasslands, scrub savannas, and orchard savannas, depending on how completely the land had been cleared for planting. Except on the occasions when succession was interfered with by frequent burning, all fields passed quickly into thickets and young forest in the wetter, higher areas. Abandoned village clearings would have been orchard savannas until succession replaced them with woodland and forest. Rates of succession have not been worked out, although experience in the recent history of Martinique indicates that regeneration of old fields to brushland and thicket occurs within seven years, and to young woodland and forest within twenty-five years, except in the dry southeast portion of the island, where changes in soil characteristics, including soil moisture retention, have proceeded toward a more xeric condition.

The species composition of these secondary vegetation types was dependent on the flora of the undisturbed vegetation, out of which the gardens had originally been carved, and on the native weeds. The openings in the natural vegetation created by destructive action of wind, flood waters, and volcanic eruptions had provided limited habitat areas for native pioneer plants. These areas had been colonized by gap phase processes passing to building phase processes with species complements associated with each. These species were given increased opportunities for colonization by human disturbance that began with the aborigines. Some, apparently native, species incidentally introduced from South America by the Amerindians, also colonized into these disturbed places.

The weedy herbaceous element in the lower and drier areas was dominated by the legumes, particularly species of *Desmodium,* the grasses *Schizachyrium* and *Paspalum* species, and malvaceous plants, especially *Sida.* In the drier regions under 1,500 mm and on heavy montmorillonitic clay soils, the malvaceous plants were more common than legumes, and grass species of *Boutleoua* replaced *Schizachyrium.* In the upper elevations abandoned gardens and settlement sites were colonized by grasses of the seasonal forest floor, the broad-leafed *Pharus, Olyra,* and *Lasiacis;* the razor-sharp sedges called "zeb rasior," *Scleria;* and composites, *Bidens pilosa* and *Ageratum* sp. Clearings maintained for a longer time and kept cleared with fire more likely, as now, to be colonized by large *Panicum* species, *Indigofera* spp., *Mimosa pudica,* and *Wedelia trilobata.*

Important shrubs, aggressively invading in lower elevations on grasslands, and later to dominate in the bushlands to follow, were *Chrysobalanus icaco* and *Lantana involucrata,* and *Solanum racemosum* and *Randia aculeata* on the drier montmorillonitic clays and the rust-brown ferruginous soils.

Croton species as well as acacias and logwood volunteered onto the shallow volcanics and the wetter ferruginous soils. Each of these elements were originally part of the dry bushland and the other coastal vegetation types. At higher elevations shrubby elements such as *Piper dilatatum, Solanum* spp., *Tabernaemontana citrifolia,* and the many *Miconia* spp. were derived from the seasonal forest and rainforest. Calabash trees, annatto, guava, and cashew trees, as well as the several species of *Annona,* frequently colonized in addition to being remnants from old gardens.

Some trees, sought after for lumber in the dry woodland and littoral woodland, are agressive pioneer trees. *Tabebuia pallida, Hibiscus tiliaceus,* and *Bursera simaruba* at the present time rapidly colonize disturbed and abandoned fields and, presumably, did so then.

Two spontaneous vegetation types began to be differentiated during this period: the leguminous scrub and several aspects of the *Randia-Croton-Lantana* thickets. As the dry bushland, dry woodland, littoral woodland, and dry areas of seasonal forest were repeatedly cleared, there developed a greater concentration of these aggressive pioneer species at the expense of more shade-demanding, less aggressive species of the vegetation. *Acacia* species began to dominate the dry bushlands from Schoelcher to Carbet on the heavy montmorillonitic clay soils. *Haematoxylon campechianum* and *Pithecellobium unguis-cati* were selectively concentrated in these bushlands by their aggressive pioneer behavior and resistance to fire. The *Croton* thicket is on the youngest and best-drained lithosols or ash from Carbet north to Prêcheur, as well as on the calcareous soils of Sainte-Anne and Diamant communes. *Lantana camara* thickets developed on the sandy soils of the leeward coasts. Here, also, accidental firing of surrounding bushlands was highly likely, and bushes were being selected for their weedy character and their fire tolerance.

Since the extent and distribution pattern of ancient Amerindian fields is obscure, the pattern of successional types is equally uncertain. If the Saladoid peoples were more conscientious about clearing fields of woody growth before planting then were the Suazoid peoples, more grassland would have been associated with their occupation than with the later inhabitants. At any one time there was probably as much spontaneous grassland as there was land in provisions gardens, and significantly more in savannas. As settlement patterns changed, there would have been a reclaiming of the land by later succession spontaneous types. Since fire is not considered to have been an important factor in maintaining the vegetation open during the aboriginal period, except possibly on the leeward Caribbean coast, the regeneration cycle was probably close to natural succession rates.

Three kinds of natural catastrophes were important during the aboriginal period and may have had increased ecological importance, adding to the disruptive effects of Amerindian economic exploitation. The first is volcanic explosions; the second, sea level changes; the other, hurricanes. After 5,000 Y.B.P. the period of the Amerindians can be said to have begun for the Caribbean Basin (see table 18). During the Archaic there were several eruptions. The first pumice episode dated at 4,610 ± Y.B.P. (Traineau 1982,69–71) was a major event by virtue of the very large volume of ejecta associated with it. On the western flank were two deposits, one on the Grand Savanna and the other along the Sèche River. Another deposit filled the Falaise Gorge. These pumice beds are from 4 m to 20 m in depth and contain numerous fragments of carbonized tree trunks. Superimposed on the pumice flows is a blanket of fine ash. In the area of Prêcheur it is almost 3 m thick, much deeper than laid down in later eruptions. It is possible that such explosions made the island unattractive to the northward-moving Archaic peoples who were not dependent upon a stepping-stone advance up the island chain. The impact was on all altitudinal vegetation types, from summit meadows to riverine facies of seasonal forest.

Three pumice events about 2,400 ± 70 Y.B.P. covered the western and southern flanks of Mont Pelée (see map 17a). This was not a major event in terms of amount of material, but, because it was likely deposited by the transport of the Trade Winds, it covered the landscape like a

Table 18. Recovered eruption events during the Amerindian period

Event	Location	Y.B.P.
Pumice event	Western flank of Mont Pelée	4,610 ± 50
Pumice event	Basse-Pointe	4,060
Nueés ardentes	Capot, Falaise, Claire rivers	3,980 ± 30
Nueés ardentes	Sèche River	3,710 ± 30
Pumice events	Prêcheur Commune	2,440 ± 70
Pumice events	Western and southern flanks of Mont Pelée	2,010 ± 140
Pumice event	Sèche River and Fond Connonville to north coast	1,670 ± 40
Missing events	—	—
Nueés ardentes	Western flanks of Mont Pelée	1,140 ± 80
Pumice events	General	650

SOURCE: Traineau 1982.

blanket rather than concentrating in the river valleys and other gorges. Roobol and Smith identified an ash cloud associated with a pumice flow at 2,560 ± 70 Y.B.P. in a quarry on Morne Ponce (cited in Traineau 1982,80).

More than a dozen other eruption events occurred between these last and a series of pumice flows around 2010 ± 140 Y.B.P. According to Traineau (1982,79–82) this was probably the last pumice paroxysm. Heavy deposits up to 15 m in depth blanketed vegetation (see maps 17b, c), completely killing all the vegetation covered during the first event. The second, directed more to the south, covered less deeply but showered a half-meter on the Carbet Pitons. This is the first covering of these high summits that has been established for the late Recent period. The third covered an almost circular area centered on the Mont Pelée summit and extending to Fond-Saint-Denis, covering Morne-Rouge, Ajoupa-Bouillon, and reaching almost to the north coast but not to the Carbet Pitons. The last event mapped over a wide extent was that of 650 + Y.B.P., or about A.D. 1300 (see map 17). This is the last pumice event recovered from the geological record. It and the previous one extended to cover Vivé, Fond Brulé, and Prêcheur and covered Amerindian sites (Roobol et al. 1976, cited in Traineau 1982,90).

The last eruption was sufficiently recent for the Caribs to call Mont Pelée the "mountain of fire." Also, early European settlers used the name "Pelée," which means bald. These eruptions caused the outright destruction of much vegetation by burial under pumice and by blasting and burning *nuées ardentes*. In fact, the pattern of vegetation change would follow the geography of destruction. Where neither the heat nor the ash fall was too severe, the effect of the ash dressing would have stimulated the growth of at least woody vegetation. Regeneration rates

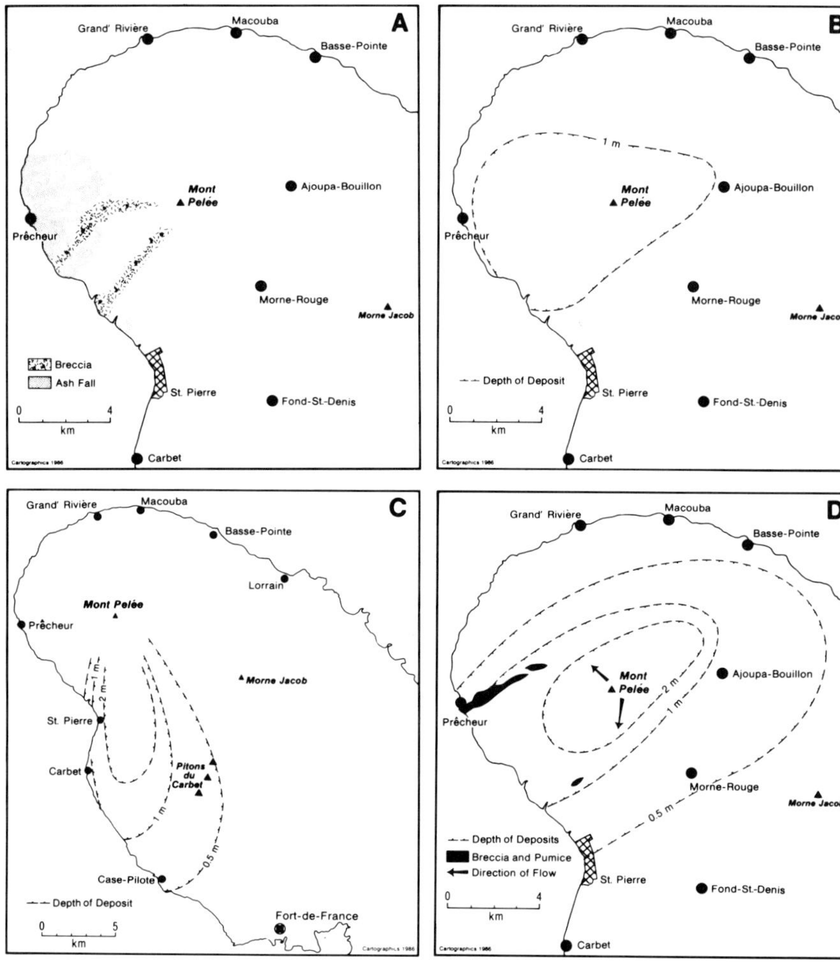

Map 17. Selected volcanic eruption events for the Amerindian period. Source: Traineau 1982,42,70, 81,86.

depended upon the degree of destruction, the altitude, the wind direction, and the exposure, as well as the new topography. The vegetation most likely damaged included montane bushland, elfin woodland, and montane forest. Some of the archeologic record was destroyed by such explosions, and probably only a few sites so buried have been recovered.

Changes in sea level caused the migration of low-level elevation vegetation types seaward during the low sea stands and landward when the sea level rose. Estimates of relative vertical motion vary 27 to 55 m (Milliman cited in Ruppé 1980,331; Blackevelder et al 1979). The island's assymmetrical submarine profile means that under lower sea level conditions, land gains would be greater along the Atlantic coast south of the Caravelle Peninsula and at Fort-de-France Bay, assuming that there was not tectonic motion. There would have been a substantial gain of mangrove and swamp woodland in Lamentin Cove and Genipa Bay and in the southern estuary boundaries. There may have been expansion of the fresh-water annona swamp along the Galion River. Behind the mangrove would have been an enlarged area of seasonal forest. The littoral vegetation types may have been reduced in area as the coastline was smoothed and the length reduced. The trend of sea-level change in the past 17,000 years has been a cumulative rise for the Caribbean Basin (Ruppé 1980,331).

The rate of infilling of Fort-de-France Bay by depositional processes varied with volcanic explosive events, but the geological evidence is for long-term filling of the bay margins. With higher sea stands, marine terraces were formed and the island's area was reduced, resulting in a loss of level land, thereby reducing the area in low-level vegetation types except for mangrove and other inundation types that are essentially coastal in distribution.

Climate change studies for the Gulf of Mexico and the Caribbean Basin for the late Recent have just begun (Bryson and Wendland 1967). Carbone (1980,123) has used a variety of data to compile a sequence for the Greater Antilles (see figure 13). Wright (1976,574) suggests that the hypothermal for the United States varied with the western United States some 1,000 years earlier than that in the southeast (6000 Y.B.P.), which might have been more in line with the Gulf and Caribbean. No information on climate change for Martinique itself has come to my attention. Much more work needs to be undertaken. There is a real need for pollen investigations from which one can work with plant evidence rather than working by analogy with climatological reconstructions, themselves derived from the geological record.

Martinique is large enough that the vegetation destroyed during the passage of a hurricane would have been a function of the path of the hurricane. Judging by the paths of historic hurricanes, one can say that no part of the island is necessarily protected from the high winds. Damage would have been more severe on the exposed and flat highland areas and less in complex vegetation types. Leeward slopes and diverse *Dacryodes* rainforests would have been least affected, as indicated by modern studies of hurricane effects in Dominica, which has some simi-

Climatic Episodes in the last 2,500 years

Episodes	Dates
Recent	1850–1960 A.D.
Xeo-boreal	1550–1850 A.D.
Pacific II	1405–1550 A.D.
Pacific I	1200–1450 A.D.
Neo-Atlantic	900–1200 A.D.
Scandic	400–900 A.D.
Sub-Atlantic	550 B.C.–400 A.D.

Sequence of Climates in the last 2,500 years

	Humid/Stable?	
End of Ostionoid Period	Dry Episode	1000 A.D.
	Humid/Stable	
End of Saladoid Period	Dry Episode	6–700 A.D.
	Humid/Unstable	

13. Two climatic sequences proposed for the period of the last 2,500 years.

SOURCES: Bryson and Wendland 1967; Carbone 1980, 123.

larities in topography and more in vegetation (Lugo et al. 1983,208). The openings created by the hurricane would have been available not only to the native pioneers but also to the exotics introduced with the Amerindians. The culturally created clearings and the naturally created disturbed areas were large enough and frequent enough that some stimulus to the expansion of the pioneer weedy component of the vegetation took place during the pre-European period of human occupation.

Notes

1. The highly fragmented and previously anecdotal work on the archeological record has been increasingly disciplined by exacting standards. This work has been presented at frequent international congresses for the study of pre-Columbian cultures of the Lesser Antilles. Ten meetings have brought Caribbean Basin and North American investigators together to share findings and discuss their implications. These presentations have been published as proceedings in French and English.

2. For differing interpretations of the migration story see Rouse 1976b, 597–612; Pinchon 1952, 305–52; Delawarde 1937b; Revert 1954; Mattioni 1973; H. Petitjean-Roget 1975, 1:90; and Allaire 1977. I believe that Desmond Nicholson's remarks are useful to consider. He says that these peoples from their start near Saladerno had "well-developed navigation skills [and] the first expansion into the Antilles must have been systematic and intended. After the maritime Saladoid had established themselves, this skill would have helped created the communication and stimulative interaction needed to support Vivé . . . [and other sites] about A.D. 500" (Nicholson 1976b,102).

3. I derived the Saladoid figure by inference from the available agricultural land in proportion by suitable coastal settlement areas on the island and by considering separately the number of recovered archeological sites presumed to have been Saladoid on the basis of pottery styles. There are no estimates of Suazoid population before colonization. Estimates for the seventeenth-century population of the Lesser Antilles vary. Vazques de Espinosa suggested 18,000 as a round figure for all the islands. Steward proposed 30,000 for the contact period. As a large, well-watered island, Martinique may well have had one-fifth to one-fourth of the total. Vazques de Espinosa [1620]1942; Rosenblat 1954,248; and Steward 1955,248. Since all the authors are biased toward low figures, I believe my estimate is reasonable.

4. Maggiola and von Altschul have presented convincing arguments for the use of cahoba among the island Arawak and for the purposeful introduction of *Peptadenia perigrina.* Maggiolo 1971, 1(1):201–16, cited in H. Petitjean-Roget 1975,294.

5. These are similar to the bird heads associated with the snuffing of cahoba, tobacco, and other hallucinogens among modern Indians in South America speaking a variant of the Arawak family language. Wassen and Holmstead 1963, 5–45; Wassen 1965,111–16.

6. The migrations of inhabitants around the shores of the island can be traced by conventional archeological dating techniques. A different method of dating the migrations is to compare the composition of the clays used in the pottery found in successive sites. This system was used by Gauthier 1974:133–39, and incorporated into H. Petitjean-Roget's typology (1975,1:157–58).

7. Nevertheless, Bouton estimated (1640,30) that there were 4,000 Caribes on the island at the time of the French colonization. Even if this estimate is accurate, it came too late to establish the aboriginal situation, for by Bouton's time the native

population of the Lesser Antilles had been disrupted by more than a century of slave raiding, reprisals, and European diseases.

8. Cited in J. Petitjean-Roget 1980,1:302; see also Halley 1654,93; Du Tertre [1667–71]1973,2:150. Du Tertre 1667–71, and Hodge and Taylor 1957 were the principal sources for the plant names used in the following section. The scientific names are used because of good descriptions that can be matched. Breton [1635]1947 was also very useful. Today in both Dominica and Martinique the Creole name *acajou rouge* is used primarily for *Cedrela mexicana.* The dry seeds of this species are not likely to have produced the oil referred to in Breton's comment, "its seed might well be used to make oil" ([1665]1892,253). This oil was used by Caliba as an insect repellent and as the vehicle for applying roucou (*Bixa orellana*). It is significant that Dominica Caribs call only *Carapa guianensis* by this folk name (Hodge and Taylor 1957,569).

9. This tree is considered native to the American tropics. The evidence is taxonomic, there being nine other species of the genus *Ceiba* restricted in the Americas, whereas *C. pentandra* has a wider distribution. The seeds are easily transported by wind, and the unopened fruit are buoyant with the kapok hairs (H. G. Baker 1965,192). It is not necessary to postulate human activity to get the tree to the Antilles. On the other hand, people may very well have brought the tree along with other plants of magical or economic utility, especially since in the West Indies these trees were believed to be the dwelling places of spirits. Schumann 1886; H. G. Baker 1965,192.

10. Some widely used names such as *maize, aji, yuca, yautia, tobacco* are Arawak words taken over and modified by the Spaniards for plants unsimilar to them when they first arrived in the Indies. Oviedo y Valdes [1526]1959,13–15.

11. Du Tertre [1667–71]1973,2:90–92. *Tannier* or *tannia* is a corruption of the Carib *taia* (Hodge and Taylor 1957,537). These are species or species segregates of the genus *Xanthosoma.*

12. Sturtevant believes that the Caribs of Dominica introduced topitambour to the French in Martinique. From there it was dispensed to the Greater Antilles and thence to the Taino Arawaks (Sturtevant 1969,190–97). See also Du Tertre [1667–71]1973,2:85 and perhaps 90–91.

13. The elaboration of words for drink and drinking suggests that liquid food may have played a relatively large role in the Carib diet. Coppier ([1664]1963,45), an early chronicler, devoted a whole chapter to their "wines." He reports a *vin d'hannanas* but claims it was less common that the other wines.

14. Breton says that the men cut down woods to make gardens and their wives planted (1963,120–21). Routes of access to the interior are offered by the principal river valleys and midelevation divides. The largest of these are the valleys of the Blanche River and Lézarde River. The Champflore Basin was accessible via Capot River. Reports of inland settlements based upon salvage materials from the Champflore Basin have not been substantiated by later work. See also Du Puis [1652]1972,91.

5 The Early European Period, 1492–1685

Shortly after their appearance in the New World and well before the formal colonization of Martinique by France, Europeans began to effect changes in the plant geography of Martinique. The Island Caribs survived the Spanish discovery and later West European incursions because they were peripheral to the more desirable Greater Antilles and because they were capable of resisting by force. Yet, their culture was subtly changed by the formation of new trade outlets, by the introduction of iron tools and cloth, as well as by new plants and animals. More substantial changes were caused by a significant decline in the Carib population through European slave raids, massacres, and disease introductions. Later, in the period before European settlement, some visitors directly modified the island vegetation by their own ephemeral plantations of high-value tobacco and indigo and partially modified it by their extraction of fuel, precious woods, and boat timbers.

During the colonization period the European took over from the Amerindian the role of ecological dominant. The attack on the plant cover was substantial and significant. In the mid-seventeenth century the imperial and mercantile designs of France replaced the earlier, essentially private activities of French buccaneers and the merchant adventurers of Normandy. The establishment of permanent settlements required a reorientation of relationships between the French and the Caribs and brought new approaches to resource exploitation. Recognizing that the Caribs inhibited the colonial development of the island, the French drove out all but a few Indians by the middle of this period and began the formation of the estate agricultural system. In this way they took over complete control of island land use, but they only added to, rather than replaced, the extant exploitative techniques.

As an *ancienne colonie,* part of the empire of Richelieu and Colbert, Martinique was expected by French officials to produce agricultural commodities to be sold in Europe. The French colonial and socioeconomic organization was gradually modified from a yeoman-farmer to a planter-slave structure, with the changes in land tenure such a reorientation required. Thus, we can see that from the time of first European settlement, ecological adjustments were made to suit the cultural objectives of the metropolitan directors of colonial settlement, rather than to make a close adjustment for local inhabitants to local environmental conditions.

The island's prescribed economic function imposed two conditions on the plant cover of Martinique: (1) a control over the dominant species in the planted vegetation types, and (2) an increase in area in export products over what might have been a more diversified planted vegetation. Because of quarrels over land tenure and the desire for explicit recognition of boundaries, all owners were required to bound their properties with living fences, a habit continued to the present day, although not all practice it. The early European settlement resulted in essentially coastal exploitation, although two transinsular roads opened some of the interior. There was some replacement of native vegetation types with cultural vegetation, a substantial enrichment of the lower-elevation flora by introductions, and a change in patterning of the vegetation mosaic.

European Visitors on the Island, 1492–1635

The first European landing on Martinique occurred on Columbus's fourth voyage. He sailed from Cadiz via the Canaries directly for Martinique, where he landed June 15, 1502, according to Diego de Poras (de Poras [1503]1899). Local tradition has identified the spot near the mouth of the Carbet River on the leeward Caribbean coast. Apparently there was no resistance to his presence, and the Spaniards may have remained several days for watering and refreshment.

Spain claimed Martinique as one of the islands of the Lesser Antillean chain. The central islands were the usual first landfall after leaving the Canaries (K. R. Andrews 1964, 159–60), but probably few landings were made by Spaniards on Martinique. Guadeloupe and Dominica on either side of the Dominica Passage were the islands usually visited after the transatlantic journey (Breton [1647]1929,76; Hakluyt 1903–1905, 191). It seems unlikely that the Spaniards on the way to the Greater Antilles would have turned as far south as Martinique for refreshment, especially as neither of the islands to the north had the poisonous snakes for which Martinique was soon notorious. This does not, of course, rule out a storm-driven vessel or one retreating from pirates. The presence of feral pigs reported for Martinique may be construed as evidence for Spanish landings, especially as the natives of other islands insisted that the Spaniards set the pigs loose (Davies 1666,69; Breton [1647]1929,30). Nothing, however, indicates that the Spaniards ever attempted settlement on Martinique (Flannigan 1844,1:3).

Slave raids took the Indian men. Slavery was legitimized in 1503 by Queen Isabella's proclamation that all Indians, especially cannibals, who resisted or refused to accept Christianity were liable to capture as slaves (Navarretti 1825–29,2:414–16). As the local Indian population declined in Hispaniola and labor problems became acute, the Spaniards raided the populations of other islands. The Lucayas of the Bahama Islands were wiped out between 1509 and 1512. The Island Caribs to the south resisted, but losses were severe in the sixteenth century, and these losses remained in the folk memory, to be reported to the early chroniclers.[1]

Diseases also took their toll. Breton ([1647]1929,48) quoted the Dominica Caribs as saying that smallpox had killed many people and that there was no remedy for the disease. Diseases were transmitted quickly within the *carbets,* and, once an infected person returned to the homestead, most of the rest could be expected to succumb to the disease. Indian massacres by the Spaniards were carried out and were in part punitive, being reprisals for Carib raids on small European settlements on the islands to the north. That these raids had the desired effect is illustrated by a report (Laudonnière [1564]1954) of some terrified Island Caribs who, when captured, talked of the Spaniards, saying they were men who castrated Indians (Hakluyt 1903–1905,9:3).

Even though none of the reports directly mentions the Martinique Caribs, it can be inferred that the shock waves reached Martinique and that populations declined there. Probably not enough population growth was gained by immigration from the more vulnerable islands to offset the losses by slaving, disease, and reprisal. The Indians claimed people of other nationalities attacked them (Breton [1647]1929,48), but the Spaniards probably bear the largest share of guilt. Bouton's (1640,30) estimate of 4,000 Caribs on Martinique fits an assumed reduction by half of the indigenous population during this contact period.

Privateering flourished within the Lesser Antilles in the sixteenth century. Three references to Martinique illustrate the nature of the contacts of Europeans with the island during this time. René Goulaine de Laudonnière, a French captain on his way to Florida in 1564, wrote, "I continued my course [from the Canary Islands] toward the West, wherein the winds favored me so well, that 15 days after our ships arrived safe and sound at the Antilles and going on land at the Isle of Martinico, one of the first of them . . ." (Hakluyt 1903–1905,9:3). During Cumberland's seventh voyage in 1593–94 the ships "fell with the Island of St. Lucya, where and at the island of Matinyno they stayed about three dayes to water and refresh themselves . . ." (K. R. Andrews 1959,24).[2] The men of the *Levrette,* when shipwrecked, found their way to Martinique and stayed nearly two years in the Case-Navire region in Schoelcher Commune (Bréard and Bréard 1889,503).

These experiences were probably duplicated many times during the sixteenth century. With its indented coastline and high relief, Martinique offered good anchorages and hidden retreats for the sailing vessels of these visitors (Hakluyt 1903–1905,5:405). The lee coast with its good roadsteads was particularly attractive to the Europeans, with their shallow-draft sailing vessels. In the seventeenth century Jean de Laet described Martinique and remarked on a spacious port where a river made watering the ships possible (Laet [1625]1640,31). Additional evidence that Martinique was used as a way station is furnished by place-names in use before Martinique was settled, such as Baie des Anglais, Galion, and François[3] on the windward side, and Case-Navire on the leeward coast.

The maritime nations of northwestern Europe were first concerned primarily with harassing and diverting part of the profitable Spanish trade. However, after the power of Spain waned with the death of Phillip II, the merchant adventurers and privateer captains became concerned with cutting dyewoods themselves and with producing some part of the commodities, especially indigo, tobacco, and ginger, that they had previously acquired by adventuring. The growing demand and consequent rising prices were the justifications of the French, British, and Dutch for investing in small, ephemeral agricultural settlements. So, like the rest of the Lesser Antilles, Martinique began to be considered as a land to be exploited. Dutch and French captains and navigators from the Bay of Biscay and Normandy were going there also for tortoise shell, tobacco, hides, and to take on fresh provisions of wood and water. Captain Girard Vaullard in 1627 contracted to cut a cargo of "bois jaune de Brésillet"[4] from the "isles de Barboude, isle de la Tortue et autres lieax circonvoisins" (Margry 1878; Anthiaume 1916,2:128).

In 1624 René Gourney left Le Havre to traffic in the islands of Martinique, Dominica, and the others nearby, and took workers who were to cultivate tobacco for three or four years (Tabellionage du Havre Registre 172.5,Oct.,1624, cited in J. Petitjean-Roget 1980,1:37). Once this system of exploitation began, a thin stream of *engagés* passed to the Antilles (Bréard and Bréard 1889,211; Delafosse 1949,241). In 1627 Thomas Hesbert contracted a three-year indenture with a La Rochelle ship captain to grow tobacco in Martinique.

Ship captains, after reaching the Caribbean, planned on settling into one of the Lesser Antilles where the agricultural situation was suitable to plant a field of tobacco, with the intention of harvesting it just before returning to home ports at the end of the season (Debien 1951,52). Although no records of tobacco planting exist for Martinique, it is known that the company of the shipwrecked *Lavrette* planted tobacco on Dominica in anticipation of their return to France (Breton [1647]1929; Barrey 1918,67).

Changes in the Plant Geography before Colonization

Selective culling of species from the spontaneous vegetation caused changes in abundance and frequency of native species. Furthermore, four additions to the flora increased the membership of the system. The introduction of pigs and other livestock introduced new consumer members into the system and, perhaps, the new process of large-animal grazing. Europeans' cropping schemes introduced two new monoculture vegetation types, which did not compensate for the losses in Carib agricultural fields caused by the reduction of population. This population loss increased proportionally the area in late succession vegetation types at the expense of early succession types.

The seagoing merchant adventurers of France introduced an easterly trade dimension and new trade goods into the trade borne on the Carib sea, thus expanding the extraterritorial dimensions of the island ecosystem. These early years saw the beginning of agricultural exports from the island. This kind of loss from the system was to dominate its history for the next 350 years.

Changes in the spontaneous vegetation caused by the presence of Europeans from 1492 to 1635 were the consequence of four activities or processes: selective utilization, grazing, clearing and burning, and fallowing and abandonment. Although no direct references to dyewood and medicinal weed exploitation on Martinique date before the early seventeenth century, Martinique, as one of the most accessible, most diverse of the non-Spanish-dominated Isles de Perou, was a likely scene of exploitation in the sixteenth century (Gosselin 1876; Bréard and Bréard 1889). *Guaiacum officinale* and Brazil woods, perhaps including *Haematoxylon campechianum,* came from the dry and littoral woodlands and the seasonal forest. Where the unidentified bois jaune[5] and bois vert[6] were obtained is uncertain since these plants are not clearly identifiable, but their properties are those common in species of the dry woodland.

Coastal forest and woodland vegetation types may have been cut out somewhat by fuel gathering, especially around good anchorages such as Case-Navire on the Caribbean side and Englishmen's Bay on the Atlantic south coast. It is also probable that trees suitable for the repair of ships such as *Cedrela mexicana* and *Tabebuia pallida* were selectively cut out of the areas around attractive shipyard or careening sites.

The Island Caribs quickly appreciated the superiority of metal axes over their stone tools for the felling of large trees, and the making of planks became feasible. The old style *ajoupa* and *carbet* continued to be constructed, but the plank floor and small household furnishings of boards made from *Simaruba amara* may well date from this time. It is uncertain whether the aborigines adopted the practice of making charcoal at this time, or whether this practice was adopted only after European settlement. In any event, the greater efficiency of the tools did not necessarily cause a more intensive use of the timber and fuel resources by the Caribs and, with the declining populations postulated above, the absolute amount of exploitation would have been reduced.

Pigs, introduced by some Europeans, were frequent enough in the surrounding vegetation that the first colonists found them a destructive nuisance in their fields (Bouton 1640,38). The Island Caribs may have acquired poultry or "poule d'Inde" at this period (they traded "hennes" with Laudonnière). But, according to Breton ([1647]1929,57), they never ate cattle, sheep, goats, or pigs, despite the large number of the latter on the island. The pigs' rooting must have modified the ground vegetation in all but the upper elevation types, but chickens probably did little damage. It is not known definitely whether cattle were present in presettlement times. If cattle had been introduced, they would have affected the woody plants by browsing and helped maintain grasslands.

There is no evidence that the buccaneers formed permanent settlements on Martinique. Those clandestine European settlements established for the raising of agricultural products were small scale, involving clean cropping of indigo and tobacco. These clearings in no way compensated for the agricultural areas lost by abandonment because of the decline in native populatons. The Island Caribs were feared enough by the Europeans that there is no reason to assume that they had to shift their settlements in retreat. Therefore, the reclamation of former agricultural lands by spontaneous vegetation types would have been generally distributed throughout the coastal areas.

The substantial reductions in Island Carib populations by disease and slaving raids in the mid-sixteenth century suggest that much of the abandoned land from the former agriculturally exploited areas was in late succession vegetation at the end of this period. It follows that relatively well-developed forests and woodlands had replaced most of the former agricultural lands, and savannas and thickets occupied an area proportional to the then-current aboriginal agricultural exploitations. The removal of the aboriginal populations may have been by *carbets,* but certainly the decline was not regionalized. It is supposed that the Caribs continued throughout to occupy the same territories on the island, but in less dense concentrations.

Fires set to clear the areas for provisions gardens that escaped into the spontaneous vegetation types continued the advance of the thorn scrub and thickets on the lee coasts and, when frequent enough, contributed to the opening of woodland understories. It is possible that activities resulting in the savanna at Case Capot took place during this period (Davies 1666,11).

Planted vegetation types were increased during this period. Bananas, sugarcane, and coconuts had entered the Martinique species complement and were encountered by the first French colonists. All had been introduced to the Greater Antilles and Brazil by the Spanish and Portuguese. However, they may have been transferred to the Lesser Antilles by Indian contacts rather than by direct European introduction.

Bananas and plantains, both *Musa paradisiaca,* were present in Martinique when the colonists first arrived (Bouton 1640,623). The cooking variety was traded in Guadeloupe in 1625 to Thomas Gage, who described them as "Those Indian fruits, the plantain above all pleasing our taste and palate" (Gage 1648,19). Breton ([1647]1929,39) spoke of two kinds, called *figues* (bananas) and *bananes* (plantains). An earlier reference to trading with the Dominica Caribs by Laudonnière in 1564 does not give enough information to distinguish between the two fruits (Hakluyt 1903–1905,10:20).

The introduction of the banana from the Canary Islands to Hispaniola by the Spaniards in 1516 is well documented in Oviedo's account ([1526] 1959,247–50). However, the plantain migration route is less clearly established (Kervégant 1936a,93; Merrill 1954,279). There had been massive Portuguese contact with Africa and Brazil as well as traffic between Brazil and the Antilles from the earliest years of the sixteenth century. The widely ranging Caribs may well have brought the plantain from Brazil in the early postcontact time. It was in Martinique by the time the Europeans colonized the island in 1635 (Bouton 1640,62–63). Certainly, once the banana had been introduced to the New World, it was rapidly dispersed throughout the islands by the aboriginal traders. Many forms and varieties of both fruits exist today. Some of the clonal distinctions found in the two types of fruits may date from this period.

By the time the colonists arrived in Martinique sugarcane (*Saccharum officinarum*) was growing "wild," and the Caribs grew it for its sweet stems (Bouton 1640,83). Breton ([1647]1929,86) described the plant for Dominica and Guadeloupe in his dictionary. Smith (1630,55) reported it for Saint Christopher before Martinique was a colony. It seems reasonable to think of the plant as a European introduction into the Greater Antilles and Brazil that soon afterward was traded by the Caribbean Arawaks and Caribs among themselves. Still, some accounts described early sugarcane plants of a stouter variety than that in the western Mediterranean and the Canaries. Labat talks of a memory of a stronger sugarcane before the Europeans arrived. However, Breton in his *Relations* (in contrast to the dictionary cited in Hodge and Taylor 1957,527) does not mention sugarcane at all. A telling linguistic argument against the presence of sugarcane before the Europeans is that the Island Carib words for sugar and sugarcane plant, *kanis* (Hodge and Taylor 1957,528), are related to European words (Taylor 1959a,306). Ogilby (1671,358) mentions that of the many fermented drinks, that made from "Sugar-Canes" is "most estemm'd amongst them [the Island Caribs]."

Sometime during this period the coconut (*Cocus nucifera*) was introduced to Martinique.[7] Breton ([1665]1892,236) found two coconut palms on Guadeloupe and planted one by his place in Basse-Terre. This may have been the result of local sea dispersal from plantings made by the Spaniards elsewhere in the Caribbean, but the history of planting of the palm in the Lesser Antilles is against this. De Rochefort (1658,65) reported that the settlers on Saint Christopher got them from Saint Barthélemy. Most of the coconut palms on beaches in Martinique are clearly the result of planting.

The papaya, despite its Arawak language name, was first described by the Spaniards from Central America. It was not mentioned by Breton (1665,242, 255, 141) for Dominica or Guadeloupe as a fruit but as an antidote to infection in wounds. Davies, translating from de Rochefort, reported that the papaya was brought from Saint Barthélemy to Saint Christopher by the mid-seventeenth century. De Rochefort himself (1658,50) described the fruit in glowing terms. Du Tertre ([1667–71] 1973,2:175–76) described and illustrated the plant. It may have been reintroduced to Martinique shortly after colonization.

If *Haematoxylon campechianum,* or campêche, as it is known in the Lesser Antilles, was not native to Martinique, it was during this period that it reached Martinique either by intentional plantings for the trade or, more probably, by natural agents of dispersal.

The tobacco and indigo planted by the *engagés* and ship captains may have been obtained from local seed sources and thus be local varieties. The tobacco may, alternatively, have been the "verraine" that was introduced by Europeans from Venezuela.

The Island Carib dooryard gardens were enriched by sugarcane, bananas, and possibly by papayas, beginning the trend toward a more cosmopolitan economic flora. The first of the river-mouth coconut trees may have started at Case-Navire, Carbet, and Fond Capot by either Europeans or Caribs. The spontaneous canafistula trees of the heights behind these anchorages may originate from these times.

New vegetation types introduced during this period were the clean-cultivated tobacco patch and the indigo field, both crops being low annuals grown in monoculture. No information on field size for this time on Martinique is available. To justify these as new units of vegetation, it is assumed that the tobacco and indigo plantings were substantially larger than those made by the Island Caribs in their gardens. Most of these plots probably were in the leeward Caribbean valleys near good anchorages.

The Founding of the French Settlements

Martinique began its French history in 1635 as a daughter colony to Saint Christopher, the first of the French colonies of the West Indies. The colony was held by one of several proprietary companies,[8] although in 1674 it came under direct governmental control and became a royal colony in 1685 (Mims 1912,60–67). However, economic control by the crown remained, whether the colony was under companies or the crown (Priestly 1939,133). All company charters provided for a seigneurial property system and commercial monopolies. These charters were implemented by the formation of subsidiary companies. Syndicates of private merchants took up portions of the company's grant in return for rents to be paid in tobacco or cotton. By 1648 these lords' proprietors were all residents of the island (Pares 1960,2). Subsidiary companies contracted for *engagés* and other immigrants. These companies began the economic development of the island and, ultimately, made the effective decisions with respect to the plant cover. However, the settlement and agricultural activities were conducted within the regulations of the *pacte colonial*, a typical seventeenth-century mercantilist framework.[9] In essence, the mother country, having founded the colonies, had to be enriched by them.

Tied to the export market, the Martiniquan producer responded quickly to fluctuations in price. The first colonists began cultivating tobacco, annatto, cotton, bois vert (Renaudot 1639,9), and cassia fistula (Du Tertre [1667–71]1973,2:164–66). When tobacco prices fell disastrously in 1638, production decreased (May 1930,108), but production increased again in 1652 when the price improved (Rennard 1929a,37). When Saint Domingue settlers began production of annatto, prices fell and the settlers stopped producing that coloring product (Delawarde 1935,94). The colonists had brought with them a knowledge of a subsistence, mixed crop–livestock economy and, during the several blockades of the period, the small cultivator of provisions expanded his production (Delawarde 1935,68).

Map 18. The island of Martinique, as sketched in the seventeenth century. Source: From Du Tertre [1667–71]1973: facing p. 98.

A small group of colonists from Saint Christopher arrived at the anchorage off Saint-Pierre and took formal possession on the fifteenth of September, 1635 (Bouton 1640,39). Pierre Belain, Sieur d'Enambuc, took possession of the island under the auspices of the Compagnie des Isles d'Amérique (Du Tertre [1667–71]1973,1:65–66). The settlement was to have existed in peace with the Caribs, but conflict was inevitable. Overwhelmed by the superiority of weapons, the Caribs retired behind a line passing longitudinally through the island, separating the French and Carib parts of the island called Basse-Terre and Capesterre, respectively. The line of demarcation was somewhat indefinite, often breached, and represented differently on seventeenth-century maps (see map 18 for one such draft). Still, for twenty years the relationship, although not friendly, was not actively hostile.

The colonists spread north and south from Saint-Pierre, filling the few suitable areas in the narrow Caribbean coastal strip and going up onto the lower gentle slopes of Mont Pelée and the flanks of the Carbet Pitons (Du Tertre [1667–71]1973,2:22). The lee coast provides easy approaches for sailing vessels, in contrast to the cliffy northeastern Atlantic coast and the reef-bound southeastern Atlantic coast. But the combination of good harbor and land suitable for a sizable agricultural settlement is rare even on the lee coast. The Fort-de-France Bay had a fine promontory for a fort but was surrounded by swamps and extensive, soggy littoral meadows. Most of the river mouths with their small amounts of sandy beach and alluvial soils were large enough for only a few households; moreover, the Caribs were then established at Carbet, occasionally called Grand Sable until 1658 (Bouton 1640,29), on the only large river with a good sandy beach of the north Caribbean coast. The Saint-Pierre roadstead with its gentle slopes watered by three small streams extending from Prêcheur to Carbet was virtually the only appropriate site for a large settlement to be serviced by a substantial number of ships. Behind the roadstead was the low pass between the Pelée and Pitons massifs leading, by way of Morne Calebasse, to the gently sloping lands of the northeast.

By 1650 the villages, more properly called hamlets, of Prêcheur and the *côte-de-fer* (Carbet, Case-Capot, Case-Pilote, and Case-Navire) communes had been established (Du Tertre [1667–71]1973,2:22–26; Bouton 1640,41), as had Fort-de-France, originally called Fort Royal. Other settlers arrived. In 1654 a boatload of Dutch Jews who had been pushed out of Brazil arrived (Du Tertre [1667–71]1973,1:457). Ten years later the colonists occupied a territory from Cape Saint Martin to the Fort-de-France Bay (Maurile de Saint-Michel 1652,26). Boats were the usual means of communication between settlements.

As the population increased, land became scarce behind the older settlements, and in 1658 colonists turned to Diamant and Trois-Ilets in the south (de Rochefort 1658,17). By 1664 there were eighty colonists with forty-nine African slaves spread out along the coast from Diamant to the Marin Cul-de-Sac (May 1930,62), and Marigot on the Atlantic had one hundred colonists and eighty slaves. Colonists began to look from the flanks of the Calebasse toward the Capesterre, which by treaty had been restricted to the Caribs. Relations between the two peoples grew worse, aggravated by the Spaniards, and a war to drive out the Island Caribs began in 1658 (Du Tertre [1667–71]1973,2:21). In 1660 Amerindians living on the northeast coast withdrew to Dominica. With this substantial removal, European settlement began on the coast of Capesterre. The immediate objectives were the region of Trinité, with its excellent anchorage in the lee of the Caravelle Peninsula, and the lower Mont Pelée slopes above Basse-Pointe.[10] Island Caribs did, however, remain on the south coast from the Caravelle Peninsula to the Salines Point.

Settlement increased rapidly. Between 1670 and 1680 all the coastal land of the north of the island was conceded and occupied (see map 19). Only the coast lands to the southeast were left because coral reefs made them less accessible and because of fears occasioned by the supposed fevers from the mangrove swamps. Most of these holdings were small; 89 percent were 70 ha or less (J. Petitjean-Roget 1980,2:1369).

The crown considered roads to be essential for the defense of the island. The first to be constructed was the Chemin de la Calebasse, which connected Basse-Pointe with St. Pierre around the southern flanks of Mont Pelée. Built soon after the expulsion of the Island Caribs, it was designed to facilitate the movement of goods to and from the colonists on the windward side, for landings on that coast were difficult. In 1680 a second transinsular road was authorized by the governor to cross the island connecting Trinité with Fort Royal (AN C^8 A^2 Blénac to Col-

Map 19. Progress of settlement in the seventeenth and eighteenth centuries. Sources: After Daney 1846; Rennard 1927; Delawarde 1935.

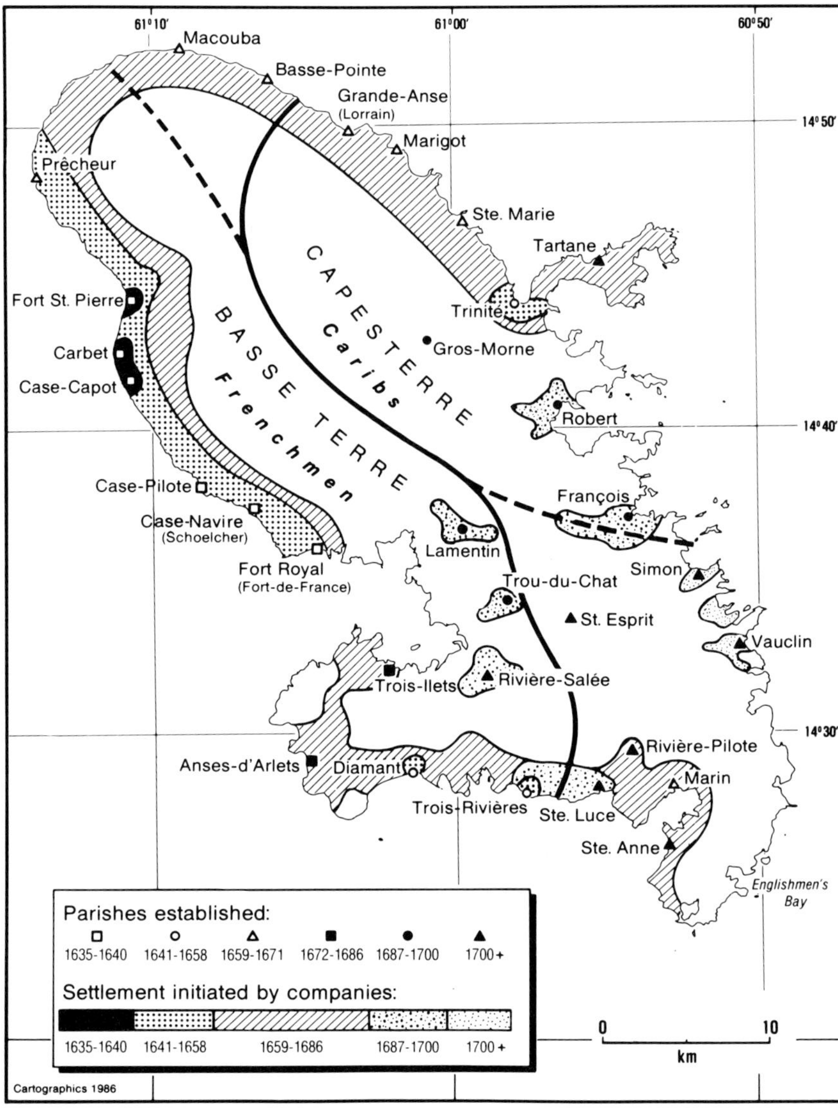

bert). Another road from St. Pierre to Fort Royal was constructed along a high interior route away from the coast. A fourth road connected the Lamentin Plain settlements with François Bay.

Colonization began along these transinsular roads. Arpent or long-lot plots were laid from the road. In order to ensure that the roads would be maintained, all *censitaires* or settlers were required to keep clear the sides of the roads, upon pain of confiscation of their territories, according to an ordinance of 1684 (May 1930,201). Small trails multiplied throughout the countryside, facilitating the opening up of these new regions. These small footpaths or trails were sufficient for the transport of the produce of the small cultivators. Cacao was carried on the heads of men or on the backs of beasts. Yet the conquest of the interior can hardly be said to have begun during this period.

During the colonization period the elimination of the Amerindians and domination by France were virtually completed. Disease, war, emigration, and assimilation with the French and African immigrants were the causes. The Europeans brought Africans with them from the beginnings of settlement. Source areas for both the European and black populations are fairly well known. The Europeans who colonized Martinique came mostly from Saint Christopher (J. Petitjean-Roget 1980,1:138,2:1429) but ultimately from Normandy, Brittany, and Picardy. They came as *colons,*[11] who paid their own passage and brought capital with them, and as *engagés* or indentured servants,[12] whose passage was paid in return for three years' service, at the end of which they were to receive a plot of land and some capital in the form of tobacco and tools (Pelleprat [1665]1857,20–22; Satineau 1928,70). Bouton (1640,8) mentioned blacks along with the first settlers, but the influx did not come until the sugar plantation economy created a demand for slave labor. The blacks came from the Cape Verde–Senegal region, the Dahomey-Guinea regions, and Angola (Du Tertre [1667–71]1973,1:462). They came as mixed lots and were sold in St. Pierre, which was then the largest slave market in the French West Indies.

The island population rose from a little more than 4,000 to 15,600 persons during the colonization period. The first handful of colonists shared the island with the native Caribs, presumed to number 4,000. By 1640 the European population of Martinique was estimated at 1,000 (Bouton 1640,96–98). In 1646 Pacifique de Provin ([1646]1939,8) gave the same estimate for the French and 400 as an estimate for the Caribs. This is a reduction of approximately 90 percent of the Carib population and more than two-thirds of the presumed population in immediate precolonial times in the first eleven years of colonization. The remaining Island Caribs were driven out to Dominica and Saint-Lucia in 1658, except for a few in the interior and the southeast.[13] The following year Maurile de Saint-Michel (1652,37) estimated the population at 15,000, certainly an overestimation. A decline in the ten years from 1671 to 1680 was due in part to the migration of Martiniquans to Saint Domingue and a decline in immigration, both the consequence of the fall in the price of tobacco (May 1930,56). The population then rose rapidly until 1685, when there were 6,200 whites and 9,400 blacks. The point at which blacks outnumbered Europeans came in 1659–60, by which time the Caribs no longer figured as a significant resource-exploiting group (Schnakenbourg 1977,44).

A dispersed settlement pattern was characteristic of Martinique for its first years as a colony. Father Bouton (1640,2) wrote: "The homesteads are now stretched out one after another without any form of town because of the mornes which separate them and because each person wishes to live on his land." Father Maurile de Saint-Michel (1652,30) wrote that there were no cities, no towns, and that each one lived in a hut near his tobacco. And again in 1657 Father Halley could say "that each island [of the French Antilles] had about 5,000 Frenchmen and the inhabitants are separated each on his land in plank huts" (Savine 1902,87–88).

Upon arrival, each colonist received several *carreau,* each equal to more than 1 ha, or about 2½ acres. The land was granted free of charge according to the *droit de hache* to occupy and cultivate, provided the colonists cleared a substantial portion within three years. At first, land distribution was without many formalities, but by 1638, to ensure order and to permit the distribution of many lots, a maximum number of *carreau* was put on parcels. The original limit consisted of a plot of 20 *carreau,* or 26 ha.[14] As it turned out, this parcel size was much too large to permit a numerous population, as Martinique was only 109,600 ha in size. The size was reduced to 10 *carreau,* or a bit over 12 ha, by cutting the plot depth in half. The exact locations of some of these sites are on Blondel's 1667 map (J. Petitjean-Roget 1980,120). From the Terrier of 1671 can be calculated the distribution of properties by size. One-quarter of the holdings were less than 5 *carreau* (6.4 ha) and one-quarter were larger than 25 *carreau* (32 ha). The other half were in between (J. Petitjean-Roget 1980,2:1,366). With the development of a sugar economy larger properties were needed, and by 1650 demands for larger holdings were being met. Areas of 75 to 130 ha became the rule.[15] Unfortunately, unexplored lands were conceded, and, in some regions, more land was distributed than existed.[16] It became necessary to formalize the taking up of land.

Very early, space was lacking for new colonists in the original Saint-Pierre Roadstead despite the fact that three *étages* or *rangs*[17] were set up (ADM, Terrier 1671,n.p.). At the same time there were settlers in the Capesterre who held large acreages, not only of cultivable lands, unused and destined for sugarcane, but also of the *chasses,* those midelevation slopes of the mountains (unsuitable for sugarcane) that remained unused even though they could have been put into subsistence agriculture (Bouton 1640,36). Colonists began to sell land by 1644 (Maurile de Saint-Michel 1652,31). They were able, however, to find free land up to 1680, but it was no longer possible to inscribe in the contracts written in France the promise of a concession of 200 by 500 *pas.* This tended to inhibit settlement and began to have a marked effect on immigration as early as 1662.

There were abuses of the system, some involuntary, others intentional. Much of the lands already conceded remained in forest because of the lack of means to bring them under cultivation (Daney 1846,1:325). The more affluent traded for lands originally obtained by concession, so larger properties were created by addition through purchase and by concession.

In order to curb speculation and to redistribute land, the Conseil de la Martinique in November of 1665 ruled that lands not in cultivation within six months were to be confiscated (May 1930,72). Confiscations were made in 1669 and in the years following, but these were all properties of small landholders or of persons without influence. The more influential continued their traffic in unoccupied lands, and the amount of land conceded was increasingly greater than the land under exploitation (Labat 1742,8:241–47).

Because most colonists continued to live on their properties, the population remained dispersed. Dwellings and industrial buildings were usually set on the coastal ends of the properties, or on the road from which the concession was laid out. Populations on large concessions devoted to sugarcane clustered in buildings near the owner's house. These consisted of slave huts and industrial buildings. The Terrier of 1671[18] shows a correspondence of many dwellings with sugar mills, indicating a tendency to keep industrial activities on local properties. The named settlements or towns were little more than sites of churches serving a dispersed population in the surrounding area. Exceptions were St. Pierre, Fort-de-France, and Trinité, which were important ports and commercial centers as well as centers for governmental services.

The land immediately adjacent to the sea was reserved after 1670 as the Cinquante Pas de Roi (de Baas [1674]1960,11). Often referred to as the *cinquante pas géométriques,* it was measured inland from the coast about 80 m from high-tide mark.[19] The *cinquante pas* was intended to hinder movement of enemy personnel, but it also guaranteed the inhabitants access to the sea and complete freedom of communication between the different thinly populated clusters along the coast. In addition, the belt set aside was intended to be sufficiently large to provide for the building of shore batteries and of forts necessary to protect the island. Captains of visiting ships might cut the timber necessary for the repair of their vessels. The land itself might be occupied by private persons or for commerce, but only at the tolerance of the governor. Artisans who had no connection with cultivated lands were often assigned small parcels of land in the zone where they could build a house, always under the condition that if there was a need for the land they had to relinquish it (May 1930,72). Merchant dwellings and business establishments occupied the zone as well (de Baas [1674]1960,11). In the earliest days agents of companies with trading ships built small huts that served as stores in this zone. Eventually, parts of the coastal strip were lost to expanding villages.

European settlers in Martinique were the major contributors to the developing agricultural system, a mixed livestock-and-crop farming tradition in which part of the produce was destined for the market. The colonists from Saint Christopher had been in the islands a while and had a chance to become accustomed to the tropical heat and the necessary work of clearing the woods and planting the native provisions, but they were not many years away from Europe. The majority of the colonists came from those parts of France where the nucleated village and open-field system prevailed, Normandy, the Ile de France, and what is now Seine Maritime. There, arable land was in unhedged fields, cereal grains predominated, and collective labor obligations were still in force. The rest came from Brittany, the Cotentin, and Poitou, where enclosures were the rule. In less-populated areas of Brittany the near fields, *terres chaudes,* were kept continuously in cultivation whereas the outfields, *terres froides* (a number of shifting plots scattered about in the waste lands) were intermittently cultivated. Animals were either pastured in herds on the stubble of the arable lands, or on meadows, or on forest mast (Grand and Delatouche 1950,272–75). Persons from southwestern Brittany were also accustomed to tethering their stock (Bloch 1966, 57). The settlers possessed a variety of agricultural strategies and concepts of the appropriate in landscape modification.[20] The majority came to Martinique to improve their economic position and may be presumed to have been prepared to develop new habits of exploitation. As migrants they were representatives of the population back home, but different in not clearly determined ways. As the ecological dominant, their role was radical and pervasive.

African contributions to the agricultural system are less clearly defined, except for crops. The Africans differed from one another, but they were all accustomed to using tree crops and starchy root vegetables, and, like the Island Caribs, they mixed many food crops in their plantings. The African pattern of life included living in villages and preparing agricultural plots by slash-and-burn techniques, including variable-term, fallow periods. Cattle were new to some Africans, but those who had kept cattle were accustomed to pasturing them on savannas and to feeding them small grains as supplemental rations.

The slash-and-burn technique the Island Caribs used to clear fields in the new land was, therefore, a new technique to only some of the colonists, and bush fallow had either been part of the European and African systems, or a variant on them. The French did not transport their old field patterns to the New World, but they did introduce monocropping of a commercial crop plant in sizable plots.

The early commercial crops except for sugar were suitable for an owner-operator labor system. One or two additional laborers, easily supplied from the indentured ranks, were all that were necessary for any agricultural activity except sugarcane growing. Some industries, particularly annatto preparation, were especially well adapted to family labor units. A dispersed population distribution was consistent with these agricultural requirements. The Indian settlement pattern, a *carbet* housing an extended family, had more impact locally than did the earliest smaller households of the Europeans.

On the other hand, gangs of relatively unskilled labor were needed for planting, cultivating, and cutting sugarcanes. The processing of sugar, even to the primitive stage requisite for export to the refineries of Europe, required considerable capital expenditure on buildings, livestock, and machinery, and some investment in skilled labor. When the price situation forced the colonists to abandon tobacco, indigo, ginger, and canafistula, they turned to sugarcane or cacao. By 1671 these shifts required some readjustment in human labor, and settlement size may have returned to that during the Indian periods.

Metal tools had been introduced by trade in the earlier contact period, but the European colonists brought with them the hoe, the plow, and probably the machete for agricultural purposes. The African, like the European, was skilled in the use of the hoe. The plow introduced was the *charrue* of simple design, effective on the soils of the island but relatively inefficient on very rough terrain. Probably its use did not continue later than 1659 (Satineau 1928,1:117). The transfer to slave labor perhaps encouraged the use of the hoe, but there was also a lack of livestock to pull the plow. Most draft animals by this time were used in the sugar-grinding operations. So long as properties were small, a single all-purpose draft animal could draw the plow then necessary, but the changeover to the larger-scale sugarcane plantation diverted the draft power and contributed to the elimination of the plow as an agricultural implement. The cutting and hacking implement known as the machete was part of the equipment of the earliest colonists in Martinique.[21] These tools made possible the stirring of the surface soils but did little to modify the horizonation or structure of the soils. Nevertheless, monocropping on slopes accelerated the natural erosion of soil.

By the end of the period, power-producing machines such as waterwheels and later windmills eliminated some need for cattle to grind sugarcane. Although the extraction rate was usually greater with cattle-driven mills, their cost was not competitive with the other mills.[22] The draft animals no longer used in mills were not used for plowing, and livestock numbers were reduced. The savannas that had been reserved for feeding the beasts were converted to sugarcane land.

Fuel requirements for cooking, manufacture of iron products, and firing the boilers for sugar making were met by the wood cut from the surrounding vegetation. Different fuels provided different flames for various stages of the reduction process (W. Barrett 1966,162,170). Charcoal was introduced at this time as a cooking fuel; however, there are no references to its use in sugar making.

Changes in the Plant Geography during Early Settlement

A property enumeration or *terrier* conducted for the Compagnie des Indes Occidentales under the direction of Pélissier provides a survey of the plant cover for the whole island in 1671, a kind of synoptic view. For each property the owner was named and the amount in crops was listed. Additionally, the surrounding vegetation was described and the neighbors were identified. The reconstruction of the cadastral picture and census figures provides a picture of Martinique settlement activities that Pélissier wished to represent to the company.[23] Pélissier omitted that part of the island between the Caravelle Peninsula and Salines Point because it was uninhabited on account of difficult access from the sea and its poor soils. It is, therefore, to be used cautiously, but it does give the best approximation of the contemporary vegetation picture available. Presumably, the amount of land reported in provisions was larger than the actual amount since the colonists were required by contracts to have some certain percentage of the land in provisions. Even these figures show that the island was by no means self-sufficient and had to depend upon food imports.

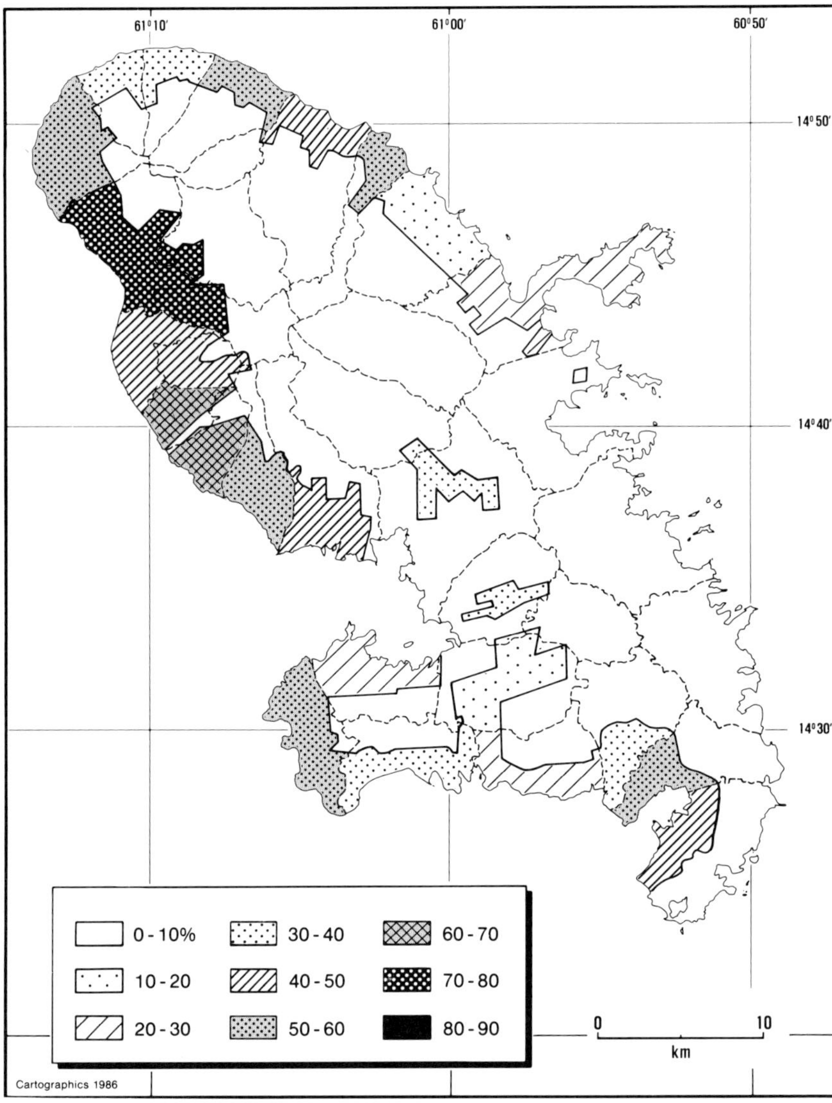

Map 20. Percentage of conceded land cleared by 1671. Source: ADM, Terrier of 1671.

On the basis of the *terrier,* a map was prepared showing the zone of lands that had been conceded by 1671 (see map 20). This document permits reconstruction of the distribution of planted vegetation types for the settled part of the island.[24] It also suggests which vegetation types were cleared and what might have been the spontaneous responses within the plant cover to the changes in land use in the colony. Pélissier wrote in his report that the heart of the island was still in forest, that the largest part was mountainous and the land was inaccessible (cited in J. Petitjean-Roget 1980,1:1,345).

Changes in the membership of the ecosystem included selective utilization and introduction of plants (and animals). Selective utilization of the spontaneous vegetation included the gathering of valuable dye-woods and dense hardwoods, and the cutting of ships' timbers. This cutting was a form of disturbance that created more gap-phase situations in the forest canopies. With the establishment of a large resident population and the associated use of lumber for housing, firewood, and charcoal making, selective inroads on the cover became increasingly important. This industrial usage is a new process active in the system since it consumed biomass associated with the production of an export commodity. That is, the replacement crop community is processed by virtue of extracted biomass for export from the local ecosystem. A new requirement introduced in this period was fuel to fire the sugar vats. Cutting for this purpose caused massive removal of spontaneous woody growth. Such activities physically removed large amounts of plant biomass. Settlement may cause the complete and permanent removal of vegetation. Lumbering and charcoal making usually create old fields, and regeneration cycles begin. The effects differed according to vegetation type, distance from the coasts, and type of settlement. Most of these areas were left to regenerate, but some were grazed and rested thereby at a succession stage.

Construction woods from the rainforests used in the colony were *Homalium racemosum, Chrysophyllum argenteum* or *Sideroxylon* sp., *Guarea ramiflora, Carapa guianensis, Dacryodes excelsa,* and *Richeria grandis* (Du Tertre [1667–71]1973,2:146–62,178). The *palétuvier* extracted from the interior may have been either *Symphonia globulifera* or *Amanoa caribaea.* The Island Caribs continued exploiting the same trees they had used until their expulsion. The only tree whose use might not have been taken over by the Europeans is the *Tapura latifolia,* for they were not interested in making paddles. As the small proprietors' houses were built of planks and squared lumber, the pressure on tree ferns and the lobster claw used in the old style *ajoupa* may have lessened. Toward the end of the period, wattle-and-daub construction using the stalks of the giant grass *Arundo donax* and sugarcane was common (Pontchartrain 1696, cited in May 1930,232).

Seasonal forest trees preferred for lumber in the colonial period were *Geoffroea inermis* (Labat 1742,3:207), *Bursera simaruba* (Du Tertre [1667–71]1973,2:225), *Simaruba amara* (de Rochefort 1658,64), *Pimenta racemosa* (Du Tertre [1667–71]1973,2:148), and *Vitex divaricata.* The latter was particularly resistant to decay, and ship captains sought it for the repair of the hulls. As the period drew to a close, more of these construction timbers were culled from the unconceded forest lands near the settlements, and, within the first decade of settlement, suitable trees were being sought for as far as "une grand demilieu dans la montagne." The wood from the *Hymenaea courbaril* was much sought-after for gun carriage wheels and ship timbers toward the end of the period but was known earlier more for its gum or copal and its fruit (Breton [1665] 1892,251). Associated with the riverine facies of the seasonal forest, these trees were growing near the coast and hence were most vulnerable to cutting. *Genipa americana* was valued for its fine-grained, easily worked wood and was useful for tool handles, barrel staves, and agricultural machines (Breton [1665]1892,251). Du Tertre ([1667–71]1973,2:155) mentioned *le courrouca,* which has not been identified. The *Prestoea montana* so desired for its cabbage is rarely mentioned after this period, and it may have been eliminated in the lowlands at this time.

Du Tertre ([1667–71]1973,2:8) noted that by 1646 on Saint Christopher there were men specializing in chopping wood and making lumber. They demanded what were thought of as exorbitant prices for their lumber, yet the return for labor was greater on the beginning plantations than in the forest. That island began to rely on a foreign supply. Martinique was close behind. Price and scarcity of the local wood was one consideration; but, when it was discovered that lumber from North America was more durable than the local timber, the demand for foreign timber increased.

Woods from the small trees in the xerophytic habitats tended to be very durable because of their impregnation with resins and essential oils that retard insect infestations and reduce water penetration (Stehlé 1945–46,314–17). There was also a tendency for these trees to have very dense, hard, and colorful woods containing dye principals. The two woods most sought-after for industrial purposes were *Guaiacum officinale* and *Krugiodendron ferreum* (Du Tertre [1667–71]1973,2:154). Both of these were used for making girders, pulleys in ship riggings, and machinery cogs. In 1683 the intendant wrote that "every year a large quantity [of gaïac] is exported" (Seignelay 1683, cited in May 1930). The Loire ports imported 35,000 tons of gaïac in 1672 and 25,000 tons in the following year (Rennard 1935b,220). By 1689 Bégon ([1689]1925, 43) reported that little of it was left in Martinique. Though not all of it came from Martinique, most of it did. Du Tertre ([1667–71]1973,2:140) mentioned that Guadeloupe had none except in a remote half of Grande-Terre. At the same time *Guaiacum sanctum* developed a reputation as a medicinal, and it was much sought-after by the pharmaceutical trade. The *G. officinale* was occasionally substituted for the latter.

The European furniture makers imported much colored wood for marquetry from the French colonies. The pink or gray semihard lumber of the *Tabebuia pallida,* the red of *Coccoloba diversifolia,* and blue-green of *Rochefortia cuneata* were well regarded when a colorful wood might be used to advantage. Bégon ([1689]1925,42) reported that the latter had begun to be very rare. *Zanthoxylum flavum* probably continued in demand. At the time Du Tertre was on the island there were a number of cabinetmakers. Trees with fine-grained, dense woods were used in the island for making barrel staves and other packaging. This served as a supplement to the legal wood imports from French Canada and the illegal ones from New England.

The principal species used in shipbuilding from these coastal forests were *Inga laurina, Bumelia mastichodendron, Bucida buceras,* and *Tabebuia pallida* (Breton [1665]1892,69). There is no indication from this period that these woods were in short supply (Du Tertre [1667–71] 1973,2:146–62). On the other hand, *Cedrela mexicana,* which was much used by boatwrights, was so scarce that it had to be gotten from Saint Lucia (Bégon [1689]1930,43).

Fruits, nuts, saponifying agents, and fiber were also gathered from the trees and plants in this vegetation. The nuts of the *Hymenaea courbaril,* the leaves of the *Pimenta racemosa,* the fruits of *Cordia collococca* (de Rochefort 1658,61,69; Breton [1647]1929,36), and *Melicoccus bijugatus,* introduced from South America, were much appreciated, but none of these uses would have affected their representation in the vegetation (Coppier 1645,54; Breton [1665]1892,25). Green tobacco (*Nicotiana tabacum*) and *carratas* were used as barbascos.[25] String and thread were made from one of the *carratas, Furcraea tuberosa* (Breton [1665]1892,40), the beach mahot (*Hibiscus tiliaceus*) (Breton [1665] 1892,36), the peelings of the several shrubby mahots, *Urena lobata, Daphnopsis caribaea,* and *Guatteria caribaea.* All of these came from the woody, spontaneous growth volunteering into savannas that developed from abandoned fields within the dry woodland, littoral woodland, as well as in the seasonal forest. Bégon ([1689]1925,46) noted that these were so common in the fields that it was not necessary to grow cordage.

The attack on the mangrove swamp began during the early settlement period with the cutting of the red mangrove (*Rhizophora mangle*). It was prized for making charcoal and for the tanning principle in its bark, which was used locally in the processing of the animal hides. Exports of hides to Europe were inconsequential.

Charcoal making was an important industry in Martinique, for it was the preferred fuel for cooking. Charcoal was made either in pits or on raised racks, which were then covered with earth after being lighted. Any wood could be used, but those that burned slowly and evenly made a higher grade product. The dense, hard woods of the dry woodland and dry bushland as well as the littoral hedge were preferred for making charcoal, but less dense woods were also used. Both *Simaruba amara* and *Bucida buceras* were well regarded as charcoal woods. The weedy guava and calabash were especially sought-after in the woodlands adjacent to the savannas. In regions near settlements, charcoal makers were often less selective and made patchy clearings of all species in the native or regenerating woodland and forest to feed the charcoal pile.

Requirements for charcoal are hard to estimate for this period, for it is not known just how much the settlers depended upon charcoal for cooking and how much it was supplemented by wood. Roughly 4 lbs. of wood are required to make 1 lb. of charcoal. Under the assumption that 1 lb. of charcoal will cook a meal for a small household (four persons), and that only one substantial meal was cooked in a day, one may make an estimate of charcoal consumption on the basis of population size. At the end of this period the population was estimated at 16,500 persons. At a consumption of 2½ tons of wood a day, the yearly total would have been 812 tons of wood for domestic fuel alone.

Increasing amounts of wood were required for sugar making. Presumably, these supplies were obtained from the surrounding woods and forests, although at the end of the period some second-growth bushlands may have been cut. The distribution of sugar mills in 1671 shows a concentration in the dry woodlands, the alluvial bottoms in the dry bushlands of the Caribbean coast, in the seasonal forest of the Atlantic coasts, and the Lamentin Plain. The leaves of sugarcane were used, but wood was needed to make the hottest flame. No figures on fuel consumed in the sugar-making process have materialized, but there must have been a real pressure on the surrounding vegetation until bagasse began to be used much later. Many of the volunteers in the savanna vegetation, including fiber plants and bulb plants, were exploited by the colonists. There was a considerable increase in savanna vegetation during this fifty-year period.

Colonists learned from the Indians as well as by experience that there were rich hunting and fishing resources. There were no large native herbivores, but the sea yielded the manatee (*Trichechus manatus*) and several large turtles (*Chelonia mydas, Caretta caretta, C. kempii,* and *Eretmochelys imbricata*), which were easily caught (Bouton 1640,76,73; Breton [1647]1929,33, [1665]1892,70,195; Coppier 1645,110–12; Du Tertre [1667–71]1973,2:216). Each of these was severely depleted. Land crabs (Bouton 1640,62) were important foods, as were sea crabs (Du Tertre [1667–71]1973,1:220). Conch (*Strombus gigas*) (Du Tertre [1667–71]1973,2:223), crayfish (*Panulinurus argus*), and oysters (*Crassostrea rhizophorae*) were among the sea resources, as well as sixteen fish species, all of which were excellent protein sources. These were early exploited by the colonists and were severely depleted by that exploitation.

The terrestrial fauna were the most easily available of meat sources. The iguana (*Iguana delicatissima*), the agouti (*Dasyprocta antillensis*), the rice rat (*Megalomys desmarestii*), and the opossum (*Didelphis marsupiales insularis*) were the most important prey after the birds and parrots. So severely were the birds and animals hunted they became scarce by the second decade of occupation and later were either exterminated or endangered. The meat supply had to come predominantly from domestic animals.

In 1640 the colonists were raising good sheep, lambs, and pigs as well as cattle and goats (Pelleprat [1665]1857,9–10). The savannas offered an "excellent pasture for the cattle." Father Halley reported that by 1657 pigs were everywhere (Savine 1902,80). By 1669 the island's livestock consisted of 500 horses and 1,600 horned animals (census of 1664 cited in May 1930,83). As there were only 118 sugar mills, not all of these animals were needed for draft purposes, and some were available for use as meat and transport.[26] As some cattle mills were converted to water and windmills, the number of animals per farm unit declined, but the conversion proceeded slowly. Even as late as 1670 four-fifths of the mills were animal powered, and the total number of livestock increased (ADM, Terrier 1671). In 1672 the island had 5,250 horned animals, 1,240 horses, a hundred or more mules and donkeys (census of 1682, cited in May 1930,83). In 1684 there were 5,483 cattle, 1,180 horses, 111 mules, and 2,166 sheep and goats (J. Petitjean-Roget 1980, 2:1386).

Grazing land became increasingly scarce; indeed, complaints were made that there was insufficient savanna for the livestock (Seignelay à Pautoulet 1680, cited in May 1930,84). Except for the Savanne de Monsieur de Vallée, which was about 44 ha in size, there were no proper pasture areas. With the gradual change to a sugar economy, more and more of the lands maintained in savanna were required for support of the beasts used in milling. However, sugarcane lands began to compete with savanna. Because sugarcane was on permanent fields, old fallow land did not accumulate as under the swidden system. "The whole coast, without exaggeration, is planted in sugar cane. There is now not enough place to nourish the beasts" (Seignelay à Pautoulet 1680, cited in May 1930,84).

The colonists were officially encouraged to develop a livestock industry as an alternative to the sugar obsession, yet no effective means were found to encourage the making of pastures. Since the savannas were not large enough to maintain all the stock, the subsistence cultivator had to be content with feeding goats and sheep on herbage and browse, and pigs on fruits and mast. The livestock of small cultivators and that on some large properties as well were permitted to run in the woodland and seasonal forest, as well as to graze on the borders of roads. This was not a new idea to the settlers, for Europeans had been accustomed to running animals in forest. Livestock, particularly cattle, appreciated the fruits of genipa, limes, canafistula, and *Pithecellobium* species, all of which have seeds that are viable after a passage through the digestive tract of these animals. Thus these animals contributed to the dispersal of these woody plants in savannas and in disturbed woodlands.

The herbaceous species members in these savannas and grasslands are a matter for conjecture. Some native grasses and sedges having a weedy character do volunteer into abandoned cultivated fields today and are found in present-day savannas. These are thought to have formed the dominant stratum in these early savannas. Ecological requirements would have segregated some species into the coastal, more xeric savannas and others into the more inland, mesic savannas.

In all savannas probably the tufted, perennial grasses *Paspalum laxum, Pennisetum setosum,* and *Andropogon bicornis* would have been found. The annuals *Sporobolus indicus, S. tenuissimus, Leptochloa filiformis,* and *Cenchrus echinatus* would also probably have been present. The stoloniferous *Panicum trichoides, Setaria geniculata,* and *Axonopus compressus* were probably in most savannas. Three sedges, *Abilgaardia monostachya, Fimbristylis dichotoma,* and *Dichromena nervosa,* have wide tolerance ranges and were probably in most pastured areas. In the savannas adjacent to coasts *Stenotaphrum secundatum* was probably an important element, whereas *Panicum conjugatum, Eragrostis tephrosanthos,* and *Spo-*

robolus poirettii were in the drier savannas of the uplands. The sedges *Cyperus odoratus* and *Scleria lithosperma* were more common in the mesic savannas, along with *Digitaria horizontalis*.

Agriculture was by far the most significant reason for clearing the spontaneous vegetation, and according to the Terrier of 1671 at least 10,000 ha had been cleared. The process used by the time of colonization, found described in planters' manuals and in accounts of the buccaneers, is the classic swidden style. What distinguished the technique from that of the Island Caribs was the generally cleared-off appearance and the regularly patterned plantings that resulted. There is no reason to suppose that fire did not get away from the colonists on occasion, but it is equally clear that any such escape was unintentional and those who permitted it were not respected as farmers.

Evidence for the use of fire to maintain savannas is curiously sparse in the printed record for Martinique. In contrast to what D. Harris (1965,100–101) found for the Leewards, there seems to have been no deliberate burning of extensive tracts of forest or savanna on the island of Martinique in the early periods. However, Du Tertre ([1667–71]1973, 2:25) did write that one could pass easily to all parts of the island on horseback, suggesting perhaps an opening up of the floor under forest canopies by fire.

Clearing for agriculture followed the path of settlement. Begun first in the dry bushlands, dry woodland, and seasonal forests of the Caribbean slope, the attack began on the northeast and south by the mid-1650s. In the south the attack was on the littoral woodland and dry woodland. In the northeast of Capesterre the seasonal forest and the littoral woodland were cleared. Slash-and-burn clearing extended into the rainforest along the transinsular route from Fort Royal to Trinité in the Saint-Joseph and Gros-Morne areas. Clearings were less common along the Route de la Calebasse. No accounts of repeated burning to maintain openings or grasslands have been found in the literature. The only large opening was that on the Morne des Boeufs above Fond Capot, where Governor du Parquet (Monsieur de la Vallée) permitted his savanna to be used communally by all in that *quartier* (Du Tertre [1667–71]1973,1:25).

Roads and clearings for new fields increased the areas subjected to the edge-effect, increasing the habitats for sun-loving weedy plant species. As settlement continued, more and more sites for native and exotic ruderals were created.

The early settlement period was one in which many plants were introduced. Most of these plants went into the planted vegetation types. Ginger (*Zingiber officinale*), jute (*Corchorus olitorius*), and flax (*Linum usitatissumum*) were introduced as field crops, but only ginger was successful, and then only when it was moved into the dooryard gardens. Indigo, mulberry (*Morus* spp.), apples (*Malus* sp.), dates (*Phoenix dactylifera*), and canafistula (*Cassia fistula*) were tree crops introduced as specialties.[27] Only the last was a successful introduction, although the crown spent considerable effort in the attempt to start the silkworm industry. The most successful bush crop introduction was cacao (*Theobroma cacao*), which may have been present aboriginally (Rofer 1699, cited in May 1930,94) but which was reintroduced about 1657 when the first of these bush plantations was made (Du Tertre [1667–71]1973,2:173).

Other economic plants were introduced into the island ecosystem. Useful food plants that found their way into house gardens were the squashes and pumpkins (*Cucurbita* spp.) reintroduced from Central America if present earlier, yams (*Dioscorea alata* and *D. bulbifera*) from Southeast Asia and via Africa, cucumbers (*Cucumis sativa*), melons (*Cucumis melo*), cabbage (*Brassica oleracea*), sorrel (*Rumex acetosa*), onion (*Allium cepa*), and lettuce (*Lactuca sativa*) transported from European gardens (Bouton 1640,46). The woody castor bean (*Ricinus communis*) from Africa and the pomegranate (*Punica granatum*) from the Mediterranean were used as living fences and hedges, along with Barbados pride (*Caesalpenia pulcherrima*), which came from the Cape Verde Islands via Barbados, introduced by de Poincy. The castor bean and the pomegranate quickly became feral, seeding into the savannas. Limes (*Citrus aurantifolia*) and lemons (*Citrus limon*) were introduced from the Mediterranean as garden and public plantings, only to go spontaneous in the savannas as well.

Most of the other introductions were useful plants that were placed in the household gardens. Some of these became spontaneous and fitted into the weed category: purslane (*Portulaca oleracea*),[28] aloe (*Aloe vera*), and bowstring hemp (*Sansevieria metallica*) (May 1930,244).[29] These introductions enriched the flora of the household gardens, the savannas, and the verges. Four plants, however, dominated the new planted vegetation types: sugarcane, indigo, cacao, and canafistula. No good information on exotic weed grasses comes from this period. Since these aliens are notorious camp followers, it is not unreasonable to assume that some common grasses had become established early in the colonization period, entering in packaging materials, the fodder of imported livestock, and the earth surrounding deliberate introductions. Likely introductions are goose grass of *Elusine indica* and *Chloris* spp.

Planted vegetation types that were present at the beginning of this period were the tobacco field and provisions fields. The primitive dooryard gardens got the complement of Old World plants and the intensity of plantings that established the dooryard garden vegetation type. The cotton bush vegetation type developed in this period. The definitely new vegetation types introduced by the settlers during this early colonial period were sugarcane fields, cacao bushlands, provisions gardens, indigo fields, orchards, and mulberry plantings.

In 1671 roughly 20 percent (23,770 ha) of the island was ceded, with about half of that cleared (see table 19). Sugarcane lands amounted to 27 percent of all the cleared land; 26 percent was in savannas; 34 percent was in provisions. There were very small areas in indigo and ginger, 73 ha and 72 ha, respectively.

According to the Terrier of 1671 the distribution of mixed food crops of provisions shows a close approximation to the settlement pat-

Table 19. Extent of selected vegetation types, 1671

	Hectares	Percentage Cleared	Percentage Ceded
Planted			
Sugarcane	2,447	26.2	10.1
Tobacco	1,019	11.0	4.3
Indigo	73	0.8	0.3
Ginger	72	0.8	0.3
Provisions	3,192	34.6	13.4
Spontaneous			
Savanna	2,406	26.0	10.1
Uncleared "bois"	11,885	—	50.0
Public land			
Urban settlement, etc.	1,675	18.1	8.1
Total ceded land	23,770	—	—

SOURCE: ADM, Terrier.

tern. The particular distribution and the size of the individual installations related to the land in indigo show the importance of individual decision. By 1671 tobacco had declined in relative importance, most markedly on the west coast. Sugarcane, by 1671, had only just begun to show dominance on the northeast coast. Its relative strength was much more on the northwest coast. Its importance in the central region, measured in terms of the percentage in cleared land, showed that the clearing of land in this region coincided with the expansion of sugarcane throughout the island. Savannas represented a major food supply source for the draft animals that turned the sugarcane plantation mills, so the amount of land in savanna was high in areas where sugarcane was important. But it was even higher in those parts of the island with a lower proportion of area in sugarcane fields, a reflection of the slash-and-burn, shifting-field agricultural techniques used.

There had been some elimination of the salt marsh territory, particularly around Fort Royal (later renamed Fort-de-France). The larger concessions granted at that time attracted planters with capital, who found the drained soil fertile (May 1930,244). By 1681 the region about Fort-de-France was partly drained and in sugarcane production, and the military camp town began to grow (May 1930,246; J. Petitjean-Roget 1980, 1347–52).

Associated with agricultural production was an increase in area in secondary spontaneous vegetation types. The opening of closed canopy vegetation permitted a more heliophytic and xerophytic flora of the dry scrub woodland to invade areas previously in seasonal forests so that the zone occupied by dry scrub woodland may well have been expanding in the *côte-de-fer*, particularly near St. Pierre. Weeds introduced into the disturbed areas were early colonizers of old fields. Some native pioneers expanded populations. It is possible that some hybridization may have taken place in these new habitats. The seasonal forest on the northeast side was being replaced by spontaneous vegetation with savannas, bush savannas, and orchard savannas, parts of which probably were counted in with the savanna figures presented under agricultural exploitation.

Even though there had been some selective cutting of the mangrove, drainage or clear cutting had not eliminated any of the mangrove areas at this time. The leguminous thorn scrub and the cactus desert areas remained constant throughout this period. The rainforest was touched only in the upper regions around Saint-Pierre and Carbet and along the road from old Fort Royal to Trinité. Some successional thicket communities undoubtedly had begun along those routes.

The hunting activities of the early European colonists reduced the number of animals that were important in the dispersal of large seeded trees, which are autopathic. How soon this lack of dispersal began to affect the species distribution is not known, but certainly by the time the animals became endangered species the effect on plant distribution must have been important in terms of reproduction.

In addition to the disturbances initiated by the European settlement, natural catastrophes continued to occur, causing some damage during this period. Traineau (1982,90,91) reports on a series of *nuées ardentes* found in the course of the des Pères River and in the Claire River (see map 21). These, from the evidence, were limited in volume and in spatial extent. Bits of carbonized wood have yielded dates from 540 ± 60 Y.B.P. to 255 ± 70 Y.B.P. (Traineau et al. 1983,267). This means that small volcanic events were occurring during the early European period as they had in the Amerindian period. These events were small and perceived as nonthreatening. They may have been the experiences that led the settlers to become blasé about the volcano and the dangers associated with it. It is certainly interesting that the southwestern flanks of Mont Pelée were the earliest densely populated areas of Martinique. Although these lands were nearest to St. Pierre, perhaps the frequent light dressing of volcanic material so increased the fertility of those light and easily worked soils that they continued to be more attractive than the heavier soils elsewhere.

Six hurricanes and windstorms hit Martinique during the early period of European occupation (Revert 1949a,514). Bouton (1640,48) was the first to describe the terrible storms. The first hit in 1642 and the others between 1651 (Pelleprat [1665]1857) and 1656, all hitting between July and mid-September (Du Tertre [1667–71]1973,2:65). One that hit Guadeloupe with particular force passed Martinique, destroying only a few small houses and leveling sugarcane and manioc plants. Another hurricane hit the night of August 15, 1666, tossing a ship onto the shore. Du Tertre ([1667–71]1973,1:492) wrote of the 1657 hurricane in Guadeloupe: "At four o'clock in the morning the great hurricane began and for 5 or 6 hours the weather ravaged terribly. As an example of the violence, it pulled up almost all the trees except for several large *Acomas* and *Courbarils* which remained denuded of branches like masts of ships. A large number of birds, chickens were killed as well as other

Map 21. Mont Pelée eruptions for 490 Y.B.P. to 320 Y.B.P. Source: Traineau 1982,89.

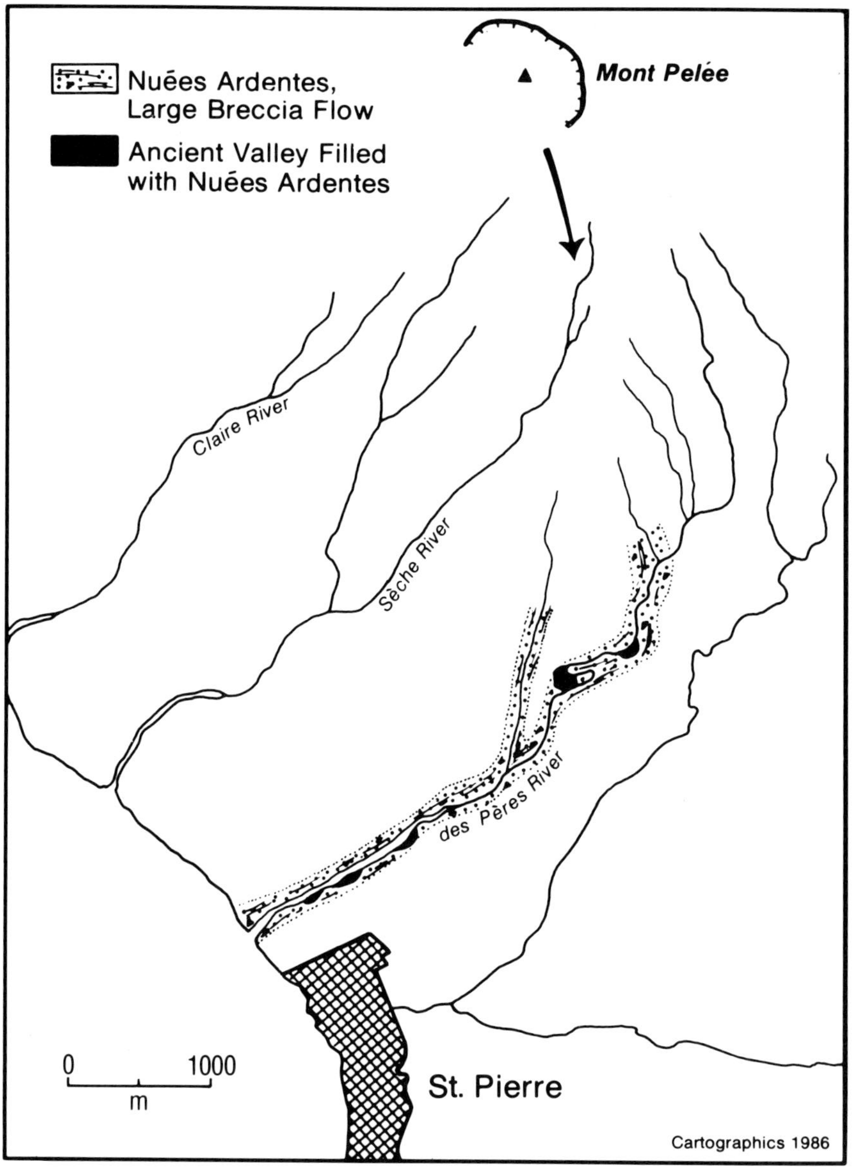

domestic animals like rabbits, dogs, pigs, and the manioc was all pulled out in all the estates which caused a great famine." Du Tertre quoted from a letter of M. Le Chevalier de Saint Laurent to Colbert ([1667–71] 1973,4:279): "the island is in a very deplorable state. Nothing remains neither my house nor my sugar house is standing and one cannot make sugar for more than 15 months. As for manioc which is the bread of the country not a single individual plant remains." Much monetary loss accompanied the vegetation destruction. A storm could put a terrible strain on the economy of the fledging colony, but Du Tertre ([1667–71] 1973,2:26) claimed that one of the several advantages of Martinique over other islands in the Lesser Antilles was that "it is very seldom subjected to hurricanes . . . the inhabitants enjoy an agreeable calm while the neighboring islands are desolate" ([1667–71]1973,2:26).

An earthquake hit Martinique in 1657. Du Tertre ([1667–71] 1973,1:493) wrote: "I threw myself on the bed but the earth trembled so hard that I fell to the floor." There is no record of what damage was caused on the island. When an earthquake followed a heavy rainstorm, landslides occurred on the upper slopes of the Carbet Pitons and Morne Jacob, as well as on smaller peaks. Although horrifying to the inhabitants and causing much monetary damage, these disturbances were factors to which the native vegetation had become adapted.

At the end of the first period of colonial exploitation the destruc-

Map 22.
Zone of vegetation modified by European agency for the end of the early colonial period.
Source: Compiled by author.

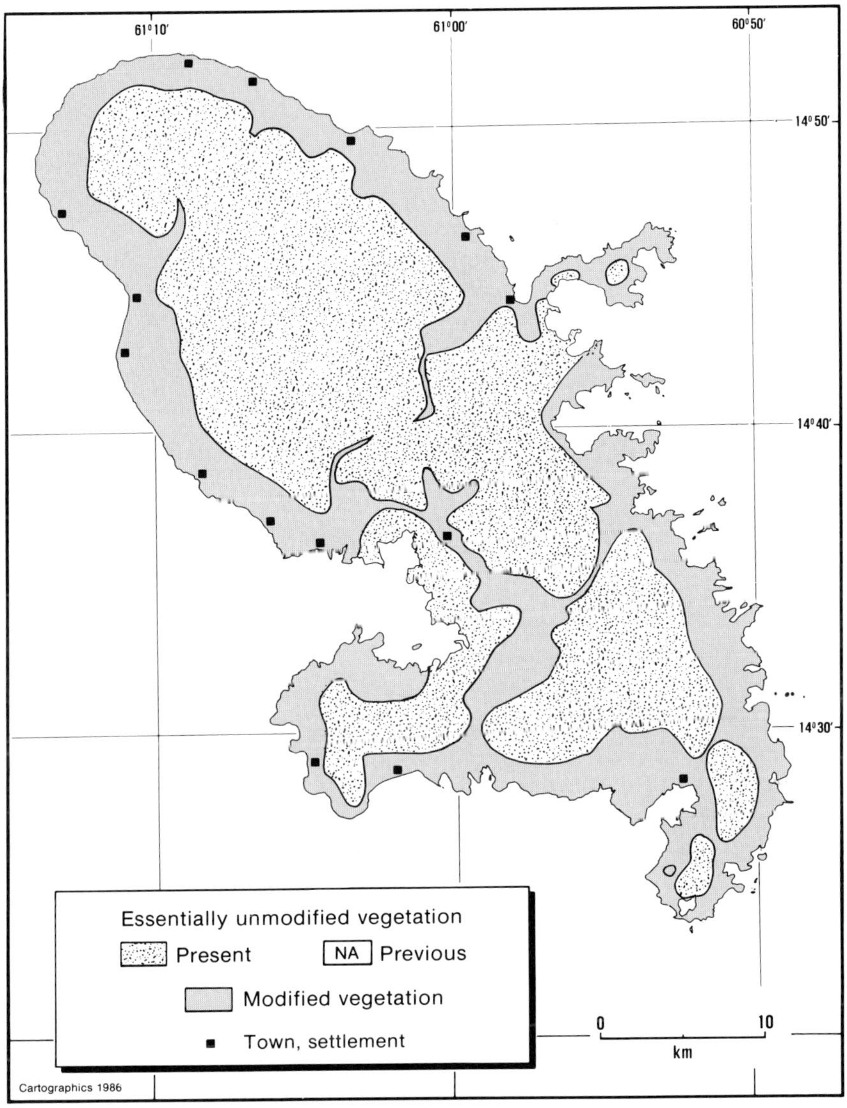

tion of native vegetation and its replacement by planted and regenerative growth had been concentrated on the coastal areas, extending several leagues inland (see map 22). The interior was, except for those disturbed areas associated with roads, still largely as Pélissier had described in his introduction to the Terrier of 1671 ten years previously: "the entire heart of the island remains in forest, the largest part of which is mountainous and inaccessible." Approximately 14 percent of the land had been cleared and was in agricultural production. Land in provisions was well distributed throughout the occupied zone, being rather more important in the northwest coast and less in the central area. Sugarcane dominated the coastal borders except in the northwest and on the south coast, where savannas were more important in areal extent, while sugarcane and savannas covered about the same area in the central portion of the island.

Associated with these agricultural lands were secondary succession vegetation types as well as islands of more or less disturbed, uncleared vegetation. The dry scrub woodland of the Caribbean coast had been cut into substantially, as had the seasonal forest, and, with regeneration, the woodlands' upper elevation border may have encroached upon the seasonal forest lands, thus obscuring the previous vegetation boundary. Elsewhere the seasonal forest and dry woodland had received the main brunt of clearing. Although there had been some draining of wet littoral meadows near Fort Royal, its mangrove borders had remained virtually stationary. Leguminous thorn scrub and cactus desert maintained their areas, although there may have been some replacement of bordering types with planted vegetation. Rainforest had receded only in the upper margins of settlement above St. Pierre and Carbet and along the transinsular roads to Trinité and Basse-Pointe.

A substantial number of cultivated plants had been purposely introduced and a number of European weeds had probably entered into the newly created ruderal habitats as well as into the savannas. Pressure had lessened on some tree forms but increased on *Guaiacum officinale* and *Cedrela mexicana,* even to the gradual eliminating of the latter in easily accessible locations.

Notes

1. Breton ([1647]1929,48) reported that when he asked why there were so few men in comparison with the women, the answer was that it was on account of the Christians.

2. Printed sailing directions or *ruttiers* from the end of the sixteenth century mention Martinique. *A Ruttier for the West Indies* contains "A ruttier that a man must keepe from Dominica to Martinino, and so to Tierra Firma" (Hakluyt 1903–1905,287–92), and the "Markes of the Isle of Maralina, or Martinino" are in *A Second Ruttier* (ibid., 309). Both of these are tentatively dated 1600 and printed in Hakluyt's collection. Mention in the second *ruttier* is made of the war-like Indians of Martinique (Hakluyt 1903–1905,10:309).

3. The ship captain François was supported by merchants of Rouen.

4. Brazilwood was the common name for several dyewoods in the sixteenth and

seventeenth centuries. The most important of these was logwood or campêche (*Haematoxylon campechianum*).

5. This may possibly be the fustic (*Chlorophora tinctoria*) of the British islands or the prickly yellowwood (*Zanthoxylum* spp.).

6. This wood was not only a dyewood but it was also used as a substitute for African ebony. Greenheart (*Ocotea rodioei*) from South America is a construction timber and not to be confused with bois vert.

7. Du Tertre (1667–71,2:185) writes, "The coconut was not yet in the French Antilles at the time of the first printing of my book (1654) and for that reason I did not speak of it. It is now very common." On the other hand, this may be a self-serving remark. Diego Lorenzo is said to have introduced the coconut in 1660.

8. The Compagnie des Isles de l'Amérique was the second of several formulated along the lines of the successful Dutch companies, receiving royal letters patent in March of 1655 (Saintoyant 1929,1:210–11). Full monopoly and rights to the new company were granted for twenty years, on condition that it should transport 4,000 immigrants. The articles of organization are summarized in Margry (1867,49–51) and Barrey (1918,201–205). For a detailed historical treatment of the colonization of Martinique and the politics responsible for that settlement, see the magisterial work of Jacques Petitjean-Roget (1980).

9. The *pacte colonial* or Régime de l'Exclusif was the name given to a series of protectionist rules begun to be stated as principle by France in 1626 (May 1930,118). See Du Tertre ([1667–71]1973,1:8–10) for details of some articles.

10. A fort at Trinité was built on land conceded in 1663 (Aff. Etrang. Mém. et Doc. Amérique V. fol. 56, cited in May 1930,62). In the same year parishes were established at Ste. Marie, Marigot, and Grande-Anse, the modern Lorrain (P. R. Dessalles 1847,3:73). The settlement at Basse-Pointe received parish status in 1665 and Macouba did so between 1668 and 1670 (Rennard 1927,68).

11. The free recruits came from the western provinces of France, especially Normandy, Brittany, Saintonge, and Gascogne (Anthiaume 1916,2:138). The most successful agriculturists were from a peasant background. Artisans were always in high demand and were permitted special considerations in passage and settlement. At the ends of European wars the crown sent soldiers, frequently without any resources, to the islands (Vignols, 1935,29).

12. Debien (1951) accumulated data in the ports of western France on points of origin of the *engagés*. There were of two types: those *engagés libres* who voluntarily bound themselves over to indenture, and the *engagés forcés* who were shanghaied. Though the majority of these also came from the western provinces, there was always a goodly number from central France as well as from other European countries (Debien 1951,89–92).

13. Yet, in 1687 Father Halley spoke of the island as being all French except for the Indians who lived in the mountains, *montagnards* (Savine 1902,76). He also estimated the population of Martinique to be 5,000, even as, listing these figures, he urged caution in estimating the island population (May 1930,54–57). However, supporting this, Rennard (1927,69), using parish statistics, put the figure at 5,003.

14. Twenty *carreau* of 200 *pas de large,* that is to say along the coast, by 1,000 *pas de haut,* or depth. A *pas* in Martinique was 3.5 pieds, or 1.137 m. This was the typical elongated plot common to all French colonies in the New World. A plot 225 m by 114 m was roughly 25.6 ha. A *carreau* in Martinique was equal to 1.2927 ha (Schnakenbourg 1977,56).

15. Or, 750 *pas* to 850 *pas de large* by 1,000 *pas* to 1,500 *pas de haut* being roughly

75 ha to 130 ha. From these figures it can be seen that Martinique did not have vast holdings in the hand of one owner or family.

16. Later on in the century it was necessary to proceed more cautiously. Not only was it necessary to obtain from the *capitaine* of the *quartier* and from the royal cadastral officer a certificate of the vacancy of the land, but all neighbors had to be present during the taking of possession and were required to declare in writing that they had no claims to this piece of property (Labat 1742,3:246).

17. *Étages* or *rangs* were developed as the land along the coast was all conceded. Behind the first *rang,* the second *rang* was then conceded. The third *rang* contained concessions well away from the coast on the flanks of Mont Pelée.

18. An enumeration was conducted by Pélissier when the Compagnie des Indes Occidentales was sold. It lists individual properties by owner within companies (*seigneuries*), giving the size of the concession, the amount of land in different crops and savanna, the number and kind of buildings, and the vegetation in the uncleared and adjacent lands. In addition, the enumerator added a recommended exploitation for the remaining land. A manuscript copy of this *terrier* is in the Archives Départementales, Fort-de-France. This copy has been used in all reports and calculations of area in this chapter.

19. It was established by decree by the Conseil Souverain de la Colonie on March 3, 1670. A further justification was given by M. de Baas, governor of Iles d'Amérique to his minister, Feb. 8, 1674, cited in Moreau de Saint-Méry (1784–86,1:272).

20. Much of the work by Duby (1954, 1968), Darby (1934, 1956), and Bloch (1966) is being used in later works by Nitz (1963), Otto and Anderson (1982), and TeBrake (1985) to underscore the influence of systems inside and outside Europe.

21. It may be derived from the buccaneer's cutlass in use toward the end of the sixteenth century. In describing the cutting up of whales in 1598, Hakluyt (1903–1905,1:414) mentioned the machette. Gage in 1648 spoke of the islanders as having no weapons but a "machete which is a short tuck" (Gage 1648,129). On the island today the implement is referred to in French as a *coutelas.*

22. The Terrier of 1671, made at the beginning of the conversion period, lists one hundred animal-driven mills, twenty-eight waterwheel mills, and two windmills.

23. A manuscript map in the office of the Service de l'Agriculture in Desclieu, Fort-de-France, was made in 1937 by Kervégant, an agent in the agricultural service. He used this *terrier* to produce a map of properties showing seigneurial boundaries and militia groupings. A more recent reconstruction made by J. Petitjean-Roget (1980) has been published.

24. Based upon a mapping by M. Sobesky of the concessions described in the Terrier of 1671. Jacques Petitjean-Roget has brought it to my attention that the assumption that all concessions described were perpendicular to the sea may have been too rigidly adhered to, and thus the geography of settlement is represented only approximately. He has analyzed the information in the *terrier* at some length in his monograph (J. Petitjcan-Roget 1980,2:1345–75). However, at the scale rendered here, such details are blurred. More critical is that the text was apparently copied or compiled by persons careless in addition. Thus, all totals needed to be checked against a resumming of the individual figures reported. Even more important for this work was the consequence of mislocation of properties when the bounding vegetation could not be as described in the *terrier.* I have used the manuscript text rather than Sobesky's map for the analysis of vegetation impacted.

25. *Carratas* was a term used for several native and introduced species that were used in stupefying fish and that furnished fiber as well (Bouton 1640,67; Breton [1665]1892,177). This method of fishing by throwing fish-stupefying agents

into the water was prohibited by decree in Feb. 1680, but to no avail. Du Tertre ([1667–71]1973,2:143) devoted two paragraphs to details of one tree and its use, probably *Piscidia piscipula* Sarg. of the Leguminosae.

26. Every family kept in its property yard a large flock of fowl, pigeons, chickens (Du Tertre [1667–71]1973,2:267), and rabbits (*Relation de Isles* 1660, cited in May 1930,82) to serve as the ordinary meat supply. These do not show up in early census enumerations.

27. The Intendant de Baas was interested in the introduction of fruit trees from France and established a model garden or *jardin du roi* in Fort Royal (*Rapport,* Feb. 24, 1970, cited in May 1930,81).

28. The proposition that purslane was already present in the New World at the time of colonization has been made (Byrne and McAndren 1955) but reintroductions from western France are almost certain.

29. The last two form conspicuous clones on abandoned house sites in the southeastern part of the island today.

Plate 1. Planted pastures, sugarcane fields, provisions gardens, dooryard gardens, savannas, dry bushlands, and dry woodland in Quartier Barrière-la-Croix, Sainte Anne Peninsula, on limestone-derived heavy montmorillonitic clay soils with 1,300 mm estimated annual precipitation. Volunteer shrubs converting former sugarcane lands (top center of photograph) are being arrested by provisions plots. Triangular portion in the lower left corner is a former plantation now in small properties. Elevation, drainage courses, and the cadastral system are the strongest controls on the vegetation pattern.

Plate 2. Clearings in seasonal forest on shallow volcanic soils of Prêcheur Commune, where there is an estimated 2,500 mm annual precipitation. Bananas are at lower elevations, provisions gardens and cleared grasslands at higher elevations, and provisions gardens and seasonal forest in canyons. North-south living fences are normal to the path of the land-sea breeze. Some carefully maintained pastures can be observed in the central shrub savanna area. Slope, terrain orientation, and elevation account for most of the pattern, although planters' decisions influence the pasture maintenance and location of banana plantings.

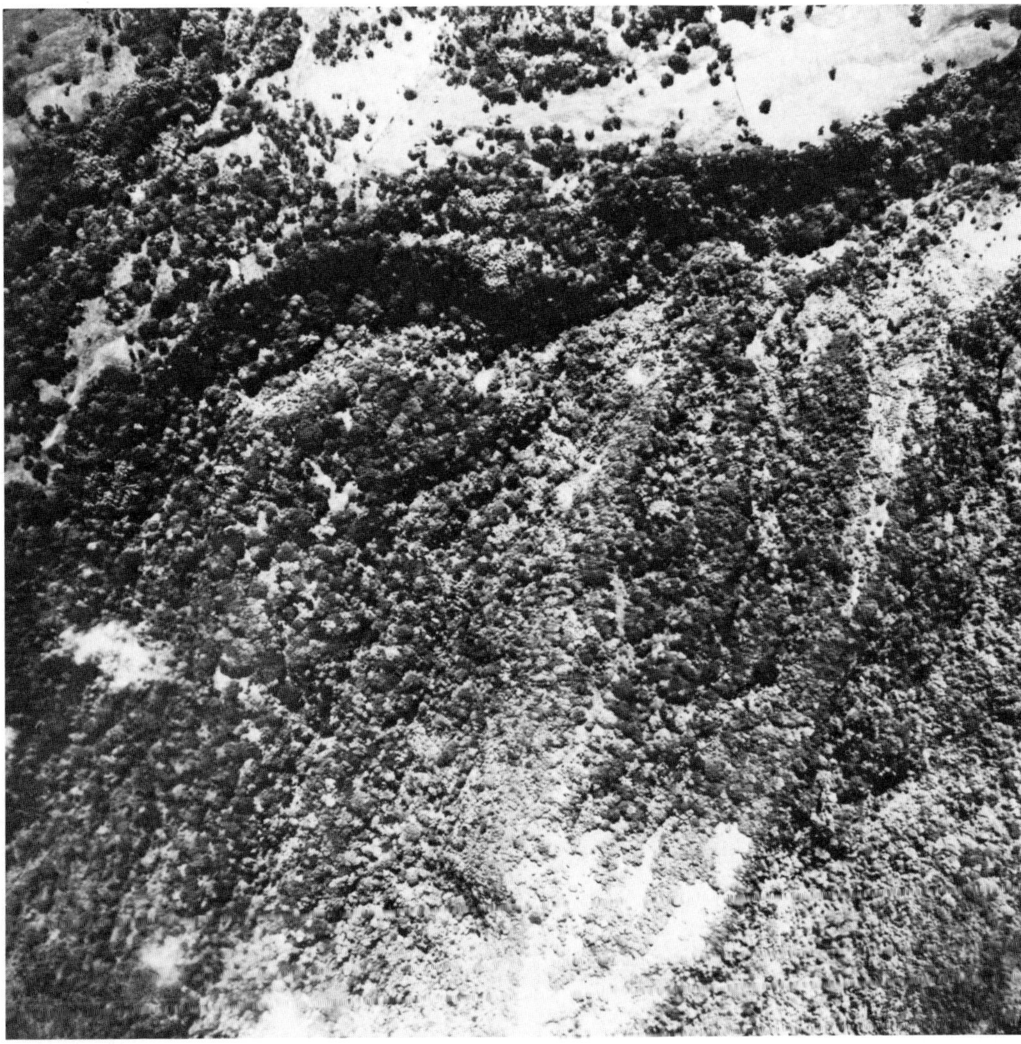

Plate 3. On Carbet Pitons, secondary rainforest on rust-brown ferruginous soils cleared for grassland pastures is visible at the top. Old pastures and provisions grounds are being invaded by pioneer plants above the canyon margin. Clumps of coconut palms identify dooryard gardens of the dispersed population in a mountain-belt area of Carbet Commune. Slope, settlement pattern, and the history of clearing and abandonment are the strongest controls on the local vegetation pattern.

Plate 4. Top. Montane bushland on Piton Dumauzé with *Charianthus nodosus, Begonia* sp., *Guzmania megastachya,* and *Prestoea montana* in the foreground. Piton Alma in the distance. Courtesy of J.-P. Fiard.

Plate 5. Bottom. Primary rainforest on the flanks of Sugar Loaf, Girou River, a tributary of Grande River. Characteristic development occurs between 300 and 500 m. Above this level, steep slopes and strong winds produce smaller trees. Courtesy of J.-P. Fiard.

Plate 6. Top. Light-colored cover on upper slopes in among the elfin woodland and montane bushland are populations of fougère calumet (*Dicranopteris bifida*), which colonizes unstable slopes. Montane forest is found on lower slopes. Remnant forest trees form the living fences in Quartier Morne Piquet, with a view of the Carbet Pitons from Morne-Vert Commune. Courtesy of E. Nurser.

Plate 7. Bottom. Young tree ferns (*Cyathea arborea*) on a cut bank above the main north-south road, near La Médaille in northern Fort-de-France Commune. Courtesy of E. Nurser.

Plate 8. Top. *Rhyticocos amara* dominates palm brake with seasonal forest species, mainly *Tabebuia heterophylla, Lonchocarpus benthamianus,* and *Zanthoxylum caribaeum,* on slopes above Cape Saint Martin. Courtesy of J.-P. Fiard.

Plate 9. Bottom. Second-growth rainforest and bamboo brake above a provisions garden in Quartier Morne Diamant (540 m), Morne-Vert Commune. Courtesy of R. J. Kimber.

Plate 10. Top. Littoral hedge and mangrove swamp, Battery Point, Caravelle Peninsula. *Coccoloba uvifera* is in the foreground. Courtesy of J.-P. Fiard.

Plate 11. Bottom. Girou River gorge after the passage of Hurricane David, showing broken trees and the debris that still blocked the streambed in 1984. Courtesy of J.-P. Fiard.

Plate 12. Top. Second-growth rainforest in an abandoned clearing near Alma in Fond-Saint-Denis Commune. *Cecropia peltata* is in the foreground, and remnant second-story trees indicate seral nature of vegetation. Montane forest is on upper slopes in the background. Courtesy of E. Nurser.

Plate 13. Below. Seasonal forest on the slopes of Morne Gardier showing a dark band of evergreen bois d'Inde (*Pimenta racemosa*). Courtesy of J.-P. Fiard.

Plate 14. Top. Elfin woodland structure, *Charianthus nodosus* and *Prestoea montana,* and canopy with the summit of the Carbet Pitons in the distance. Courtesy of J.-P. Fiard.

Plate 15. Bottom. Riverine facies of seasonal forest composed primarily of savonnette (*Lonchocarpus* spp.) trees in swale between fallow land and sugarcane fields. On the crest of the morne between Rivière-Salée and Diamant communes are elements of dry woodland. Courtesy of J. Barkley.

Plate 16. Left. Bois blanc (*Simaruba amara*), in undisturbed lower montane rainforest, with M. Guy Pitula, a forest guide, on Sugar Loaf. Courtesy of J.-P. Fiard.

Plate 17. Right. Southern slopes of Cape Salomon, Anses-d'Arlets Commune in January at the end of the rainy season. Dark trees are *Acacia muricata,* a species typical of the driest rocky slopes, and *Hippomane mancinella.* Courtesy of J.-P. Fiard.

Plate 18. Top. Dry woodland on limestone of Morne Caritan, Sainte-Anne Commune, with red-barked *Bursera simaruba* and *Capparis cynophallophora,* still existing in 1982. Courtesy of J.-P. Fiard.

Plate 19. Bottom. Understory of the dry woodland, Caravelle Peninsula, showing fine specimens of ananas bois (*Bromelia* sp.). Note the light passing canopy and dense shrub layer. Courtesy of E. Nurser.

Plate 20. Top. Elfin woodland showing the characteristic wind pruning of the trees. Summit slopes on Chapeau Nègre in northern Schoelcher Commune on an exceptionally clear day. The trees in the foreground are *Myrcia deflexa,* a species often forming typical associations with *Prestoea montana* in upper montane forests. Courtesy of J.-P. Fiard.

Plate 21. Bottom. Tree fern colony on the 1902-blasted and ash-covered Plateau Palmistes, Mont Pelée. Courtesy of R. J. Kimber.

Plate 22. Top. Mangrove-swamp woodland dominated by black mangrove (*Avicennia germinans*) in Lamentin Commune. Note the conspicuous pneumatophores thought to bring air to the root system. Courtesy of J.-P. Fiard.

Plate 23. Bottom. *Pterocarpus officinalis* in swamp woodland, Robert Commune. Courtesy of E. Nurser.

Plate 24. Top. Littoral meadow, mangrove swamp woodland, and second-growth dry woodland in the south end of Marin Point in Sainte-Anne Commune. Courtesy of R. J. Kimber.

Plate 25. Bottom. Littoral woodland remnants on a small beach north of Englishmen's Bay, Sainte-Anne Commune. *Erithalis fruticosa* and *Hippomane mancinella* fringe the shoreline. Courtesy of J.-P. Fiard.

Plate 26. Top. Interior view of a twenty-year-old mahogany (*Swietenia mahagoni*) plantation on Morne Mitan in Morne-Vert Commune. Wide spacing and poor maintenance resulted in poor configuration of trees. Courtesy of R. J. Kimber.

Plate 27. Bottom. Primary rainforest canopy with a view toward Plateau Clark (550–650 m). The largest trees are *Sloanea, Talauma,* and *Dacryodes.* Courtesy of J.-P. Fiard.

Plate 28. Top. Mixed cactus scrub and desert rock pavement border at Vauclin Point north of the town. *Furcraea tuberosa* and *Cephalocereus* are conspicuous among the *Lantana croton* and *Borreria laevis* in the bushland. Courtesy of J. Barkley.

Plate 29. Bottom. Grazed clearing in second-growth cactus scrub bushlands, with *Heteropogon* sp., the most palatable of the volunteering grasses. Candle cacti rise slightly above the tops of *Cordia* and *Randia aculeata* crowns on Vauclin Point. Courtesy of E. Nurser.

Plate 30. Top. Landscape of desert rock pavement on Petrifications Savanna in Sainte-Anne Commune. Cliffs of Point d'Enfer in the mid-distance about 30 m high. Courtesy of N. Marshall.

Plate 31. Bottom. Dry season aspect of an island of cactus desert on the Petrifications Savanna in Sainte-Anne Commune. Raquettes (*Opuntia dillenii*) and bois genou (*Lantana camara*) are the largest individuals. Courtesy of J. Marshall.

Plate 32. Top. East coast littoral of Englishmen's Bay: *Hippomane mancinella* on sea slope, *Cephalocereus nobilis* and *Calliandra slanei* (right) in foreground. Piton Crève-Coeur in the distance. Courtesy of J.-P. Fiard.

Plate 33. Bottom. Windswept littoral hedge on Martineau Point, Sainte-Marie Commune, and bananas in cleared littoral woodland on the near side of Anse du Four. Mesophytic ruderals in the foreground. Courtesy of N. Marshall.

Plate 34. Top. Detail of leguminous thorn scrub with climbing cactus (*Cereus* sp.) and gommier rouge (*Bursera simaruba*), coast road north of Fort-de-France in Schoelcher Commune. Courtesy of J. Barkley.

Plate 35. Bottom. Floor of the *Cordia* thicket dry bushland during the dry season on Morne du Cap, Case-Pilote Commune. Courtesy of R. J. Kimber.

Plate 36. Top. Poyo banana planting near Basse-Pointe at 100 m in middle climate belt on dry, shallow volcanic soils. Exposed to the prevailing trade winds, leaves whipped on outside plants. Close spacing reduces danger from windfall. Courtesy of J. Barkley.

Plate 37. Bottom. Sugarcane fields and second-growth dry bushlands in Sainte-Anne Commune. Old stone pier for windmill on the right. Courtesy of J. Barkely.

Plate 38. Top. Coconut grove and mangrove swamp woodland border in Marin Point, Sainte-Anne Commune. Coconut grove is grazed by *créole* cattle, and the nuts are not harvested. Replaced by Club Med facility. Courtesy of J. Marshall.

Plate 39. Bottom. Coastal bluffs 3 km south of St. Pierre in Carbet Commune. Thorn scrub on the borders of coconut plantation on uplands. *Croton* and *Asclepias* in the beach vegetation. Courtesy of R. J. Kimber.

Plate 40. Top. Sugarcane fields on the alluviums of Lamentin Plain south of Quartier du Bac, Ducos Commune. Landscape created on extensive sugar-factory-owned lands. Courtesy of R. J. Kimber.

Plate 41. Bottom. Peanut (*Arachis hypogaea*) fields divided into *tâches* by maize (*Zea mays*) on shallow volcanic soils in Prêcheur Commune. Second-growth dry woodland and bushland begin at the break in slope. Courtesy of J. Barkley.

Plate 42. Top. Cotton bush in the dry bushlands of Quartier Morne Blanc of Diamant Commune. Pastured clearings in the bushlands can be seen on the mornes in the background. Courtesy of E. Nurser.

Plate 43. Bottom. Remnant cacao plantation with breadfruit (*Artocarpus altilis*) trees. Fallow grassland slope grazed by cattle is in the near ground and a provisions plot in the rear, Canton Suisse, Morne-Vert Commune. Courtesy of E. Nurser.

Plate 44. Left. Provisions garden of manioc subtype in Quartier Caplet, Morne-Vert Commune. Manioc (*Manihot esculenta*), sweet potato (*Ipomoea batatas*), carrot (*Daucus carota*) complex. This photogaph was taken about four months after planting. Courtesy of E. Nurser.

Plate 45. Right. Provisions garden of the yam subtype in Grande Savanna in Ajoupa-Bouillon Commune. Yam (*Dioscorea alata* or *D. cayennensis*) on the bamboo poles, dasheen (*Colocasia esculenta*) and barbade (*Maranta arundinacea*) below. Courtesy of E. Nurser.

Plate 46. Top. Pineapple planting, fallow grassland, and dooryard garden near Vert-Pré in Robert Commune. Coconuts and gliricidia are used to delimit properties. Courtesy of E. Nurser.

Plate 47. Bottom. Planted pasture on coastal stretch on Hon. Paquemar in Vauclin Commune. Portion of dooryard garden on the right, littoral-zone fence weeds in the foreground. Courtesy of R. J. Kimber.

Plate 48. Top. Improved pasture with remnant dry woodland species on rock outcrop in Hon. Paquemar, Vauclin Commune. Clones of cabouya (*Sporobolus indicus*) and petit foin (*Heteropogon contortus*) were hand-planted two years before this photograph was taken. Courtesy of N. Marshall.

Plate 49. Bottom. Grazed savanna in former littoral woodland on Caracoli Point, Caravelle Peninsula. *Tabebuia pallida* in rear center. *Croton niveus* shrubs volunteering from the secondary dry bushland in the background. Courtesy of E. Nurser.

Plate 50. Top. Citronella grass (*Cymbopogon nardus*) hayfields, leguminous thorn scrub, and second-growth dry woodland in Fonds-Bourlet, Case-Pilote Commune, during the rainy season. Courtesy of R. J. Kimber.

Plate 51. Bottom. Cleared, cattle-grazed, wet montane meadow at Plateau Boucher, Fort-de-France Commune. Montane forest on the east wall of one of the Carbet Pitons in the distance. Cloud level approximately 800 m. Courtesy of R. J. Kimber.

Plate 52. Top. Dry season aspect of regenerating dry bushland in old fields of different ages of abandonment of the *côte-de-fer* of Case-Pilote Commune. Courtesy of R. J. Kimber.

Plate 53. Bottom. Bromeliads, lycopods, and ferns of the 1929 ash deposit near the summit of Mont Pelée (1,200 m). The mist is characteristic of afternoons and is present many mornings. Courtesy of R. J. Kimber.

Plate 54. Top. Morne Saint Croix, 733 m. *Prestoea montana* and *Myrcia deflexa,* an association typical of montane rainforest on relatively undisturbed slopes of Mont Pelée and Piton Mont Conil. Courtesy of J.-P. Fiard.

Plate 55. Bottom. Mont Pelée from the southeast on a rare clear day. Banana plantings are in the foreground. Courtesy of C. de Lucy de Fossarieu.

Plate 56. Top. Garden areas divided by living fences on ridges surrounding Morne Petit Pain (532 m) in Morne-Vert Commune. Canyon walls of the Carbet River in rainforest, seasonal forest, and bamboo brakes. Courtesy of E. Nurser.

Plate 57. Bottom. *Philodendron giganteum* in the understory of seasonal forest, crestline forest, north slope of Piton Mont Conil. Courtesy of J.-P. Fiard.

Plate 58. Top. Secondary succession in seasonal forest in Plateau Fofo, Fort-de-France Commune. *Acacia* spp. and *Swietenia mahagoni* the more conspicuous woody volunteers. Courtesy of E. Nurser.

Plate 59. Bottom. Guavas (*Psidium guajava*) invading a fallow grassland 3 km east of Fond-Saint-Denis. Second-growth rainforest species colonize into the old field after the guava bushes. Courtesy of E. Nurser.

Plate 60. Right. Montane forest on Morne Rose, 600 m in Case-Pilote River drainage system, Schoelcher Commune. Courtesy of J.-P. Fiard.

Plate 61. Below. New plantation of *Pinus caribaea* in rainforest clearing in the southern part of Morne-Rouge Commune. Courtesy of N. Marshall.

Plate 62. Top. Coconut planting behind black sand beach on the north coast. Anse Lévrier, Prêcheur Commune. In the distance Morne à Lianes. Courtesy of J.-P. Faird.

Plate 63. Bottom. Degraded littoral woodland on sand, back of beach 1 km east of Diamant. Sea grape (*Coccoloba uvifera*) and poirier (*Tabebuia pallida*) trees dominate, with an occasional exotic East Indian almond (*Terminalia catappa*). Courtesy of N. Marshall.

Plate 64. Right. Hurricane recovery in seasonal forest between Anse Couleuvre and Grand Rivière, northwest of the island. *Ochroma pyramidale,* a pioneer weedy tree, is in the foreground. Trees of this species that were seedlings after Hurricane David (1979) were 10 m high in 1986. Courtesy of J.-P. Fiard.

Plate 65. Below. Low rock walls dividing orchard savanna area cleared out of littoral woodland on Atlantic coast of Sainte Anne Peninsula. Volunteer plants in both fences include *Optuntia dillenii* and *Argemone mexicana* as well as species of *Croton.* Courtesy of E. Nurser.

6 The Colonial Plantation Era, 1686–1847

During the colonial plantation era the Europeans cleared and replaced much of the remaining vegetation with cultivated crop lands. The avowed policy of the crown to develop a free peasantry in Martinique, with its attendant emphasis on small-scale, diversified agriculture that would provide tropical commodities for the mother country, had given way to a policy that not only recognized but aided the accumulation of large estates manned by a large, permanent slave class devoted to the production of an export commodity, sugar. This began in the last part of the seventeenth century, continued throughout the eighteenth century, and into the mid-nineteenth century. Paralleling this trend was the establishment of smaller upland holdings devoted to specialty bush crops such as cacao, coffee, and cotton as well as the herbaceous crop tobacco. Although the term *plantation* has been traditionally used for the agricultural systems developed, it is more accurate to refer to these as *estates*, and to the system as *estate agriculture* (cf. a long argument by J. Petitjean-Roget 1980).

With this change in land use came regional changes in field patterns and in settlement morphology. Strategic considerations, population growth, and the dispersed settlement pattern all caused the proliferation of secondary roads. Technological changes in the sugar industry had had few effects on the vegetation, but engineering works such as flumes, drainage ditches, and dams built during this period were designed to provide the settlers with some independence from strictly local physical environmental controls and therefore facilitated the attack on native wild vegetation.

The mercantilism implicit in the *pacte colonial* was modified to the exigencies of the colonial situation during the earlier part of this period by nonenforcement and during the latter by a more liberal official policy. These conditions drew the Martinique economy to the North American markets as well as those of Europe, besides France. At the same time increased population growth provided an internal market for provisions. The effect of the course of settlement and exploitation was such that by the end of the second European period, roughly three-fourths of the island was either in cultivation or in spontaneous vegetation of successional or much degraded types and only one-fourth was in largely undisturbed vegetation. That undisturbed part was in the mountainous center of the island. The flora meanwhile had been enriched by many exotics.

Even more than in the early colonial period, Martinique during the years 1686–1847 was influenced by decisions and conditions external to its territory. In effect, the forces leading to change in vegetation were the decisions implementing French policy in Paris. Agricultural activities, forestry, and local settlement changes were carried on within a changing economic and political context that divides roughly into three parts: 1686–1762, a period of general prosperity during which Martinique was governed directly as part of the Domaine d'Occident; 1763–1802, a period of national exchanges and fiscal reform in which social and economic conditions were under active debate and which the momentous upheavals of the French Revoluton hardly influenced; and 1802–1847, a period of economic disquiet resulting from the competition of French sugar beets with Martiniquan sugarcane and to active discussions about the emancipation of slavery begun with the French Revolution.

During the first period (1686–1762), the Martiniquans developed a real dependence upon the British northern colonies[1] for flour, wood, dried salt cod, beef, and slaves (May 1930,162), while those colonies became dependent upon the French Caribbean colonies for sugar, molasses, and slaves (Pares 1932,69). There was also some commerce between Martinique and the rimlands of the Caribbean. The island's merchants traded French liquors, oil, and candles as well as Martinique coffee, annatto, and sugar with the Spanish colonies (Rufz de Lavison 1850,1:12). By 1699 the Martiniquans began to import from the northern coast of South America the mules, horses, and oxen[2] so necessary for running the sugar mills (May 1930,159; Chanvalon 1763,132). Trade continued well into the eighteenth century with the people of the South American coast (Venezuela and Cayenne) despite the discouragement by the Spanish crown. Traffic with neighboring European countries was permitted to French metropolitan merchants, but the colonials were not so free, so their greatest export and import traffic was with the French Atlantic ports of Nantes, La Rochelle, Bordeaux, and with the Mediterranean port of Marseille (Daney 1846,3:115).

During these eighty years the agricultural system itself was permitted to develop indigenously in the Antilles without the central bureaucratic control that existed in European France. The innovations associated with the agricultural revolution in France did not pass to the Antilles. Indeed, efforts by one governor to control production by limiting the number of sugar mills resulted in his rejection by the Martiniquans, and he had to return to France. There was little debate over the institution of slavery in this period.

The official economic policy of Paris changed dramatically in 1763 when the old policy of strict mercantilism was abandoned for a more liberal policy. The government, following another return of Martinique after a short occupation by the British, undertook to reform the civil and military administration as well as the commercial regime. In an effort to solve the planters' mounting debt problem, caused by controlled prices and comparatively high costs, the government permitted foreign buyers to buy syrups and rum in exchange for essential foodstuffs and other necessities when these were in short supply (Martineau and May 1935, 224). There were no *engagés* to be found. The cost of obtaining slaves from interior Africa was rising, and, within French government circles, there was speculation as to whether the blacks might be better protected, live longer, and bear more children if they were free. Strong humanitarian concerns current in French society were added to these arguments with the emergence of the emancipation movement.

The port of St. Pierre was declared a free port in 1784, and colonial assemblies were formally promulgated in 1787, but the disorganization associated with the French Revolution and yet another English occupation virtually eliminated any social gains. Indebtedness continued to plague the planters. However, the French loss of Saint Domingue at the end of the eighteenth century together with the disruption of its plantation economy eliminated those lower cost sugars from the French market, thus providing the Martinique planters with better markets and prices (Parry and Sherlock 1956,172).

The last period beginning in 1802 was transitional in every respect. During the Napoleonic wars, the English isolated the Antilles from France. Costs mounted prohibitively while revenues declined. The use of sugar beet, to which French refiners had been forced to turn during the blockade, was officially encouraged by the emperor after 1812. This dealt a severe blow to the sugarcane industry. The refineries in France found using a local supply of sugar to be more economical than importing partly refined products, and there was little sympathy in France for the planters' plight, a feeling fed by increasingly negative attitudes toward the institution of slavery.

The Settlement Pattern and Process

The vast majority of the individual estate properties were established in the eighteenth century (Dauvergne 1948,242) and were the result of the assembling of numerous small land grants. The grain of the topography in most of Martinique is such that these new, larger properties contained different soils and varying exposures. Consequently, the planters grew accustomed to planning for wide ecological variations within their properties and to maintaining areas of undisturbed vegetation on rough or steep land within the exploited lands. Only on the lower western slopes of Mont Pelée, in Basse-Pointe Commune, and in the Lamentin Plain were the sugarcane fields nearly identical in size and physical characteristics.

This process of enlarging holdings was most pronounced from 1667 to 1740, although individual properties continued to expand and others to be cut up during the whole of the colonial plantation era. The big planters continued to enlarge their properties, and many middle-sized property holders were forced to leave with their slaves for other islands and the Carolinas. The exodus was particularly marked during

1716 and 1718.[3] Its consequence was a net diminution of area in provisions and other subsistence crops.

In time, areal specialization took place so that the lower, flatter lands were taken up in sugar estates and the lands on the steeper slopes and the midelevations were taken up in the smaller holdings of the secondary export commodities and in provisions gardens. The average sugar estate was 170 ha, but reports mention some at 1,000 ha, and even one at 1,500 ha in size (Lavollée 1841,111). The classic land tenure pattern of a few large holdings and many small holdings was constituted during this period.

The sugarcane estates became a distinct settlement form as well as a landholding system. The four essential parts to a sugarcane habitation or estate were the headquarters, the *plaine* in sugarcane, the grazing lands, and the slave lands. The estate headquarters area, referred to during the eighteenth century as the *savanne,*[4] consisted of an open space surrounded by hedges and sometimes by a wall within which were the buildings necessary for the functioning of the estate.[5] Included were the *grande maison* or big house, a mill or two activated by oxen or water or wind for the crushing of the sugarcanes, various buildings associated with the boiling down and processing of sugar, the bagasse buildings, slave housing, huts for indentured artisans, a pottery-making building, a storehouse, and lime kilns. Sometimes a sheep run and an artificial pond called a *mare* were built.

Sugarcane lands were marked off in rectangular patches except when a morne or steep-sided ravine cut through the property. Between these fields were access roads, which often occupied as much as a tenth of the area in fields.[6] Cutting across the fields might be a canal or aqueduct. The ravines, where possible, were planted in plantains and bananas for provisions. By and large, the initial investment in fences and ditches discouraged changes in field size and pattern (W. Barret 1966, 152). Labat was most specific that individual fields be identified for evaluation of production and specific management concern (Labat 1724, 2:231). After several ratoons, the field would be converted to pastures (also commonly called *savanne*) or left fallow (*en frisch*), in both instances to be used for grazing livestock.

On the slopes of the mornes were the slave lands called *places à nègres,* where the slaves produced most of their own provisions. Where there was not enough of this rough land, territory suitable for sugarcane was expected to be given over to provisions, since a precisely defined allotment was required by law for the use of each slave: in 1724, 500 manioc plants (Revert 1949a,317) and in 1736, 25 plantains.[7]

The development of the big sugar estates led to larger rural population agglomerations. Labor requirements caused slave populations to increase rapidly. Père Labat, who between 1698 and 1702 managed a plantation for the Dominicans and is the best source on management strategies, pointed out that for a sugarcane holding of only 30 ha, 120 blacks would be required. By 1825 estate labor requirements had doubled (Revert 1949a,280).

The cacao plantings, the coffee plantings that largely replaced them in 1727–28, and the cotton bushes involved the cultivation of the mornes or hilly land for export produce. These holdings were also called *habitations,* or estates, but were smaller and without the complex industrial aspect of the sugar plantations. According to the census of 1788, there were 948 coffee estates averaging 6 ha, 101 cacao estates averaging 11 ha, and 233 cotton bushes or plantings averaging 7 ha (Rufz de Lavison 1850,2:20). By 1845 the number of coffee estates had fallen to 715 but averaged 26 ha, indicating that the smaller ones had not been economic units (Rufz de Lavison 1850,2:119).

The modest labor requirement together with the fact that most owners were operators meant that on coffee estates, compared with sugar estates, fewer slaves needed housing. As a consequence, smaller population aggregates were spaced more closely together in those areas dominated by three bush crops. The slaves housed on these estates lived in huts surrounded by their gardens rather than in slave quarters. Consequently, the vegetation mosaic was finer in grain in these areas than in sugarcane areas.

Usually, the coffee estates were on the crest of mornes or on the steeper, upper slopes. The commercial plantings surrounded the headquarters area. This area contained the big house, some small service buildings, a vegetable garden, a mill operated by animal power, slave huts, and a *cour à nègres.* A characteristic item of coffee estates was the *glacis,* a large rectangular area in front of the big house that was used for the drying and preparation of the coffee grains (Dauvergne 1948,250). Narrow roads or paths that provided access to the surrounding bushes frequently became water-filled gullies during rainstorms. Occasionally a small aqueduct ran to the property.

By 1717 the cultivable land of low elevation on Martinique was entirely ceded (Dauvergne 1948,250), but lands continued to be registered for newcomers and not all were in the mountainous interior. The Norman inheritance system in use, which required an equal division among heirs,[8] tended to make properties decline in size (Chanvalon 1763,3). However, Daney (1846,2:113–14) reported that strategies had been developed to circumvent this practice, making possible the maintenance of large sugar properties by the middle of this period (Durand-Molard 1807–11,1:227).

The course of population growth increased at a slower rate than earlier, but by the end of the period the inhabitants numbered over 120,000. In 1685 there were 9,000 blacks and in 1690, 13,000. In spite of periodic emigration of those settlers crowded out by the land acquisitions of the big planters, the population increased modestly until the 1720s. Then it began to increase rapidly by immigration until the 1740s, at which time the wars of the Austrian succession reduced the number of white immigrants. The total population peaked in the eighteenth century in 1740 at 85,360, to level off to slightly over 80,000 before gradually climbing again. The white population peaked in 1742 at 16,071 and then declined somewhat, but the black population continued to mount.

By 1787 there were roughly 15,000 whites, 80,000 slaves, and 2,000 former slaves. In 1818, after disturbances of the revolution and the empire, there were 9,800 whites, 77,000 slaves, and 11,000 freedmen. More and more blacks received freedom either by purchase or by gift from the master. On the eve of emancipation in 1845 there were 77,447 slaves, fewer than 10,000 whites, and 30,000 to 32,000 blacks and mixed-blood freedmen, giving a population density of 108 per sq km.

At the same time that the rural population agglomerations in the lower elevations were becoming larger and farther apart, penetration continued into the undeveloped parts of the island, and new towns were formed. In 1710 there were twenty parishes representing as many towns (Daney 1846,2:205). Even though the total number of towns increased into the early nineteenth century, the estate was still much more important as a center of population than were most towns. The Monnier map of 1825[9] illustrates the settlement pattern as it had developed near the end of the colonial plantation era. Larger estates such as La Montagne above St. Pierre and Beauregard in Sainte-Anne Commune had many more buildings than did the smaller towns, although the actual groupings of the symbols may be an artifact of the cartographer. This same map shows the farm arrangements of a large number of small cacao and cotton estates.

The island still had many isolated farmhouses or *cases à demeurer.* These, for the most part, were areas colonized by liberated slaves and free mulattoes. Most mulatto men were set up by their fathers as petty artisans in the towns, but a significant number were established on small properties. This was a common and accepted practice in the earlier half of this period, and it continued long after governmental policy had assigned slave status to these products of irregular unions. Areas having dispersed dwellings labeled as Fond-Mulatre or Fond-Gens-Libres tell their origins. These names are still in the toponymy of modern maps.[10] At an accelerating rate, particularly after 1830, slaves were freed by their masters, and they tended to congregate. The Saint-Esprit area was recognized as a region of "rich mulattoes" by the beginning of the nineteenth century; Rivière-Pilote, southwest of the Marin Cul-de-Sac, was known as a region with many former slaves.

The increasing population's need for land supplemented the continuing strategic and commercial reasons for building roads into the interior of the island. Intendant Blenac arranged for the building of a new road connecting Fort Royal with St. Pierre that did not follow the coast but passed high above Case-Pilote and avoided the coastal town of Carbet as well as the roadstead of Grande-Anse of Carbet. At the same time, a road was put through to François from Fort Royal by way of the Lamentin Plain, cutting the overland route to the windward coast via Gros-Morne–Trinité by at least a third. Petite-Anse and Grande-Anse of Diamant were connected to Ste. Luce and Marin Cul-de-Sac as well as with the road to Fort Royal (May 1930,201).

In 1718 the Conseil de la Marine charged the administrators of the colony to put roads into the interior in order to facilitate "settlement by small colonists."[11] The heights above the windward coast where provisions, cacao, and later coffee did better than sugarcane were particularly dependent upon roads. In 1734 more roads were made to join the towns to interior uplands and valleys in order "to locate a very large number of new colonists" (May 1930,204). Road conditions were poor; nevertheless, it was possible in the eighteenth century to move overland to all principal settlements.

As new roads were built, planters established themselves where wagons and pack horses could take out produce for export. Lacking the competitive advantage of a monopoly of transport, the soil-impoverished coastal lands became less attractive to new settlers than did the virgin lands of the interior, and the population became more evenly distributed.

Management and cultivation techniques in sugarcane remained generally the same throughout this long period, but innovations in milling begun in the earlier period diffused rapidly and became common (Chanvalon 1763,129). New varieties of sugarcane were introduced. The small-plot farmer continued to use slash-and-burn techniques, having perhaps a tenth or less of his concession in cultivation at any one time (Dauvergne 1948,250). The large estate holders sometimes used field rotation, but later in the period more and more of the estate land was continuously cropped until exhausted and then put into long-term fallow. New planting of sugarcane fields took place only where the ratoon canes began to reduce production (Dauvergne 1948,259). Multiplication of industrial agricultural enterprises was a function of increased land under cultivation, and their distribution mimicked the settlement pattern.

Despite innovations in metropolitan France, cultivation techniques in Martinique remained the same[12] except in the heavy soils about Lamentin, where the plow was reintroduced (Chanvalon 1763,127). The spread of the plow was limited by the availability of pasture and green feed and by some doubt about its usefulness under some Martinique conditions.[13] Even Chanvalon was doubtful of its utility in areas heavily infested with ants. In 1818, in an effort to stimulate the use of the plow, the French government sent five models to Martinique and offered to send agriculturists to demonstrate them (Rennard 1954,74), with little result. No significant changes in the processing of coffee or cacao are reported. Cotton mills were installed beginning in 1763 (May 1930,90). The census of 1783 twenty years later lists 181 such facilities (Rufz de Lavison 1850,2:308).

There seem to have been few provisions against pests. It is not even clear whether the practice of burning cane fields to get rid of rats and the *fer de lance* was common. Suggestions that contouring with small terraces might be useful in retarding erosion do not appear to have been taken up by many cultivators (Chanvalon 1763,123). There was no manuring reported except for the piling of collected weeds about coffee or cacao bushes (Daney 1846,4:197).

The dependence upon animal power for the grinding of sugarcane was lessened by the introduction of water-driven mills and of windmills capable of more cost-efficient pressing of the sugarcanes (W. Barrett

1966,156–58). Neither of these ever surpassed the number of animal-driven mills for the whole island, although waterwheels became more common in the northern part of the island, where many streams are perennial. Windmills were important in the windward coast and Lamentin Plain, but animal mills remained the only ones in significant numbers in the south. After a sharp increase in the first third of the period in waterwheels and windmills, the proportion remained relatively constant.[14]

With the construction of engineering works the settlers began to be somewhat free of local environmental controls on the siting of industrial enterprises as well as on agricultural production and stock raising. Canals, flumes, and aqueducts, dams and their impounded waters, drainage and irrigation ditches, cisterns and water tanks all were designed with considerable skill, and quite a few of these remain in use today.

Seasonal water flows were controlled for use in mills by the canals and flumes that tapped the stream courses above the dry belt or diverted water from one watershed to another. In the Carbet-to-St.-Pierre leeward slope were a number of elaborate engineering works. The Delajus Canal carried water from the Carbet River along the 35 m contour for 1.6 km to a point due north of the Lajus Distillery, then passed under Morne Charlotte for 100 m to empty into the estate reservoir on the south-facing slope of Lajus Ravine. The building of the tunnel and flume took three years of slave labor.[15] Farther north, a longer canal was constructed to deliver water from the upper Carbet River along the margins of the spurs to the Desfontaines estate. Called the Fond-Mascret Canal, it, too, passed part of its course underground.[16] The Usine Vivé was powered and was served until 1962 by a canal taken off the Capot River. On the north coast a canal diverted water from a tributary of the Grande River, carrying it along the top of the spur to the Beauséjour Estate sugar presses. The French usually put the sugar works on hillsides so that gravity could be used to maintain flow from one stage in processing to the next (W. Barrett 1966,151).

Drainage and irrigation ditches substantially altered local soil conditions, making it possible to use previously useless land, as in the lower-lying portions of the Lamentin Plain, or to use more intensively land already in cultivation, such as the agricultural land in seasonally dry climatic regions. Even the heavy alluvial soils of Lamentin provided excellent harvests of sugarcane for small investments in canals and drainage (Chanvalon 1763,14). Lacking pottery drains, cultivators often used bamboo (by now grown in swales in the foothills) with good results (Satineau 1928,1:134). Both the large and small planters irrigated other crops, especially coffee and cotton.

Dams were built in the upper ravines to store water of the rainy season in lakes or ponds to be tapped during the dry season. Above the tidal reach on the Salée River, the river was "stopped in its course by the tanks (*digues*) made for the sugar mills" (Chanvalon 1763,20). In the southern half of the island, low areas were excavated or earthen walls were built up to form ponds (*mares*). At least some of these reservoirs were made close to or in the headquarters of the estates, since those ponds were used for domestic water supply as well as for watering stock.

Even so, surface waters were never plentiful or good enough to satisfy domestic needs in the drier regions of the south. Cisterns were built on the roofs of churches and large houses to hold the rainwater of the rainy season. Chanvalon (1763,22) wrote, "In all the sections of the island where there is bad water, the water is collected in the rainy season. The inhabitants too poor to have appropriate vessels to conserve it in, gather the water from ponds. . . . There is general joy when enough can be collected for the year because of the regular drought which subjects the two peninsulas of Sainte Anne and Diamant." The Caritan cistern dates from the eighteenth century.

The amount of livestock and hence grazing pressure fluctuated with political and economic developments in the Caribbean. The number of horses and mules increased about five times during the colonial plantation era (see table 20). Horses were important as riding animals and for draying, but some were used in the mills, as were the mules. The mules were often the only draft animal for settlers of the upland, where the secondary roads in use were not much more than trails. Cattle numbers increased threefold in this time. Most were used to haul cane and to turn the mills. Old animals were butchered for meat. Sheep and goats are not reported in significant numbers until 1769, when 1,049 were enumerated. According to the census of 1783 they were well distributed throughout the island, although the northeastern part of the island had fewer animals, and Vauclin had the most animals of any commune. The concentration of sheep and goats in the southeast quarter of the island had begun. Pigs were reported for the first time in the census of 1769, but domestic pigs were certainly kept as dooryard animals from the time of earliest European settlement.

Sugar refining, begun in 1680, was prohibited in 1695[17] in the French West Indies, leaving to Martiniquans the production of raw sugar

Table 20. Census of livestock for the colonial plantation era

Year	Horses and Mules	Cattle	Sheep and Goats	Pigs
1687[a]	1,736	5,718	2,011	—
1697[a]	3,608	8,084	—	—
1701[b]	3,971	10,311	—	—
1735[c]	9,327	15,436	12,296	4,324
1769[d]	7,651	13,307	12,571	1,049
1783[e]	9,396	9,359	11,209	1,598
1789[f]	11,055	10,852	11,150	8,400
1845[g]	8,064	14,133	15,192	3,905

[a]Schnakenbourg 1977, 79.
[b]Rufz de Lavison 1850, 1:103.
[c]Schnakenbourg 1977, 80.
[d]Rufz de Lavison 1850, 2:200.
[e]Rufz de Lavison 1850, 2:308.
[f]Boston *Gazette*, July 5, 1790.
[g]Rufz de Lavison 1850, 2:308.

to supply the refineries of France. The colonies were, however, permitted the making of clayed sugar, since this was not considered refining.[18] Production of sugar steadily increased until 1753, when a decline began because of the wars and confused political situations in the Caribbean,[19] only to rise after the end of the Napoleonic wars to even higher levels than that of ninety years before (Pares 1932,68–72). The number of sugar mills continued to increase until there were 592, although there were marked fluctuations in production in response to the market situation (Rufz de Lavison 1850,2:199).

Several changes in the industrial process had an impact on labor requirements on the sugar estates. The three-roller vertical mill[20] still used in midcentury (Chanvalon 1763,134) was improved by the invention of the *doubleuse* (Dutrone 1790,104), a shield that guided the once-crushed cane into position for the second crushing, thus eliminating the need for one mill operator. The mill was also later made of iron rather than wood, which led to greater dependability and length of life. At the end of the period bagasse began to be used as well as wood to fire the sugar, just as Chanvalon (1763,18) had recommended fifty years earlier. The use of green bagasse was discovered just after the invention of the multiple-effect evaporation system in 1840 in New Orleans. At the end of the period it was not yet used on the island (Chapman and Chapman 1911,39).

With steam power available, the horizontal mills came into greater favor and by the end of the colonial plantation era had replaced many of the older types. The first of the steam-driven mills was built during the British occupation of 1813. By 1827 there were 184 water mills, 27 windmills, 211 cattle mills, and 13 steam-driven mills (Dupontés 1979,204–207). Open-fired batteries for boiling down the syrup continued in use, but the application of steam power increased the efficiency of the operation.

Changes in the Plant Geography

The flora of Martinique expanded dramatically during the colonial plantation era. Some two-thirds of the post-European introductions took place during those years, while there was virtually no loss in native plant species members of the ecosystem. However, many animal and bird species were eliminated. Selective culling modified further the forest and woodland composition, and most of these vegetation types were significantly reduced in area. Six new planted types were assembled with their weedy complements. An unidentified cacao plague drastically reduced the cacao plantings and accelerated the change to other specialty crops, especially coffee.

Burning began to be used as a management tool on sugarcane estates. Grazing measures varied in importance as livestock numbers fluctuated. Arrested succession on the Caribbean littoral and the southern peninsulas caused stability and accounted for the larger area for vegetation types that were previously much more ephemeral and restricted in area. Such process management led to greater control over the ecosystem territory.

As a consequence of severe erosion, the moisture-retaining capacity of soil in different regions was reduced, and plants in those places were subjected to greater drought stress than the rainfall warranted. Drainage of wet meadows modified habitats so that sugarcane and provisions might grow where previously the saturated soils supported swampy meadowlands and mangrove. The effect of natural catastrophes on livelihood activities was made considerably worse and longer lasting by economic losses that often pushed the management decision toward abandonment of any agricultural enterprise. The total area in spontaneous vegetation regained area from planted vegetation.

Never again would the island system be independent of decisions made in France; the dimensions of the system had been permanently changed. The effects of the *pacte colonial* varied throughout the period as European quarrels made tight or loose control over Martiniquan exports and imports desirable for the mother country.

At the mid-eighteenth century roughly the interior half of the island was still in native vegetation, although there had been some disturbance within it (see map 23). The coastal half was in planted and secondary spontaneous vegetation but contained islands of relatively undisturbed vegetation. These islands were largely on land topographically difficult for cultivation, such as steep valley sides and cliffs, or edaphically unsuited, such as swampy areas in heavy alluvial soils or stony ground. Chanvalon's (1763) account makes possible a synoptic view of what the vegetation fabric was like.

The north Caribbean coast was covered with modified vegetation. The small areas of level and gently sloping lands such as the crowns of spurs and alluvial valley bottoms were in continuous cultivation, principally in sugarcane and provisions. Water-driven mill wheels had reduced the need for livestock, and the previous grasslands now were in sugarcane. Steeper slopes rimming Mont Pelée were a mosaic of dooryard gardens, provisions gardens, especially fields of manioc (Chanvalon 1763,16), and volunteer second growth. Leguminous aliens and the native *Croton* species were increasing in importance, as were the pioneer elements of the seasonal forest and dry woodland on the steep slopes of the valleys connecting the bottoms with the level ridges and spurs.

The lack of fertilizer and long periods of fallowing after the exhaustion of continuously cropped land meant that many of the landholdings may have been in secondary vegetation types derived from the bushland, woodland, and forests. Daney (1846,2:262–63) described the landscape immediately north of Fort-de-France, which is illustrative of the *côte-de-fer* as a whole: "A morne extends from the northwest to the southeast toward Fort Royal, which it dominates. The interval which separates the place where the enemy was camped from the morne itself was almost covered with brush and some tall trees which had been saved in the midst of fields are covered today with diverse plantings." On the foothills

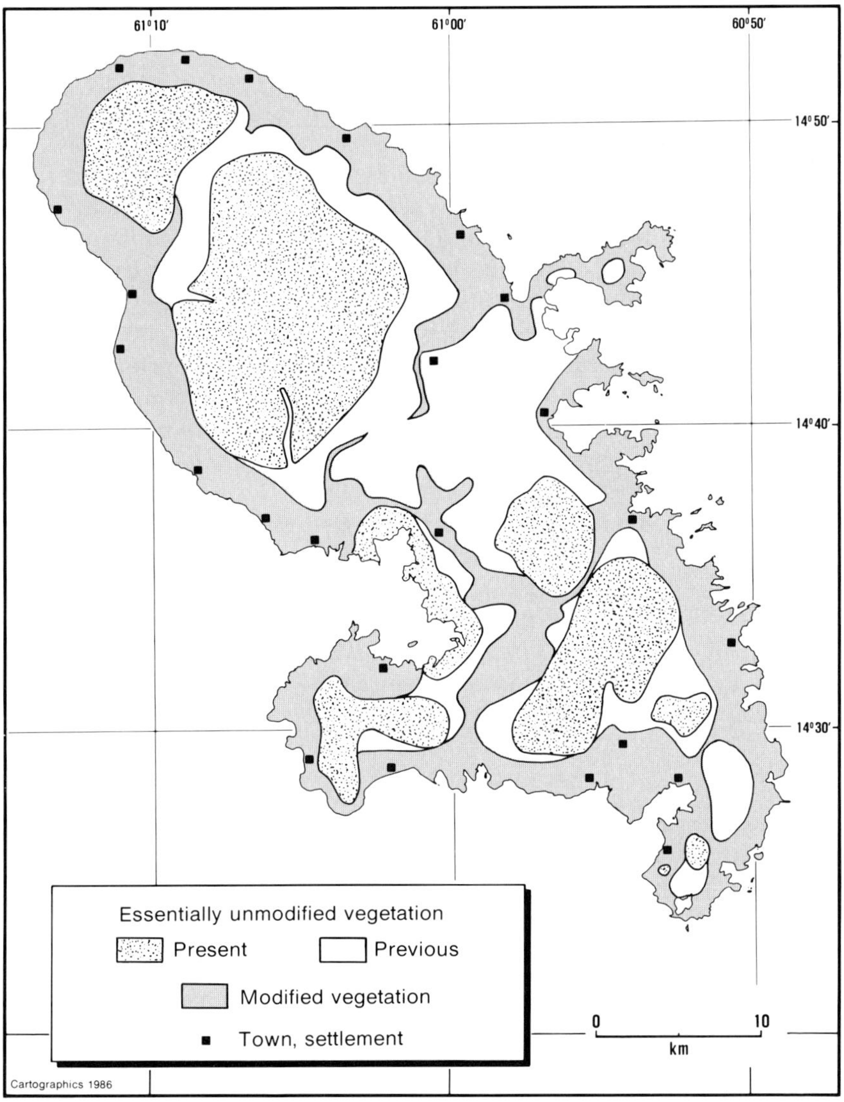

Map 23. Zone of vegetation modified by human agency in the mid-eighteenth century. Source: Compiled by author.

of the Carbet Pitons in Morne-Vert and, perhaps, Fond-St.-Denis, coffee plantings were being established on the steep slopes (Chanvalon 1763,12).

The north Atlantic coast still had a rarely invaded inner fringe of seasonal forest and lower-level rainforest. The coastal flatlands were in sugarcane. Coffee and tobacco were in the midelevations, mixed with the dooryard gardens and provisions gardens. As Chanvalon (1763,29) wrote, "occasionally some settlers established farther inland, cultivating there coffee and tobacco." There were cacao plantings, remnants from the previous period, on the valley sides of the Capot River and in the midreaches of the northern streams draining Mont Pelée. Tobacco plantings were most important around Macouba (Chanvalon 1763,17). Alluvial valley bottoms from Trinité north to Basse-Pointe were almost exclusively in sugarcane.

The steep canyon walls of the lower courses of eastward-flowing rivers were in riverine facies of seasonal forest, culled of the more useful timber species. In these narrow canyon-like valleys were small plots of bananas and plantains. Clumps of bamboo were taking hold in the ravines and gullies. The Caravelle Peninsula, especially in the central portion, was in sugarcane fields except for the mangrove swamp areas, which were as extensive as ever, and in the severely browsed littoral woodland and dry woodland of the point, as well as the severely eroded montmorillonitic clay soils, which were abandoned to savanna.

The south Atlantic coast still had a few indigo plantings, but sugarcane and pasture land dominated the lower elevations, and coffee estates and seasonal forest covered the foothills of the Vauclin massif (Chanvalon 1763,12). The size of rock pavement areas remained the same, as did the cactus desert. Some second-growth bushland had taken over on abandoned sugarcane fields.

Cotton plantings dominated the Sainte Anne Peninsula, covering most of the hills as well as the flat lands (Chanvalon 1763,17). Dooryard gardens were concentrated on the hills of the individual plantations along the rest of the coast. Chanvalon (1763,16) reported that in the *quartiers* around Rivière-Pilote where the soil quality of fersiallitics varies conspicuously from excellent to mediocre, a patchy land-use pattern gave a moth-eaten appearance to the hilly landscape. In the western peninsula, in the hills of Anses-d'Arlets, coffee plantings were prominent, but cotton was a locally important crop, as was sugarcane. Much of the surface of Anses-d'Arlet, Diamant, and Trois-Ilets communes was still in dry woodland vegetation. The mangrove swamp had not retreated much from its aboriginal area in Marin Cul-de-Sac and seaward of the Salines Lake. In the corridor strip connecting the south coast with the Lamentin Plain and its bordering hills, sugarcane fields formed long reaches interrupted by bamboo-filled swales and riverine facies of seasonal woodland.

The central portion of the Lamentin Plain, improved with drains, and its bordering hills were largely in sugarcane fields and littoral meadow. Almost all the land bordering the bay was still in mangrove swamp (Chanvalon 1763,14,20). Coffee plantings were on the hills and higher ground, together with small settlements and dooryard gardens. The largest-scale, most regularly patterned agricultural landscape in Martinique was the Lamentin Plain.

The mountainous interior had been breached significantly in only two regions. The Gros-Morne area, previously in cacao plantations, had become known for its coffee. The Champflore Basin around Morne-Rouge and the Ajoupa-Bouillon area were in provisions gardens and savannas. These midelevation forage areas were known to produce livestock for meat. Much of the grazed savanna was weedy, overgrown, abandoned swidden lands. The Carbet Pitons and Morne Jacob, how-

Table 21. Martiniquan furniture woods of the eighteenth and nineteenth centuries

Species names and habitat	Color and other characteristics	Use
Acajou *Cedrela mexicana* Present on all the island at low elevations	Red	Wainscoting, furniture
Acajou blanc or bois blanc *Simaruba amara* Average height, very widely distributed in seasonal forest and its margins	Milky white with pearly reflections in the varnish	Interiors, fine furniture, and chests
Acomat *Mastichodendron foetidissimum* Dry forest, became rare because of overuse	Yellowish white	Most commonly used wood of 18th and 19th centuries
Balata *Manilkara bidentata* Average height, seasonal forest	Deep maroon, chestnut	Construction, ordinary furniture
Balata rouge *Pouteria hahniana* Large tree, rainforest	Reddish brown	Construction
Bois amer *Picrasma antillana* Rainforest, stream courses	Whitish wood with hint of yellow; very hard, takes a good polish	Wainscoting
Bois bander *Richeria grandis* Average height, rainforest, common	Dark red	Paneling
Bois cannelle *Cannella winterana* Large tree, rainforest, rare	Reddish brown, aromatic	Small boxes, small pieces of furniture
Bois d'Inde *Pimenta racemosa* Average height, dry woodlands, and seasonal forest	Light brown	Turned wood, chests and boxes
Bois Madame *Styrax glaber* Small tree, rainforest, and seasonal forest	Yellow with darker marbling	Marquetry, small pieces
Bois mahot *Trichilia simplicifolia* Small tree, rainforest, and seasonal forest	Brown	Construction, furniture
Bois pelé *Myrcianthes fragrans* Small tree, rainforest, rare	Brown	Small furniture

Table 21. *continued*

Species names and habitat	Color and other characteristics	Use
Bois pistolet *Guarea ramiflora* Small tree, rainforest, rare	Reddish with little venation	Small furniture
Bois rivière *Chimarrhis cymosa* Rainforest	Yellow	Wainscoting, paneling, turned feet for furniture
Bois de rose *Cordia alliodora* Dry woodland	Red, very precious wood	Furniture, wainscoting, paneling
Bois violet, bois de sept ans *Meliosma herbertii* Rainforest, very rare	Very lovely dark wood with yellow highlights	Quality marquetry and furniture, very fine furniture
Bois de Chine *Murraea paniculata* Introduced shrub, cultivated in gardens	Very clear yellow	Turned wood and small pieces of furniture
Cannellier *Cinnamomum zeylanicum* Small tree, lowlands	Brown, aromatic wood	Bookcases, chests
Catalpa *Thespesia populnea* Small tree, lowlands	Light gray	Turned objects, small pieces of furniture
Café grand bois *Drypetes glauca* Average height, rainforest and seasonal forest	Red	Small pieces of furniture
Citronnier *Citrus divers* Small tree predominant in the zone of cultivation	Yellow, light yellow, or golden yellow	Marquetry, small turned objects, decoration
Citronnelle *Triphasia trifolia* Shrub of the littoral woodland	Light yellow	Decorations, small pieces of furniture, marquetry
Courbaril *Hymenaea courbaril* Large tree, seasonal forest and dry woodland	Reddish brown, hard wood; takes a good polish	Furniture, large poster beds; much abusive cutting
Gaiäc *Guaiacum officinale* Small tree, dry woodland Became very rare at the end of the 18th century	Black, very dense wood; takes a magnificent polish	Elegant furniture; furniture sellers in Europe get an incredible price for it
Giroflier *Syzygium aromaticum* Small tree, Indonesian origin, not widespread	Reddish brown, aromatic wood	Very much in demand for turned wood

Table 21. *continued*

Species names and habitat	Color and other characteristics	Use
Goyavier *Psidium guajava* Small tree, native to glades and wastelands	Yellowish gray	Small pieces, furniture
Laurier fine *Ocotea leucoxylon* Average height, rainforest and seasonal forest	Grayish brown	Very much in demand for paneling and veneering
Magnolia *Talauma dodecapetala* Large tree, rainforest	Very dark	Furniture, popular from the end of the 18th century through the 19th century
Mancenillier *Hippomane mancinella* Average height, found behind sandy beaches	Grayish black with golden highlights; takes a good polish	Paneling, wainscoting, chests, caskets
Poirier-pays *Tabebuia pallida* Average height, dry areas	Light yellowish gray	Furniture, mainly in the 19th century; mills and ships in the 18th century
Pois doux marron *Catalpa longissima* Large tree, extremely rare	Brown	Furniture
Sapotillier *Achras sapota* Average height fruit tree, predominant in the area	Reddish	Furniture, construction
Tendre à caillou *Acacia muricata* Average height, dry woodland, became very rare in the 19th century	Gray-black, very hard and precious wood	Wainscoting, some furniture

SOURCE: Darmezin de Garlande and Poupon 1982, 48–52.

ever, were still in rainforest and fringing seasonal forest. Most of the Pelée massif was still in forest (Chanvalon 1763,13).

Selective cutting of construction materials, furniture woods, and dyewoods out of the spontaneous forests and woodlands continued into this period. Even as late as 1694 all the houses in St. Pierre were made of local wood except for two stone sugar refineries and the Jesuits' House (Daney 1846,1:330), but by the turn of the century most of the lumber used in town homes, commercial establishments, and sugar works was imported from Saint Lucia (Daney 1846,2:296). During the eighteenth century more and more buildings were constructed of stone, but most building was still in wood, susceptible to the frequent fires, which only intensified the need for timber. After the very first settlement, good, even fine, furniture was made locally (Darmizen de Garlande and Poupon 1982). Some thirty-one different species were used to make furniture, boiserie, and for veneers (see table 21). Lumber imports rose, suggesting that local supplies of lumber were not adequate, either because good lumber trees had been culled or that the local sawmills could not keep up with demand (Kopp 1930,4). However, in the pioneer settlements building materials would have continued to come out of the local spontaneous vegetation type, less desirable species or not, because the settlers would have lacked the capital needed to purchase and haul lumber.

By the mid-eighteenth century, officials began to express concern over water and wood supplies. Intendant Phélipeaux requested the appointment of a Grand Maître des Eaux et Forêts, who would have control over hunting and fishing as well as over water supply and forests.[21] Among his duties would be the identification of saplings of suitable construction species to be reserved in the new settlement clearings, as well as the making of a recommended list of fast-growing, easily maintained trees that would be required planting in previously exploited properties, preferably in the savannas and the hedges. This plan was the first to incorporate planted wild forest tree species outside of forest preserves and woodlots.

The harvesting of wild fruits and plant parts for food continued during this period, but at a much-reduced intensity since the imported domesticated species were more productive and more accessible. The gathering of medicinal plants continued unabated, if the recipes for potions of the modern Martiniquans are any indication (Ayensu 1981; Morton 1981). The cutting of fuel wood for charcoal making increased as the population increased.

Mature colonies of tree ferns (*Cyathea arborea* and *C. muricata*) in the rainforest were destroyed on Morne Calebasse by woodsmen who brought the trunks into the settlements for house construction (Chanvalon 1763,176). The first species is more weedy than the other, so probably it increased generally in importance on the cut-over lands during the eighteenth century. Phélipeaux recommended the preservation of saplings of acajou (*Carapa guianensis* or *Cedrela mexicana*) and *Homalium racemosum* (Rennard 1935b,126), suggesting that these two species had become scarce. After Fort Royal was established as the naval harbor for the French fleet in the West Indies in 1823, ship timbers regularly were taken out of the rainforests of the southern slopes of the Carbet Pitons.

The seasonal forest tree le cipre (probably bois-Chypre, *Phoebe elongata*, or possibly *Cordia alliodora*), with its brownish yellow, very resistant wood, was also mentioned by Phélipeaux as a timber tree to be conserved (Rennard 1935b,171). Another much-used timber tree was *Hymenaea courbaril*. According to Chanvalon (1763,165) the slaves liked the strong-tasting floury substance surrounding its seeds. When these seeds matured in the month of November, large groups of slaves would assemble in the forests to gather the hard, woody siliques. This collecting may have aided in the tree's dispersal outside of forested areas, while logging for naval timbers reduced its numbers within them.

In the earlier years of this period, domestic and industrial fuel requirements subjected the surrounding seasonal forests to severe depletion. Although the use of coal for firing the sugar vats is mentioned in 1708, it was not until the mid- to late nineteenth century that coal may have replaced wood in the Lamentin Plain, Sainte-Marie, and Trinité, where transport costs from the port were not too high (May 1930,151).[22]

Guaiacum officinale or lignum vitae was increasingly cut out of the dry woodland and littoral woodland, for it was used in making gears and cogs for the sugar-mill machinery. It was impossible to enforce a ruling of 1701 that confined its use for pulleys in naval ships' rigging (Daney 1848,2:296). This understory tree became progressively more scarce during this period, although it appeared on official price lists as late as 1777 (Chambon 1777,174). "Bois jaune" and *Acacia muricata,* being similarly resistant to wetting and not susceptible to termite damage (Stehlé 1945–46,317), was also culled. Bois jaune was on export lists to the mid-eighteenth century (Chambon 1777,174).

Tabebuia pallida, a dry woodland species, was one of the fast-growing woods that Phélipeaux suggested for planting in the lower elevations (Rennard 1935b,171). In the early eighteenth century the settlers were still using boughs of *Erithalis fruticosa* and *Zanthoxylum spinifex* for torches (Labat 1724,2:239–40).

Leather and hides were regularly exported to Europe during this period (Chambon 1777,1:175). The leather was processed with local tanning agents derived from the bark of three mangrove species, and from the roots of logwood (*Haematoxylon campechianum*). Mangrove trees also make excellent fuel, and without doubt they were cut out of the swamp woodland near Fort Royal, François, and Marin. There were four tanneries in 1763. Of the three on the west coast, the most important was in St. Pierre, where mangrove bark was brought from as far south as the Fort-de-France Bay. The Rivière-Pilote tannery was near the mangrove swamp of the Marin Cul-de-Sac. Since mangroves sprout and grow rapidly, the area under mangrove retreated only where drainage programs were undertaken. The drainage of wet littoral meadow areas probably modified the regional drainage pattern somewhat and, thus, the adjacent mangroves.

Several activities reduced the bushlands slightly. There are scattered references to logwood cutting in official reports. In 1842 a thousand kilograms of this dyewood were exported to France (Martinique, *Bulletin officiel* 1842). The dry scrub species of these bushlands make excellent fuel, in part because of their high densities and heat conductivity, in part because of the resins and gums they contain. These species were much sought-after for charcoal as well. The effect of cutting in these bushlands is described by Chanvalon (1763,38), who said, "the tree (or shrub) is cut to the ground . . . after which [it] throws out numerous sprouts forming a kind of coppice." Since these vegetation zones were close to population concentrations on the west and south coasts, exploitation must have resulted in thicketizing this vegetation.

Naturalized and native heliophylls volunteered into recently abandoned old fields, providing useful food and fiber plants along with herbs and simples. Used for fruits were guavas, mangos, tamarinds of the earlier introductions, and breadfruit, jack fruit (*Artocarpus heterophyllus*), oranges and lemons (*Citrus* spp.) of the later period. *Calotropis procera* and the local mahot species were used for fiber. Plants with real or supposed pharmaceutical or magical properties commonly came out of these areas. Exploitation of this sort would tend to encourage rather than suppress the used species since the majority are pioneer species.

Grazing continued to be an important stress factor on the vegetation, with both the domestic and the rare domestic-gone-wild animals using savannas, bushlands, and woodlands. Feral animals were probably no longer important within the spontaneous vegetation. There are no references to hunting wild animals for meat. Sheep and goats, as well as some cattle, did roam freely into the dry bushlands and woodlands near the pasture savannas as well as into remnants of the seasonal forest. However, most cattle and horses fed on the savannas or fallow grasslands. Beginning with 1764, pasture areas, listed as savannas and grasslands, appear as sizable areas in the census, but they were probably merely the uncultivated spontaneous, herbaceous vegetation on inaccessible or rough land rather than true planted grassland pasture.

Livestock, savanna, and cropland figures for 1783 indicate that cattle numbers are higher in those communes with more than the average percentage of land in gardens, except in Lorrain and Prêcheur. Communes with many cattle generally bordered the Caribbean. The southeastern communes, having 21 percent or more cultivated land in sugarcane, ranked high in horses. The four southeastern communes held 35 percent of the sheep and goats. The lowest density of animals was to be found in the communes Trou-au-chat (Ducos), Saint-Esprit, Rivière-Salée, Diamant, and Anses d'Arlets, all of which have less than one grazing unit per ha.[23]

By the end of the period two-thirds to three-fourths of the island had been cleared at some time, yet in the early years clearing progressed too slowly, according to the crown. Directives were sent continually to the governors and intendants, encouraging the exploitation of already conceded territory. According to a directive in 1713, one-third of the allotted land was to be cleared within three years, a second third in the three following, and the first year had to see the settlers making an *établissement* on pain of losing title to the concession (May 1930,74). However, speculation in land continued, and it was not until a declaration of 1722 that clearing of new holdings was actually enforced.[24]

Land clearing for sugarcane estates virtually eliminated the seasonal forests, except for areas in steep slopes in the south and in the middle, seaward slopes of Piton Conil and Mont Pelée. Clearings for coffee and cacao estates were made in rainforest lands as well as in the upper elevations of the seasonal forested lands. South of the Calebasse

Road settlement was made into the Champflore Basin. Sometime during the second half of the eighteenth century the wide gentle spurs of the adjacent mountains were cleared of rainforest for agriculture and pasture. In 1850 Rufz de Lavison (1850,2:43) referred to the savannas of Champflore as "those lovely plateaus which are dissected having sufficient altitude to correct for their latitude."[25] Clearings in the rainforest for coffee plantings pushed the border up and inward in Basse-Pointe, Macouba, Trinité, Gros-Morne, and modern Saint-Joseph communes. Clearings were were also made for the dooryard gardens and provisions gardens.

By the 1730s the *cinquante pas* reserves in drier areas were referred to as coastal savannas, difficult to cross because of the many spiny cacti *Opuntia antillana* or *O. dillenii* (Durand-Molard 1807–11,1:375). Elsewhere along the coasts complaints multiplied over the difficulties of maintaining grasslands, or *haie.* "The vegetation is fast growing, it is only with a great deal of attention that one can have good grasslands. . . . The shrubs as well as all the other plants shoot up" (Chanvalon 1763,28).

Occasionally, fires got away from the charcoal burners in the dry bushlands, leading to a concentration of more fire-tolerant plants such as *Randia aculeata* and species of *Capparis,* certainly along the Prêcheur-Carbet coast and perhaps in the Caravelle Peninsula.[26] The same cause may have led to development of relatively pure populations of *Croton flavens,* as in the Schoelcher–Case-Pilote region of the west coast, on the south coast in Diamant, and, later, the establishment of *Croton* and *Lantana* bushlands in the early nineteenth century in Marin and Vauclin communes.

However, runaway fires did not pose a management problem on the windward coast. Labat (1742,1:203) said the burning of stumps in fields newly prepared for sugarcane was practical because the vegetation surrounding the field was not flammable and would prevent the fire from running away from the blacks tending it. The burning of the weeds and brush was recommended because the ash was valuable fertilizer for the crops (Raynal 1777,3:259). Burning also was generally supposed to inhibit the growth of weeds in the first year of planting.

The vast majority of the useful and weedy species introduced from abroad came into Martinique during the colonial plantation era, particularly during the last half, when some two-thirds of the identifiable introductions took place. The majority of these plants were minor landscape elements, being either food or ornamental plants, but several plantation crops did enter. Examples are Arabian coffee, new varieties of sugarcane, cotton, and cacao, and some timber and construction species, which today are being exploited as crops. There were also a large number of woody and herbaceous weedy exotics that colonized rapidly into the disturbed habitats created by clearing, agriculture, and abandonment.

Those primarily influential in making introductions were the planters and their agents, although the crown had some concern with plant introductions.[27] During the eighteenth century the ministry of colonies had young trees and seeds sent in order to expand the economic base of the island. The Jardin Colonial des Plantes at St. Pierre[28] was established in 1803 and began receiving plants from the older gardens in Saint Vincent, Jamaica, and Cayenne, Surinam, and from gardens in Mauritius, Réunion, and Paris (Martinique, *Bulletin Agricole* 1918,14–16). Nurseries were set up in the 1820s to provide trees for planting along the roads and along public promenades in towns (Martinique, *Lois et statutes* 1827,43,no.14). Chambers of agriculture were established late in the period, primarily for the purpose of holding discussions on agricultural techniques and marketing, but they had some responsibility for plant introduction (Martinique, *Bulletin officiel* 1845,15:394–96). Only a few of the many introductions can be mentioned.

Two cinnamon trees were sent to one of the estates in 1720 (AN F^3 251 1767). An introduction of more consequence resulted in significant area in cropland; at the period of greatest extent, coffee covered 5 percent of the cropland. Several attempts to introduce the fragile seedlings of coffee (*Coffea arabica*) were made in the beginning of the eighteenth century (Kervégant 1937,57), but it was not until 1721 that Desclieux succeeded in bringing live plants from Paris. Other seeds came in by way of Surinam and from Ile Bourbon (Réunion) at the same time or perhaps a few years before (May 1930,101). In the nineteenth century superior varieties were again imported from Ile Bourbon, including café Moka in 1839 and the café de l'Yemen in 1840 and again in 1842 (Rennard 1946,74).

Higher yields from sugarcane fields were made possible by the replacement of the old creole cane variety by the more rapidly maturing varieties of Otaheite cane[29] introduced from Mauritius in the 1780s (Légier 1901,510ff.). In 1782 Cossigny introduced some sugarcane varieties from Java to Mauritius and Réunion. They were later brought to the French West Indies together with Otaheite sugarcane. A second introduction of Batavian canes was made in 1789 by Guyet-Duclos, who imported them to Martinique at the request of Intendant Foulon d'Ecotier (Boyer de Peyreleau 1823,2:25). With these introductions came the possibility of local selection of varieties, for these new sugarcanes set seed, in contrast to the sterile creole cane (Deerr 1949,25). Introductions were made repeatedly from the Indian Ocean realm in 1791, from Cayenne in 1818, from Panama in 1837, Egypt in 1840, and India, Madagascar, and Bourdon in 1842 (Deerr 1949,18; Rennard 1946, 73);[30] but Batavian was not accepted as readily as the Otaheite, which became the predominant variety grown in Martinique during the nineteenth century (Deerr 1949,18).

In 1742 a cacao from Cayenne, more vigorous and more disease resistant than the previous variety, was a significant factor in reestablishing the diminished cacao industry (May 1930,96). Growth and ginning experiments were conducted on several cotton varieties and species coming from other islands and from the Carolinas of the United States (Satineau 1928,145; Kervégant 1936b,34). Sea Island cotton was the type grown most commonly. Chanvalon (1763,159) reported that he was growing "coton du Mississippi" in his experimental garden.

Of the construction materials, only bamboo (*Bambusa vulgaris*) became widespread during the period. Bamboo was introduced from southeast Asia in 1747 by Mahé de la Bourdonnais (J. Petitjean-Roget 1953, 15). Mahogany (*Swietenia mahagoni*) was introduced to Martinique sometime between 1750 and 1806 as a specimen and windbreak tree (Hughes 1750,79), but it was not mentioned as a timber tree in Martinique in the 1880s, although it was noted for Guadeloupe by de Lanessan (1886a,172) in his catalog of useful plants.

Labat (1742,1:355) reported bringing the avocado (*Persea americana*) from Saint Domingue in 1694 to plant in his garden at Macouba, some distance north of the sugar plantation in Fond-Saint-Jacques near Ste. Marie. Black pepper grains were sent to the intendants of Martinique and Guadeloupe (Archives nationales, Series F^3,19:II:1774). The Mexican avocado (*Persea drymifolia*) was introduced at the end of the eighteenth century (Kervégant 1937,54). The sapodilla (*Achras sapota*) was introduced in the mid-eighteenth century (Hughes 1750,132; Kervégant 1937,130). The mango (*Mangifera indica*) may have arrived in Martinique by the end of the seventeenth century (D. Watts 1966,67), but ten years after the 1782 introduction to Jamaica it had certainly reached Martinique (Dancer 1793). Named varieties of grafted mangos were sent to the Société d'Agriculture et d'Economie Rural in 1842 (Martinique, *Courrier de la Martinique,* Nov. 15, 1842). Both the breadfruit and châtaignier (*Artocarpus altilis*) and the jack fruit (*A. heterophyllus*) were brought from Mauritius to Martinique in 1792 (Fosberg 1960,106). However, the breadfruit tree had been sent by the crown in 1790 (Archives nationales, Series F^3,89,Jan. 25, 1790). In the same shipment carried by Fournier and Martin were the carambola or carambolier (*Averrhoa carambola*), the mangostene (*Ziziphus mauritiana*), the Bitimbier (?), and the Ravensana (?) (Archives nationales, Series F^3,89,Jan. 25, 1790). Captain Bligh brought the otaheite-apple or prune cythère (*Spondias cytherea*), along with another introduction of the crambola (Howard 1953,88–94). The akee or ris de veau (*Blighia sapida*) from West Africa was brought to Jamaica in 1793 and thence to the French Antilles (Kervégant 1936a,88). The Chinese lemon (*Triphasia trifolia*) was probably also introduced at this time.

Labat (1742,1:384–87) mentioned a series of garden vegetables that he arranged to have grown in the kitchen garden at Macouba. Many are the same as mentioned by Bouton (1640,8) when referring to the vegetables raised by the French settlers in the earliest period of settlement, so these plants of Labat were doubtless reintroductions. New listings were the three onions—chives (*Allium scheonoprasum*), leek or poireau (*A. porrum*), and garlic (*A. sativum*); the parsnip (*Pastinaca sativa*); salsify (*Tragopogon porrifolius*); and beets (*Beta vulgaris*). It is probable that the social distinctions, by then already developed, would have tended to permanently maintain these and other European vegetables in the gardens of the planter class. Also introduced early into gardens were the American chayote of chrystophine (*Sechium edule*) and granadilla or barbadine (*Passiflora quadrangularis*), which, though not common, were distributed generally and not on a class basis. The tomato (*Lycopersicon esculentum*) was probably introduced at this time, since it was in Barbados by midcentury (Hughes 1750,148). This was probably an introduction via Europe rather than directly from the Americas.

Useful plants were sent to the French colonies often at the instigation of the local agricultural organizations. In the *Courrier de la Martinique* for November 15, 1842, is a report of a shipment of plants received from Mauritius. Some forty-four species had been packed in three cases. Black pepper was again sent and arrived in good condition. Several sweet oranges had been sent, but only two plants survived. Ornamentals such as *Cycas circinalis* were introduced, as was *Rapsis excelsa* from China. These plants are still found in gardens, although they are not widespread. Other plants in the same shipment were *Curcuma domestica* and *C. amada*. Shipments such as these were being sent from one agricultural society to another in the nineteenth century. Such exchanges replaced those of the crown of the eighteenth century.

The African slave trade brought with it African and Asian food plants, which were used as middle-passage foodstuffs. True yams were important root crops that carried well and that were appreciated by crew and the slave cargo alike. The Portugese yam (*Dioscorea cayennensis*) and couche-couche douce (*D. esculenta*) were added to other yams that had come into the gardens in the previous period. Dasheen, taro (also called choux de Chine [*Colocasia esculenta*]) from the Pacific, and okra (*Hibiscus esculentus*) from Africa became important elements in the slave diet, as did pigeon peas or pois d'Angole (*Cajanus cajan*), probably brought from Africa, as Labat (1742,1:375) indicates, although the latter may be a native of India. Pigeon peas, like other plant introductions, may well have been introduced several times, thus accounting for some of the variability that Duss (1897,1:206) found in the species on the island.[31] Two other beans were important articles in the slaves' diet, the horse bean or fève (*Vicia faba*), introduced from the Mediterranean (Chanvalon 1763, 190), and the hyacinth bean or pois boucousou (*Lablab niger*) (Ballet 1890–99,1:524) from the Old World tropics.

Numerous herbaceous species introductions came voluntarily into the spontaneous vegetation. The raspberry or framboise du pays (*Rubus rosafolius*), originally from Asia (Kervégant 1937,19), today is naturalized along roadsides in the midelevations. It had been part of that shipment reported in the *Courrier* of 1842. The four o'clock flower or belle de nuit (*Mirabalis jalapa*) from South America, originally introduced as an ornamental, had naturalized in open associations at low and midelevations (Chanvalon 1763,192). Similarly, pepper grass (*Lepidium virginicum*), which is a common herb in the grasslands and savannas, came in during this period (Kervégant 1937,27). *Stachytarpheta* sp. came in sometime during the first half of the eighteenth century (Hughes 1750,86), as did *Vigna sesquipedalis* and *Clitoria ternatea*. Both have become naturalized in the savannas and grasslands of the drier parts of the island. Cress or cresson d'eau (*Nasturtium officinale*), introduced in the mideighteenth century, escaped into the wet savannas and grasslands in cultivated areas (Chanvalon 1763,189,192). *Emilia sonchifolia* from South America and its close relative *E. coccinea* had arrived by the mid-century and were in the manioc gardens and in sugarcane fields (Ballet 1890–99,1:524–25). *Calotropis procera* is from the Old World and tends to be found in coastal areas and to be relatively tolerant of drought (Ballet 1890–99,1:524–25). *Asclepias curassavica,* introduced to Jamaica, was probably sent to Martinique as well as Barbados during this period as a cultigen curiosity (D. Watts 1966,59). It has escaped into the spontaneous vegetation.

Several exotic trees also entered the spontaneous vegetation. The Indian almond or almondier (*Terminalia catappa*) was introduced to Jamaica from the East Indies in 1790 (Broughton 1794,34) and was listed as "very common" in 1886 on the littoral by de Lanessan (1886a,163). The East Indian rose apple (*Syzygium jambos*) was introduced as a nurse tree for coffee bushes and as a windbreak for plantations along the windward side of the island. Its fruits are rarely used, but it is now found in selected areas of rainforest (Kervégant 1937,221) and very commonly in degraded forests and gallery forests (Fournet 1978,873). In 1816 Pierre Baudin introduced into the botanical garden in St. Pierre the Asian loquat or neflier du Japon (*Eriobotrya japonica*) and the litchie (*Litchi chinensis*). He also introduced the governor's plum or merisier de Madagascar (*Flacourtia indica*), which was well regarded as a hedge plant like the other merisiers for which it was named (Kervégant 1937,92).

The bilimbi or cornichon (*Averrhoa bilimbi*) came in at the beginning of the nineteenth century (Ballet 1890–99,1:522). The kenette (*Melicoccus bijugatus*) was in the botanical gardens and Saint-Pierre by 1829 (Kervégant 1937,93). However, the mandarin orange (*Citrus nobilis*), imported from Indochina, and the Cayenne cherry (*Eugenia uniflora*), coming in at the same time from South America, became better distributed in Martinique than the bilimbi and akee. Frangipani (*Plumieria rubra*) from Mexico or Central America was an early garden introduction (Labat 1742,1:379).

A number of small trees and shrubs that were to become important elements in the bushland vegetation also came in at this time. The New World *Acacia macrantha, A. farnesiana,* and the Old World *Albizia lebbeck* were very common at the end of the nineteenth century (Duss 1897, 249,251), and their presence elsewhere in the West Indies earlier supports this (Maycock 1830). The rare South American *Caesalpinia coriaria* also was introduced during the colonial plantation era for its useful tanning properties.

Decorative elements in the spontaneous vegetation of today which were brought in as ornamentals to the botanical garden were the flame tree or poinciana (*Delonix regia*) from Madagascar (Duss 1897,230), the East Indian vine, *Thunbergia fragrans* (Ballet 1890–99,1:524–25; Duss 1897,2:428), the *Crotalaria juncea* and *C. quinquifolia* (Duss 1897,1:194). The *Pandanus utilis,* introduced from the Pacific (Duss 1897,2:485–86), naturalized rapidly from the garden introductions throughout the lower and midelevations. The ornamental hedge plant *Leucanea leucocephala* and its relative *L. brachycarpa* came in at this time as well (Duss 1897, 1:247–48).

Upland rice or riz de montage (*Oryza sativa*) was introduced from Manila to Martinique in 1820 but was only grown as a minor food plant in the Gros-Morne area and the hill lands above Fort-de-France (Kervégant 1936a,88–89), although Chanvalon (1763,191) had it in his garden and paddy rice had been urged by the crown since 1686 (May 1930,244). Luffa, citrouille de Guinée (*Luffa cylindrica*), was in Martinique by the mid-eighteenth century (Chanvalon 1763,189).

A number of pasture plants—legumes and grasses, primarily—were introduced purposely to improve animal feeds and accidentally as weeds. Chanvalon (1763,190) experimented with lucerne (*Medicago sativa*) and yellow sweet clover or melilot (*Melilotus officinalis*), both of which flowered and seeded. *Desmodium latifolium*, a native of the East Indies, naturalized out of the botanical garden in St. Pierre, as did *D. girans* (Duss 1897,1:200,204). Guinea grass (*Panicum maximum*) and Para grass (*Brachiaria purpurascens*) were both listed as introduced in this period by Ballet (1890–99,1:526–27). Guinea grass, introduced from Africa to Jamaica in 1741 (Edwards [1819]1948,1:254), probably reached Martinique late in the eighteenth century, as it did into the British Leewards (D. Harris 1965,119). Para probably came to Martinique from Barbados, where it was present by the mid-eighteenth century (Hughes 1750,240). Ballet (1890–99,1:527) has Para originating in South America, which agrees with Hitchcock (1936,247), who has it arriving in the West Indies from Africa via Brazil. Nonetheless, Lavollée (1841,50–51), who was sent to the Antilles on a fact-finding mission for the Ministère de la Marine et des Colonies, complained that the pastures were neglected and recommended that *Panicum maximum* be planted for the eight to ten hay cuttings it would give a year.

Eragrostis pilosa was probably introduced from the Midi of France during this period (Duss 1897,2:502) for, according to Grisebach (1864, 532), it was in Saint Vincent. *Dactyloctenium aegyptium* had become common throughout the lower elevations of the island (Duss 1897,2:508). The ornamental *Setaria geniculata* was probably introduced from France (Duss 1897,2:524); and, since it was in Saint Kitts, Antigua, and Trinidad in the mid-nineteenth century, it was in Martinique by the early nineteenth century along with its close relative *S. indica* (Grisebach 1864, 554). This survey of introductions is incomplete, but it accounts for the majority of the important naturalized common species. Those interested in a more complete listing should consult Fournet's (1978) flora.

During the colonial plantation era new vegetation types, both planted and spontaneous, developed; their areal extent can be documented and geographic shifts can be established with the aid of several agricultural censuses of the eighteenth and nineteenth centuries. The majority are not published in their entirety, but parts of the censuses of 1682, 1697, 1701, 1769, 1785, 1789, and 1845 are reported in specialized publications. The total area in cultivation was reported in a number of censuses and enumerations. In the eighteenth century the high was

Table 22. Extent of selected vegetation types, 1783

Type	Hectares	% of island area	
		99,137 ha	105,682 ha[a]
Planted			
Sugarcane	14,555	14.7	13.8
Provisions	5,426	5.5	5.1
Coffee	4,505	4.5	4.3
Cacao	1,107	1.1	1.0
Cotton	873	0.9	0.8
Subtotal, planted	26,466	26.7	25.0
Spontaneous			
Plaines (rotation fallow sugarcane land)	10,733	10.9	10.2
Savannas and grasslands	13,588	13.7	12.9
Mornes	33,652	33.9	31.8
Forests	14,698	14.8	13.9
Subtotal, spontaneous	72,611	73.3	68.8
TOTAL	99,137	100.0	93.8

SOURCE: Rufz de Lavison 1850, 2:308–309.
[a]The area of Fond-Saint-Denis and Saint-Joseph added to the 99,137 ha enumerated = 105,682 ha, not far from the total of 109,600 for Martinique.

38,702 ha for 1789. That figure was not even approached until 1835, when the total area was 38,320 ha (Schnakenbourg 1977,55–56).

The complete census of 1783 has been published and gives an excellent base from which to reconstruct a map of cultivated land, as it was probably meant to be exhaustive. It provides data showing variation in plant cover by communes (Rufz de Lavison 1850,2:308–309). Nine classes of vegetated surface were enumerated for twenty-six communes (see table 22). These were *plaines, mornes,* and areas devoted to sugarcane, coffee, cacao, cotton, savanna, provisions, and standing timber.

The first two categories were new in this census and have been interpreted as representing secondary spontaneous growth. *Plaine* is interpreted as fallow sugarcane land. This is justified by the contemporary vernacular use of *plaine* on plantations for the land in sugarcane. The sugarcane land was presumably in fallow rotation and not abandoned to spontaneous vegetation permanently; it was included in the arable land. The savannas and grasslands totals were not in the official island statistics. *Mornes* are interpreted as hill lands formerly in cultivation and now abandoned and in bush fallow. If the steeper slopes, where erosion presumably had destroyed productivity, were in fallow, this term must have represented land in a fallow period longer than usual in the farmer's agricultural system. Therefore, this is a substantial portion of this category in the census that can be considered as secondary spontaneous vegetation in midsuccession stages. Cattle and sometimes goats roamed in these areas, which may have actually resembled the savannas that were more strictly delimited. All but two of the thirty-four communes of the present can be thought to have been subsumed in this census. The two that probably were not enumerated are Saint-Joseph and Fond-Saint-Denis communes, which were at that date still probably largely in native forest vegetation. When those areas are added to the enumerated area for the census, a figure of 105,682 ha results, approximately the area of the island.

The proportion of land in cultivation plus in *plaines* and savannas for each commune in 1783 is shown on map 24. Close to two-fifths or more of the land was in cultivation in all but four communes: François, Marigot, Macouba (including Grand' Rivière), and Prêcheur. In each ex-

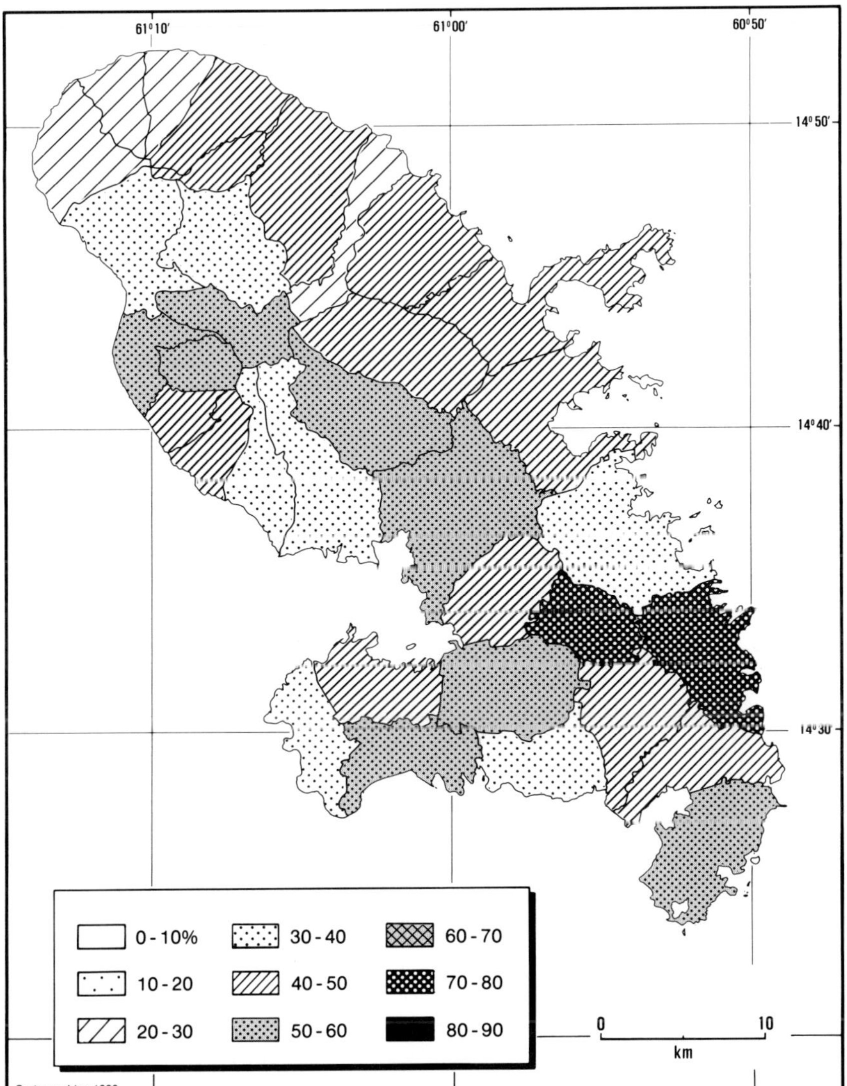

Map 24. Percentage of land in cultivation, census of 1783. Source: Rufz de Lavison 1850,2: 308–309.

cept François rugged mountains occupy a large portion of the commune area. François contains much of the modified hill-country vegetation. The very high values indicated for Saint-Esprit commune probably result from censusing an area greater than the modern area of Saint-Esprit. However, this area is one of small properties in which most of the land is in cultivation. It may have included part of modern Ducos, then called Trou-au-Chat, which was probably also intensively cultivated. However, since there is no evidence for other boundaries, no changes have been made on the maps. Vauclin and Trois-Ilets had extensive sugarcane lands associated with much land in provisions.

At the mid-eighteenth century Martinique was described as an island cultivated only in sugarcane and coffee (Chanvalon 1763,18). The trend toward sugarcane's dominance of planted vegetation had begun in the early colonial period. Coffee did not make its appearance until the 1720s (Chambon 1783,1:280–86; Raynal 1798,4:325) and became an important crop only after the cacao disaster and the economic depression caused by the War of the Spanish Succession. Planted vegetation types present at the beginning of the period were the dooryard and provisions gardens, orchards of remnant mulberry and canafistula trees, cotton bush, cacao bushlands, low annuals of indigo and tobacco, and sugarcane plantings. New types developed during this period were the bamboo plantings, coffee bushlands, which replaced much of the cacao bushlands, and perhaps hayfields of introduced grasses (Raynal 1798,4:339).

The dooryard gardens were greatly enriched throughout the period by food and ornamental introductions from South and Central America, Europe, Africa, and Asia. The sapodilla tree, the avocados, the mango, and later the breadfruit, jack fruit, otaheite-apple, and bilimbi went into the tree stories. New citrus plants, cacao, and coffee were also in gardens, adding to the shrub and small-tree levels. To this group can be added the red frangipani (*Plumieria rubra*), which came in very early along with Barbados pride (*Caesalpinia pulcherrima*), and later, the screwpine.

The horse bean, luffa, christophine, citronella grass (*Cymbopogon nardus*), and possibly okra and eggplant (*Solanum melongena*) were introduced to the dooryard garden. The last two were probably in the provisions gardens as well. An occasional taro and arrowroot or envers blanc plant were also among the herbaceous plants in the garden. Weeds in the dooryard garden by this period included goose grass and purslane (*Portulaca oleracea*) introductions from the earlier period as well as nut grass, crowfoot grass (*Dactyloctenium aegyptium*), and *Setaria geniculata*. The red-flowered *Emelia fosbergii* must also have appeared in the upper elevation gardens as weeds (Duss 1897,2:374).

With the large increase in population, the dooryard gardens increased dramatically in areal significance. Their distribution reflected the settlement pattern. Some regional concentration may have taken place in the Rivière-Pilote Commune and in the Mont Vauclin region. Perhaps parts of Saint-Esprit Commune may have been marked by a high pro-

Table 23. Provisions figures for selected dates, 1719–1835

Date	Manioc Hills[a]	Hectares	Plantains Hills	Hectares	Total Hectares
1719	2,221	140[b]	—	—	—
1731	22,771	—	3,051	209[c]	—
1742	43,756	—	3,782	520	—
1751	49,440	—	5,434	269	—
1762	22,562	—	3,698	84	—
1767	17,930	—	3,509	528	—
1773	38,670	—	5,858	407	—
1784	37,878	—	3,090	638	—
1785	—	6,238	—	—	6,238
1789	—	—	—	—	10,154
1816	—	—	—	—	7,645[d]
1826	—	—	—	—	9,402
1829	—	4,585	—	2,398	6,983
1832	—	—	—	—	10,097
1835	—	—	—	—	13,389

Source: Schnakenbourg 1977, 60, 70, 73.
[a]In thousands.
[b]Calculated by May (1930).
[c]Censused area for this year was 1,966 ha.
[d]Lavollée (1841) gave a figure of 8,856 ha.

portion of the land surface in dooryard gardens, anticipating the concentration found in the mid-twentieth century.

The provisions gardens, which had maintained themselves in proportion to the population during the early colonial period, began to fall behind relative to sugarcane as imports of flour from France rose to fill the food requirements (Labat 1742,1:367), only to build back later in areal importance (see table 23). The lack of manioc was particularly felt, and in the first years of the eighteenth century the local government was charged by the crown to require the planters to grow enough manioc to feed both whites and blacks (May 1930,80), but these regulations were relatively ineffective. In 1719 there were only 2,221,000 hills, or 140 ha for 43,000 inhabitants (May 1930,80). In 1723 a new ordinance put the matter more forcibly, and an annual census was ordered. Increase was slow. In 1731 there were 48,126,000 hills, and the number of hills remained high until 1762, when it fell; in 1767 it fell again. By 1773 production areas had increased to subsistence levels (Schnakenbourg 1977, 69–70).

Meanwhile, the administrators also encouraged the growing of plantains. In 1730 all those having slaves were required to put some land in plantains (Rufz de Lavison 1850,2:289). More than 3 million stems were planted, amounting to an area of 1,966 ha. An ordinance of 1736 outlawed the *pratique du samedi* and required the planting of twenty-five plantains for each slave (Durand-Molard 1807–11,1:399). In 1734 there were enumerated 5 million plantain stems, roughly 3,278 ha; in

1752 there were 1½ million, corresponding to 4,197 ha; and in 1751 there were 5 million, or 3,280 ha in plantains.

During the same period, and for the same reasons, the growing of sweet potatoes and yams (*Dioscorea* spp.) was also encouraged by regulation. In 1731 there had been 210 ha censused of these provisions grounds.[32] In 1740 an ordinance imposed a regulation of 1 *carreau* for every thirty slaves, which seems to have been enforced (Durand-Molard 1807–11,1:236).

These regulations essentially required the distribution of provisions to correspond generally with areas of high population density, as noted above. Regulation encouraged the field cultivation of mixed provisions, although provisions were not always in consolidated areas. Labat (1742, 3:142–43) even suggested sowing the European string bean (*pois de Rome*) in the verges along the roads, together with sweet potatoes and manioc, in order not to preempt land from sugarcane. One reason for the enumeration of plants or hills and stems in the various censuses of the eighteenth century was that these plants were not located in easily enumerated areas but were scattered in borders and patches.[33] Contemporary descriptions of estate layouts put plantains in draws or ravines, sugarcane and the other commercial crops on the upland or crowns of spurs. Such a distribution was both convenient from a labor management point of view and ecologically suitable. This practice continues today on large holdings.

The Old World taro and New World maize were not regulated. During this period they were not significant elements in the food patterns of the islanders because, first, they were novel and, second, they were not accepted by the population. Maize has remained a chicken feed.

Although the regulations were effective in building up the production of provisions, the total area in provisions in the last part of the period did not expand in proportion to the population. Rather, it tended to remain stable, except during the blockades of the several wars of the period, when the administrators attempted to modify the purely industrial direction of agriculture.

Canafistula orchards appear to have increased some in area, although the export figures show only a minor increment. By the third quarter of the eighteenth century there were about 113 ha in canafistula (Rufz de Lavison 1850,1:200). Some 23,129 trees were in the *département* of Fort Royal, 716 trees in Trinité, and 5,507 trees in Marin, in addition to the 47,561 trees in the zone of earlier concentration along the west coast of Prêcheur. The low figure for Trinité is not surprising since the tree does not bear well in the middle belt, under mesophytic conditions. On the other hand, it naturalizes easily in the dry regions of the island. These areas would have been at the expense of dry woodland and littoral woodland except in Trinité, where seasonal forest land was used. In only a few cases would the borders of canafistula orchards be against the spontaneous vegetation. Most often, the orchards would have existed as islands in the sugarcane lands, provisions gardens, and fallow grasslands and savannas. Mulberry orchards continued to be urged, but no new plantings appear to have been made. The black mulberry was reported in the dooryard gardens.[34]

Bamboo was introduced as a construction material and was planted in dooryard gardens as well as in groves on commercial estates. Its use spread throughout the island; it was used for small aqueducts and drains as well as in building construction. It was used to stabilize the downslope created by road building. Currently, it forms considerable groves or brakes as a consequence of its mode of vegetative reproduction. The growth habit causes it to shade out other plants, making it an effective competitor. Although now locally considered spontaneous, in most instances it has been possible to identify an original planting or to suspect an original plantation.

Cotton continued as a planted vegetation type and expanded after industrialization occurred in France proper. There were 56,000 bushes in 1719 (Schnakenbourg 1977,61). Sometime during the eighteenth century Sea Island varieties were introduced from North America. The number of plants by 1731 was 2,900,000, occupying an area of 1,900 ha. Plantings averaged about 2 million plants throughout most of the eighteenth century, although there were more than 3,223,000 censused for 1735. In the last two decades of the eighteenth century especially, with the disruption caused by the French Revolution and the Napoleonic wars, production fell substantially. In 1807 there were 791 ha in cotton. In 1826 there were 491 ha and in 1835, 178 ha (Schnakenbourg 1977,65).

Colonists of the second quarter of the nineteenth century reported that low prices were not the only problem. There had been a plague of a caterpillar that multiplied so fast that a whole harvest could be lost. This infestation was probably the cotton worm or cotton caterpillar (*Aletia argillacea*), the most serious pest of cotton in the West Indies (Monkhouse 1910,262). Further decreases continued, and by 1845 there were only 155 ha in cotton, and virtually all of this production was concentrated in the southern peninsulas and littoral (Chanvalon 1763,17; Lavollée 1841, 37–38; Kervégant 1936b,100). As the cotton lands were abandoned, they were allowed to revert to spontaneous bushlands, for the soils were eroded and no longer fertile enough to compete for sugarcane with lands in other parts of the islands, especially those in the Lamentin Plain.

Cacao, which had been only a small item in the planted vegetation at the end of the previous period, increased in importance with the turn of the century. The Terrier of 1671 contains many notations to the effect that a particular piece of property could be exploited in cacao, but no really significant cacao plantings were yet present. By 1727 about 9,477 ha (Revert 1949a,343) were in cacao trees, and according to May (1930,95) the cacao planter had a greater opportunity for riches than three-fourths of the sugar planters of the day.

Cacao plantations, earlier planted under sheltering trees left from the seasonal forest such as *Calophyllum calaba* and later planted under nurse trees, were intertilled with provisions in the early years of the plantings. Cacao takes six years to come into bearing, and the planters fell back on provisions growing to support the new plantings. Manioc provided a suitable nurse shade for the young seedlings, and other crops such as pumpkins, cucumbers, and sweet potatoes on the ground were compatible and tended to keep down the weeds (Labat 1742,6:396) and to help control surface erosion. After the plantain came into bearing, the dense shade cast by the large leaves tended to keep the ground bare of weedy growth, and the nurse trees were tall enough to shade the bushes.

The cacao bush is strictly tropical, a demanding plant in terms of soil nutrients, and much was written about soil exhaustion under the primitive techniques of the first settlers and the consequent need to clear new land to replant the artificial bushland because of the lack of manuring practices. More important, probably, was the physical stripping away of the upper soil layers, exposing the shallow roots of the cacao tree by rapid erosion caused by heavy rainfall, steep slopes, and friable soils. Cacao reacts unfavorably to rapid fluctuations in humidity even though it will tolerate a dry season if soil conditions are correct. Delawarde (1935, 104) was right in considering erosion as a primary cause for the "exhausting" of these ferralitic soils. The territory about Gros-Morne, with its fertile, unexploited young allophanes, good drainage, adequate rainfall, and cool temperatures as well as protected aspect (protection from drying winds was absolutely essential), became the principal cacao region. The other region was in the valley of the Capot River, the boundary between Basse-Pointe and Lorrain communes (May 1930,95).

In 1727 a severe earthquake and a violent storm destroyed 6 million of the 14 million cacao trees (Chanvalon 1763,17). In the ten years from 1722 to 1732 export figures for cacao fell to less than 5 percent of previous years.[35] A root fungus attacked the surviving trees, almost completing the ruin. A few trees survived, but it was not until the introduction of a South American clone in the 1740s that the area under cacao began to expand once more. Chanvalon (1763,17) recorded that cacao plantations were found in a few places and were very small. By 1775 the area under cacao expanded to about 1,000 ha and continued to rise until 1789, when it amounted to 1,191, only to fall again during the first half of the nineteenth century, when the area fluctuated between 1,076 and 492 ha (Schnakenbourg 1977,65). Cacao plantations generally were situated adjacent to forest or to and within seral communities derived from bush fallow.

The distress and emigration caused by the cacao disaster led to the promotion of Arabian coffee as a plantation crop. Expansion was rapid once the seedlings were available. The 100,000 bushes of 1727 became 1,802,000 in 1731, occupying more than 1,113 ha (Schnakenbourg 1977,61). Trinité Commune, adjacent to the unfortunate Gros-Morne, was the location of 1,150,000 bushes, or 753 ha of that total (May 1930, 103). In 1734 coffee covered 5,246 ha, and in 1737 it covered 8,316 ha. Thereafter, coffee bush numbers fluctuated around 10 million bushes or 6,556 ha for the rest of the eighteenth century (May 1930,103). There were fluctuations, with a peak in 1773, when 9,300 ha were under coffee (Schnakenbourg 1977,62), and a low of less than half that in 1783. During the first half of the nineteenth century the area under coffee declined steadily from 5,149 to 3,082 ha (Schnakenbourg 1977,65). In 1820 there appeared an unidentified infestation,[36] and by 1845 coffee bushlands occupied only 1,921 ha (Kervégant 1936b,100).

Coffee and, to a lesser extent, cacao were grown under nurse trees. Several trees, both introduced and native, were used for this purpose. At higher elevations the exotic *Syzygium jambos* and the natives *Calophyllum calaba* and *Inga laurina* were used, and at lower elevations coconuts were used. Sometimes the plantings took a rectilinear pattern, as on Mont Vauclin. In others, hedges were spaced perpendicular to the Trade Winds and separated some 25 to 30 m. In other cases, windbreaks of *Calophyllum calaba* surrounded the plantings on all sides, and *Inga laurina* was spaced regularly throughout the plantation (Lavollée 1841,35; Revert 1949a,342). The herbaceous ground layer was controlled by the cutting of weeds with the machete and naturally by shading from the tree canopy above.

According to Chanvalon (1763,17) coffee was cultivated in most communes. It replaced cacao and cotton plantings and caused new clearing to be made in the interior rainforests of the island.[37] While Arabica coffee is a rainforest understory species, under properly shaded conditions it will grow in lowland and even rather dry habitats.[38] The differences in environments created beans having different flavors. Although production was higher in other parts of the island, the coffee produced in the dry Anses-d'Arlets and Diamant communes (usually grown above 100 m) was the most highly regarded (Chanvalon 1763, 63). Coffee bushlands, therefore, bordered all vegetation types except the montane types and the xeric cactus scrub, desert rock pavement, and probably mangrove swamp woodland.

Tobacco had become a minor member of the planted vegetation because of poor prices that were the result of the saturation of the French market from overproduction. When demand again grew to the point that English tobacco was being purchased to satisfy demands in the mother country, efforts were made to encourage tobacco in the colonies, but with little success. Between 1733 and 1741 there was an increase from about 3 to 5 ha (May 1930,89). Tobacco plantings were always maintained in the Macouba Commune, where the dry, rust-brown ferruginous soils and the well-ventilated but humid atmosphere produced the much-appreciated "violet" flavor in the tobacco (Chanvalon 1763, 17; Saint-Croix, 1820,187). In the Tardieu atlas of 1817 annotations include references to tobacco in Macouba. In 1832 Sainte-Marie and Macouba communes exported some 700 kg of tobacco to France (Rennard 1946,74). In the eighteenth-century census materials examined, tobacco is not listed in the tables, indicating a lack of significant area and value for the colony.

Indigo was being encouraged by the government, and good prices stimulated production at the beginning of the period. In 1690 there were some twenty indigo plantations, but by 1700 the number had fallen to five enterprises, and exports were on the order of 50 to 100 quintaux per year.[39] There is no record of indigo after 1726[40] until 1775, when Raynal

(1798,4:338), quoting from ship registers, reported 1,147 quintaux loaded from Martinique, Saint Lucia, and Guadeloupe. Indigo again appears on price lists beginning in 1829 but after 1841 it is again not listed (Martinique, *Bulletin officiel,* vols. 2–15). However, the Marquis de Saint-Croix (1820,187) reported that indigo had been abandoned for a long time. The nineteenth-century production was small and confined to the wild indigo (*Indigofera suffruticosa*), which satisfied the dyers of the colony.

Ginger increased in export value slightly, but only an occasional figure gives evidence of continued production, and the rhizomes were probably derived from plantings in dooryard gardens and specially designated field plots. Experiments with cloves made by M. Gautier on Morne Bon Repos came to nothing because even though the trees produced well, France would not accept the harvest (Lavollée 1841,39).

Sugarcane surpassed tobacco in area in 1650, and thenceforth did not lose its primary position as a planted vegetation type until the second half of the twentieth century. Between 1764 and 1789 production areas covered more than 14,000 ha, except for 1778, when the total fell to 13,507 ha. In 1789 the area in sugarcane was 19,108 ha. After 1789 production of all crops declined substantially, although areas under sugarcane were reduced less, proportionally (Kervégant 1936b,98; Deerr 1949,233,235). For all years of record, sugarcane amounted to more than 40 percent, and in most years, more than 50 percent of the planted vegetation area, and it ranged from 10 percent to 20 percent of the total island area.[41]

Sugarcane fields generally were in the gently sloping lands and rolling hills at low elevations; they were located on a variety of soils and in regions having rainfall from 1,200 to 2,300 mm. These conditions obtain in the principal part of the low elevations of Martinique, which bears out that, where it was economically feasible to grow cane, it was grown. Sugarcane does not do well above 300 m in this latitude. Widely distributed, it was bordered by all types of vegetation except the montane types and the desert rock pavement. It occupied the principal portion of the cleared seasonal forest and much of the littoral and dry woodlands. Some dry bushlands as well as littoral meadow lands had been occupied by sugarcane. Some abandonment of land in previously dry woodland areas resulted in ultimate replacement by dry bushland.

Both Guinea grass and Para grass are grown in hayfields to supplement the spontaneous savannas and grasslands today. There are no contemporary accounts that refer to hayfields as such in the colonial plantation era. Lavollée (1841,51) suggested strongly that Guinea grass hayfields would be very productive. However, the remark in Duss (1897, 2:519) that Guinea grass was introduced for feed for the horses of the gendarmerie leads to the supposition that it was grown in pure populations properly called hayfields. Since, according to the proprietor of Hon. Paquemar,[42] the horses of the gendarmerie in the early nineteenth century were raised mostly on the estates of coastal Vauclin Commune, the hayfields may have been located there.

It is probable that the plant associations that constituted the different planted vegetation types found on the island today were assembled in virtually their present weed composition during this period. The introduction of exotics and the clearing of much of the land eventually under cultivation provided different habitats for the weedy introductions and native pioneer species that could be selected for association with each crop plant vegetation. There are enough references to weeding to provide documentary evidence supporting the inference that weedy growth was present in planted vegetation.

Common to all sugarcane fields with low-density planting and most ratoon fields would have been the introduced *Eleusine indica,* the several feather or finger grasses of the genus *Chloris,* the native *Centrosema virginianum,* the introduced *Calopogonium mucunoides* and *Crotalaria juncea,* the hairy *Euphorbia hirta,* and the several *Phyllantus* species, both natives and introduced. In dry areas, particularly on the montmorillonitic clay soils, were *Andropogon condensatus, Heteropogon contortus, Leptochloa virgata,* and *L. domingensis* and the introduced *Cynodon dactylon.* The woody *Lantana involucrata, Macroptilium lathyroides,* as well as herbaceous *Belchum pyramidatum* and *Ruellia tuberosa,* would have seeded into poorly maintained fields. In wet areas caused by either high rainfall or seasonally poor drainage the common weeds were the sedge *Fuirena umbellata, Lindernia microcalyx, Bacopa monnieri,* and *Aeschynomene sensitiva.*

In the coffee plantings and in the more open cacao plantations the native commelina or herbe grasse (*Commelina cayennensis*) may have been found, as were *Eleusine indica,* mentioned earlier, and *Heliotropium angiospermum.*

Weeds common in the sugarcane fields and in the coffee and cacao plantations that were capable of enduring the putatively greater traffic in the provisions and dooryard gardens were, for example, *Eleusine indica, Euphorbia hirta, Phyllantus* spp., and *Heliotropium angiospermum,* as well as the sedges *Fuirena umbellata* and *Abilgaardia monostachya.* All of these are common elements in the weed associations in modern times.

Clearing and abandonment were still part of the agricultural system. By the early nineteenth century local cultivation practices had led to substantial soil erosion and soil exhaustion. Sugarcane yields were 50 percent of those in Cuba (Lavollée 1841,iv). Depleted lands, unable to maintain production, were allowed to revert permanently to spontaneous vegetation; there were also the short-run fallows, continuously disturbed areas, and areas under regeneration. The composition of the vegetation varied with this cultural control and with physical factors as well. In addition to the shrub and orchard savannas vegetation types, large silverleaf balsa stands developed for the first time in abandoned mid-elevation gardens.[43]

Experimental clearings and plantings had quickly located the patches of good soil in the settled zones. These were gradually put into continuous cultivation until exhausted, when they, like the poorer soils, were allowed to revert to bushland and woodland and then were exploited for wood and charcoal. In the woodlands, and especially the bushlands, regeneration frequently involved stump sprouting, which produced a coppice-like thicket within a few years (Chanvalon 1763,28). As the population density increased, the rotation period was reduced and the coppice growth never completely regenerated to woodland.

Such pioneer, fast-growing trees of the woodland as *Bursera simaruba* and *Tabebuia pallida* greatly increased in proportional representation, as did *Croton* spp. and species of Rhamnaceae and Leguminosae. In the forest areas the coppicing habit was not so characteristic, and the stumps disintegrated so that regeneration by seedlings was more characteristic; such weedy tree species as *Geoffroea inermis, Ormosia monosperma, Ilex cuneifolia,* and the shrubby species *Citharexylum spinosum* and *Daphnopsis caribaea* increased in relative importance. In the rainforest areas the *Inga laurina, I. ingoides, Sterculia caribaea, Simaruba amara, Cecropia peltata,* and *Chimarrhis cymosa* are among the earliest trees to enter after the colonizing shrubs of the genus *Miconia, Bunchosia nitida, Boehmeria ramiflora, Piper,* and members of the Lauraceae family.

In the littoral woodlands, where permanent clearings among the trees had been made as in the *cinquante pas,* the stoloniferous *Cynodon dactylon* grew in patches. After continual abuse under sugarcane cultivation, even the better dry, shallow volcanics of the hill lands became eroded and colonized by pioneer species. Invading the abandoned lands were *Croton* and *Randia* spp., acacias, *Lantana* spp., and especially logwood and wild tamarind. Mahogany later proved an aggressive pioneer in the savannas of the Caribbean littoral and may have been so during the latter part of this period.

The Caribbean littoral and the southern peninsulas were particularly affected by this shift toward early successional stages and arrested succession. *Cassia emarginata* became a persistent invader of the pastured savannas of the southeast. The *Erythrina corallodendron, Adenanthera pavonia, Pithecellobium unguis-cati,* and the introduced *Tamarindus indica* became important tree species in the poorly maintained but grazed savannas.

Bush plantings of coffee and cotton were not soil conserving. The limestone-derived vertisols, especially those deteriorated by erosion, had their moisture-retaining capacity reduced, probably permanently. Associated with the transition to a drier, more xeric condition was an expansion of dry bushlands and the mixed cactus scrub in Diamant, Anses-d'Arlets, Sainte-Anne, and Marin communes.

The flatter, more humid landscapes of the Capesterre retained their fertility longer and remained in continuous cultivation throughout this period. Sugarcane lands once taken into cultivation were not abandoned but put alternately into grass fallow. The abandoned provisions gardens were invaded first by *Solanum* spp., *Piper* spp., and *Miconia* spp. Later the pioneer trees of the seasonal forest entered, forming the silverleaf-balsa stands. This association, whose members were part of the native flora, became areally significant with the increased clearing and abandonment of seasonal and rainforest lands. Left to regenerate further, they were ultimately replaced by more shade-demanding members of the seasonal forest and rainforest.

The destroyed cacao lands of Gros-Morne were in part claimed by the seral bush species of the rainforest, the crécré or *Miconia* spp., and so on, mentioned above. These successional communities were later more often cleared for the growing of provisions than were the forested lands proper.

During the colonial plantation era natural hazards caused much misery and occasionally resulted in the abandonment of crop lands. Renouard (1820,97) reported a flood that rose 30 to 40 feet around Morne de St. Croix in Saint Pierre, which would have damaged the coastal houses and gardens. Severe earthquakes with their attendant landslides struck Martinique in 1727, 1736, 1767, and 1839. An eruption of Mont Pelée in 1792 caused little damage. Two disastrous hurricanes in 1767 and 1839 and a very severe storm in 1788 caused much damage in the plant cover. It was, however, the catastrophic plague affecting the cacao trees in 1727 that had the longest lasting effect on the plant cover. Following immediately the earthquake of 1727, which had consumed most of the free capital for recovery, the plague wiped out whole plantations. Its apparently endemic presence deterred replanting.

Earthquake damage in the countryside primarily followed from landslides and from the mudflows resulting from the derangement of drainage systems. In each case some areas of secondary succession were set in train. Occasionally, fires in the industrial establishment reached cane fields, but this was rare. A disastrous earthquake in 1839 destroyed Fort-de-France (Garaud 1895,115). According to the American consul in St. Pierre, of the 10,000 buildings in that town, only seven were left undamaged.[44] The rebuilding of Fort-de-France put a severe strain on the economy, for much of the timber was imported. However, there was considerable use of native timber during the rebuilding. The previous, less disastrous earthquakes had also put strains on local lumber supplies.

The eruption of Mont Pelée in 1772 released some sulfurous gases and was accompanied by several minor earthquakes. Trees were burned and the surface about the vent was covered with sulfur. However, there was little concern over this event except on the part of inhabitants of the immediate vicinity (Lacroix 1904,30).

Hurricane damage, on the other hand, was severe. The damage caused by a high wind that hit the Rivière-Pilote area was similar to the damage of hurricane winds elsewhere, as described by Chanvalon (1763,153):

> Defoliation and death accompanied this storm. Its tracks are like those of a fire, everything disappears in its path and the change is so prompt and terrible . . . the plantation buildings are destroyed all at once, nothing of their debris remains. Trees perhaps older than the building of our Colony . . . are uprooted, lifted from the earth and deposited upside down;

> those which remain are whipped like fragile reeds; the plantings of all kinds are destroyed and turned over; the grass even is torn up and dried as if burned. Everywhere the eye can see crevasses and gulches are filled with branches and mud brought down by the rains.

In most cases the damage was repaired by the recovery of the vegetation and by replanting. But when the destruction was greater than the proprietor was able to finance, plantations fell into ruin, not to be exploited for many years. The plantation of M. De Pagerie (father of the Empress Josephine) in Trois-Ilets was destroyed in 1767 and never rebuilt (Deerr 1949,30). The hurricane of 1766, the most severe to that date, was reported by Raynal (1798,4:336): it "destroyed the provisions and harvests, rooted up the trees, and even overthrew the buildings. The destruction was so general, that scarce a few inhabitants remained able to administer comfort to so many unfortunate people, and to relieve so many miseries."

Such general calamity put terrible strains on the balance of trade and created severe deficits. The food scarcities caused by these hurricanes and by earthquakes resulted in famines and much suffering. These tended to influence the authorities to open the island ports to foodstuffs brought by foreign ships (McCloy 1966,20), which took in exchange sugars and molasses from the sugar plantations. During these catastrophes the strict regulations of the *pacte colonial* were modified.

Plant introductions had greatly enriched the local ecosystem membership (see map 25). Settlement activities tore at the plant-cover fabric, cutting up the large areas of native spontaneous vegetation into smaller units, creating many more edge effects and boundary communities. Settlement processes were supplemented by increased grazing pressure on the available grass and shrub lands. Significant erosion had occurred in both the humid central region as well as in the drier southern penin-

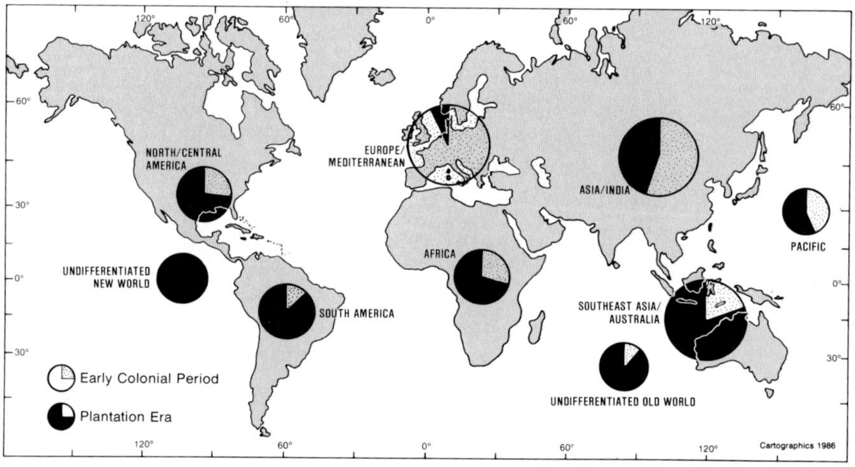

Map 25. Provenience and origins of mid-colonial plant introductions, 1689–1850. Source: Compiled by author.

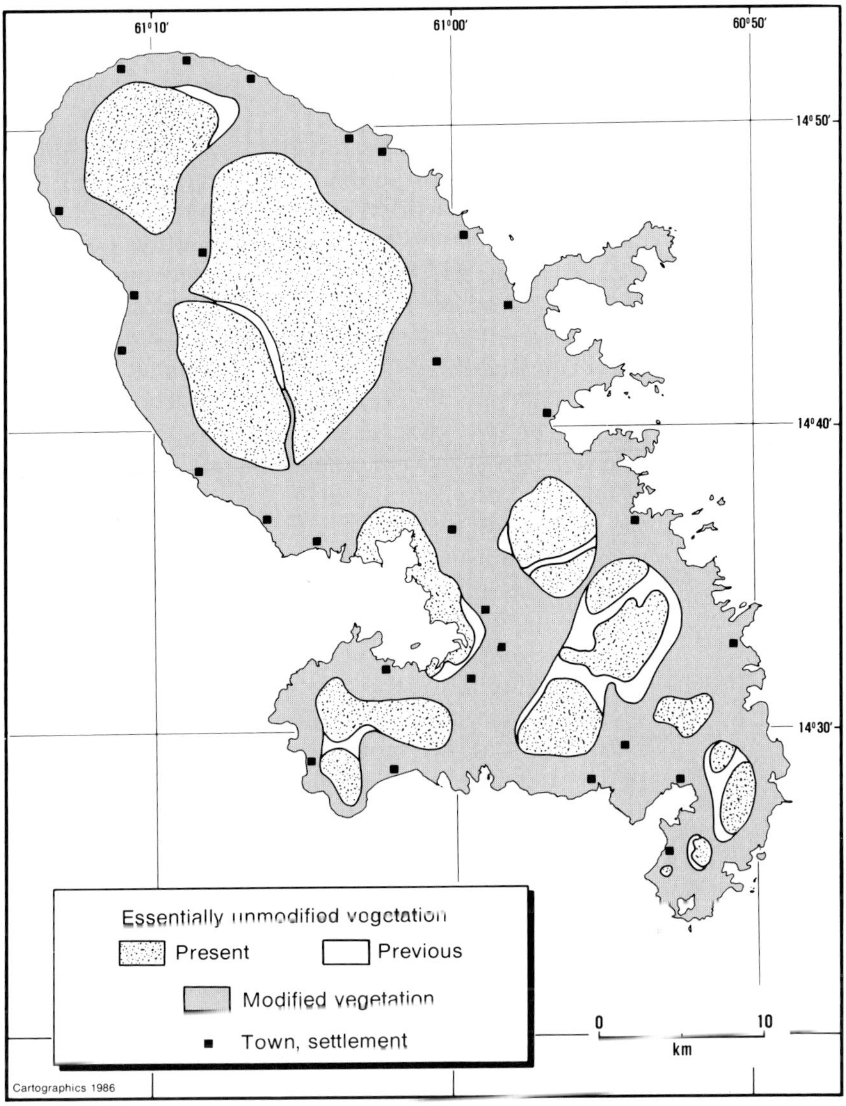

Map. 26
Zone of vegetation modified by human agency in the first half of the nineteenth century. Source: Compiled by author.

sulas, thus changing soil moisture regimes in those areas. The impact of human occupation was profound and, in parts of the island, irreversible.

Between two-thirds and three-fourths of the island plant cover had been cleared, burned, or otherwise destroyed beyond recognition by 1848. Only in higher elevations and on steep slopes were there significant areas in native, spontaneous vegetation (see map 26).

The Pelée massif was still capped in montane bushland, elfin woodland, and montane forest, and was fringed by a much-attenuated belt of rainforest on the Caribbean slopes. The Carbet Pitons and Morne Jacob massifs were also still dominated by those native vegetation types; they also had islands of native laural woodland. On the western flanks of the Pitons some midelevation reaches of the river valleys were in unexploited riverine facies of seasonal forest. Borders between the vegetation types were essentially unchanged by humans, except where the montane forest and rainforest had been destroyed along the Route de la Calebasse.

Some of the southern hills and mountains were still in native vegetation. In the western hill areas and in the Sainte Anne Peninsula, areas of reasonably intact dry woodland such as on Morne Bigot and Morne des Pères remained. In the central portion containing Mont Vauclin, Morne Pitault, and Morne Caraïbe there were several islands of seasonal forest in an otherwise domesticated landscape. The area of mangrove swamp woodland and littoral meadow about the Fort-de-France Bay had been slightly reduced by drainage activities. Swamp woodland in Marin, Vauclin, François, and Robert communes had been partially drained, but few changes in the position of the borders had taken place. The desert rock pavement was intact, but the cactus desert may have retreated somewhat in Vauclin Commune, as those areas were cleared for headquarters of plantations. On a larger scale there were islands of relatively undisturbed vegetation in areas awkward topographically for farming and in situations remote from roads.

Secondary spontaneous vegetation types were important in the midelevations of the Capesterre and on the steeper slopes of the lower elevations bordering the Caribbean. Crécré bushlands, *Miconia* thickets, silverleaf-balsa stands, and candlewood copses were all to be found scattered on abandoned lands in communes from Macouba to Gros-Morne. Dry bushlands of *Croton* and *Lantana* types had developed over the dry shallow volcanic stretches abandoned or in long-term bush fallow of the *côte-de-fer.* On the vertic montmorillonitic clay soils the leguminous bushlands had colonized in the dry south and Caribbean coastal borders. Fallow lands and pastures under poor grazing management had been invaded by unpalatable woody species and by those species with particularly attractive fruits whose seeds were unharmed by being passed through the digestive tract of the grazing animals. This degeneration of planted grasslands to savannas was most marked in the dry areas of the southeast. The abandoned bush plantations passed quickly to bushland and woodlands or forest.

Planted vegetation types dominated most of the land under 100 m and much of it below 300 m. A little over a third of the island had been censused as agricultural land. Some portions of the lower border of the rainforest on the windward coast had moved upslope as a result of its replacement by planted vegetation. Sugarcane fields were the most important planted vegetation type and had occupied most of the land suitable for the sugarcane varieties then available. Of the low annuals, only tobacco is listed as occupying any area in fields, a mere trace in terms of the island surface area. Cotton had been declining in the first half of the nineteenth century and remained on a few hectares in the Diamant, Vauclin, and Sainte-Marie communes. Coconuts had been planted along the beaches fronting the towns in small bays and valley mouths, and the old canafistula orchards with their spontaneous offspring continued to be harvested and the produce exported. The majority of these were on the north Caribbean coast. About 10 percent of the planted island surface was in provisions gardens.

Notes

1. News of this island was regularly reported in the commercial press of New England and the Carolinas. See the eighteenth-century editions of the Boston *Gazette* and similar newspapers.

2. The record from the slaughterhouse in St. Pierre gives a breakdown by source for creole cattle. For the year 1845 source areas were Puerto Rico, Angustura, and America [northern British colonies and Canada] (Rufz de Lavison 1850,2:79).

3. In the sister colony of Guadeloupe the governor wrote to the king that "the colonists having medium-sized holdings [*moyen colons*] who are the defense of the colony are gone completely" (Archives Nationales, C^8A^{19}). Again in the late 1720s, 250 colonists, their families, and their slaves immigrated to Saint Lucia and Saint Domingue because of the cacao crisis (Mémoire sur la culture du café aux îles du vent, June 28, 1731, cited in May 1930,261).

4. Not to be confused with the *savane à betail.* One might better translate *savanne* as "central gathering place." Many towns today have an open place, sometimes grassed, which is referred to as *La Savanne.* This is not restricted to Martinique. Port-au-Spain in Trinidad has such an open space.

5. May (1929,37) is of the opinion that the disposition of these buildings resembled certain farms of Normandy, whence came many of the early settlers to Martinique.

6. *Lizières* (a word used in the French West Indies meaning "boundary roads" or "fences") were usually only casually maintained and were a refuge for weedy plants.

7. A detailed royal ordinance collecting previous regulations into a systematic statement was published in 1685 as the *Code Noir.* Together with certain later supplementary enactments, it regulated the life of the slaves until emancipation. The regulatons and ordinances are published in Isambert et al. 1821–33,29:494–504 (cited in McCloy 1966,16).

   Under the *Code Noir* the slaveowner was required to provide a standard allotment per slave in provisions. This was subvented by the system referred to as the *pratique du samedi.* Saturday was the day that was left to the slave to cultivate a plot of soil. This "free" time put the responsibility for food and clothing on the laborer. This practice often led to malnutrition, substandard clothing, and high mortality rates among slaves.

8. This is a feature of all *coutumes* of France north of the Roman law areas (see Ferrière 1770).

9. A photograph copy of this manuscript map in twelve sheets is in the Archives Départementales de la Martinique, Fort-de-France. The photographic resolution is poor, but the representation of buildings is reasonably good.

10. In addition to the freed and free blacks, there were a small number of runaway slaves who were called locally *négres marrons.* These sought refuge in the still-uncleared seasonal forests of the higher mornes and in the heights of the mountainous country. How many there were in Martinique has not been established, although Roncière (1929–33,1:456) says that there were probably never more than 800 *marrons* in Martinique. McCloy (1966,38) is of the opinion that not more than 400 ever lived in Martinique at any one time. In contrast to Saint Domingue and Jamaica, which had many *marron* settlements, Martinique is a

small and easily penetrated island. In 1833, after the emancipation in the English islands, most *marrons* attempted to escape to Dominica and Saint Lucia.

11. A cadastral map made in 1705 as part of an intended census for Intendant Vaucresson shows new clearings in arpents fashion aligned along the roads in what had been forest (Archives du Ministère des Colonies, G^1 cited in May 1930,202).

12. Chanvalon (1763,5) is sharply critical of the local agriculturalists, doubtless because of his association with the *agronomes* and other scientists of the Jardin du Roi and the Royal Academy of Sciences, who were examining French agricultural practices from a scientific standpoint. See Duhamel's *Nouveau System* in his translation of J. Tull's work (Bourde 1953,5–6).

13. There was considerable argument in the British Leeward Islands as well about the use of the plow. See Peterkin 1790,27–35.

14. The increase in mills between 1769 and 1783 from 311 to 342 is distributed proportionally among the three kinds. The dry south, however, kept to the animal-driven mills, especially in Rivière-Pilote Commune (census data from Rufz de Lavison 1850,2:200,308).

15. Plantation records quoted in 1963 by M. Bally, proprietor of Lajus Estate, Carbet Commune.

16. Both of these canals are still in use and can be picked up on large-scale topographic maps of the region.

17. The first sugar refinery in the West Indies was installed by Marquis de Maintenon at the Chateau de la Montagne near St. Pierre (Deerr 1949,230). In spite of continuous controversy led by the French refining interests, who were supported by Colbert, there were eighteen refineries in operation in Martinique in 1695 when orders were finally given to destroy all refineries (Deerr 1949,232).

18. The production of clayed sugar (*blanc*) was introduced by Martin le Fol (Josa 1931,80). At the point in the manufacturing process that the crystallized sugar is to be separated from the molasses surrounding the crystals, the sugar mixture is put into conical molds, which are then capped by a layer of moist clay or earth. Rouen clay was most often used in Martinique. The moisture from the wetted clay percolates through the sugar, washing away the molasses. The crystals are left relatively free of impurities and comparatively white in color, hence the term "plantation white," which was often used for the product (Chapman and Chapman 1911,40).

19. See Pares 1932,68–72. Martinique changed hands between France and England a number of times. It was one of the key items in the negotiations over Canada.

20. A Sicilian invention of 1449, it was common throughout the Caribbean and considered the best arrangement, although horizontal and two-roller mills were in use (Deerr 1949,536).

21. Official concern with water supply and forests had existed in France since Colbert's Ordinance of 1669, which included provisions for planting forests (Bourde 1953,8,122). It is here suggested, however, that local pressure for timber protection was behind the official request (Mémoires de Phélipeaux Jan., 1763, cited in Rennard 1935b,171).

22. It is not known just when coal began to be used in Martinique; however, it and wood were used to the end of the nineteenth century (Rolph 1917,221). May (1930,151), during a discussion of early eighteenth-century commerce, poses the question of what sources other than France were available to the Martiniquans in times of distress. Among the items listed as being in short supply was

"coal for the sugar mills" (Mémoire sur le commerce des îles, 1708,C^8B^2, cited in May 1930,151). I have found no other reference to coal in that century.

23. Horses and cattle were equated, and sheep and goats were calculated at the ration of 6 to 1 of the large animals, after the custom of the grazing unit used by the U.S. Department of Agriculture.

24. Declaration of Aug. 3, 1722, cited in May (1930,75). This was a regularization of the custom followed during the early settlement period (see Du Tertre 1654,452 and Labat 1742,1:199).

25. Revert (1949a,192) reports a colony of Alsatians who at the end of the eighteenth century were unable to succeed in Champflore Basin because of the lack of roads.

26. Lightning generally occurs during heavy rainstorms, so it is thought that lightning-set fires were very few.

27. Surinam and the botanist Plumier went to the West Indies on royal orders in 1684 to make a floristic inventory and to determine what useful plants might be gathered and what might be introduced from elsewhere (Bégon [1689]1925, 42). In the manuscript collections of the archives of the Ministère des Colonies is a collection labeled F^3, Nos. 26–38, which concerns the history of Martinique. Another is F^3, No. 39, which concerns a geographic description of the island. Photocopies of these materials were graciously given to me by Bernard Petitjean-Roget and Emil Hayot.

28. M. Artaud, secretary of the Society of Agriculture and Rural Economy, was given charge of the garden. The purpose of this establishment was "to naturalize in Martinique the plants of the East Indies, principally the species, to furnish the botanical gardens of metropolitan France those which would be lacking there; to assemble according to a botanical system the indigenous plants; and to form a depot of medical plants for the use of the poor." By decree of the governor in 1840 the garden was made the responsibility of the Agricultural Society (Martinique, *Bulletin Agricole* 1918:14–16).

29. Creole cane takes fifteen months to mature, whereas Otaheite matures in twelve. Otaheite demands greater care in cultivation (Goveia 1965,127). Traditionally *Saccharum officinarum* has been the botanical name associated with sugarcane, but as Deerr (1949,12) explains, it is better to think of the sugarcane varieties as being derived from *S. officinarum* and hybrids with *S. spontaneum* and *S. robustum*.

30. Deerr 1949,20–21. Bourbon or Otaheite cane was taken from Tahiti to the island of Mauritius by Bougainville in 1768, from which it was distributed to the Caribbean (*Annales du Museé d'Histoire Naturelle* [Paris]8:479). Du Valdailly's letter of Nov. 24, 1842, to the minister contains an appended note listing the contents of five boxes of sugarcane pieces of five different varieties of Otaheite sugar.

31. It is also possible that it was introduced in the previous period since Bouton (1640,51) reports it as common in Saint Kitts, although Kervégant (1936a,83) considers that this name belongs to pois de sept ans (*Phaseolus lunatus*).

32. It is not known which of the yams were involved, but probably in early introduction *Dioscorea alata* was planted most.

33. The censuses through 1769 report plantains and manioc by individual plants or hills, and areas have been calculated for what was a dispersed population. Conversion factors were taken from Kervégant (1936a,97), who appears to have carefully examined the variations in planting patterns. His conversions tend to give a conservative translation to surface area, compared with those of Satineau (1928) and Revert (1949a).

34. The black mulberry rather than the white mulberry (*M. alba*) is common in modern Martiniquan gardens. The white mulberry is less hardy in the island environments.

35. In 1722, 14,000 quintaux of cacao was exported; in 1732, 454 quintaux was recorded (*Statistiques du Commerce,* cited in May 1930,96).

36. This was actually two plagues, according to Revert (1949a,341): an unidentified fungus and a leaf-eating caterpillar. The latter was described by Lavollée (1841, 36) in some detail.

37. "The forests on all sides are being cleared, plantations are being made in the heart of the island . . . even on the mountains themselves" (Chanvalon 1763, 17). Leblond (1813,35), who toured the Antilles and South America from 1767 to 1802, in describing the Caribbean coast of Martinique, mentioned the coffee plantations as occurring below the forest and above the sugarcane and being distinguished by color from them both in the landscape. Lavollée (1841,35) reported that up until fifteen years earlier, coffee was the second most important crop by value.

38. *Coffea liberica,* a robust, tall-growing, West African lowland species, grows without benefit of shade in Martinique.

39. *Statistique du Commerce,* cited in May 1930,90. There were nine indigo plantations in 1700, according to the Abbé Raynal (1798,4:326).

40. Chanvalon (1763,17) wrote, "Indigo had also been abandoned for a long time. Some had begun to cultivate it but the true indigo (*Indigofera tinctoria*) has been lost. The false indigo (*Indigofera suffruticosa*) is used. It grows everywhere abundantly especially in arid and worn-out soils." Although an exotic, as is true indigo, it has found a niche in the vegetation of Martinique and is quite naturalized on the drier parts of the island today.

41. These figures are based upon the modern figure for the area of Martinique of 109,600 ha. I make the assumption that areas in commercial crops were probably very accurately known, whereas the other areas such as forest were estimated, thus accounting for the much smaller totals for the island derived from the enumerations in the few complete censuses for the island made during the period.

42. André Asselin, plantation operator of lands in Vauclin Commune, possesses family records going back to the early nineteenth century.

43. This is assuming that the Amerindian settlements were all coastal. Two interior lithic sites reported by Allard and Mattioni may be the first indications that the pre-Columbian settlements got into the midelevations. If these people had practiced agriculture, then the *Cecropia-Ochroma* type would have merely reappeared. If, on the other hand, they were merely lumbering sites, as has been suggested, an occasional gap may have been filled by these trees but not a really significant area.

44. Paton Gay to Secretary of State, Feb. 7, 1839, Consular Dispatches St. Pierre, Martinique, vol. 2, U.S. National Archives, cited in Mitchell 1963.

7 The Attack on the Interior, 1848–1920

Emancipation of the slaves in 1848 led to the settlement of much of the remaining lands of the interior. The migration of freed slaves also stimulated the growth of towns. The labor shortages that emancipation created brought about the reintroduction of indentured labor, and some 90,000 *engagés* from Africa and Asia came to Martinique between 1852 and 1887. Prosperity had begun to collapse before emancipation, and the planters, hard pressed by the sugar-beet growers and the French sugar industrialists, barely survived the cash burden of hired labor. Relief in the form of credit issued by the formation of state-funded institutions with banking functions provided for the resale of bankrupt estates to Martiniquans. Land was thus kept in the hands of the local population rather than overseas investors. These newly constituted estates were then converted to a labor-saving central factory system.

The steam-powered *usines,* or sugar factories, of the new technology consumed enormous quantities of sugarcane, which was bought on contract from surrounding estates. Most minor industrialists with their small sugar works and distilleries became purely planters. The industrial labor saved was released for other activities elsewhere, and a large portion of the field labor supply was redundant during the six-month-long dead season. On sugar estates not producing rum and too far away from the sugar factories to bear transport costs, land was put into other commercial crops or allowed to revert to spontaneous vegetation.

Concern over timber and water supplies, which had been expressed earlier, was increased by the consequences of clearing and burning associated with the new upland settlements in the hill lands. The setting aside of the Forêt Domaniale and the organization of the Service des Forêts implemented a plan to maintain lumber and fuel supplies on a sustained yield basis and to provide for watershed conservation. The formation of the forest guard within the Service des Forêts made the enforcement of regulations possible.

The explosion of Mont Pelée in 1902 killed some 29,000 people and destroyed the town of St. Pierre. Most of the native white population kept town houses in the capital, and because an election was scheduled in a few days, many of the white elite were in town and were killed. Thenceforth, the white obligarchy lived on estates; urbanism on their part was a thing of the past. The permanent reconstitution of the intelligentsia and the commercial sector of the population resulted since mulattoes replaced most of the whites in these roles. The official resettlement of refugees in the interior mountain forest lands was the last of the concerted settlement attacks on the native vegetation.

Newer commercial relationships with France, although never releasing Martinique from a commitment to French markets, did permit some legal exchange with North America and Britain. World War I briefly stimulated the sugar industry, but the combined efforts of the beet and wine interests pressed for protection for the homeland and, in the end, Martinique production was controlled but also supported by a quota system. Events during this period led to the largest expansion of planted vegetation in the history of exploitation. The settlement process went to the locally recognized limits of cultivation. Continued new plant introductions further enriched the planted and weedy floras.

Social and Economic Developments

The Revolution of 1848 established the Second Republic, whose constitution provided for the direct representation of Martinique in the French National Assembly (Renard 1973,19). This voice, however, was not strong enough to determine policy for Martinique, and the French administrator in Martinique continued as the dominant power even in local government (Parry and Sherlock 1956,250).

As political structural relationships with France changed during this period, leading to incorporation and governmental structure similar to that of a mainland *département* (S. H. Roberts 1929,6–9), the administrative policy tended to minimize the special interests of Martinique agricultural producers and processors to the advantage of the industrial and agricultural interests of the mother country. However, changes in the political structure of France itself and the ideologies behind those changes did have significance for the land use and, consequently, the plant cover.

Up to 1848 the growth of towns had been slow, and only St. Pierre, Fort-de-France, Marin, and Trinité were of any consequential size. Land tenure was largely freehold by this time, the average units being over 50 ha for sugarcane exploitation, 10 ha or less for bush crops, and 3 ha or less for producers of provisions exclusively. Most towns were connected by road, and virtually all freight was hauled overland rather than by coasting vessels.

The newly freed slaves slowed work or refused to work, using the slogan "La Liberté." So difficult were these years for many landowners that they left Martinique entirely, some for New Orleans, others for the English islands. Indeed, in one area near Case-Pilote it was reported that "all the landowners . . . have left the colony" (Adelaide-Merlande 1981b, 4:52). However, new labor systems were constructed and indentured laborers were added to the thinned ranks of permanent workers. Nevertheless, the sugar estate remained the basic economic unit.

The Second Empire supported the interests of the white planters. The newly defined agricultural proletariat did not become politically active and maintained a position of inferiority. However, the reestablishment of republican government in 1870 with the Third Republic inspired a major uprising in the south of the island; it was put down effectively, but not until many plantation headquarters had been destroyed and fields had been burned (Pago 1981,229–30). Strikes elsewhere in the island, similarly inspired, compounded the economic problems of planters faced with increased competition for the sugar market (Mitchell 1963,22–23).

For most of the period markets for its produce, not production, was Martinique's chief economic problem. The concept of Martinique as an integral part of France prohibited its concluding separate trade treaties with other nations such as were being made by neighboring British islands (S. H. Roberts 1929,641). But by the last quarter of the nineteenth century, ships from several nations were trading in Martinique ports. Customs figures for 1879 show that about half of the imports were brought in French ships into the ports of Fort-de-France, St. Pierre, Trinité, and Marin. The number of foreign ships entering Martinique in 1879 and 1884 was greater than the number of French ships, but the tonnage was about the same (de Lanessan 1886b,746). However, one advantage to producers on the island came out of assimilation; that was a nonrestrictive migration of French West Indians to the mother country. The enlargement of a Caribbean element in Paris served eventually as a specialized market for Caribbean produce.

During the late nineteenth century the sugar industry generally became one of the most competitive industries in the world. Competition drove out the small and less efficient producer except where supported for political or strategic reasons. Martinique survived as a sugar producer only by virtue of its highly protected market (Jackson 1958, 108) and the special banks set up by the French government to issue credit (de Lanessan 1886b,972). Planters, reduced in number, became through intermarriage a tightly knit oligarchy that controlled the economy of the island. Sugar production rose and fell through the period, peaking in 1875 and again with the sugar crisis of 1884 (Guy 1900,168–69). Twentieth-century production to 1920 did not again reach these levels,[1] although sugarcane plantings continued to expand.

The American Civil War interrupted the main supply of cotton to world markets, and Martinique, like many other countries where it could be grown, temporarily expanded the area under cotton cultivation. However, with the reentry of American green-seed, mechanically ginned cotton, the island's hand-picked cotton lost its market. With the short-term decline of coffee production in Brazil, caused in part by emancipation of slaves in 1888 (Imbart de la Tour et al. 1900,358), Martinique coffee found an expanding market. In addition, markets for cocoa and chocolate in France opened up with changes in food habits and with newly developed industrial uses for cocoa butter. Sales of rum expanded with the shortages in brandy created by World War I, but at the close of the conflict rum was put on the quota system to satisfy the wine industry of France.

The emancipated field slave went either to unexploited, unoccupied hill lands or to the towns, since to work as a hired laborer at jobs previously handled in a slave capacity was regarded as demeaning and not befitting free status. Only in the Lamentin Plain did the former slaves remain to serve as hired labor. The labor shortage created by this exodus led to the legalization of a system of indenture, beginning in 1852. In the nineteenth century no Frenchmen could be induced to come to Martinique under labor contracts, but some 16,000 Africans did come, as well as more than 78,500 Asians, of whom 77,000 were Indians, 1,000 were Chinese, and 500 were Annamese (Leiris 1955,27; Revert 1949b,61).

The African indentured laborers were recruited in Africa under conditions not different from earlier slave trading, but they were indentured upon their arrival in the colony (Adelaide-Merlande 1981a,4:125–26). While blending into the black creole population, they were distinguished for a time by being referred to as *congos* or *nègres de Guinée.*[2] Trade in African *engagés* was supposed to end in 1860; however, the *Annuaire de la Martinique* for 1863–64 reported several thousand Africans introduced each year. Some, after they had worked off their indentures, found assimilation difficult and were repatriated; others formed enclaves in the settlement fabric. The district of Morne l'Africain in the northern hills of Diamant Commune and some hamlets in François Commune originated in this manner (Leiris 1955,28). For most, assimilation took place within a generation. In 1901 there were 5,341 Africans, whereas by 1910 only 838 were enumerated (Revert 1949a,242).

A few East Indians entered Martinique in 1852, but it was not until the contract with the Compagnie Général Maritime in 1853 that they were brought in any numbers. By 1857 the company was to bring in 15,000 Hindus (Revert 1949a,241). Between the years 1863 and 1869 the number of Indians averaged well over 8,000 (Adelaide-Merlande 1981a, 4:146). Some of these indentured servants came from French India, which became French Indochina, but the majority came from British India (Leiris 1955,28).

Most of the latter were from the south of India, Dravidian in stock and speaking Tamil. Although by pigmentation dark and thus, it was thought, easily assimilated, the East Indians remained largely apart from the rest of the population. They did not settle in villages but remained scattered on or near the rural properties they were imported to service. In Martinique today the *coolies* or *Indiens* are most conspicuous in the populations of Basse-Pointe, Macouba, and Saint-Pierre communes, being descendants of the 10,000 Indians who elected to remain after the system of indentures was stopped.[3]

Indentured workers came also from China proper and have become assimilated. A few surnames and some physical characteristics remain, but otherwise there is little to distinguish their descendants from other Martiniquans (Horowitz 1967,13). Prior to the introduction of indentured labor from Africa and Asia, some 300 contract laborers came to the island from the Madeira Islands (Adelaide-Merlande 1981a,4:

125). They were severely hit by disease in the first year, and no more laborers were sought from those islands. The population by the middle of the period was, thus, considerably more mixed than at the time of emancipation.[4]

The population growth for the island varied, with immigration providing only a part of the growth during the twenty-five years that indentured labor was legal (Martinique, Conseil Général, Dec., 1884). Natural increase, in contrast to the period under slavery, was high. In 1888 there were 158,660 persons counted in the official census. The 1902 eruption of Mont Pelée killed one-half of all the whites on the island and about 29,000 persons in all (Revert 1949a,242). Yet, the very high birth rates rebuilt the population to 229,839 by 1921.[5]

The former slave population quickly colonized the hill country not yet occupied by the freed slaves of the former period. This settlement activity represented an increase in small, individually owned subsistence properties. It was not a new kind of settlement in a radically different ecological zone but rather a filling in and extension upward of the limits of settlement. It was during this period that the limits of settlement, as they are currently understood were reached.[6] Delawarde (1937a,25) wrote: "There exists an altitudinal limit beyond which human settlement may not pass. It is a function of the zone of continuous cloud cover and of exposure to wind and varies, therefore, with respect to these. On the slopes of the Capesterre where the rainy Trade Winds blow, it is located between 350 and 450 m, where the cold is perceptible; while on the western slopes, man has installed himself up to 600 or 700 m in the lee of the winds and on a drier soil."

From the beginning, these squatter properties were small. With equal division among heirs the rule under the Code Napoléon, which merely reinforced the previous *coutume de Paris*, the majority of these properties became very quickly less than 1 ha. After a while, because such small properties were difficult to exploit, some were not physically divided with living fences but were left as a single piece to be exploited by one of the heirs, with the proceeds to be divided equitably at harvest time.[7] Occasionally the small parcels were sold to one of the heirs.

The inheritance pattern became greatly confused by the third generation. Landmarks no longer were in existence, and boundaries were imprecise. Family lands became the cause of much litigation on the part of the small proprietors. Revert (1949a,265) quotes as a sample of cadastral confusion a deposition in the case of a transfer of property: "area indeterminate, boundaries uncertain, the seller declares that he does not have title to communicate, the notary declines all responsibility on the subject of this real estate which is the object of the present act." Indeed, the cadastral confusion by the twentieth century was so great that roughly twice as much land was owned by deed or common acceptance as is contained in the island.[8] Although this condition did not hold for all lands at this time, it was true of the majority of the smallest holdings.

The heights in the south were rapidly colonized to their summits.

The surface of Mont Vauclin was completely divided and settled, though not entirely cleared. The two branches of Pilote River served as access routes into the Morne Vert of the southeastern hill country, and the valley flanks were completely taken up in holdings. Unoccupied land in the northern half of the commune of Sainte-Luce was colonized. The upland windward north was colonized from the sugar estates in the lowland. Gros-Morne Commune entered a second cycle of exploitation, this time in small subsistence properties. The *étages* of Trinité, Sainte-Marie, and Lorrain were established at this time. The settlement of the *hauteurs* such as the Hauteurs Bourdon above Macouba and those above Basse-Pointe, as divisions of old estate lands not suitable for sugarcane, began in this way.

The release of the slaves to colonize the interior lands permitted the reconstitution of a peasantry or yeoman-farmer class and a consequent economic disengagement of a large proportion of the island population from the plantation world (Mintz 1967,93). Small landholders had had a place in the land tenure system, but after emancipation this class grew in importance. This was furthered by the government-instituted policy of offering state land divided into parcels of less than 6 ha as settlement plots to the laborers who were pushed out by the expansion of the central factory system (Martinique *Conseil Général Mémoires,* 1868). In 1860 the government assigned the last of the concessions near Morne-Rouge in the Champflore Basin[9] to laborers displaced by the setting up of the sugar factories on the eastern slope.

During the insurrection of September, 1870, thirty-four estates were burned in Vauclin, Marin, Sainte-Anne, Rivière-Salée, Sainte-Luce, and Saint-Esprit (Rennard 1929a,106). The majority of these were not rebuilt. Some of those estates reverted to the Crédit Foncier Colonial, as did others in financial distress elsewhere on the island. These repossessed lands were either sold to local capitalists or, failing to attract a big buyer, were divided up into small parcels, collectively referred to as a *quartier,* with the name of the former sugar estate attached.[10] Other lands reverted to the colony and were redistributed in small lots.[11] The depression of 1880 caused an additional number of estates to fail and be redistributed in this fashion (Adelaide-Merlande 1981c,4:323–27).

Out of the tragedy of the 1902 eruption came government resettlement schemes. The settlements were to be on "the domaine imperfectly utilized" according to Governor Lamarie. It was the policy to reestablish these people around important towns or sugar factories where labor could be used. However, the government made an effort to relocate persons rather than merely to offer land. Some consideration was given to matching the old physical environments to the new environments in these relocations.

These rural resettlement schemes were countered by the rural-urban migration that was beginning. After 1830, when manumissions became more and more frequent, there was a slow migration to towns. Emancipation accelerated a sizable influx of newcomers to St. Pierre and Fort-de-France. Housing was scarce and slums developed on the margins, especially in Fort-de-France and along the waterfront on unoccupied *cinquante pas* lands of the coastal strip.

The road network was enlarged to connect not only towns but also the growing number of rural districts of high settlement density. The really substantial road-building achievement was the extension of the road called La Trace in 1850, connecting Fort-de-France with Fond-St.-Denis via Colson and Alma (de Lanessan 1886b,372). The amount of improved road had risen to 927 km in 1886.[12] Maintenance, however, remained a problem. By 1901 Lasaulce (1901,3) reported that there were 617 km in the colonial system and many roads had been permitted to fall into ruin, including the northern road from St. Pierre to Grand' Rivière, which had still been passable in the 1890s (Geoffroy 1897,n.p.).

Railroads were restricted to the large plantations, although a number of suggestions for a public railroad net were put forward.[13] These short-run railroads permitted the transport of sugarcane by steam engine, thus eliminating need for some of the draft animals. Unfortunately, the steam engine threw sparks, which caused fires in the sugarcane fields or drew the blame for deliberately set fires. These railroads were most important in the Lamentin Plain and on the windward coast.

Motorboat connections between west coast and south coast towns were available on a regular basis. Motor launches connected St. Pierre with Case-Pilote, Fort-de-France, and the south coast ports of Anses-d'Arlets, Diamant, Poirier on Point Borgnessee just east of Sainte-Luce, Marin, and Sainte-Anne. Fort-de-France was connected by different lines to Lamentin and to Petit-Bourg. These offered regular scheduled runs during the week (Geoffroy 1897,appendix).

The application of the steam engine to the process of sugar refining was a major technological revolution. The first steam-operated mill in the Caribbean was in Cuba in 1819. The effect on Martinique was to increase the competition for markets (Deerr 1949,235). By 1860 it was recognized that a central factory system was essential to maintaining a competitive position in the sugar market. The first steam-driven plant was installed sometime between 1804 and 1815 in François at Habitation Vapeur (Buffin 1981,4:190), but it was John Thorp who really started the steam plant at Fort-de-France in 1845 (*Sugar,* 1888–92,2: 435). The sugarcane was collected coastwise and hauled to the sugar factory in barges drawn by steamboats (*un bateau à vapeur de la force*), according to Deerr (1949,235), this being the first application of steam to draying in Martinique. Such transport reemphasized coastwise traffic in internal trade.

John Thorp established a large sugar factory at Point Simon in 1845, the Usine Sinson, and on the windward coast in François Commune in 1846,[14] and a third at Larienty in Lamentin Commune (Buffin, 1981,4:185). By 1884 there were sugar factories at Galion in Trinité Commune, Soudon in Lamentin Commune, at Dillon in Fort-de-France Commune, at Petit-Bourg in Ducos (the old Tour-au-Chat) Commune, at Rivière-Salée, at Robert, and at Trinitè. Four sugar factories were built in 1884 at Basse-Pointe, Vivè, Lorrain, and Ste. Marie. Eventually twenty-one usines were established: twelve from Basse-Pointe to Vauclin, two on the south coast, six in the Lamentin–Rivière-Salée, and one at St. Pierre (B. Petitjean-Roget 1981,218).

The transformation of sugar making into a concentrated industrial activity would have been impossible without capital furnished by the land-credit banks. Small merchant banks were not in a position to provide the large sums necessary. The Société de Crédit Colonial was established in 1860, later to be transformed into Crédit Foncier Colonial (Buffin 1981,4:181; B. Petitjean-Roget 1981,210). Money could be lent for the construction of steam-driven mills and for the renovation of existing machinery such as the replacement of the old vertical, three-roller mills by the horizontal mills, which were more easily powered by steam engines (W. Barrett 1966,156). The growth of the large central factories was paralleled by increasing mechanization of field operations with plows, important in all but remote sugar plantations.

The effect, besides that of concentrating the industrial processing to a few large sugar factories,[15] was to lead to substantial indebtedness on the part of many planters, who were finally eliminated in the depression of 1880. During the period 1883–90 most sugar factories were divested of their agricultural lands upon the advice of the Crédit Foncier Colonial (B. Petitjean-Roget 1981,4:218).

The use of green bagasse as fuel was made possible by inventions in the sugar region of Louisiana. In 1854 Moses Thompson developed a new furnace for burning wet bagasse. It was this innovation that made practical the modification of machinery in the Antilles (Sitterson 1953, 153). All these modifications in technology facilitated the concentration of the industrial activity and helped separate the agricultural from the industrial activities as well as set the course for the concentration of sugarcane production areas. At about the same time the *pacte colonial* was abandoned for the *loi de libre échange,* and restrictions on the manufacture of high-quality white sugar were lifted.

The use of chemical fertilizers became more widely known during the last half of the nineteenth century largely through the efforts of the agricultural societies. The fertilizers were used primarily on sugar estates, which had the capital to invest. The evident exhaustion of lowland soils throughout the island but particularly in the southern montmorillonitic clay soils led to the recommendation that fertilizers be used more and more (O. Hayot [1881]1938,69). At the turn of the century there was a fertilizer factory in St. Pierre employing people "according to the market demand" (France, *Annuaire Colonial* 1902,54).

Other innovations favored the expansion of pineapple and banana growing and thus of land in the cultivation of those crops. The first pineapple-canning factory was built in Denel in Gros-Morne Commune sometime before World War I (Sylvestre and Lotaut 1947,2:275). The building of ships that could carry bananas was urged in 1910 on the Compagnie Transatlantique, which held the monopoly on freight carriage to France (*Journal d'Agriculture Tropical* 1910,103; 1911,260). In 1919 plans were laid to build a special fleet of banana boats.

Changes in Plant Geography

Modification of the plant cover continued during the final attack on the island interior. Species membership changes resulting from selective cutting and habitat destruction reduced the populations of some plant species to dangerously low levels. Several species were reported as extinct by the time Duss prepared his *Flore phanérogamique des Antilles françaises*. These species losses were offset by the introduction of new crop species, exotic pasture grasses, and ornamentals as well as by the reintroduction of weedy plants, especially the woody legumes. The retreat of spontaneous vegetation, as agricultural exploitation penetrated deep into the previously uncultivated central massif, eliminated virtually all undisturbed seasonal forests and the lower margins of the rainforest, simplifying the vegetation mosaic below 300 m. The small plots of spontaneous forest were reduced to areas smaller than could accommodate the species complexity, and the relative importance of species varied from place to place as an accident of clearing. Modifications in habitat continued as more open herbaceous cultivated cropland replaced forest lands; erosion destroyed previously fertile and water-retention areas. The extraterritorial dimension to decision making continued, and Martinique suffered from local and world market disruptions.

Not until the end of the period was a rational system for the exploitation of woody vegetation implemented. The selective cutting of lumber, fuel, and dyewoods was conducted on a purely individualistic basis and without government subsidy. Most construction in Martinique during the nineteenth and early twentieth centuries continued to be of wood. According to one estimate roughly nineteen out of twenty of the town houses, business establishments, warehouses, and the like were built of wood imported from North America, valued particularly because of its superior durability (Lasaulce 1901,20). On the other hand, much of the construction in the more rural areas was done with local lumber (Lasaulce 1901,26). Consumption of such local lumber for construction purposes was estimated at the turn of the century to be 600 cubic meters per year (Lasaulce 1901,23). Various industrial uses of wood (e.g., for furniture and cabinet work and wagon and wheel manufacture) were satisfied from the local forests, as were most of the demands for poles, stanchions, and firewood. Staves for barrels continued to be imported from North America. Stave wood was duty exempt at the time (France, *Annuaire Colonial* 1902,63). Small furniture and cabinetmaking establishments continued to use local hardwoods, but these crafts gradually lost ground during this period, and imports took the place of the local products (Martinique, *Almanach Martinique illustré* 1908).

The yearly volume of cabinet and furniture wood consumed in the colony by the end of the nineteenth century amounted to 140 cubic meters, of which about one-half was derived from local sources. Poles, sticks, and firewood amounted to about 130 cubic meters per year, all from local timber sources. As the trackage of railroads operated by the sugar factories expanded, the consumption of wood in the form of ties increased. In 1902 there were 182 km of track on the plantation and factory lands. Assuming an average service life of five years per tie, the annual consumption was estimated to be 44,000 ties, or 2,200 cubic meters (Lasaulce 1901,20).

Charcoal was the only domestic fuel of consequence, although firewood was used for certain home industries such as the making of manioc flour. Most of the charcoal was produced in the colony, and all of it during this period was consumed on the island.[16] *Homalium racemosum, Chrysophyllum argenteum, Sideroxylon foetidissimum, Manilkara bidentata, Pouteria hahniana, Ceiba pentandra, Simaruba amara, Cedrela mexicana,* and even *Dacryodes excelsa* were used to make charcoal (Hearn 1890,291). As the population rose, the charcoal consumption increased accordingly. In 1901–1904, 993 tons or 34,673 cubic meters were consumed in the colony (Lasaulce 1901,19). According to Lasaulce (1901, 20): "But today the island exports no more hardwoods: it has even been found necessary to import such from neighboring islands; and yet the destruction of the forests still goes on. The domestic fabrication of charcoal from forest trees has been estimated at 1,400,000 hectolitres per annum. Primitive forest still covers the island to the extent of 21.37 percent; but to find precious woods one must go into the interior."

Industrial fuels were coal, wood, and bagasse. The use of coal by the sugar factories is reported by Rolph (1917,241) for the latter part of the nineteenth century and the twentieth century, but one coal merchant of St. Pierre appears to have handled the requirements of the island (*Annuaire Colonial* 1902,93). Distribution would have been fairly simple with the coastal shipping then developed out of St. Pierre. Systems using green bagasse were incorporated into the sugar factories built in the second half of the nineteenth century (Josa 1931,142), but the energy conversion was not sufficient to handle the requirements of the industry, and wood continued to be used to fuel the sugar-evaporating process (Rolph 1917,241). The use of wood was extensive enough that comments were made in the contemporary literature about the stripping of woody vegetation from the mornes around the sugar factories, hence, the creation of bushlands or savanna from woodlands and forests.

The dyewood of commercial significance during this period was logwood. On average, 1 million kg of the heartwood and roots of this tree were exported annually to France throughout the nineteenth century. Individual years fell considerably below this level, as in 1898 when only 728,668 kg were exported, or rose above it, as in 1895 when 2,234,000 kg were sent to France (Imbart de la Tour et al. 1900,326). Lasaulce (1901,21,28) estimated that 1,928 metric tons of logwood had been taken from the southern Martinique forests on the dry, eroded vertisols.

As has every forester who has commented on the problem, Lasaulce (1901,23) insisted that instead of being an importer of wood, Martinique should export a considerable quantity. Part of the problem lay in the fact that there was in France no large, central lumber market for tropical woods, although some tropical woods went through the ports of Marseilles, Le Havre, and Bordeaux[17] and, according to reports,

Table 24. Average annual consumption of wood products

Derived	Commodity	Amount in m³	
Local products	Charcoal	34,673	
	Construction timbers	600	
	Railroad timbers	2,200	
	Cabinet and furniture woods, wheel and carriage woods	70	
	Logwood	3,680	
	Other[a]	2,000	
			43,223
Imported products	Construction timbers	11,229	
	Industrial woods	3,821	
	Other	67	
			15,117[b]
			58,340[c]

SOURCE: Lasaulce 1901, 23.
[a] Includes faggots, firewood, stanchions, poles, etc.
[b] The original work had the figure 16,117.
[c] The original work had the figure 59,340.

a considerable quantity of West African woods were distributed from Hamburg (Imbart de la Tour et al. 1900,323–24). Many of the species in Martinique forests were recognized at this time as being excellent for inlay cabinetry, parquet flooring, and paneling (de Lanessan, 1886a,128). But there was no system organized for the export or even the extraction of these widely spaced forest elements. Some of these less common hardwoods were used by local craftsmen, and the "colonial furniture" of Martinique may have many woods employed in the manufacture of a single piece after the Empire and Regency fashions.

Lasaulce (1901,21) estimated the average annual productivity of Martinique forests to be 6 cubic meters per hectare on an area in forest of some 22,000 ha, making a total of about 132,000 cubic meters per year. The annual consumption was less than half of this production, yet, even so, a significant portion of that consumed was imported. His estimates are arranged in table 24.

From the first year of the period official concern for the conservation of useful woods was expressed in government decrees. Such a decree by the governor in 1848 set forth regulations for the exploitation of construction timbers,[18] but there is no evidence that it was ever enforced (Lasaulce 1901,32). In 1852 regulations for the selling of charcoal were issued, but these apparently were not enforced (Arrêté du Gouverneur, July 7, 1852, Martinique, *Bulletin officiel*). In the same year the Service des Ponts et Chaussées was given the responsibility of guarding the waters and forests of the colony, but enforcement was well-nigh impossible. There had been no delimitation of the woodlands and forests, and the ownership of forest land by different interested parties—the state, the colony, and private individuals—though legally established was not surveyed.[19]

In 1869 water conservation and erosion problems caused by the active clearing of the interior forests were debated in the Conseil Général. Rivers were choking with sediment eroded from deforested slopes. Fort-de-France Bay received so much sediment that the interior bays were filling in. A commission was charged with drawing up a forestry code based on that of France to regulate exploitation and to provide for an administrative structure to enforce the regulations (Martinique, *Bulletin officiel* 1869). In 1896 another commission was appointed. Lasaulce, inspector of water and forests, was charged by the Ministère des Colonies to prepare a proposal for a forestry program and a budget for implementing it. In 1901 he published his *Rapport*. By the end of the period the Forêt Domaniale had been constituted and forest guards appointed (Martinique, Préfecture, *La Martinique* 1963,48).

Rainforest hardwoods were not exploited to the extent that they might have been because of the lack of access roads. Many of the tree species that had been first culled from the seasonal forest were untouched in the rainforests because of inaccessibility even at the end of the period. According to Lafcadio Hearn (1890,278), "There are plenty of *gommiers* (*Dacryodes excelsa*) still, but the difficulty of transporting them to the shore has latterly [1880s] caused a demand for the *gommiers* of Dominica." Those trees best regarded for construction from the rainforest and culled from it were *Talauma dodecapetala, Ormosia monosperma, Nectandra saguinea, Sloanea* spp., *Pouteria hahniana, Homalium racemosum, Chimarrhis cymosa,* or *Mimusops balata*. The *Prunus dussii* and *Talauma dodecapetala* were important for wagon axles and wherever a hard, dense wood was needed. *Homalium racemosum* was considered almost indestructible and was used for mill machinery and where construction was to be kept under water. *Guatteria caribaea* was considered a superior canoe wood. *Ocotea leucoxylon,* a very common tree, was useful for flooring though not otherwise a valuable tree (de Lanessan 1886a,158). As late as the 1880s the trunks of tree ferns were still used to build small huts in the Morne-Rouge area (Hearn 1890,191). Of this list only *Chimarrhis cymosa* was mentioned as one of principal construction woods in the *Annuaire Colonial 1902* (France, *Annuaire Colonial* 1902,51), probably because the others had become rare. Lasaulce (1901,6–18) listed seventy-six trees in twenty-seven families. These are listed in appendix C.

The best regarded of the construction timbers of the seasonal forests appears to have been *Manilkara bidentata,* for it is mentioned by all writers as a superior wood. Hearn (1890,277) writes that it was rapidly disappearing in the 1880s. By 1902 it had become rare (France, *Annuaire Colonial 1902* 1902,51). *Hymenaea courbaril,* which was also to be found in the rainforest and the more humid seasonal forest, was being aggressively culled in the 1880s (Hearn 1890,277). The courbaril had become rare in 1901 (Lasaulce 1901,9). Also reported rare was *Side-*

roxylon foetidissimum (Duss 1897,386), which had still been common in Prêcheur, Rivière-Pilote, and Lamentin in the late nineteenth century (de Lanessan 1886a,162). *Cedrela mexicana, Geoffroea inermis, Lonchocarpus* spp., and *Pimenta racemosa* were all important in construction and in the making of wagons or machinery (de Lanessan 1886a; Duss 1897,1:165), but *Sapindus saponaria* was becoming rare (Lasaulce 1901,213). Lignum vitae, *Guaiacum officinale,* which had been so prized in the previous period, had become scarce outside of cultivation (Duss 1897,1:137).

The seasonal forest of Sainte-Luce, which had remained as a sizable "island" throughout the nineteenth century, was described with rapture by the traveler Garaud (1895,248–53), indicating that much of the vegetational physiognomy was still intact, but Lasaulce (1901,26) estimated that in all of the southeastern part of the island, including Mont Vauclin, not more than 500 ha remained and much of that was severely damaged. The destruction of the forests must have been intense between 1886, when de Lanessan compiled his catalog,[20] and 1901, when Lasaulce wrote his report.

The cabbages of *Roystonea oleracea* and *R. regia* were in high demand in St. Pierre as a fresh salad vegetable. There were cabbage hunters, *chasseurs-de-choux,* who lived by furnishing the hearts of young palms to the city markets (Hearn 1890,279). Although the mature palm tree may grow to more than 30 m in height (Duss 1897,2:488), Hearn noted that by the 1880s "few palms reached a height of even 12 feet." Still, Duss (1897,2:488) could write that the cabbage hunters climbed regularly into the high *grand bois* to return laden with cabbages that sold for a good price.

The railroad ties to supply the tracks of the sugar factories were made from wood cut in Diamant and Sainte-Luce dry woodland (France, *Annuaire Colonial 1902* 1902,51). *Tabebuia pallida* continued to be regarded as an excellent wagon wood and lumber tree (de Lanessan 1886a, 167). It was listed as one of eight important construction woods in 1902 (France, *Annuaire Colonial 1902* 1902,51). *Bursera simaruba* was used in the making of boats and oars and paddles (de Lanessan 1886a,161). *Melicoccus bijugatus,* also found in the seasonal forest, was still uncommon (Lasaulce 1901,13). *Crescentia cucurbitina,* which had been in the forests of La Régale and Trois-Ilets, was, at the time of Duss (1897,1: 119), represented only in the gardens at St. Pierre. *Catalpa longissima,* previously abundant in the low elevations of Martinique (de Lanessan 1886a,167), had been virtually eliminated because of its demand in construction work (Duss 1897,2:419). *Tecoma leucoxylon,* a good furniture wood, had been also virtually eliminated from the seasonal forests and dry woodlands (Duss 1897,2:420). *Vitex divaricata* of the seasonal forest had become less abundant because of its high demand among wheelwrights and carpenters (Duss 1897,2:469).

Rhizophora mangle was highly prized as a wood to fire bakers' ovens (France, *Annuaire Colonial 1902* 1902,51). Convenient to the rapidly growing capital, the mangrove swamps of Fort-de-France Bay were culled of this and other tree species for the making of domestic charcoal fuel.

Most of the logwood exported was drawn from the leguminous thorn scrub bushlands of the communes of Saint-Esprit and Rivière-Pilote (de Lanesson, 1886a,154), presumably those trees colonizing the abandoned sugarcane and planted bushland plantations. Because these were probably coppiced, the exploitation might be considered a kind of harvest. Since logwood had also been a favorite living fence, some quantities from the fences may have been a cash crop for the peasantry. Much of the charcoal manufactured during this period came out of the thorn scrub as well as the *Croton, Lantana,* and *Randia* bushlands. The introduced mango, which was occasionally emergent from these bushlands as well as present in the savannas of the period, was considered a superior charcoal wood (Lasaulce 1901,18).

As the small landholders moved into the dry lands of the south, they exploited the mixed cactus scrub for the inflorescent stalks of karatas or langue boeuf (*Furcraea tuberosa*), which by the time of Duss (1897,559) was also in use as hedgerows in dry Carbet, Case-Navire, Diamant, Anses-d'Arlets, Caravelle, and, curiously, in the interior commune, Gros-Morne. The wood from the savanna *Acacia muricata* was in demand, as was that of *Leucaena glauca,* for inlay cabinet work (de Lanessan 1886a,156). *Psidium guajava* has a strong, durable wood that was used in machinery and in light construction (de Lanessan 1886a,165).

Travel accounts during this period have remarks about grazing animals. By and large these animals were in pastures made by clearing and burning in the middle and mountain belts, although pastures were being established in the southeastern part of the island. These pastures were largely orchard savanna. If early a grassland, the land quickly became covered in scrub savannas without the constant grubbing out of the woody invaders. After the refugees from St. Pierre introduced cattle into the bushland and dry woodlands of the south, many cattle became feral in the area around Diamant. Hunting expeditions for wild cattle were part of the folk memory in Diamant Commune in 1963, although the cattle had disappeared.

The clearing of virgin and second-growth spontaneous vegetation during this period caused the retreat of spontaneous vegetation to its greatest extent. All areas ecologically suitable for agricultural production had been brought into cultivation at one time or another, except in the middle and upper reaches of the valley of the Lorrain River. Even there, some clearings were made for bush plantations (Lasaulce 1901, 26–27). The products went out on mules or on the heads of porters on the step mule trails. Woodland clearings made in the twentieth century for crop plantings in the southern hill country frequently revealed foundations of *établissements* that had been abandoned during the economically troubled period of the last third of the nineteenth century (Delawarde 1937a,39). There was clearing in the rainforest and montane forest of the Ajoupa-Bouillon and Morne-Rouge communes for pasture to accommodate the increasing cattle herds (de Lanessan 1886b,744). Even though the land under cultivation decreased after 1895 (Kervégant 1936b,100), clearing continued to bring new land into production. According to Lasaulce (1901,30), the worst attack on the virgin forests was clearing for planting: "But clearing in the present period is being pursued in a blind rage. At all points of the forest, in the interior as on the perimeter, one finds clearing, small, large, in all dimensions and, something extraordinary which one cannot see without being profoundly saddened, it seems the preference is for the steepest slopes where all remunerative agriculture is impossible." He further estimated that some 400 ha per year were being clear-cut in the forests for agriculture or other exploitation. This is a rate that could not have lasted for long.

In 1894, according to the *Annuaire de la Martinique,* forests occupied 24,695 ha. By 1901 the remaining forested areas of the island were estimated to have an area of about 22,000 ha. About 27 percent of the northern part of the island was still forested. Forests had virtually disappeared from the littoral, and there were no sizable plots below the 400-m contour (Lasaulce 1901,25). Only about 4 percent of the surface remained in forest in the southern part of the island (Lasaulce 1901,26). The resettlement schemes, except for those at Colson and Alma in the rainforest, were carried out on land previously in cultivation, although the spontaneous vegetation had recolonized many plantations, and bushlands or young forests had to be cleared by the new settlers (see map 27).

Clearing techniques had not improved over those at the time of Labat. "It is not rare to find in the interior massifs a battery of trees cut a meter above the earth whose corpses lie unused on the ground" (Lasaulce 1901,29). Efforts to control erosion, which Chanvalon (1763,28) had suggested, do not seem to have been taken seriously until after 1880 under the supervision of O. Hayot as director of the Service de l'Agriculture (Martinique, *Bulletin Agricole* 1936,107). Wooded land in the upland of Diamant Commune was reported cut by a new landowner in order to pay for the recently purchased landholding (Delawarde 1937a,39). From 1913 on, after all danger from the volcano appeared to be over, some evacuees returned to their original holdings near Mont Pelée, taking up the cultivation of the raw, new soils. These were few, however; dense human settlement on the Carribean slopes of Mont Pelée was a thing of the past.

New plants were continuously introduced to Martinique. The directors of the Jardin des Plantes were energetic collectors of native plants and exotics. The gardens at Tivoli, opened in 1869 in the rainforest, permitted the acclimatization of plants demanding a mountain-belt environment.[21] Agriculturalists also were influential in the exchange of better varieties of domestic species, especially of the commercial crops. Visiting civil servants and botanically inclined travelers also aided exchange. The indentured laborers, particularly those from India, brought with them useful plants and weeds. The Hindus also brought a taste for rice, a food habit that was quickly transmitted to the rest of the population (Parry and Sherlock 1956,203). This use of rice changed somewhat

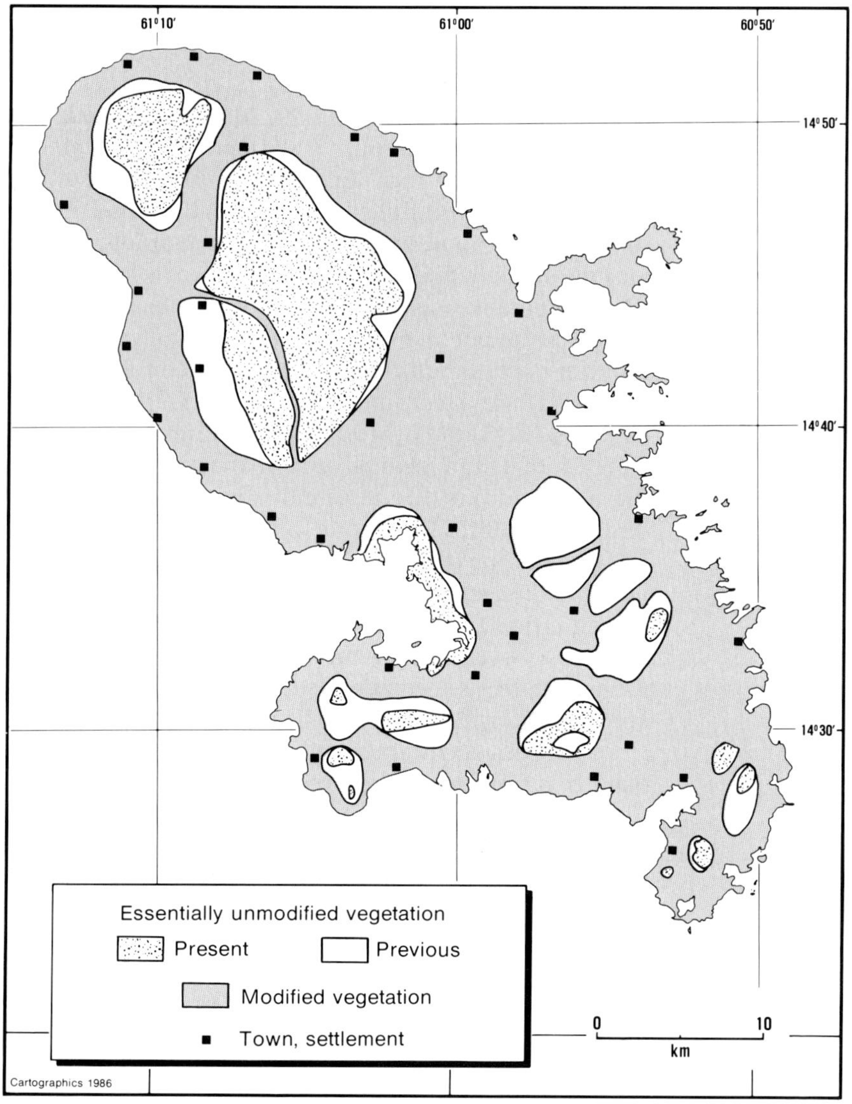

Map 27. Zone of vegetation modified by human agency in the late nineteenth century. Source: Compiled by author.

the traditional dependence on the *gros légumes* (e.g., manioc, yams, sweet potatoes) and led to the importation of large quantities of rice because the colony did not develop lowland rice cultivation.

In the Archives d'Outre-Mer there are numerous references to the introduction of plants to Martinique. Sometimes these were the result of direct requests; sometimes the introduction seemed to be part of a system of exchange among the colonies. In a letter of June 27, 1862, three eucalypts were discussed: *Eucalyptus globulus, E. gigantea,* and the species preferred by the director of the Jardin des Plantes de St. Pierre, *E. robusta.* Again in 1875 the Société d'Acclimitation in Paris sent seeds of *E. globulus* (Archives du Ministère de la France Outre-Mer, Series F, box 11g, dossier 1057).

In a letter from the minister to Van der Siwaelmen of June 6, 1884, note was made of seeds of three species of quinquina that should be sent to colonies. They were sent to the botanical garden in St. Pierre at the end of the next month (Archives du Ministère, Outre-Mer, box 11g, dossier 1057). Perhaps one of these was *Quassia amara,* which is today found in gardens and is semispontaneous about settlement sites in low to mid-elevations (Fournet 1978,1070).

In 1860 the governor of Martinique reported to the minister of Algeria and the colonies that he had caused a number of plants to be sent to Gabon and hoped that other plants might be sent back to Martinique in the same vessel. Attached to the letter was a list of nineteen requests prepared by the director of the botanical garden, along with his list of the plants sent to Gabon. Another list of plants to Algeria underscores the fact that the exchanges may have been frequent.

Thierry in his capacity as director of the botanical gardens at St. Pierre introduced the cultivation of Venezuelan indigo (*Indigofera disperma*) in 1887 (Geoffroy 1897,13). This indigo is a shrub growing to 2 m in height and was a more vigorous species than the true indigo (*Indigofera tinctoria*) or indigo anil (*I. anil*), which had been commonly grown on the island. The seeds were distributed generally among agriculturalists, but the species was grown only in a few estates on the island.

Improved varieties of cacao and coffee were imported from Trinidad and St. Vincent by Nollet, a later director of the botanical gardens, for experimental plantings. *Carillo* and *Forastero* varieties of cacao were introduced, and some crosses were made with the local seedling varieties. Those introductions that proved successful were grown in nurseries for distribution to growers (Geoffroy 1897,13). In spite of a small staff, 50,000 coffee and cacao plants were made available in 1890 to the planters. At the turn of the century immigrants from the Bahama Islands brought with them some new long-staple cottons of South Carolina and Sea Island strains (Imbart de la Tour et al. 1900,488).

Otaheite sugarcane remained the standard sugarcane in Martinique until the end of the nineteenth century (Deerr 1949,20). Batavian or Transparent cane had been introduced earlier, but it did not become important in field plantings. In 1871 Rouf reported finding self-sown seedlings on moss growing over a waterwheel on an abandoned estate (Deerr 1949,26). Experimentation with seedling canes was carried on in the island, and exchanges were made by Littee, a local experimenter, with Barbados and other sugarcane-producing islands to develop disease-resistant varieties (Stehlé 1954,n.p.).

Some new or potential crop plants were introduced with considerable difficulty. Kola or goora nut (*Cola acuminata*) was introduced at the instigation of the botanist Hecket in 1882 (Geoffroy 1897,15). A native of Africa, this tree, whose fruits yield a kernel used for medicinal purposes and the making of a stimulating drink, was introduced from the Guianas, where it had been introduced from Africa. The habit of chewing the nut was not introduced. Nursery seedlings were distributed free for test purposes to planters working rainforest clearings.[22] Seeds of coca de Perou (*Erythroxylon coca*) were also set out in nurseries at St. Pierre (Geoffroy 1897,18).

Liberian coffee (*Coffea liberica*) was introduced in the 1880s. Among the plants introduced by the laborers from India was turmeric (*Cucurma domestica*) (Duss 1897,583), betel nut palm (*Areca cathecu*), the decorative *Asystasia gangetica* (Stehlé 1944,26), and the fast-growing softwood *Agati grandiflora* (de Lanessan 1886a,155). Perhaps *Rheedia macrophylla*, a native of India that superficially resembles the mangosteen, also came in under their auspices (de Lanessan 1886a,165). The horseradish tree or ben aile (*Moringa oleifera*) was also introduced in this period (de Lanessan 1886a,431), perhaps from India via Africa. The balsam pear, so associated with the East Indians that it is called pomme coolie or liana paroka (*Momordica charantia*), had come into Martinique in the previous period (Grisebach 1964,287) and was incorporated into the gardens of the East Indian immigrants. The plant took hold quickly, and harvests were collected from the spontaneous wayside growth. Betel (*Piper betel*), originally brought from the East Indies, was grown in the gardens to use with the palm nut for chewing (Duss 1897,489).

Hibiscus abelmoschus was imported from India and grown both for its attractive yellow flowers and for seeds that yield a musk scent (de Lanessan 1886a,435). Also introduced for ornament was *Polyscia fruticosum* and *Nothopanax cochleatum* (de Lanessan 1886a,463; Duss 1897, 323). *Quamoclit pinnata*, introduced in the earlier period, had become spontaneous and was especially common in cemeteries (Duss 1897,440). *Oplismenus africanus*, derived from Africa, had also become spontaneous, although it was still grown as an ornamental in parks and gardens (Duss 1897,514).

Not previously mentioned for Martinique but possibly an introduction, as it was in the British Leeward Islands earlier (D. Harris 1965, 144), is *Prosopis chilensis*, whose fruits were used for cattle feed (de Lanessan 1886a,424). *Acacia farnesiana* had become the most common of the thorny legumes (de Lanessan 1886a,425). *Albizia lebbek* was considered a useful introduction on account of its gum, which resembled that of gum arabic, as did that of the new imported *Acacia vereck*, which was still rare in 1886 (de Lanessan 1886a,425). *Leucaena leucocephala* had become widely distributed throughout the island by the end of the century (Duss 1897,248).

A dramatic specimen tree, never common but always conspicuous, was the baobab (*Adansonia digitata*), introduced from tropical Africa (de Lanessan 1886a,436). An important lumber tree, the tulip tree (*Liriodendron tulipifera*), having a light lumber quite resistant to insect damage and hence useful for interiors and cabinetry, came in during this period.[23] The graceful saman (*Samanea saman*) was rare at the time of de Lanessan (1886a,157) but increasingly common by the end of the century (Duss 1897,1:254). It became naturalized in northern forests during the twentieth century. The Australian pine or filao (*Casuarina equisetifolia*) was considered naturalized in Martinique by 1886 and was a useful interior construction timber (de Lanessan 1886a,170), but Duss (1897,1:190–91) comments on its attractive foliage and its locations—the Port des Transatlantiques in Fort-de-France and in Morne Rouge and Trinité, all artificial plantings. There is no evidence that the species is spontaneous today. According to Duss (1897,1:191) two other species, *C. quadrivalvis* and *C. tenuissima,* were grown in the gardens of plant collectors. Eucalyptus was a late introduction. There is no listing in de Lanessan's catalog, and Duss (1897,2:276) merely lists three species—*Eucalyptus robusta, E. amygdalinda,* and *E. grandiflora*—as present in the French Antilles, with the comment that the first was rare in Martinique. *Tektona grandis* was introduced into Barbados from southeast Asia in 1799 (Schomburgk 1848,614) and was probably present in some gardens by the end of the period, although it, too, was neither listed in de Lanessan nor noted for the island in Grisebach's flora.

Some of the weeds common today entered at this time. *Chloris barbata* was found only about the harbor where the ships of the transatlantic steamers docked in Fort-de-France and in Trois-Ilets across the bay at the time of Duss (1897,2:508). Another introduced species, *C. cilitata,* was frequently found in the verges in Rivière-Pilote, Marin, and Ste. Luce.[24] Sorghum or *sorgho à mil* (*Sorghum bicolor*) was introduced and grown in small quantities for bird seed. Duss (1897,531) points out that one variety, which he called *Sorghum vulgare* var. *effusis,* had become spontaneous in different parts of the island.

Vétiver (*Vetiveria zizanioides*), originally from India, was used as a specimen plant in gardens and about houses.[25] The aromatic roots were used in linen closets (Duss 1892,2:529–30). There is no evidence that the oil was distilled from the roots during this period. Strawberries (*Fragaria* spp.) from different parts of the world were tested in gardens about the island (Kervégant 1937,79). The strawberries introduced in the preceding period had become spontaneous in open areas at higher elevations. In describing his climb up Mont Pelée, Garaud (1895,61–62) tells of coming out into a stretch covered with wild strawberries, an indication that dispersal mechanisms were locally effective for this plant.

Immediately after emancipation the area in planted vegetation dropped 15 percent and only the area in provisions increased between 1845 and 1850 (Kervégant 1936b,100). The area under sugarcane was reduced by the greatest amount, as might have been expected, with the abrupt loss of field laborers. Even so, sugarcane remained the dominant planted vegetation type, always constituting more than half of the land in cultivation.[26] The amount of land in cultivation peaked in the year 1879 with the stimulus of government-sponsored supplies of labor and capital. The depression of the late 1880s and early 1890s and the renewed labor shortage was reflected in a general decline in the exploitation of agricultural land. Planted bushlands declined during this period, never to regain their former area. At the end of the period indigo plantation fields were no longer represented in the planted vegetation on the island. A few new types had made their appearance. Forest timber plantations, coconut groves, citrus orchards, the truck gardens of European vegetables, and planted grassland pastures began to be vegetation types recognizable in the landscape.

Changes in dooryard gardens during the period included the incorporation of new plants into the species composition list. The horseradish tree (*Moringa oleifera*) and the tulip tree, two recent introductions, and the previously introduced mahogany[27] were brought into the gardens as elements in the tree flora. The horseradish tree has a thick, soft root that may be used as a condiment, a substitute for the true horseradish (Gooding et al. 1965,171). The seed yields a fine lubricant, but whether this was exploited is not known. The other two are lumber trees. The introduction of timber plantings having a long maturation period into the dooryard garden strengthened a developing commitment to place, turning the small landholder into something resembling a peasant farmer. The betel nut palm was probably restricted to gardens of horticulturally minded planters and gardens of the East Indians (Gooding et al. 1965, 171).

The musk mallow probably was found most often in the dooryard gardens as a specimen plant or two. Other ornamentals were the romarin and the cypress vine. The balsam pear and sorghum were present in gardens in varying degrees. Turmeric and betel nut were in all the gardens of the East Indians throughout this period.

The significant effect of increasing the number of rural households, each on its own plot of land, the result of several generations of property division, was that the trees of the gardens surrounding the huts provided a wooded aspect to the landscape. There were breaks in the canopy and sunlight reached the ground in parts of the gardens, but often it was only the species mixture that revealed the presence of human habitation from a distance. The conspicuous life form of the coconut palm was always a sure indicator. Its wide distribution below 300 m probably dates from this period.

Not all dooryard gardens had been removed from estate lands, however, as might have been implied in the settlement discussion above. Since planters supplying the sugarcane factories required seasonal labor, the planter let some small cultivators remain in the midst of his land so that there might be a supply of casual labor and a nucleus of cane cutters during the dry season (Delawarde 1937a,40). Those few workers who held year-round laboring positions on estates and who had the usufruct of estate lands also had gardens. Hearn (1890,198), writing at the end of the 1880s, describes a row of such gardens: "each [little hut] was in a little garden planted with bananas, couscous, camanioc, choux caraibes, or other things and hedged about with roseau d'Inde and various flowering shrubs."

The distribution of dooryard gardens probably still mirrored the settlement pattern of the period, especially the rural settlement. Dooryard gardens gained at the expense of dry bushlands and some relatively undisturbed dry woodland on the old fersiallitic soils at the southern end of the island. They gained at the expense of relatively unexploited rainforest and successional communities: crécré (miconia) thickets, silverleaf-balsa stands, savannas that recolonized the abandoned cacao bushlands, and some coffee lands in the north. Dooryard gardens were interspersed within the commercial agricultural lands in the south, whereas in the north these dooryard gardens tended to be above the zone of commercial agriculture or in the ravines.

The herbaceous planted vegetation types were the old provisions gardens and the new truck and flower gardens. The vast majority of former slaves who established their own holdings made provisions gardens in which the *gros légumes* (provisions) were grown both for the table of the household and for sale to the immigrants in the rapidly growing towns. With the increase of a population no longer having access to produce from their own gardens, there developed a proportionally greater demand for provisions in the towns than had existed previously. In the resettlement schemes following the Mont Pelée explosion, the island government considered the location of potential markets for the products of the evacuees' provisions gardens. Still, the area recorded as being in provisions gardens did not expand to the extent it might have. The area in provisions in 1920 was only 12,300 ha compared with 11,731 ha for 1845, although there had been a gradual increase to 15,710 ha in 1889.

Two developments may account for the lack of expansion: first, the decline in provision areas on estates with the increasingly ignored requirement for a set amount of provision land per slave (the worker), and second, the intensification of farming techniques. In any case, foodstuffs continued to be imported as always, with flour and rice becoming more important. "For all, it [the steamer] signifies the arrival of food. The island does not feed itself: cattle, salt meats, hams, lard, flour, cheese, dried fish all come from abroad, particularly from America" (Hearn 1890,310). The developing taste for wheat bread and rice reduced the demand for manioc. Hearn (1890,380) stated that by the end of the 1880s "more banana trees are yearly being cultivated."

Prêcheur Commune continued to have the largest percentage of land in provisions gardens. Fort-de-France became a large market, as St. Pierre always had been for provisions. Gros-Morne in the 1880s was described as having many provisions gardens (de Lanessan 1886b,738). The large covered market in Fort-de-France was built in the 1890s and rebuilt at considerable cost after a fire and earthquake. After the volcanic eruption of 1902, Fort-de-France had the largest market on the island. All commune capitals had marketplaces, and most southern communes had built covered markets by the end of the period. Pineapple exported to France at the turn of the century (Imbart de La Tour et al. 1900,471) probably came out of provisions gardens in much of the upland area. However, considerable numbers of pineapple plantings must have been concentrated in Gros-Morne Commune in order to supply the cannery at Denel.

With the increasingly large town populations and the changes in food habits associated with urban life, demands for European vegetables increased. This change in food habits was led by metropolitans staying on the island for short civil service assignments. Gradually, the country-people learned to plant these vegetables for the urban market.[28] Gardens in which these vegetables were grown were called *jardins maraîchers.* They were generally best developed in the mountain climatic belt because of the demands of individual species. The gardens were considered "European" in morphology because the vegetables and flowers were planted in pure populations of a few rows each. The mixed planting scheme of the provisions garden was never practiced in these gardens.[29] Fond-Saint-Denis was the location of most of the truck-flower gardens supplying St. Pierre, although a few were located in Morne-Rouge and Morne-Vert (Geoffroy 1897,15).

The uplands about St. Joseph and about Camp Balata in Fort-de-France Commune supplied Fort-de-France. Geoffroy (1897,15), a visitor from Guyana, wrote wistfully, "The truck garden vegetables of Europe do well on the upland plateaus: the market at Fort-de-France is adequately supplied by melons, strawberries, radishes, cabbages, cauliflower, artichokes, asparagus, etc. which we eat in Guyana only in the preserved state." Elsewhere the truck-flower gardens were very minor elements in the landscape and the economy. Indeed, this vegetation type did not cover a significant area during this period. The land devoted to these gardens was taken from the lands of provisions gardens, which had the required superior soils.[30]

Tree plantings included the old canafistula orchards, which continued throughout the period. Increased exports may have resulted not only from the increased area in trees, but also from greater care of the bearing trees. Exports at the beginning of the period amounted to about 100,000 kg. Between 1860 and 1880 the average export was two and a half times that. The peak export year was 1887, when 437,209 kg was sent to France. At the end of the century, which is the last date for which there are figures, 134,000 kg were exported (Imbart de la Tour et al. 1900,400). However, after the eruption of Mont Pelée, the refugees planted a number of orchards in Diamant Commune.

New tree plantings were coconut groves developed as a plantation crop in Grand' Rivière; citrus orchards in Carbet and Diamant communes (France, *Annuaire Colonial 1902* 1902,101); and some experimental timber plantings of *Khaya senegalensis, Swietenia mahagoni,* and *Carapa guianensis,* all in the state forest land of the interior mountains (Lasaulce 1901,17). The coconut groves were to provide copra for a proposed soap and margarine factory. The timber plantings were an effort to reduce eventually the island's dependence on North American timbers.

Travelers passing through Martinique during this period waxed eloquent over the bamboo brakes found in the rainforests and in the cultivated lands. Along the Capot River to the west of the Champflore Basin, Garaud (1895,45) noted bamboos and tree ferns, indicating that bamboo perhaps was migrating spontaneously into disturbed streamside habitats. Again in the territory between St. Joseph and Gros-Morne, Garaud (1895,83) noted bamboo along midelevation stream courses. In describing the road called La Trace near Deux-Choux before it turns toward Fond-St.-Denis, Stuart (1891,97) wrote, "feathered bamboo shot up their bright green varnished stems piercing the foliage in search of sunlight." In the Champflore Basin grasslands pastured by goats and sheep had scattered clumps of bamboo, diversifying the landscape (Garaud 1895,352).

Planted bushlands were cotton bush, cacao, and coffee plantings. Cotton plantings were declining in area at the beginning of the period and fell to a low of 10 ha by 1850 (Kervégant 1936b,100). In 1865 the area in cotton bush expanded abruptly to 388 ha in response to the markets made available by the absence of American cottons because of the Civil War. In the four census years from 1860 to 1880 the area remained above 200 ha. It then fell abruptly below 100 ha and, in 1895, to 18 ha (Kervégant 1936b,100). The progressive lowering of cotton production accompanied an increase in sugarcane areas, although a few former cotton bushlands went into sugarcane. However, as Imbart de la Tour and associates (1900,487) pointed out, much of the decline in the modest production can be assigned to the cupidity of the growers, who mixed poor quality cottons in with the better to increase the quantity, thereby losing a reputation for supplying good quality cotton. When it took place, expansion of area in cotton bush was in Sainte-Anne and Vauclin communes. Decline in area took place in Anses-d'Arlets. Only a small area in the hill country of Diamant continued to grow cotton. Consequently, when the decline came at the end of the century, it was in the southeastern communes.

The area under cacao gradually increased from a low of 212 ha in 1860 to 1,800 ha in 1920 (Kervégant, 1936b,100). The major impetus to this increase was the sugarcane crisis of the 1880s (Imbart de la Tour et al. 1900,352). The increasing use of more disease-resistant cacao varieties led to an increase in production greater than might have been expected by the amount of land in cultivation. Practically all the steep-sided valleys of the northern coast from Prêcheur to Macouba to Ajoupa-Bouillon had cacao trees protected by shelter trees usually left standing from the seasonal forest that had been cut for the plantations (Geoffroy 1897,12). Occasionally, *Calophyllum calaba* was planted as a shelter belt tree in exposed areas.

The land under coffee plantings dropped erratically from 1,256 ha in 1850 early in the period to 150 ha in 1913 (Kervégant 1936b,100). The significant break in plantings occurred with the outbreak of a disease caused by *Hemileia* sp., which causes the plant to form a witch's broom (Geoffroy 1897,17; Imbart de la Tour et al. 1900,358). The introduction of Liberian coffee, which is resistant to the disease, helped some, but the flavor of Liberian coffee beans did not find favor in France. The demand for coffee had been climbing,[31] and officials continued to demand diversification, suggesting coffee as an alternative commercial crop in the Antilles.

In spite of the decline in these two bushland crops, travelers continued to notice the cacao trees and to mention coffee bushes. The local consumption of cacao and coffee was satisfied by local production. The taste for a mixture of Arabian and Liberian coffee, which was so pronounced in the mid-1960s among the country people of Martinique, may have dated from this transition period.

Shelter trees for the coffee plantations that are recorded for this period include those used earlier, such as *Inga* spp. *Calophyllum calaha,* and some not mentioned previously. Two natives, *Cedrela mexicana* and *Sloanea truncata* (châtaignier du pays) (Imbart de la Tour et al. 1900, 363), and the introduced leguminous tree *Erythrina corallodendron* (Stuart 1891,90) are the new shade trees mentioned. Much of the production for the local consumption did not come out of the bushlands but out of the dooryard gardens and the upland savannas maintained by grazing on land formerly cultivated in coffee.

At the end of nineteenth century Blérald (1898,n.p.), writing about tobacco cultivation in Martinique, reported that Martinique produced a "cigar fair in aroma and burning" but that the annual production did not keep up with local consumption. Tobacco and other specialty crops considered as low annuals were enumerated under provisions throughout most of the latter part of the nineteenth century (Kervégant 1936b,98). However, in 1896 "tobacco and diverse crops" amounted to 2,379 ha for the whole island (Imbart de la Tour et al. 1900,516).

Almost singlehandedly Thierry revived indigo as a commercial crop. Beginning in 1889 production exports amounted to 43 kg. By 1895 it had grown to 1,621 kg, still a feeble showing compared with the old days (Imbart de la Tour et al. 1900,477). This production apparently came from plantations concentrated in Thierry's holdings in Grand' Rivière Commune (France, *Annuaire Colonial 1902* 1902,100). Finally, competition from the synthetic dyes proved too much and, with this last effort, indigo ceased to be a vegetation type in Martinique.

Vanilla, which grows in Martinique on stakes or on trees planted for the vine to clamber over, was just beginning to be a crop plant at the turn of the century. Demanding a hygrophytic environment, such as in the mountain climactic belt, it was grown commercially in large scale at Camp Balata along La Trace and in Morne-Vert (Imbart de al Tour et al. 1900,384).

By the turn of the century agitation for the production of fresh banana fruit for France led to a small export of 5,000 kg in 1904. In 1907 the island exported 20 metric tons and in 1910 some 13 metric tons. World War I's disruption of markets caused planters to lose interest. Bananas were used as an intertilled crop in the new plantings of coffee and cacao since the bananas would shade young plants and at the same time supply a cash crop, but only a very few agriculturalists listed themselves as banana growers in the *Annuaire Colonial 1902.*

Sugarcane expanded to its largest area yet during this period, when in 1920 it reached 28,000 ha. As in previous periods it was the only crop dominating truly large areas of the island. This expansion was marked by a certain selectiveness. Sugarcane had been widely distributed in the earlier periods. By the end of this period the crop was concentrated on those soils most productive in sugarcane (fortified in the late part of the period by fertilizers) and in those areas linked to the central sugar factories by railroads or short road distances. Estates located on the coasts with good boat landings had an advantage. By 1895 the Usine Marin was the only sugar-processing plant in the dry south, and it served the plantations rimming the Marin Cul-de-Sac (Geoffroy 1897,15).

New varieties of sugarcane, including seedling types locally developed, were beginning to be selected for the requirements of different soils and climates on the island. In the late nineteenth century "rind disease" caused the disappearance of Otaheite sugarcanes (Deerr 1949,29). The 1916 outbreak of mosaic disease[32] in Puerto Rico was the first in the Antilles and led to the breeding of other strains of *Saccharum officinarum* that were resistant to the disease.

A distinction between sugarcane fields for the production of sugar and industrial alcohol and those destined for rum was made. Desirable combinations of sugarcane varieties and ecological conditions are different for sugar and rum. The large estates attached by contract or ownership to the central factories used sugarcane varieties yielding the highest sugar production for their region. With the increased mechanization of the sugar industry, a series of smaller estates formerly used for the making of sugar with small works devoted their full production instead to rum. These small stands were found higher and in smaller fields than that common on the sugar-producing estates, except for the Hon. Lajus[33] in the Carbet region, which was as large as any estate found on the island. Increased appreciation of Martinique rum made possible the utilization of small sugarcane properties.

Throughout this long period *savane* refers to grasslands and savannas. The evidence for plowed and sown grasslands has not been found. References to hayfields are similarly scarce. Hearn (1890,269,272) noted "meadow and cane and cacao" in the Champflore Basin and "guinea-grass run mad" in the savannas along the Morne-Rouge road. On the road from Fort-de-France to Gros-Morne, a view of the Lamentin Plain showed "vast savannas where the herds of cattle wallowing in Para grass to their stomachs grazed in liberty" (Garaud 1895,81). The grasses introduced during this period were primarily ornamental and weeds, but some did enrich pastures.

The decline in areas under coffee in the humid mountain belt and cotton in the littoral and dry belts provided significant areas for recolonization by spontaneous vegetation. As noted above, some of the sugarcane lands were allowed to disappear with the economic problems of the 1880s. There was also, particularly toward the latter part of the nineteenth century, much abandoned land within areas dominated by planted vegetation types. Some of this land was not reclaimed for many years, permitting a sequence of successional spontaneous vegetation types to develop. As has been pointed out, these periods of spontaneous regeneration were cut short by the settlement activities of former slaves and indentured laborers, by resettlement schemes of Crédit Foncier Colonial, and later by the resettlement schemes of the government for the refugees from the Mont Pelée eruptions. However, not all of these abandoned lands were recolonized, particularly those most severely eroded during cultivation and those bypassed by the improved wagon roads.

The successional communities described for the preceding period recolonized these abandoned lands in sequence during this period. Since more mountain-belt areas, formerly in rainforest, were among the lands abandoned, the mesophytic secondary communities particularly rich in melastomes became important. Former cotton bushlands, ordinarily dry woodland areas, were colonized by more xerophytic associations, either leguminous thorn scrub dominated by species of *Acacia* and *Haematoxylon* on the old volcanic-derived soils, or bushlands dominated by *Croton* and *Lantana* on the limestone-derived soils.

A number of pioneer woody shrubs passed into the savannas that were maintained by the grazing of animals in the Vauclin, Marin, and Sainte-Anne communes. The most important of these were the native *Crotons, Acacias,* introduced *Acacias, Leucaena leucocephala,* and *Ziziphus mauritiana.* Mesquite and *Acacia tortuosa* both expanded under grazing, the first because of the palatability of the fruits and the second because of the unattractiveness of the plant to grazing animals. Citrus, particularly lemons and limes, were spread by cattle in the Diamant pastures. Guavas and bamboo were noted in the savannas of the more humid north. Garaud (1895,312) speaks of "bouquets of guavas" in the Champflore savannas.

Earthquakes, two volcanic eruptions, a major fire in Fort-de-France, and a hurricane were physical afflictions that are known to have affected the vegetation of Martinique during this period. Epidemic diseases in Arabian coffee and Otaheite sugarcane spelled ruin for many individual planters. Though there are no estimates of total costs to the island from these catastrophes, they were in the millions of francs.

The volcanic eruption of 1851 was associated with a rent in the western flank of Mont Pelée. The explosion was almost coincidental with the last of a series of earthquakes that had begun in May and ended in August (Hearn 1890,273). On the morning of August 6 white ash rained down to cover Morne-Rouge, Carbet, Fond Carré (a suburb of St. Pierre), and Prêcheur (Heilprin 1903,278). The ash also whitened the countryside (Hearn 1890,274). The eruption caused panic among the inhabitants, but it turned out to be merely a nuisance in the towns, and the countryside was enriched by the ashfall.

Several earthquakes between 1883 and 1885 served only to intensify the economic crisis in sugar and the economy in general (de Lanessan 1886b,733). On June 22, 1890, a great fire destroyed Fort-de-France (Garaud 1895,231). Municipal records and public archives and hundreds of buildings were consumed. Another fire in April of 1891 only added to the misery. The town, rebuilt almost entirely of wood after the earthquake of 1839, had been ripe for a conflagration. The next rebuilding proceeded slowly and was again in wood, largely imported.

Then, in mid-1891, a hurricane or tropical cyclone hit the whole of the island hard. Garaud (1895,269) gave a graphic description, loosely translated as follows:

> The church and gardens of Morne-Rouge were destroyed. At St. Joseph only the town hall was left standing. Marin was flooded, Vauclin mowed down, Robert left for dead, and François annihilated. Ducos was buried under debris. Almost 40 small hamlets were utterly destroyed. All the breadfruit trees were decapitated. Not a mango was saved. The manioc harvest was lost. The sugarcane fields were severely damaged and most of the *usines* ruined. Tall trees of the Caribbean coast, the ceibas, and courbarils were uprooted. The wreckage was indiscriminate, but the rural population suffered from want of food from their destroyed gardens and the flattened towns could only compete for the limited food supplies. Relief ships from North America and France were dispatched, but the suffering was intense.

The hurricane severely disrupted the economy and destroyed most of the isolated trees and many in groves and gardens. Even in the forested lands emergent trees were torn loose and overturned, taking the associated epiphytes and lianas with them. Cacao plantings were badly damaged by the hurricane wind. The winds destroyed most of the Liberian coffee bushes planted in the 1880s and many of the still-bearing Arabian coffee bushes (Geoffroy 1897,13,17).

The single most dramatic natural disaster to affect the plant geography of Martinique was the 1902 eruption of Mont Pelée. For two weeks prior to May 8 the volcano had been spewing clouds of ash and vapor. With increasing activity the ash began to accumulate, burying the ground-level flora and covering all the upper-level plants with the heavy coating. No real assessment of the effect of this first ash shower can be made except to record that observers noted that the ash stimulated new growth. Vegetation was wiped out by the mud flows from the Dry Lake at the summit of Mont Pelée following a series of discharges of volcanic water before the first major blast, and by the series of explosive blasts on May 8, 20, and 26, June 6, July 9, and August 30, 1902. Boiling mud advancing as an avalanche destroyed the Usine Guerin at the mouth of the Blanche River (see map 28). Torrents of volcanic water carried debris that wiped out the town of Prêcheur and much of Basse-Pointe (Heilprin 1903,271). As in the previous explosions of 1851 and 1782, no lava flowed, but ash, pumice, and incandescent bombs were thrown out with great force (Heilprin 1903,281). None of the large pieces of ejecta were thrown far—the upper levels of the cone received these. Pieces found farther away were moved by stream action or were dislodged by earthquakes, rolling downslope. Some boulders 20 to 30 ft in diameter were reported to have rolled two or three miles. How much these boulders crushed and tore up forests is only to be guessed at. As for much of the fine material, it was picked up and retransported by the wind in the following months. The area receiving ash by wind transport is shown on map 28. Besides the Caribbean towns of St. Pierre and Prêcheur, the interior towns of Basse-Pointe, Ajoupa-Bouillon, and Morne-Rouge were wiped out, and the communities of Morne Capot and the Hauteurs du

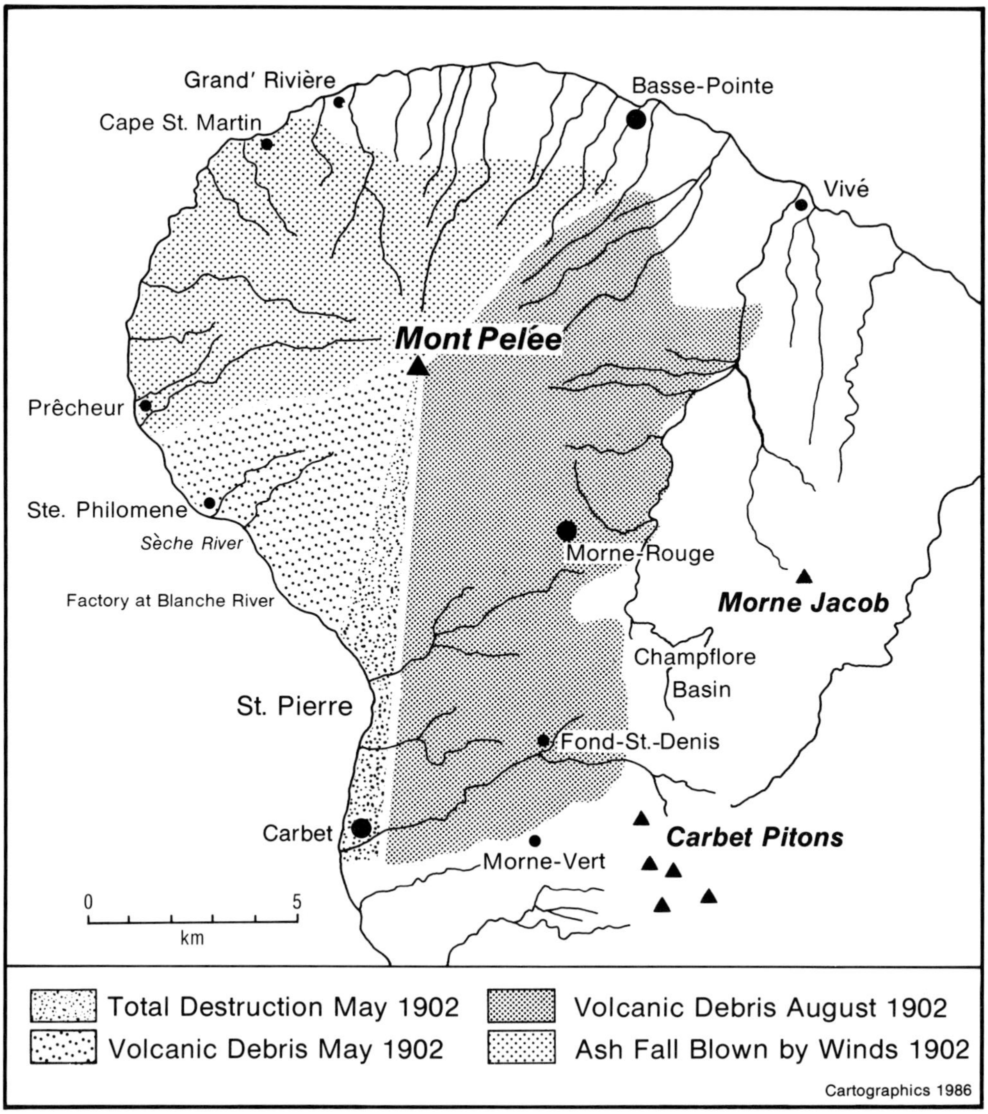

Map 28.
Areas substantially affected by the eruption of Mont Pelée of 1902. Source: After Heilprin 1903.

Bourdon were wrecked or entirely wiped out (Heilprin 1903,232). Fauquet (1912,158) estimated that there was a loss of 35,000 lives, although lower estimates are considered more accurate.

In two minutes the explosion of May 5 blasted a wedge of absolute destruction between eight and nine square miles in an arc from just north of Carbet to north of St. Pierre. Considerably beyond this wedge extended an area over which most vegetation was destroyed. Beyond that was an area in which singeing ash fell and minor fires caused by flaming cinders damaged the surface vegetation. Four more eruptions continued the destruction in the same area, extending its borders somewhat. The explosion of August 30 extended over a much larger area and stopped only a few kilometers from the Atlantic coast. More than one-half of the rainforest areas of Martinique were affected by these explosions. Slightly less than half of the montane forest was severely damaged. The elfin woodland on Mont Pelée was virtually destroyed, as was the montane bushland of the peak. The blast destroyed one rum-producing area and halved the total rum production, leading to a redistribution of quotas (Mitchell 1963,23).

On the nature of the gas explosion, Lacroix (1904,672) said, "There is no reason to seek for a combustible gas; the trees are not burned and the palms from which the leaves have not been forcibly torn show these to be simply dried out." Not even the slightest indication that vegetation had been burned with a flame is reported for the Morne Bajai and Ajoupa-Bouillon region. Hovey (1902,347) in his preliminary report said, "in many places the line of demarcation passed through single trees leaving one side scorched and brown while the other side remained as green as if no eruption had occurred." The trees left standing within the blast region were largely stripped of leaves and were completely dry. No sap was discernible in the twigs. Where the blast damage was less severe, the "leaves hung to the branches, shriveled up as though having been rapidly passed through a dry heat furnace or scorch blast" (Heilprin 1903,313). The rich riparian forests, clothing the walls of the deeply incised radial streams on the mountain outside the blast zone, were torn away by torrents of mud and volcanic water or were blasted dry by the hot gases.

Recovery outside the zone of total destruction was rapid. On his August ascent of Mont Pelée from Hon. Leyritz on the east flank of the mountain, before that month's explosion, Heilprin (1903,218) reported "a light growth of grass had begun to cover the arid slope of ash and cinder and the blackened forest of the ravine slopes was also touched on the crown with green. The beautiful tree fern gave evidence of new life. . . . It was evident that the burned forest was not absolutely dead." The sugarcane growers and peasants with bananas noted the spurt of growth on land that had received an ash dressing (Heilprin 1903,210).

The topographic modification of the summit affected recolonization, in part. The slopes of the inner and outer caldera were colonized by grasses and lycopods, but succession was slow. The flora, although small in number, possesses a high capacity for recolonization and is able to recover without loss of species (Stehlé 1937,256–60). The colonization was largely by air from propagules from the herbaceous communities below and from the upper slopes of the nearby mountains. On the highest peaks of Mont Pelée *Gaultheria sphagnicola, Sphagnum* spp., *Tibouchina, Guzmania plumierii,* and *Gesneria ventricosa* community is mixed with tree ferns. After the explosion the first colonizers were the weedy species *Sauvagesia erecta, Desmodium canum, Erigeron canadensis,* and *Stenotaphrum secundatum,* followed by the pioneer fern *Dicranopteris pectinata.* Later, Stehlé observed the arrival of *Tibouchina chamaecistus, Pitcairnia bracteata, Baccharis cotinifolia,* and then *Clusia* sp. and *Charianthus nodosus*. However, the full elfin woodland had not yet been reconstituted in 1937.

Lacroix and his associates (1902,382) noted that only about one-third of the mountain was ravaged by the first blasts. They note, "The north and east sides have barely suffered, or at any rate the deposit of ash on them was so thin that when we were in Martinique [early June] it had been mostly washed away and the beautiful verdure which characterizes that lovely island flourished to all appearances as if no eruption had taken place." These comments must be considered in the light of the controversy over the governor's not evacuating the town. The account may have minimized the effects. However, after the August eruption the damage was more general, though not as severe as that which occurred in the path of total destruction in May.

Ajoupa-Bouillon and Gros-Morne were resettled, and the heights (600 m) at the foot of the Carbet Pitons were settled with government funds. The central road La Trace was rebuilt in case of the need for evacuation.

At the end of World War I the Martiniquans were interested in just what opportunities for rural settlement existed on the island (Buffon 1920). By 1920 the area of unmodified vegetation was smaller than ever before, probably considerably less than 20,000 ha (see map 29). The valley walls between Mont Pelée and slopes of the pitons and mornes to the north of the massif were in virtually undisturbed forests. The central mountain core had much undisturbed or minimally disturbed montane and rainforests. In the south only isolated patches of forest remained. The seasonal forests were relatively undisturbed only in the Préfontaine region of Morne Caraïbe. These forest islands, none of which exceeded 120 ha in size, were bordered by a variety of planted vegetation and secondary spontaneous plant associations. Relatively undisturbed dry woodland had retreated to an all-time minimum extent. However, there had been little disturbance of the montane forest, the elfin woodland, or the montane bushland. Borders among these and between montane forest and rainforest were relatively undisturbed. The lower border of the rainforest had retreated about the Carbet Pitons and Morne Jacob massifs. The lower margin was also moving upslope along the windward coast, in Gros-Morne, Saint-Joseph, and Sainte-Marie communes, and about the Champflore Basin.

The area devastated by the 1902 eruptions of Mont Pelée was slowly being recolonized by secondary vegetation. The borders of the higher altitude types maintained their historic positions except where they had been severed and obliterated by the 1902 explosion.

Secondary woodlands and forests were regenerating on abandoned coffee and cacao areas on the seaward flanks of the Piton Mont Conil from Ceron around to Grand' Rivière on Dominica Channel. Leguminous scrub and *Croton* thickets dominated the land in steep slopes along the Caribbean coast. Remnants of the seasonal forest and dry woodland occupied river courses and steeper slopes above 250 m except for the broad crowns of the interfluves (Buffon 1920,19–20).

Map 29. Zone of vegetation modified by human agency after World War I. Source: Compiled by author.

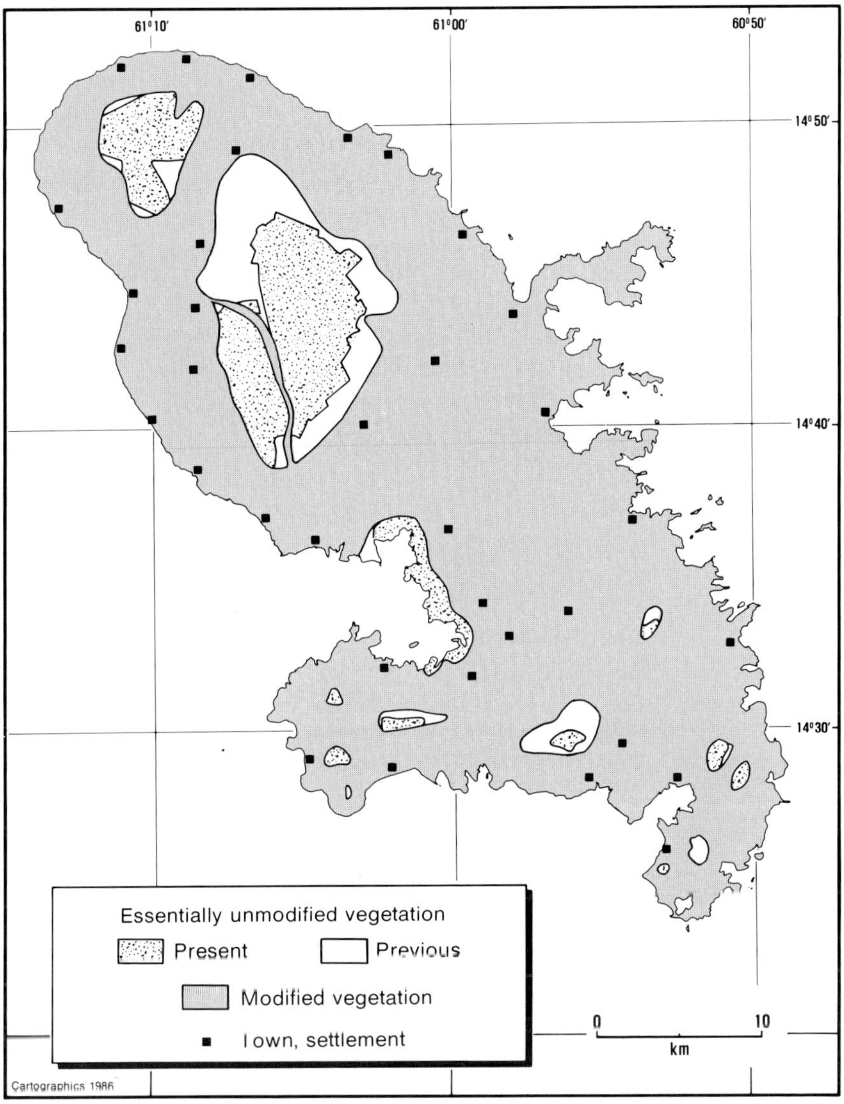

In the Sainte Anne Peninsula the evergreen dry woodland had reclaimed those eroded hilltops abandoned from farming for the longest time, and deciduous dry woodlands existed on those that had never been cleared. The rest of the severely eroded sloping land was in leguminous thorn scrub and mixed cactus bushland. Shrub and tree savannas had taken over the untilled lower elevations. In the southwestern communes of Diamant and Anses-d'Arlets, leguminous thorn scrub and dry bushland dominated by *Zanthoxylum* and *Croton* species covered the land in slope that was browsed by feral animals. Orchard and shrub savannas had colonized the abandoned sugarcane lands of the Vauclin interior.

According to Buffon (1920,20), the planted vegetation constituted a continuous belt occupying most of the land in gentle slope up to 250 m on the north windward coast, to 400 m on the Caribbean slope, and over the majority of the island south of the line from Fort-de-France to Robert. There was roughly 55,000 ha in planted vegetation, including sugarcane fallow land but excluding savannas (which included some uncensused improved, planted pastures). Provisions and dooryard gardens covered some 13,000 ha about the major towns and in Prêcheur Commune. Tree crops, including citrus, amounted to 600 ha, found mostly in Ajoupa-Bouillon and Diamant communes. Bushland crops amounted to 2,050 ha with 1,800 ha in cacao plantings, almost all of which were in Ajoupa-Bouillon Commune, and 250 ha in coffee plantings, most of which were in Vauclin Commune. Sugarcane occupied roughly half of the planted acreage. Pastured grasslands and savannas occupied 12,000 ha, and fallow grasslands, also pastured, amounted to between 6,000 and 7,000 ha, so that these together occupied roughly a scant third of the planted vegetation.

Notes

1. The destruction of beets in World War I caused an increase in sugarcane production in 1914 and 1915 for export to France, but the quotas instituted strictly limited production.

2. Leiris 1955,23. Both *Congo* and *Guinée* were synonyms for the continent of Africa and were not being used in their more restricted political sense. However, Debien (1951) has sifted the evidence in the slave trade and established that demands of the planters often made members of one tribal group more marketable than those of another group.

3. Although assimilated linguistically by the 1960s—they all spoke creole and all were nominally practicing Catholics—they kept the worship of village deities, a persistence of the rural South Indian form of Hinduism practiced by their ancestors (Horowitz 1961,340), and they kept certain Indian food habits such as the liberal use of turmeric.

4. One indication that labor was not limiting for long is the fact that the fall in exports of 1848 was overcome by 1860 (Chemin-Dupontes 1909, chap.8).

5. France, INSEE 1955–60,38. Revert (1938,34) quotes a figure of 244,369 for 1921, which may be closer to the real population size than the official figure because of the depression of official figures caused, as he says, by the nonenumeration of those wishing to evade taxation.

6. There is little doubt that these are self-imposed, and the creoles resist changing their housing, food, and clothing habits. The local whites will say that the blacks are unable to live in the cooler temperatures of the higher elevations, but they do not live there themselves. Chanvalon (1763,13) said the same. White and black doctors on the island today point out that blacks are very susceptible to respiratory infections and sinus trouble.

7. This was a practice old in France by the time of Napoleon. This was the practice under old French *coutumes*.

8. Personal communication 1963, from J. le Blanche, head of the Service des Travaux Cadastraux. A detailed cadastral survey was made in the 1960s so that it would be feasible to introduce the taxing of real property. The aerial photographs taken for this cadastral survey were made available to the author, who used them for the preparation of a manuscript vegetation map.

9. Archives Topographique, Enregistrement, Fort-de-France.

10. The modern large-scale topographic maps in which the dispersed dwellings are located often show the small huts contained within a rectangular or quadrilateral space. The Quartier Mamelin and Quartier Plateform along the Belleview River in Marigot Commune are two examples in the north.

11. Arrêté of April 22, 1877, in France, *Annuaire Colonial 1902* 1902,60. The proclamation covers the size and nature of lands to be conceded, the time allotted for bringing into exploitation, the payment, and so forth. The concessions of domain lands, authorized by the deliberation of the Conseil Général, decreed that individual lots could not exceed 6 ha. They applied exclusively to cleared lands, and at least five-sixths of the concession had to be exploited in secondary products such as coffee, cacao, vanilla, ramie, annatto, to the exclusion of sugarcane or provisions (Arrêté of April, 1777, article 1, France, *Annuaire Colonial 1902* 1902,60).

12. A sign was discovered in the twentieth century near Colson commemorating the opening of 15 km of road in thirteen days by a detachment of the 2nd Infantry Regiment of the Marine in 1850 (Delawarde 1937,176).

13. The earliest plan was that of Jourdon, submitted in 1882. It would have linked all the important towns north of the narrows just south of Lamentin. Even though the Conseil Général approved the plan, it was never implemented because of a disagreement between the construction engineer and his financial associates (Revert 1949a,404).

14. According to Buffin (1981,195) there were two important innovations associated with Thorp's enterprise: the establishment of colonage and payment according to the amount of sugar in the sugarcanes rather than by weight.

15. Even in 1897 Geoffroy (1897,10) found some of the old-fashioned sugar works, but these were associated with the agricultural rum distilleries rather than sugar and industrial rum producers. In Martinique it is the convention to call rum for human consumption "agricultural rum."

16. It is not known just how much charcoal was imported. Lasaulce (1901,19) stated categorically that none was imported, yet charcoal is listed as a duty-free import when derived from North America (France, *Annuaire Colonial 1902* 1902, 63). These listings may, of course, relate to imports of an earlier time and may be anachronisms.

17. Customs declaration cited in Imbart de la Tour et al. 1900,323.

18. Arrêté du Gouverneur, 1848 *Bulletin officiel,* cited in Lasaulce 1901,32.

19. State lands would include those on the *cinquante pas* and small wooded areas used for special purposes, such as that used by the different military services. These amounted to about 300 ha. The colony had two large areas: the larger on the Carbet Pitons, the smaller on Mont Pelée. The first (5,000 ha) was much disturbed by timber cutting and shifting cultivation. The Mont Pelée holding contained about 2,000 ha and was even more disturbed than the Carbet forests. There were a few small wooded holdings in the south and some farmer estate lands acquired by the Crédit Foncier. Lasaulce estimated that 7,500 ha belonged to the colony. The rest of the forest, estimated at 14,200 ha, belonged to private individuals (Lasaulce 1901,28).

20. Hearn (1890,277), visiting in the late 1880s, wrote that hundreds of the fine great trees had been cut down by the charcoal makers "but the forest is still grand." As a nontechnical observer Hearn would have been unaware of the damage by selective culling unless the physiognomy had been drastically modified or local people had complained to him of destruction; therefore, we can presume that the "forest" remained.

21. A permanent demonstration garden was created by decree on Jan. 2, 1854, and was established in 1869 in Tivoli (France, *Annuaire Colonial 1902* 1902,94).

22. In 1897 the garden at St. Pierre had distributed more than 600 small plants to proprietors in Gros-Morne, Fond-St.-Denis, Case-Pilote, and Blanche River in Prêcheur (Geoffroy 1897,15–16). The growers who received the plants represent a list of the big sugar planters, indicating a continued interest in experimental horticulture on the part of this class of farmers.

23. de Lanessan 1886a,153, 421. Identified in two places in the catalog by de Lanessan, this tulip tree probably is not a misidentification of the African tulip tree or bois flambeau (*Spathodea campanulata*), since de Lanessan discusses the latter species when listing the trees of Senegal and makes no mention of the introduction to America.

24. Duss 1897,2:508. Gooding et al. (1965,45) call *C. barbata* a common weed of dry ground from the West Indies and Mexico to Argentina. *C. ciliata* is not widely distributed but exists also on dry soil in the same areas. Hitchcock and Chase (1950,527) treat the latter as if native.

25. The genus and common names are derived from a Tamil word (Hitchcock and Chase 1950,172), but there is no good evidence that it was brought to Martinique by the East Indian indentured laborers themselves. It was probably introduced by horticulturists.

26. Calculated from Kervégant 1936b,100.

27. Occasionally a timber tree was planted for each child in the household, a kind of investment for the children, dowry for the girls and cash for the boys. These were arranged as a windbreak at the bottom of the garden or as a property line row or scattered throughout the garden (Vauclin and Morne-Vert communes in 1963).

28. Stuart (1891,102), who was traveling and botanizing in Martinique during the late 1880s and early 1890s, recounts his instructing his hostess at Morne-Rouge in how to cultivate radishes. He also told her of various European vegetables that could sell well at St. Pierre and that could be raised in the cool mountain climate. Caution should tell us that he may have been more pleased at the telling of it than effective at plant introduction.

29. When vegetables derived from abroad are grown in a mixed complex such as the string bean, eggplant, cucumber combination common today in Ajoupa-Bouillon, the garden has been called a provisions garden, according to local usage. The term *jardins maraîchers* (market garden) is used only for gardens having this morphology in the 1960s and 1970s. In the 1980s no regular distinction in morphology could be made.

30. All statistical data from this period combine the provisions and truck-flower gardens areas in a single figure.

31. In 1825 the Antillean colonies had produced 30 percent of the coffees consumed in France. In 1897 the proportions had been reduced to 1.3 percent (Imbart de la Tour et al. 1900,321), a reflection of the lack of appreciation of the quality produced as well as an actual reduction in area planted in coffee and an expansion of production in other French overseas possessions.

32. This disease was, according to Deerr (1949,29), associated with an insect vector *Aphis amidis*.

33. *Hon.* is the French Caribbean abbreviation for *Habitation,* which translates as "plantation," or, as it is now used, "estate."

8 The Modern Period, 1921–1985

Modern developments in the plant cover have reflected rising levels of professional competence in agricultural production, an increasing effectiveness in implementing French government programs, expanding participation by the local government in policy formation and regulation of internal affairs and, finally, a loss of competitiveness of export crops on the world market. During this period the damage caused by renewed volcanism and by several hurricanes was carefully documented, and informal studies were made of the natural recovery of destroyed vegetation.

The French centralization policy survived the dramatic upheavals associated with World War II. Centralism, begun in the nineteenth century with the direct representation of the colonies in the French parliament in 1846, was made more effective when, in 1946, Martinique was again made a *département* of metropolitan France, this time one of three Départements d'Outre-Mer (DOM). Its civil service was integrated with that of other *départements*.[1] Social assimilation after World War II continued, as a greatly improved and expanded educational program was carried out by teachers and administrators trained in metropolitan France. Then in the late 1960s, partially as a response to the growing sense of nationalism or autonomism in the other French islands and the British-dominated islands, a more effective implementation of the link with Paris was made by the governments of de Gaulle and Pompidou. Although the independence movement is much discussed, it is espoused by few.

The rising level of education paralleled an apparent increase in receptivity to new techniques of production and economic organization, although many small proprietors were still conservative. The commercial policy continued to be one of market quotas and price subsidies, which, under rising costs of production that were high already for the West Indies, tied Martinique to France as effectively as had the *pacte colonial* in earlier periods.

All types of landownership remained. During the 1930s large estates were both assembled and reassembled, but land reform in 1959–60 and again in 1963 attempted to favor the small cultivator. Sharecropping increased in importance up to the 1960s (Horowitz 1967,31, 107). Land reforms actually made it possible for the larger landholders to dispose of lands not suitable for mechanized agriculture. The formation of cooperatives increased the capital that could be marshaled by peasant farmers to further their own interests. However, mechanization increased the production-cost gap between those controlling large operations and the small proprietors, resulting in economic tensions for which no good solutions have materialized. Squatters have been tolerated recently, not only in the coastal margins but on some larger estates. No recent government has been willing to move against this form of land redistribution even when given the authority.

Sugarcane no longer dominates the economy. By 1963 banana exports were slightly larger by value than sugarcane products, and by 1968 the area in bananas surpassed that in sugarcane. By 1981 substantially more area was in banana plantings than in sugarcane but well down from the high in 1978 (Martinique, Chambre de Commerce 1984,16). Pineapples had the highest return per hectare of the three principal exports (France, DDA 1975,60). The exploitation of the forested lands came to be largely under regulation by the Service des Eaux et Forêts, more recently by the Office National des Forêts (ONF). Even as selective cutting continued to be practiced occasionally, a policy of sustained timber yield has been gradually implemented with systematic plantation schemes for mahogany and Caribbean pine on both ONF and private land.

Sizable population increases have taken place, although the rates of increase have declined because of birth-control programs and emigration. The increased urbanism of the population has been associated with increased internal circulation, the result of a better road network and more vehicles. Radio has become ubiquitous and television is commonly available. Several local newspapers have wide circulation.

Increased population density, as well as unemployment and underemployment on the island, has stimulated considerable discussion as to the relative merits of the different agricultural enterprises in terms of work-days consumed in production and in value returned per hectare. Although the social benefits of production processes with high labor requirements under a minimum wage law are not locally denied, the employment of labor-saving techniques on large and medium-sized holdings increases. Industrialization, proposed as an alternative use of redundant labor and as a way to reduce the imports and thus improve the balance of payments, began to get serious attention in the 1970s (Martinique, Préfecture 1977). A generous social welfare program has increasingly supported the underproductive and nonproductive labor pool. However, the French government has begun to look less kindly on the huge demands made on the national treasury by the DOM.

During the modern period new plant species continued to be introduced and ameliorated in government research stations. Planted vegetation gradually again reached the proportions of its greatest nineteenth-century extent, and then began a decline. With improvements in water distribution, the traditional restriction of bananas to the more mesophytic and hydrophytic environments no longer obtains. They are grown in some dry parts of the island and have been introduced into a crop rotation scheme in others. Pineapples have taken over former banana areas at upper midelevations on the windward coast. In the 1980s the decline of sugarcane and bananas as export commodities refocused attention on vegetables and flowers for export and on a variety of polycultural agricultural systems to produce food for local consumption. The expansion of planted vegetation was chiefly in new pastures and timber plantations. Forest lands are increasingly managed for sustained yields, and the tree plantations are gradually increasing in size. Several hurricanes caused severe damage and some change in the vegetation pattern largely through economic repercussions. Severe droughts in the early 1970s substantially reduced the area of agricultural vegetation types, although the value of agricultural production continued to increase.

Social and Economic Developments

Despite all efforts of local politicians, colonial subordination has remained the central economic fact. Special interests of the *département* were minimized, and the balance of trade continued in favor of European France even after decentralization led to the setting up of economic regions in metropolitan France and the overseas departments in 1979. Martinique is one of the regions; it has two economic and social councilors. The census of 1921 reported 229,839 persons, with women outnumbering men (France, INSEE 1956–57,38). The gross density was 210 persons per square kilometer. The towns and rural *quartiers* of agglomerated settlement and much of the road pattern had become established in its present form. The population continued to expand under continuing high birth rates and declining death rates due to improved medical services and public health programs. Population increased steadily from 1921 to a peak in 1946 of 265,950 (France, INSEE 1956–57,38). It fell and then rose again, reaching a censused 328,566 in 1982, giving a gross density of 300 persons per square kilometer (France, INSEE 1983,50–51).

Beginning in 1960 the growth rate began declining. By 1974 estimates were that zero population growth would be achieved by 1980 (Martinique, CENADDOM 1975,23). Although this is not yet accomplished—the birth rate in 1985 was 3.457 or 5.9 percent—the actual population figures have declined since 1982 due to emigration. The percentage of the population less than fifteen years of age has declined from 43 percent in 1967 to 33 percent in 1982. The rise in population between 1946 and 1960 was slowed somewhat by immigration to Panama, Guyana, and the other Antilles, and from 1970 on by immigration to Europe.[2]

During the modern period the previously rural population became increasingly urban. In 1935 one-eighth of the population was urban. By 1954 a third of the population was urban, and by 1974 the urban sector had almost half the people (47 percent).[3]

The population of all communes increased until after 1963, when only eleven increased and, of those, only five increased by more than one hundred people (France, INSEE 1956–57,154, and 1975,n.p.; Martinique, Préfecture 1977,2). The rest were losing population. The regional urban centers of Ste. Marie and Lamentin increased proportionally over other towns. Ste. Marie had begun to compete with Trinité for the role of regional center, and Lamentin had become the urban focus for much of the manufacturing and industrial developments within the Lamentin Plain. The phenomenon of rural-urban migration had arrived in Martinique, and urban ways of life passed to the country. As Roman Paquette (1982) put it, *morne paysan* had become *morne urbain*. The application of minimum-wage law, together with the various family assistance programs, has constantly raised the buying power of the population. New habits in food consumption tended toward more expensive items such as meat, milk, and so-called European vegetables. By 1975 imports exceeded exports by four times (Martinique, Préfecture 1977, n.p.; Martinique, Chambre de Commerce 1984,14).

The internal economy is divided between commercial agricultural estates and essentially subsistence agriculture; analyzed another way, the economy falls in three categories: the working population (11 percent), a slowly expanding manufacturing sector (18 percent), and a greatly expanding tertiary employment sector (71 percent) (Martinique, Chambre de Commerce 1984,7). A few local processing industries include agricultural product processing such as chocolate making, pineapple canning, sugar refining, beer brewing, and rum distilling. Other industries include a refinery for motor oils and a bottling station for butane gas. Industrial zones begun in the early 1970s have increased in area, but the proportion of persons employed in such industries is relatively low and has declined by 1 percent since 1974 (Martinique, Chambre de Commerce 1984,7).

Industrialization has lagged in spite of official encouragement, and even in the late sixties some economists thought that linkage had been a deterrent. "Integration as a department has been an obstacle to, rather than an encouragement, of industrialization" (Lasserre 1961,1066). However, two new industrial sites have been planned by the Chamber of Commerce and Industry (CCIM), which points up the continuing faith in industrialization. A study conducted in 1978 tested fifteen possible industrial projects, including the packaging of detergents and the manufacture of pressurized PVC (polyvinyl chloride) pipe and electric batteries. The agro-industries suggested for possible development include the processing or production of poultry, cold meats, banana chips, crackers, and candy (Martinique, Chambre de Commerce 1984,43). Participation by France in the European Common Market certainly did not serve Martiniquan interests.

From the end of World War I until 1946 the attention of France was directed primarily toward her African colonies and those in the Far East rather than toward her old colonies of the Caribbean. In Africa cacao production was encouraged, as was banana production, both of which were in direct competition with Martinique production. Still, Martiniquan sugar remained on the quota system, and during the worldwide economic depression of the 1930s quotas were revised in an attempt to alleviate the distress on the island. Over the long run, however, the period between the wars was one of prosperity for the Martiniquans, and a rising standard of living provided for new local markets in meat, milk, and margarine products (Sylvestre and Lotaut 1947,1:53). Markets in metropolitan France developed for bananas and canned pineapple, and later developed for fresh bananas, mangos, avocados, and other fruits.

The establishment of the Vichy government after the fall of France during World War II resulted in a blockade of the French islands by Britain and the United States. The island turned to self-sufficiency strategies. Under the de facto government of Admiral Georges Robert during World War II, sugarcane land was converted to provisions land (Revert 1949a, 318). Animals were grazed wherever there was browse, but their numbers declined. Widespread famine was eventually averted only by food gifts from the United States (Horowitz 1967,14). During this period political cleavages split the island community profoundly, and social and economic relationships reflected the wartime politics long after the end of the conflict.

After the war, the need for food and fuel imports rose as acreages under export commodities expanded again. Increasingly large amounts were supplied by the United States (Revert 1949,526), as France was involved in reconstructing her home economy. Fuel oil and kerosene were increasingly used in domestic cookery. By the early 1970s these fuels had largely replaced charcoal, and foresters and conservationists began to see forests relieved from attack by the makers of charcoal. However, with the oil shortages and increasing price of petroleum products in the 1970s, charcoal making began to regain importance, and by the mid 1980s it had become an important consumer of spontaneous forest biomass. No statistics, however, were being collected.

Machinery and spare mechanical parts were imported from the United States during the midcentury years. Exports to the United States included industrial molasses, grains, vanilla, goat and sheep skins, and, after 1961, increasing amounts of sugar (Sanderson 1964,22). The expanding tourist trade of the Caribbean opened a new market on cruise ships for provisions and expecially truck crops. The hotels, after a difficult beginning period, began to be an important labor consumer.

After World War II, Martinique began to reassess her traditional isolation from her West Indian neighbors. In 1946 Martinique joined the reorganized Caribbean Commission. This gave Martinique continuing opportunities for dialogue, increased the informal exchange of expertise in agriculture, and facilitated the pooling of research activities. Participation in other regional organizations in the last quarter of the century brought the French Antilles, and Martinique specifically, much closer to effective interaction with other West Indian states. Passage of a law greatly strengthening the autonomy of the regions of France (France, *Journal Officiel de la Republique Française* March 3, 1982,730) led to renewed commitment to intra-Caribbean collaboration and cooperation with international organizations (Martinique, Conseil Régional 1985, 83–86). The French law and the regional plan to foster economic, social, and cultural development provide the first legal and operational basis for effective interaction with other states of the Caribbean.

Political integration brought the island into closer economic relations with the mother country. One local consequence was an increase in the cost of labor, for French civil legislation, including the minimum-wage laws, became operative in the island (Mestres 1956,1947). Another effect was the immediate application, with some slight modifications, of the French social security system. Essentially free medical and pharmaceutical care is provided the population under the provisions of the *assistance medicale gratuité.* Child allotments are provided to the population by the *allocation familiale* for families on salaried incomes and by the *assistance à la famille* for agricultural tenants, small landowners, and others not on fixed incomes (Slater 1977,97–98). These allotments had been part of a governmental policy to increase population in France after World War I. When Charles de Gaulle visited Martinique in 1963, he denied that they would be withdrawn in an effort to control the population growth rate of the island, and when President Giscard d'Estaing visited in 1975, he promised even more liberal social benefits. Politically, no leader feels strong enough to modify the system substantially.

Small- and medium-sized rural properties remain the most numerous on the island, but big estates have grown larger (France, DDA 1975, 12, 146; Linares 1984,9). In 1934 the famous plantation of Père Labat, Hon. Fond–St. Jacques, developed by the Dominicans in 1658, was cut up into small lots for distribution to small farmers. The 342 ha was divided into 203 lots (Delawarde 1937a,136). But the reassembling of properties began at once, and after World War II, local observers could write that the Hon. Fond-St. Jacques was reassembled into sugarcane and banana plantations (Sylvestre and Lotaut 1947,1:325). However, some of the smaller plots in the *hauteurs* remained in peasant hands.

The Conseil Général in the late 1930s charged the agricultural service with the task of finding suitable land within the state holdings for agricultural exploitation. The motives were mixed, but among those stated was the need to relieve the congestion of towns, to increase the production of provisions, and to reestablish secondary export commodities. Between 1939 and 1941, when this lending of state lands for agricultural production was most important, one hundred lots, or 129 ha, were rented for 49,609 francs. Theoretically, the land was rented for three years at a price of 5 francs per ha (Sobesky 1940,247). The majority of these lands were in the central mountain area (86 ha in sixty-six parcels). A smaller area was let out in the forests north of Deux-Choux (40 ha in thirty-one lots). In the south, only three lots were let—two in Sainte-Luce and one in Vauclin for a total of only 3 ha distributed (Sobesky 1940,249).

Direct governmental intervention in land reallocations involved action by the Société Assistance de la Technique et du Crédit (SATEC). A regular program of land redistribution was undertaken (see map 30). Plantations were purchased, divided into parcels, and allocated (S. Graham 1969,3–4). Land-reform bills were passed between 1961 and 1964.

Map 30.
Location of areas affected by land-reform legislation.
Source: Martinique, Chambre de Commerce et d'Industrie de la Martinique 1985.

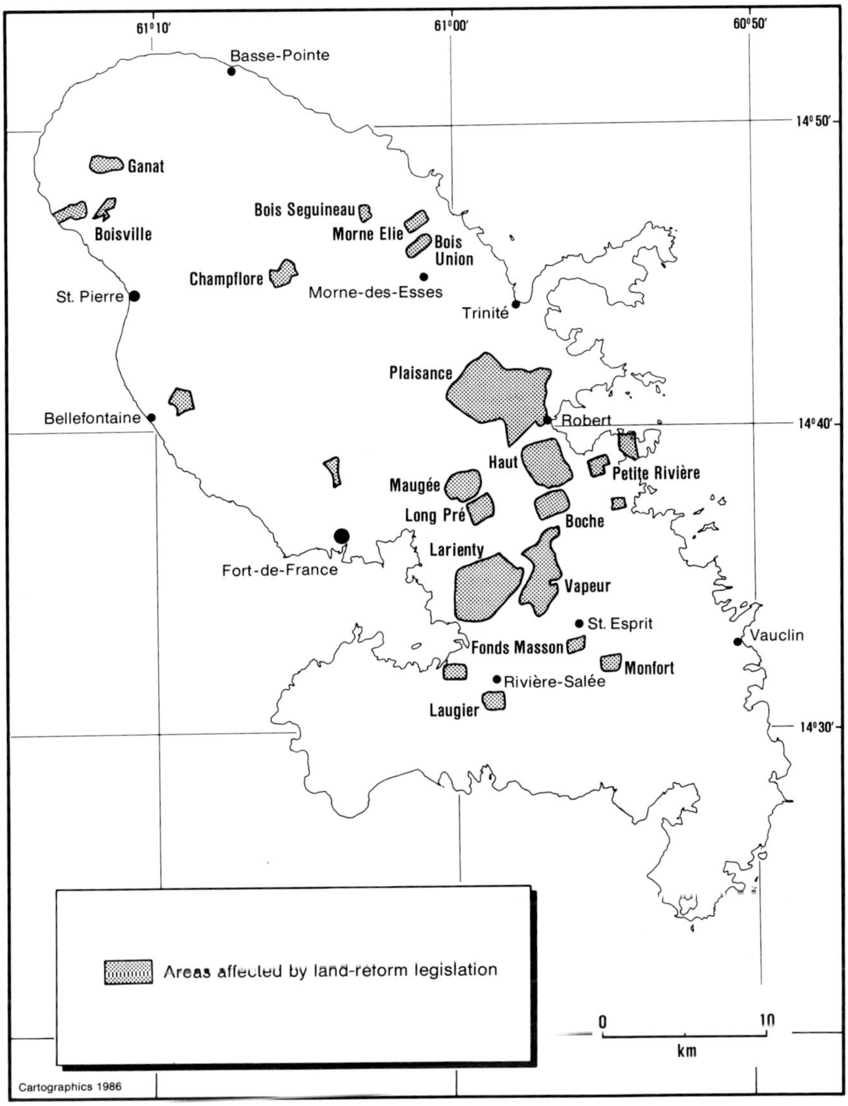

The law of 1961 amended in 1962 was administered by the Société d'Aménagement Foncier et l'Establissement Rural (SAFER), which was created for the purpose of widening the agricultural tenure structure (Massé 1979,33). However, since funding was nonexistent and purchase was contingent on the owner's willingness to sell, progress was slow. Indeed, between 1969 and 1975, 800 ha was ceded back to SAFER and other lands were sold for nonagricultural uses (Massé 1979,133). The Société d'Intérêt Collectif Agricole d'Accession à la Propriété du Larienty (SICAAP) saw to land distribution into fifty-four small parcels (Graham 1969,4). Later, the 900 ha of the Larienty estate was divided into 226 small farms and 226 housing plots.[4]

In spite of these efforts at land reform, 66 percent of the agriculture units exist on 10 percent of the agriculturally exploited land, while 3 percent of the agricultural units contain 54 percent of the land (Linares 1984,3). Since 1980, small numbers of squatters have taken over land, building houses and making provisions plantings as "witness" to their "occupation" of the land. No legal action has been successful against these squatters.[5] The distribution of small properties coincided well with the areas of high density, which have been relatively constant for forty years.

However, countering the continued effort to provide small properties for the rural population is the coalescing of larger holdings into ever-larger units. The consolidation is vividly illustrated by the large number of abandoned estate headquarters falling into disrepair. In a tier of communes from Trinité to Sainte-Luce there were more estates that had been abandoned since 1920 than there were estates operating in 1963, and even fewer existed in the mid-1980s (Benoist 1968,130–59; Linares 1984,3). The response to overseas competition by the Martiniquan sugarcane producer was to increase the size of the production unit and to plant increasingly productive strains of sugarcane. Some consolidation was accomplished de facto by the formulation of harvest contracts between smaller planters and the sugar factories. In other cases, the planters were bought out completely. Mechanization accompanied the enlargement of units and the abandonment of the hilly parts of plantations.

The road network improved during the twentieth century, with two notable exceptions. The old mule trail from Prêcheur to Grand' Rivière, still serving the cacao growers at the end of the nineteenth century, was in ruins by the end of World War I due to landslides and washouts. It has never been rebuilt, and today the adjacent lands belong to La Société pour le Conservation de la Littoral. A similar road into the valley of the Lorrain River from Ste. Marie via the Quartier Beuzaudin has also been allowed to fall into ruin. For the rest of the network there was a significant improvement of roadbeds and surfaces as well as widening and realignment of many previously existing roads. In the 1980s access to regional centers on the coast is easy and cheap for people of the interior.

During the 1980s congestion had become so great in the capital that freeways and bypasses were designed and constructed to facilitate the north-south movement of traffic as well as the transinsular traffic. Banana growers and SATEC personnel have constantly agitated for the improvement of roads in the interior to facilitate the export of bananas from the small properties in the uplands and the mountains. Sociologists joined economists in insisting that the secondary road network was inadequate (Edmond 1963,7). According to foresters, timber extraction has been slowed by the lack of roads into some of the best timber-producing forested areas.[6] Agitation for more and better roads is a political constant in all of the Lesser Antilles; Martinique is no exception. Not only did the amount of surface in roads increase, withdrawing this area from vegetative cover, the access to land they provided changed the land used in these relatively opened-up areas.

In the 1920s converted trucks and auto buses, locally called *taxis,* began to take advantage of the roads, making trips daily or thrice weekly from the main centers of the different communes to Fort-de-France, St. Pierre, and Trinité. These privately owned vehicles carried passengers, newspapers, and produce. The improved road system also facilitated the delivery of milk derived from increasingly large and numerous herds of cattle and goats. The slowly increasing tourist trade began to be served by an expansion of the fleet of conventional taxis. The number of large cars increased, and, in spite of higher fares, the local population began to take advantage of their superior comfort and speed, so now much interregional as well as intraregional traffic is by collective taxi. The number of private automobiles and trucks has increased enormously, so that traffic congestion is chronic, particularly during the sugarcane harvest.[7] This of course is not exceptional; it is the case on any sugarcane-growing island. But concerns about air pollution from motor vehicle exhaust systems are limited, and no real debate over this environmental issue is taking place.

Travel by water has declined. The canal of Petit-Bourg still carried much sugar and rum out of the Lamentin Plain to Fort-de-France in the late 1940s, but motor boat transport from Fort-de-France to Rivière-Salée was discontinued in 1947, and that to Grand' Rivière, perhaps the most isolated of all principal settlements by road, in the early 1960s. The St. Pierre–to–Fort-de-France connection by sea no longer functions, and there are no plans to reactivate intracoastal transportation.

Appreciation of nature as part of the national patrimony came late to Martinique. Fraser (1956,25–28) reported that there was a conservation attitude developing in the British Windward Islands, but it did not influence the French islands, which came to this view independently. Its direct antecedents consisted of the work of a few avid naturalists and some legislation in metropolitan France. Fathers Delaware and Pinchon stand out as the preeminent naturalist popularizers; they published many articles concerning the biota of Martinique as well as its early history. Henri and Madelaine Stehlé, who worked in both Guadeloupe and Martinique, published many botanical and plant geographical works beginning in the 1930s. Lucienne Maurice, professor at the Lycée Schoelcher, was one of the most active of the amateurs, whose influence on students, amateurs, and visitors to the island was profound. For many years the natural history research was carried on by talented amateurs who were disciplined historians, such as Jacques Petitjean-Roget and Emil Hayot. Their courtesies to visiting researchers were remarkable and invaluable. Jacques Portecôp produced a sophisticated ecological study and mapped the plant communities in 1978 for a dissertation in plant biology. Scientific competence is joined with popular concern.

Concerted efforts at conservation education in Martinique began in the early 1970s. Informed amateurs as well as the professional staffs in the departmental services wrote popular accounts to sensitize the local population to ecological concerns (Fiard 1979). Efforts to set aside certain areas as reserves for their biological and aesthetic values began, perhaps as a consequence of improved communications with France and the neighboring islands, as well as a significant rise in literacy and educational opportunity.[8] By 1974 two very important political acts rewarded the work of ecologically minded local conservationists. These were the Délibération du Conseil Régional de la Martinique of September, 1976, and the Arrête Ministeriel of August, 1976. They provided for an organizational charter for the Parc Naturel and charged the local authorities, the director of the Rural and Urban Mission and the *préfet* of Martinique Region, to carry out the order.[9]

In 1982 members of a syndicate supervised the organization and running of the park. These were l'Etablissement Public Régional, Le Département de la Martinique, La Ville de Fort-de-France, Le Syndicat Intercommunal d'Aménagement Touristique du Nord (SIATNO), Le Syndicat Intercommunal à Vocations Multiples du Sud (SIVOM), La Chambre de Commerce et d'Industrie de la Martinique, La Chambre d'Agriculture de la Martinique, La Chambre des Métiers de la Martinique, and l'Office National des Forêts. The park has a variety of functions and uses. Five objectives have been identified as being of prime importance: (1) identifying and protecting the natural environments, the historical items, and architectural patrimony; (2) promoting rural development and local products; (3) stimulating outdoor activities; (4) improving access and rest lodges according to local abilities; and (5) undertaking an information program and developing public sensitivity to the problems of the environment. Conservation education, public opinion, and enabling legislation changed the conservation movement into an active force for monitoring vegetation and, indeed, for change.

Soon after World War I the question of credit for agricultural innovations found favor among the island population. The Crédit Agricole à la Martinique was begun in 1928 and has been reorganized and refinanced several times. Functioning rather like a credit union, it has an administrator of the bank for the whole of the island and several councilors. Each commune has a local committee. In two cases they were begun by and have served tradesmen, but the rest are for small farmers and planters.

In August of 1956 the Société d'Assistance de la Technique et du Crédit (SATEC) was formed as a quasipolitical national agency to give credit but more particularly technical assistance of the sort given by agricultural extension agents in the United States to small farmers (Martinique, SATEC 1963,20). Spraying services to those who formerly could not afford the expense on their small plots, pick-up of banana stems, and marketing were done or arranged for by SATEC personnel. The Société d'Intérêt Collectif Agricole (SICA) is a government cooperative formed under the Code Rural and Code Civil in August, 1961, to aid in the improvement of agricultural techniques of the small agriculturalist, the marketing of the produce of its members, and the furnishing of supplies such as fertilizer, herbicides and pesticides, and facilities. However, it was not until the Seventh Plan (1976–80) that one stated objective was to promote local production of food products in order to reduce massive imports of edible products (*Promotion, Martinique* 1980,13, cited in Paquette 1982,37). Additionally, the plan put island self-sufficiency in food on an equal plane with export agricultural commodities and provided plans for the development of infrastructure to support this change in emphasis. In May, 1985, a colloquium was held on the campus of the Université Antilles-Guyane on the subject of Caribbean farming systems and alternatives for development. It was sponsored in part by the Conseil Régional, and investigators from throughout the Caribbean were invited to participate.

Beginning early in the period refrigerated boats for the shipping of bananas came into use, improving the quality of fruit at destination; however, the expansion of the industry was in major part tied to adjustments negotiated via customs regulations. Pineapple-canning factories were built to process pineapples grown in St. Joseph, Gros-Morne, and the Champflore Basin. After World War II, packaging and fast transport problems were solved, and Martinique began to supply a substantial amount of fresh fruit to metropolitan France.[10]

The Centre de Recherche des Antilles et Guyane Françaises was established in 1947 primarily as a plant-breeding center, but also to investigate plant pathogens and possibilities for biological control. Herbicides, pesticides, and biological control came into general use. Hoeing and weed chopping are no longer necessary on large estates.[11] Both the agricultural research center and the agricultural cooperatives began to promote the use of fertilizers by the small cultivator. Until the second half of the modern period they had been restricted to the industrially organized agricultural enterprises. The agricultural research station introduced green manures and conducted experiments with chemical fertilizers on various soils. The quasigovernmental cooperatives and credit agencies either provided fertilizer supplies to members or made them available at low cost. Local manufacture of fertilizer has not kept pace with demand, and imports continue to increase.[12]

The mechanization of agriculture began in 1935 with the use of tractors for plowing on large sugarcane estates (Robequain 1958,409). Cane continued to be hauled in the fields by mule and cattle almost exclusively until after World War II, when trucks were used to move the cane to railroad cars or to barges. By 1975 most large sugarcane estates had been mechanized. Land too steep for tractor plowing was no longer in sugarcane. Much land preparation on estates devoted to bananas was done by bulldozers. Mechanization was perceived as the means by which labor costs could be reduced to make the larger-scale enterprises economical.

Disease control, the object of considerable research, has been generally successful with the major export crops. A locally developed spray for the sigotoka disease of bananas brought this disease imported from the Pacific under control.[13] The cane borer disease is largely controlled by an imported parasitic fly. However, no really good control has been developed for cacao diseases, so breeding and selection goals are for disease-free varieties. Similarly, the coffee parasites, *Heterodera radicicola,* which attacks the root system, and *Ceneostoma caffeola,* a moth whose larvae eat the leaves (Sylvestre and Lotaut 1947,1:304), were not controlled in Arabian coffee. Hence, virtually all commercial plantings of the post–World War II era were of Liberian coffee, resistant to both infestations. It is, however, no longer economically feasible for Martinique to compete with Colombia and Brazil in coffee. Most locally consumed coffee is imported.

Ditch irrigation and later sprinkler irrigation made possible the move of banana plantings to the south windward coast and into the lower elevations. During the 1970s the large operators shifted dramatically to irrigated banana fields in what had been sugarcane lands. This put tremendous pressure on the small farmers, who could not compete in quality control or production guarantees. Drainage activities increased the amount of land available for agriculture in Lamentin and Robert communes.

After World War II the town and commune of Rivière-Salée undertook a number of drainage activities. Ditching and the planting of red mangrove or paletuvier (*Rhizophora mangle*) were used in reclamation. A 60-ha area of littoral meadow in the Lamentin Plain was drained by a canal (Sylvestre and Lotaut 1947,2:295). Large-scale ditching and drainage activities begun in 1974–75 cleared almost 1,000 ha of mangrove, which was turned into improved pasture, sugarcane, and industrial parks. The Blanche River, a tributary of the Lézarde River, was dammed in the central region of Gros-Morne Commune in the early 1960s. The Manzo Barrage built on the border of Ducos and François communes is used primarily for domestic water supplies. A small hydroelectric installation produced electricity for the southern and central part of the island until the large plant in Fond Laillet was built in the early 1980s. That plant now serves the whole island.

Livestock rearing and fishing expanded under the influence of diversification programs and the change in food tastes made possible by the relative affluence of the population. Cattle herds were improved, first by importing blooded bulls from the United States and France and later by artificial insemination. At first available only to larger planters, frozen semen was made available to the small landholder. In fact, the frozen semen is today more important to the small cultivator, who almost always has a cow or two, than to the large holder, who has large herds including bulls. These developments stimulated interest in pasture grasslands (Cléach 1974).

The availability of ice, and particularly refrigerators, in the interior made feasible the keeping of fresh fish in markets, stores, and homes. In 1963 SATEC offered seagoing fishermen aid in acquiring outboard motors for their boats and modern line and net fibers, particularly nylon (Martinique, SATEC 1963,21). Motor-driven launches were replacing the older *gommiers,* local ocean-going canoes that had a capacity of only 50 kg each. Plans for enlarging the functions of the organization were based on the hope that the expanding fishing industry would absorb some redundant agricultural labor as well as provide food to raise the nutritional level for the island population, but these have been only partly successful.[14] In the *Plan Régional 1984* several agroindustrial objectives have been identified in aquaculture and in pelagic fishing (Martinique, Conseil Régional 1985,79–81). These include efforts to cultivate conch, a highly prized shellfish used in a variety of creole dishes.

Changes in Plant Geography

Four representations of the vegetation patterns are possible for the modern period. One for 1935 followed a decline in sugarcane areas and preceded the expansion of bananas and pineapples. The second, for 1963, followed some twenty years of incorporation as a French *département,* diversification in commercial crop production, modifications of the wage structure with its influence on consumption patterns, and the appearance of a long-term timber-planting program. The third, for 1975, reflected the effects of modernization of the infrastructure, rural-urban migration, expansion of the public sector (especially increased social benefits), and modifications in technology and labor requirements. The fourth, for 1985, was witness to an urbanized, service-employed population, the impact of several devastating hurricanes, a reorganization of land tenure, and a focus on the small plot farmer following drastic declines in demand for Martiniquan sugar and a rise in demand for provisions and other vegetables. Perhaps even more important was the commitment to centralized planning for regional growth informed by regional planning technologies and chastened by a newly developed conservation ethic.

From 1921 to 1985 the area in censused planted vegetation went from a little less than one-third of the total surface area to a bit more than one-fifth of that area (see table 25). Abandoned farmland had gone into spontaneous savanna, bushlands, and reconstituted forested lands. The growth in planted pasture is a post–World War II development, with the greatest expansion in the period 1960 to 1973, when it reached 25,366 ha (France, DDA 1975,103). Other changes in the planted vegetation were the substantial increase in timber plantings, the expansion of banana plantings at the expense of sugarcane, the establishment of pineapple plantings as large-scale commercial efforts, and the modification of some previous ecological limitations on the distribution of commercial species. New to the planted vegetation types were truck gardens of traditional European vegetables. Eggplant had a short-lived preeminence in the mid-1970s when 3,540 tons (1974) were produced, but by the 1980s there were few fields, and only 439 tons were produced in 1982 (Martinique, Chambre de Commerce 1984,16). Avocados and mangos joined citrus as orchard crops, indeed replacing most commercial acreages of grapefruit. In 1982 the Office National des Forêts published a map showing the forested areas by type, *Carte Forestière de la Martinique,* and an accompanying interpretive booklet (Poupon 1982).

Table 25. Island surface in selected vegetation types during the modern period

	1920[a]		1935[b]		1964[c]		1975[d]		1981[e]	
	Ha	%	Ha	%	Ha	%	Ha	%	Ha	%
Planted										
Sugarcane	28,000	25.6	13,782	12.6	13,500	12.3	7,652	7.0	4,545	4.1
Bananas	—	—	2,141	2.0	7,700	7.0	9,094	8.3	7,215	6.5
Pineapples	—	—	200	0.2	800	0.7	858	0.7	652	0.6
Provisions	—	—	5,850	5.3	—	—	5,932[f]	5.4	6,882	6.3
Truck crops	—	—	—	—	—	—	221	0.2	310	0.3
Bush plantings	2,050	1.9	935	0.9	500	0.5	—	—	—	—
Planted pasture	—	—	422	0.4	1,000	0.9	3,049	2.7	2,191	2.0
Tree crops	600	5.5	—	—	6,000	5.4	925	0.9	1,613	1.5
Spontaneous										
Spontaneous grasslands	12,000	10.9	27,334	24.9	20,000	18.2	11,130	10.1	11,405[g]	10.4
Savanna	—	—	—	—			11,187	10.2	6,741	6.1
Bois	14,000	12.8	14,301	13.0	8,635	7.9	23,470[h]	21.4	19,494[i]	17.8
State forest[j]	7,000	6.4	7,000	6.4	7,000	6.4	10,030[h]	9.2	11,426[i]	10.4
Mangrove and semi-inundated land	6,000	5.5	8,698	7.9	3,000	2.5	2,000[h]	1.8	1,920	1.8

NOTE: Percentages are based upon an island area of 109,600 ha.
[a] Buffon 1920, 20.
[b] Martinique, *Bulletin Agricole* 5(2): 95–112.
[c] Martinique, Préfecture 1964, 24, 45; France, Service des Eaux et Forêts 1964, n.p.
[d] France, DDA 1975, 2:12; Paquette 1982, 33.
[e] France, DDA-RAG 1983, cited in Linares 1984, 36.
[f] Paquette 1982, 33 gives 14,025 ha. This difference is probably in dooryard gardens.
[g] Spontaneous grasslands in the agricultural census plus the area in mountain summit grasslands.
[h] Tillon 1975, n.p.
[i] Derived from figures in Poupon 1982, 2–3.
[j] Does not include forested land on private estates.

According to this text some 44,740 ha were in forest, a considerable difference from the usual figure, which had fluctuated between 20,000 and 30,000 ha since the 1920s (Poupon 1983,3).

The census of 1935 was conducted by personnel of the Service de l'Agriculture and the Service des Eaux et Forêts; as such, it is the first professionally conducted agricultural census.[15] Areas of twenty-two different crops and other vegetation types were given for each of thirty-two communes. From the planted vegetation types reported, data were compiled in the following categories: land in sugarcane, banana plantings, pineapple plantings, cacao bushlands, coffee bushlands, tree crops, citrus, cotton, low annuals, and provisions.[16]

The spontaneous vegetation reported in the census consisted of *savane, bois,* mangroves, and semi-inundated lands. The *savane* figure for each commune included very small amounts of planted pasture as well as spontaneous grasslands and savannas. To the *savane* figures for each commune in the census were added those for sugarcane fields in grass fallow. The *bois* figures accounted for most of the woody vegetation. They included all forests, dry and littoral woodlands, and at least in Anses-

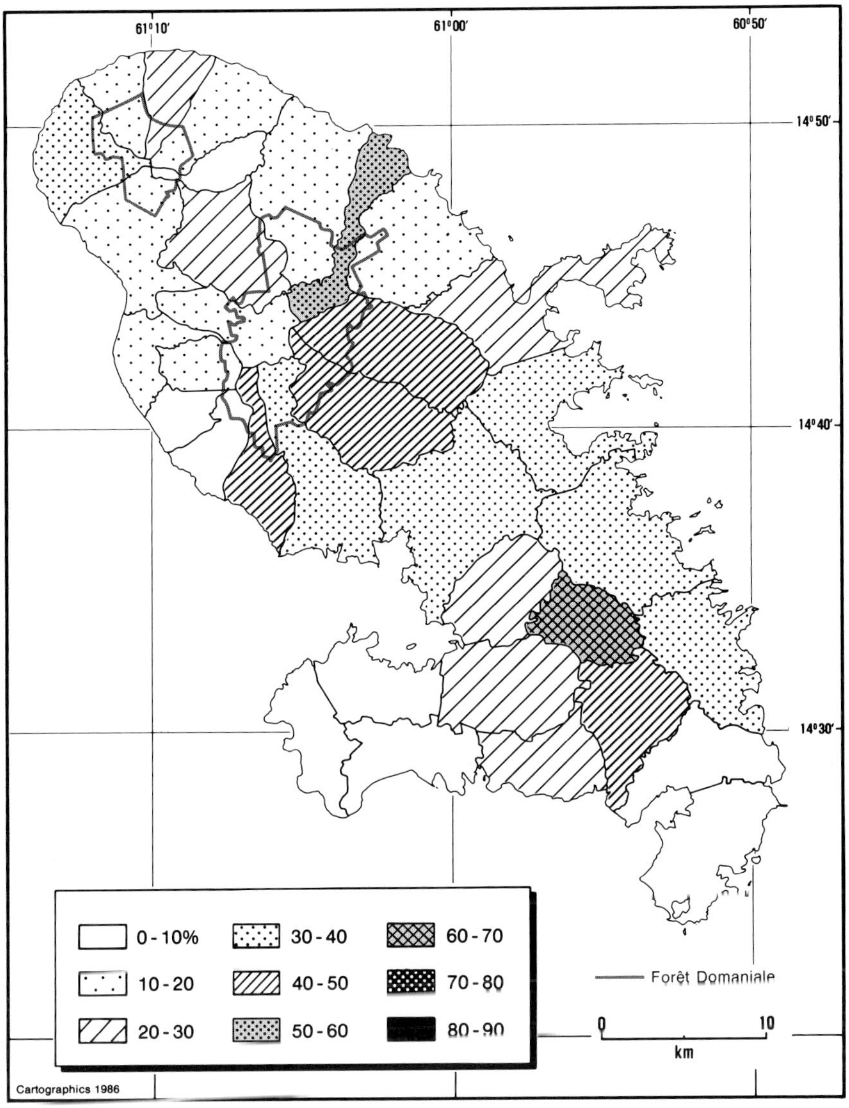

Map 31. Percentage of land in cultivation, 1935. Source: Kervégant 1936b.

d'Arlets, Diamant, and Sainte-Anne, leguminous thorn scrub, *Lantana,* and *Croton* bushlands. Probably the Prêcheur figure includes bushlands as well. The mangrove and semi-inundated lands figures as reported are ambiguous in distribution but appropriate in size.[17] At least 56 percent and probably approximately 65 percent of the island surface was in spontaneous vegetation (see map 31).

There was no census for 1963, but the actual geographic patterns of vegetation for the island were mapped at different scales in some detail. During the field sessions, the spontaneous vegetation types were identified and characterized prior to mapping. The presumed native spontaneous vegetation types are listed in table 7. Those derived from the human exploitation of spontaneous vegetated areas were the two

Table 26. Spontaneous vegetation resulting from human manipulation

Woodlands: continuous woody cover at a level of 4 m or more in height. Now considered spontaneous although originally planted, and some downslope areas are planted as a part of highway maintenance.

Bamboo brake: bamboo clumps virtually homogeneous in species composition.

Bushlands: generally continuous wood cover at a level below 4 m in height. Dense communities varying in species dominants.

Randia-Croton-Lantana thickets: dense, small stem communities, species composition varies. These thickets include *Myrcia citrifolia* and occasionally *Coccoloba pubescens,* which can reach to 6 m. Other *Croton* complexes are deciduous part of the year and may contain colonies of a large Agavaceae.

Leguminous thorn scrub: more or less closed canopy of microphyllous, thorny bushes and small trees, larger stems than in the above, containing *Myrcia* sp.

Savannas: continuous herbaceous ground cover with scattered woody growth.

Shrub savanna: scattered low, woody growth not exceeding 4 m in height. May form small thickets in key areas.

Orchard savanna: scattered trees over 4 m in height. Shrubs may or may not be conspicuous.

Herbaceous communities: continuous herbaceous cover with no important woody growth.

Fallow grassland: spontaneous herbaceous growth grazed by animals. Part of field rotation system in agriculture.

Pasture land: spontaneous and planted elements in herbaceous growth, permanent pasture.

NOTE: This table supplements Table 7 in chap. 3 for spontaneous vegetation.

savannas, fallow grassland, the *Randia-Croton-Lantana* thickets, leguminous thorn scrub, and the bamboo brakes (see table 26).

The planted vegetation is divided like the spontaneous into categories on the basis of structural criteria (see table 27). These classes are subdivided into eighteen types on the basis of the useful crops that dominate the aspect of the vegetation. Occasionally, the weedy growth may be considerable, thus confusing the aspect, and in most coffee and cacao bushlands the nurse trees are conspicuous. Gardens have a complex morphology, including many life forms. Dooryard gardens differ from provisions and truck-flower gardens in having a more complex, less formal patterning of the plants and by being located about the dwelling. Provisions gardens are dominated by tropical staple vegetables, whereas truck-flower gardens are dominated by midlatitude exotics. Until recently, provisions grounds tended to have polycultural morphologies, but the truck-flower gardens consisted of small monocrop patches (Kimber 1969). This distinction is no longer so useful (Etifier-Chalono 1985).

The tree plantations and bush plantations separate easily on the basis of species dominants, although the physiognomy is distinctive in each case. The herbaceous communities divide first between the clean cultivated crops and the herbaceous sward of the planted pasture. The low, soft-leaf annuals such as peanuts, tomatoes, tobacco, and indigo are easily segregated from the perennial crops: bananas, pineapples, sugarcane, and hay. The first three of the latter category constitute different

Table 27. Structural organization of the planted vegetation of Martinique

Gardens: many life forms or different useful species.
Dooryard gardens: heterogeneous structure and composition, trees common.
Provisions gardens: mixed plantings of vegetables and root crops.
Truck-flower gardens: "European" plants. Generally monocropped and planted in rows.

Tree plantations: dominant trees over 4 m at maturity.
Mahogany forests: timber plantation forming two-story forests.
Teak-pine-maho plantings: timber plantations, mature plantings, two-story, differentiated by species.
Coconut groves: palms commonly in rows. Occasionally useful plants grown between rows forming a second story.
Canafistula, mango, and avocado orchards: large fruit tree plantings. Only young orchards interplanted with other species.
Citrus groves: evergreen trees not exceeding 5 m.

Bush plantations: dominants less than 4 m high.
Cacao bush: regular planting pattern commonly with nurse trees. Cauliflory, large-leaved shrubs.
Coffee bush: regular planting pattern with nurse trees. No cauliflory, smaller leaves on crop shrub.
Cotton bush: no nurse trees. More heterogeneous plantings pattern.

Herbaceous crops: woody growth absent or incidental and isolated.
Low annuals: tobacco, peanuts, tomatoes, indigo grown in pure populations. Field crop pattern.
Bananas: large, broadleaf plants to 4 m.
Pineapple: terrestrial bromeliad, planted in rows.
Sugarcane: perennial field crop, reaching over 5 m at maturity.
Hayfields: fodder plants mowed for hay. Annual, rarely permitted to reach 5 m.

Herbaceous sward: low grasses. Earth rarely visible.
Planted pasture: continuous sward rarely over 1½ m.

types because of their distinctive morphology and the large areas devoted to their cultivation. The last of these crops, which during part of the year physiognomically resembles sugarcane, is separated from it according to its different use, as well as by species composition.

A medium-scale map at 1:100,000 for the whole of the island was compiled from aerial photography and field reconnaissance. Where parts of the island proved inaccessible for checking in the field, information furnished by the forestry service personnel was relied upon. From this manuscript map was calculated the areas used in the statistical treatment. Twelve sample areas were mapped at 1:10,000 to gain familiarity with the larger scale mosaic, which led to a better understanding of the vegetation patterns and the human decisions largely responsible for those patterns (Kimber 1969).

There was more land in cultivation in 1963 than was censused in 1935 (see table 25 above). Some 27 percent of the land surface was in crops, including sugarcane, bananas, pineapples, provisions, and bush and tree crops, including almost 1 percent in timber plantings. Eighteen percent of the island area was in savanna and spontaneous grasslands.

Bois, a unit carried over from earlier census, amounted to only 8 percent, the low for the modern period. State forests are reported as 7,000 ha, the same figure of 1920– 35. The mangrove swamps had been reduced a third by drainage works in the Lamentin Plain.

The island was revisited during the summer months of 1975. The areas intensively studied in 1963 were reexamined, and maps were prepared of the actual vegetation. Field mapping was supplemented by maps prepared by Colmet-Daage and associates. The census of agriculture taken later was available in manuscript form.[18] For the island as a whole, nineteen land-use categories were censused by area and number of cultivators. These are much the same units as in 1935 and permit some direct comparisons. Within the communes the breakdown into categories was less detailed, and the system used combined several vegetation types that previously had been distinguished.[19]

The planted vegetation covered a quarter of the island,[20] the forested land amounted to a fifth of the island (see table 25), and the second-growth grasslands, savannas, and woodlands are estimated to have risen to a quarter of the total land surface (see map 32). The urbanized surfaces, roadways and verges, and rural industrial complexes had expanded. Sugarcane lands had dropped to half the area they had been in 1935 and accounted for only 7 percent of the island surface.

The area in bananas was 8.3 percent of the island, well over 1,000 ha, more than that in sugarcane and more than four times the area noted in the previous census. Pineapple areas had increased slightly. Figures for cotton, cacao, and coffee areas are not recoverable, and only isolated clumps of these bushes were observed in the field. Fruit trees of all kinds increased very slightly, to 0.8 percent of the land area from 0.5 percent. Planted pasture land had risen five times.

Over a third of the forested land was in public land. Mangrove swamp areas had been reduced significantly in the Lamentin Plain borders, and the total for the island was about 2,000 ha, or a little less than 2 percent of the total. Indeed, enough of this vegetation type had been lost that the loss of bird habitat was noted, and conservation efforts to save the mangrove had begun (Fiard 1979).

The most recent census data available is that of 1981. In that census are documented a number of consequences of the change in processes and decision making of the previous ten years. One of the profound changes was that the area in cultivation was almost 10,000 ha less than in the 1973–74 census (see map 33). Sainte-Marie was still the commune with the highest proportion of land in cultivation and was the only commune with more than three-fifths of the land in cultivation. Twelve communes had two-fifths of the land in cultivated vegetation. Lamentin had dropped conspicuously from 65 to 47 percent of commune area in cultivation, largely as a consequence of the industrial and urban developments. Fort-de-France had a big drop, from 26 percent to 9 percent, again caused by suburban development, road building, and new areas in industrial and warehousing uses. The Atlantic coast communes lost except for Vauclin, which held its own, and Marigot, which more

Map 32. Percentage of land in cultivation, census of 1973–74. Source: France DDA 1975.

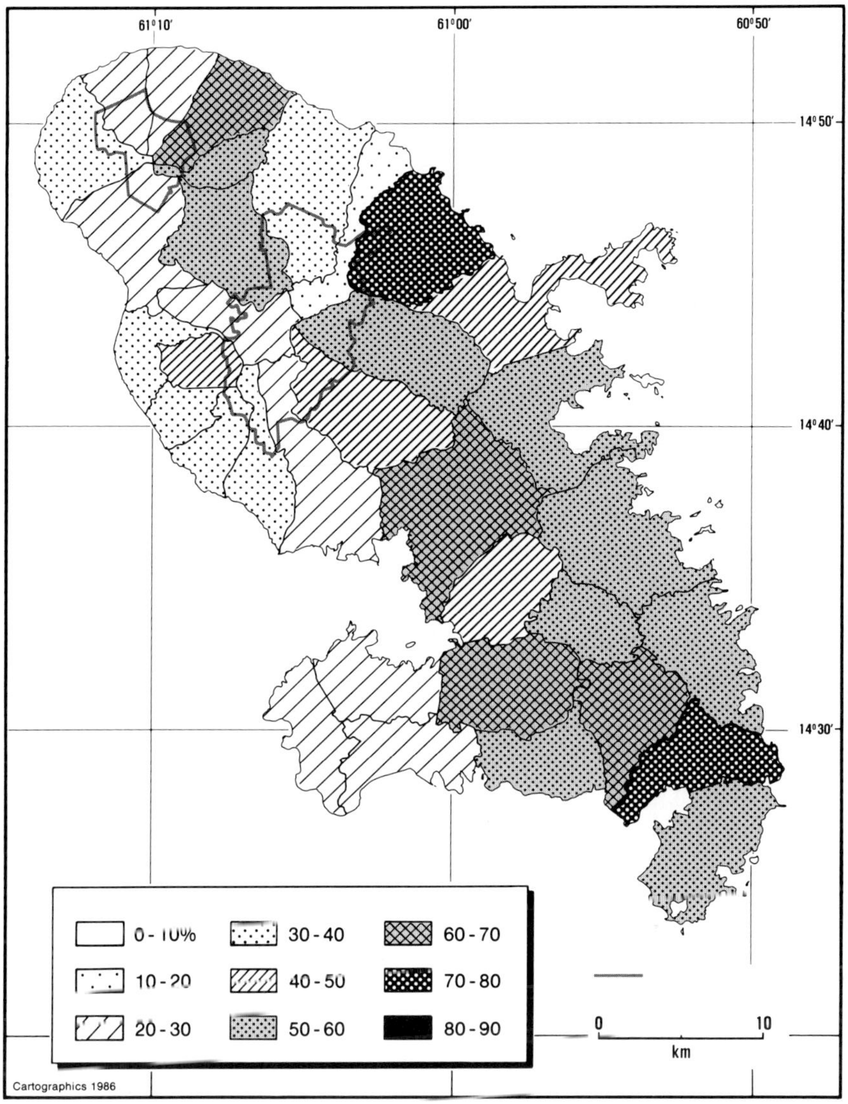

than doubled its cultivated land area. The lowland central communes of Lamentin, Ducos, and Rivière-Pilote had about the same proportion of land in cultivation, dropping to that of Ducos in 1923–24, only a little more than two-fifths of the land in cultivation. The Caribbean coast was stable, with the same proportion as in the early 1970s. The reduction by half in Case-Pilote and Schoelcher was an artifact of the method of enumeration. The southern communes all had less than 15 percent of the agricultural land in cultivation.

The planted forest areas had developed dramatically in structure and species composition as the dominant timber trees reached normal canopy heights. The Forêt de Montravail was particularly interesting physiognomically. A spontaneous second story had become well estab-

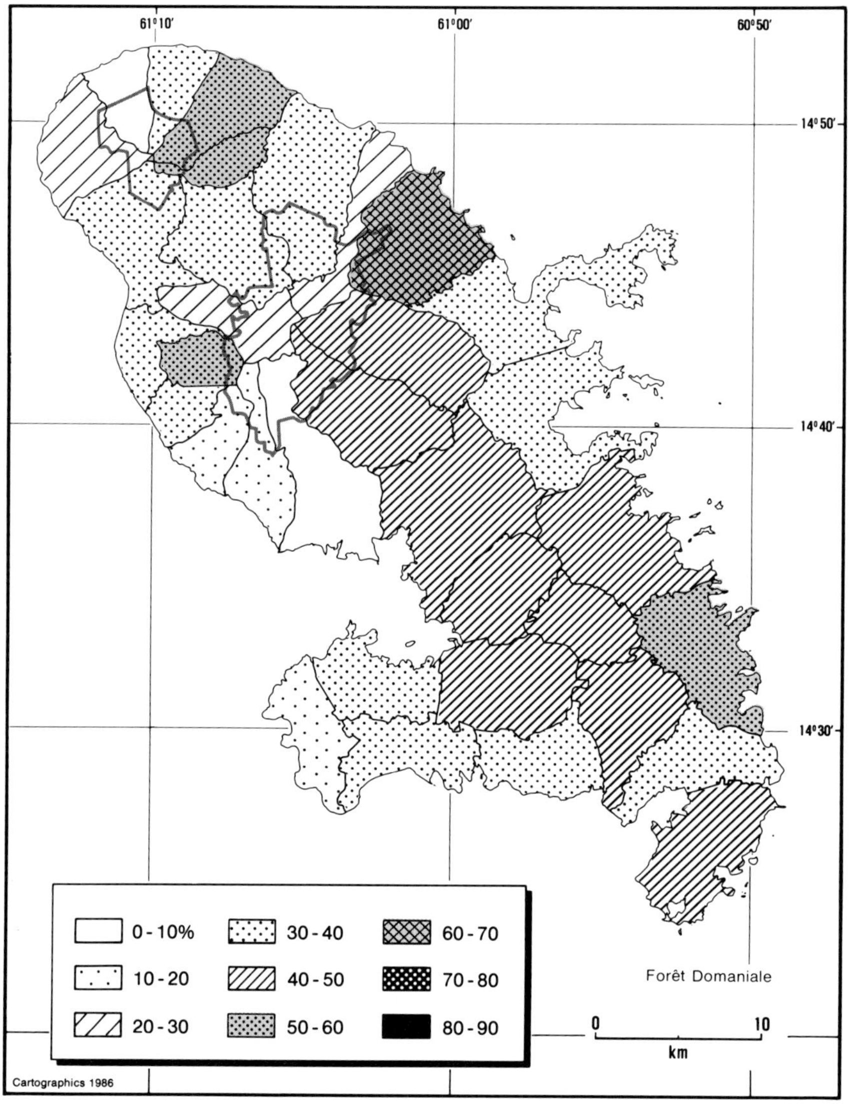

Map 33. Percentage of land in cultivation, census of 1981. Source: France, DDA 1981.

lished. The mahogany planting above the Hauteurs Bourdon was interesting, both physiognomically and floristically.

The land in spontaneous vegetation was not increased by the 10,000 ha lost to cultivation because of industrialization and urbanization. However, some part of it was abandoned land. These areas, on lands in steeper slopes away from the growing urban centers, had gone into bushlands and savannas.

The dry woodland of the Caravelle Peninsula and of the other smaller peninsulas to the south are more disturbed, even cut out, than in 1975. However, the trend observed was that the spontaneous vegetation was undergoing successional development to the extent that very good examples of all native vegetation types may well be constituted by the year A.D. 2000. The national park movement, a more explicit conservation ethic, and a more enlightened citizenry may encourage and preserve the spontaneous vegetation to an unprecedented degree. The Gallerie Géologique and the Gallerie Botanique are but two examples of the concern of the Conseil Régional about the physical, environmental patrimony of the island population.

All four ways in which the plant cover of the island may be changed were important in the modern period. Membership changes were particularly important in reconstituting the herbaceous vegetation. Culling activities declined and then expanded again. Abandonment of farmland to spontaneous vegetation caused by market loss led to a net loss of cultivated plant cover. A new emphasis on polycultural systems may lead the small farmer to make a finer grain to the cultivated landscape.

Selective utilization of the spontaneous vegetation continued to range from trees for lumber to herbs for medicine. Lasaulce's pleas for the rational control of the island's forests became policy, and forest guards were generally successful in its implementation. In March of 1922 the Conseil Général was authorized to legislate concerning state forests, and within the year the forestry code had been drawn up and passed (Martinique 1922). State, communal, and private holdings came under its provisions. In 1982 the Office National des Forêts estimated that the forested area amounted to 44 percent of the total island area (see table 28). This estimate, arrived at by aerial photograph interpretation, is probably closer to the real figure than earlier calculations represented in previous census figures.

Most of the clandestine cutting of state forests for lumber has been curtailed. Lumber for general construction, cabinet work, and furniture

Table 28. Area in forests in 1982

Type of Forest	Ha
Mangrove swamp woodland	1,920[a]
Coastal formations included in the 50 *pas géometrique* including land behind mangrove swamps; manchinel and seagrape woodland on sand beaches; dry woodland on volcanic coasts; northeastern cliffs in *Tabebuia pallida* and *Erithalis* spp.	1,700[b]
Dry woodland[c]	5,710
Dry forest degraded to *Croton* thicket	7,000
Semideciduous forest with *Simaruba amara*[d]	15,280
Rainforest with *Talauma* and *Dacryodes*	11,850
Degraded forest of high altitude[e]	1,280
Total area in forest	44,740

SOURCE: Poupon 1982, 2.
[a] Of which 1,160 is in Fort-de-France Bay and 726 in a more-or-less linear pattern, disturbed and discontinuous in the southern half of the island.
[b] As of June 20, 1982.
[c] *Forêt sèche à gommier rouge* in the original work.
[d] *Forêt semi-humid à bois blanc* in the original work.
[e] Elfin woodland and upper mountain thicket of this work.

continued to be cut, but at a much slower rate than in the previous period, something like 550 cubic meters per year—just a third of the previous rate. By midcentury, house construction even in the rural communes was increasingly of cement and concrete block, and the majority of rural buildings still of wood or wattle-and-daub in 1963 had been replaced by concrete-block houses by the 1980s.

Firewood and charcoal consume the largest quantity of wood from the spontaneous vegetation. Important at the beginning of the period, this cutting had been more or less brought under control by the application of the regulations by forest guards by the middle of the period. The principal uses of firewood in the 1950s were for firing the brick and tile kilns and for starting the fires in the sugar mills and the distilleries, which are then continued with bagasse (Martinique 1961,54). The use for the latter purpose was much reduced with the introduction of fuel oil in the sugar factories after World War II and later with the reduction in the number of those factories. Charcoal making increased directly as the population increased until the late 1940s, when kerosene began to be available in quantity and at prices the rising level of living could afford.[21] In 1947 charcoal making was still an important commercial enterprise in Fond-St.-Denis, Ajoupa-Bouillon, and Basse-Pointe communes in the north and in Sainte-Anne in the south (Sylvestre and Lotaut 1947,2:224, 193, 201). The principal market was Fort-de-France, although St. Pierre had begun to recover some importance as a consuming center. The average volume per year of firewood and charcoal extracted from forests controlled and supervised by the state was 2,606 cubic meters (Martinique, 1961,55), and much more was cut from the bushlands or *haillier* than from the forests. Then, the cutting virtually stopped. By 1963, charcoal making for the Fort-de-France market was important only in Morne-Vert and Diamant communes (Sylvestre and Lotaut 1947,2:212). During the late 1960s and the first half of the 1970s it was insignificant, but by the early 1980s the increased costs of petroleum products made charcoal and wood attractive fuels once more.

Minor selective uses of spontaneous vegetation were the gathering of fiber plants for the making of hats, baskets, and nets. The *bacaua,* the indispensable work hat of the agricultural laborer, is made from *palmier à balai* (*Thrinax morrisii*) fibers. Baskets are made from veteveria; trousers, of sisal and sansiveria fibers. The ropes used on fishing boats as well as the nets are made of the various mahot fibers, although nylon is increasingly being used for the nets. In Anses-d'Arlets during the 1940s egg baskets were made of roseau, which are species of *Gynerium* and *Arundo* (Sylvestre and Lotaut 1947,2:198) taken from the roadside ditches. In the upland community of Morne-des-Esses the old "Carib weaving" style of basketry using fibers and barks was redeveloped in the 1960s and 1970s, and some of the materials were reported to be difficult to find. However, the development of outlets for crafts has seemingly had a limited impact on these natural resources in recent years.

The old practice of collecting medicinals from the forest continued, chiefly in attempts to cure venereal diseases and to lower fevers.[22] In the

1940s the greatest number of herbalists were from Saint-Joseph Commune. In 1963 and 1975 the largest number of herbalists in the great market of Fort-de-France came from the same commune, a change from the late nineteenth century, when Ste. Marie was the center of the herbal collectors.

Assessment of the exploitation by vegetation type reveals that the rainforest stands were exploited extensively in the first part of the twentieth century and last part of the nineteenth. During the 1920s and 1930s rainforests on the east-facing slopes of Mont Pelée in Basse-Pointe and Macouba communes were selectively cut for construction timbers, cabinet woods, charcoal, and for making *gommiers* (Delawarde 1937a, 119). The timber was cut into boards and carried out on the woodsman's head. Since World War II, most of the lumber has been cut in sawmills, but even as late as 1963 forest guards would permit a single tree to be cut down and sawed in the forest and the lumber to be carried out to the small lot buyer. By 1975 the cutting of lumber in the Forêt Domaniale was all done by government contract.

In the 1930s the tree ferns were still in demand for the making of small huts on the agricultural clearings in the forest where the cultivator could rest during the day or remain overnight (Delawarde 1937a,119). There are no reports of using tree fern species (*Cyathea* spp.) in the later years of the period for building purposes, but they have been increasingly used for the nursery and horticulture trade. Since these are effective colonizers of landslides, there seems to be no excessive pressure on these species.

According to Sylvestre and Lotaut, most of the highly prized furniture and building woods of the mid-1940s came from regenerated forest islands scattered throughout the zone originally in seasonal forest. *Cedrela mexicana* was sought in Macouba, Robert, and Vauclin communes (Sylvestre and Lotaut 1947,2:260). *Guaiacum officinale*, still in use for the manufacture of mill machinery for the sugar factories, was found mostly in the forests of Vauclin and François communes during the 1880s and 1890s. Some was exploited during the 1930s. It is virtually extinct today; a few mature trees on Morne Malgré Tout west of Englishmen's Bay have produced a number of seedlings. Shortly after World War II there were forty-eight men listed as furniture makers in the capital, and there were cabinet workers in Schoelcher and Vauclin communes (Sylvestre and Lotaut 1947,1:323–48). The Bois de Beaudella and Bois de Prefontaine in Rivière-Pilote Commune, about 100 ha in area, furnished wood for cabinet work and local construction (Sylvestre and Lotaut 1947,2:282). Besides *Cedrela mexicana, Tabebuia pallida* came out of the wooded areas in Robert Commune. In the state forest areas of Gros-Morne the exotic *Litchi chinensis,* as well as the native *T. pallida* and *Simaruba amara,* was cut for furniture wood (Sylvestre and Lotaut 1947, 2:237). In Sainte-Luce the Forêt de Montravail was selectively cut of *Simaruba amara* and other unidentified lumber trees (Sylvestre and Lotaut 1947,2:295). By the end of the 1970s this spontaneous forest had been largely replaced by timber plantations, which were ready for culling in the mid-1980s. The first lumber cuttings by the Office National des Forêts of mahogany were in the Forêt de Montravail in 1985.

From the dry and littoral woodlands *T. pallida* was the most important construction wood. Some charcoal was made from *Coccoloba uvifera. Erithalis fruticosa* and *Fagara microphylla* continued into the 1930s to be cut for torches (Delawarde 1937a,130). During the World War II blockade, all lime used was manufactured on the island. The lime kilns were all located in littoral woodland or littoral hedge areas, and these were clear-cut for fuel to fire the kilns; however, most of the trees had grown back to cover the kiln sites by 1963. The poisonous *Hippomane mancinella* has been cut out of woodlands on beaches where *Thespesia populnea* and arborescent sea grape *Coccoloba uvifera* are abundant enough to provide shade for bathers. Sites where this had been done with particular thoroughness because of the tourist trade were Anse l'Etang on the Caravelle Peninsula and on the Diamant Grande-Anse. *Myrcia citrifolia,* used so intensively during World War II, was probably made more important by the disturbance activities of those who gathered the wild cherries for fresh fruit and jam.

Recreational uses of the coastal areas caused thinning of the vegetation. The understory has been destroyed in some littoral woodlands, and regeneration of both canopy and shrub species has been inhibited. Grasses colonized into open areas, especially *Cynodon dactylon* in favorite bathing beaches.

Leguminous scrub and other bushlands furnished the wood for charcoal in Sainte-Anne Commune. *Haematoxylon campechianum* (logwood) makes a high-quality charcoal that continued to be made available at outlets in Fort-de-France. Logwood for dye remained an object of exploitation until 1929. A dye factory had been installed at St. Pierre, which was supplied by small boats bringing the cut wood from the southern end of the island and also from Guadeloupe (Revert 1949a,347–48). Destroyed by the Mont Pelée eruption of 1929, the factory was never rebuilt and logwood is no longer utilized for dyestuff. It has been used continuously as a cabinet wood and for inlay work (Sylvestre and Lotaut 1947,2:295). *Cassia* sp. is a cabinet wood derived from the bushland. Ébène vert, either *Tecoma leucoxylon* or *Rochefortia cuneata,* derived from the dry bushlands on limestone soils, has been used in inlay work for furniture.

Rhizophora mangle was collected for tanning and for fuel from the mangrove swamps in Lamentin and Robert communes. All tanning in the island uses local tanning agents from the bark of spontaneous trees, most of them red mangrove.[23] Efforts at draining littoral meadow by planting mangrove trees is reported, but the extent of this planting has not been determined. However, construction of large drains in the Lamentin Plain has systematically reduced the surface in mangrove. In 1963 the Service des Eaux et Forêts calculated that 3,000 ha were left in mangrove, and by 1981 there were only 1,920 ha left of the woodland (Poupon 1982,2).

The high-altitude elfin woodland and other communities of Mont

Pelée have begun to be affected by tourism. As tourism has increased on the island, the relatively easy climb of the famous volcano has attracted the more vigorous. As has been pointed out by Claude Sastre (1978,90–91), many of the woody and herbaceous species have lovely flowers, which visitors not only pick but pull up by the roots. Trampling has also caused an undetermined amount of soil compaction on the trails and at the relatively level rest stops along the route. *Charianthus nodosus, Pitcairnia spicata,* wild thyme (*Tibouchina chamaecistus*), and *Gaultheria swartzii* are all vulnerable shrubby species because of their attractive foliage and flowers. The herbaceous *Polygala antillensis,* which had been thought extinct after the 1902 eruption, was collected in 1977 (Sastre 1978,85).

Grazing continued in the spontaneous vegetation, although the area in planted pasture increased. Livestock numbers increased during the modern period except during the blockade of 1943 and 1944, when the islanders were forced to kill off a large number for meat (Revert 1949a,335). Cattle numbers increased by more than 100 percent (see table 29). Most of this was in dairy cattle, although all beef breed herds have been expanding (France, DDA 1975,115). The encouragement of livestock raising by governmental subsidies and the provision of credit had raised local supplies in the early 1960s to the point that there was a decline in import of animals on the hoof in the customs statistics for 1961–62 (Martinique, Conseil Régional 1964,11). With the Sixth Plan of the early 1970s setting forth goals for local production to satisfy 46 percent of meat and egg requirements and 26 percent milk products, the government began a real emphasis on livestock raising. This led to a greater concern with scientifically increasing the productivity of natural range and pasture. This in turn stimulated an interest in productive pasture grasses and legumes to be used as planted pastures.

Semiwild livestock were reported in several parts of the Island after

Table 29. Livestock in the modern period

Year	Horses and Mules	Cattle	Sheep and Goats	Pigs
1922[a]	—	25,000–30,000	30,000	20,000
1935[b]	12,680	30,551	35,380	13,019
1942[c]	10,480	44,097	33,476	14,549
1943[d]	10,480	41,577	14,678	7,886
1959[e]	14,475	63,000	33,509	45,000
1968[f]	4,500	40,000	35,000	30,000
1973[g]	4,500	44,958	44,975	29,377
1981[h]	504	44,285	47,888	22,173

[a]Estimate by Bassier, quoted in Revert 1949a, 331.
[b]Census of 1935 in Martinique, *Bulletin Agricole* 1936, 5:122.
[c]Revert 1949a, 331.
[d]Revert 1949a, 331.
[e]Martinique, *Annuaire* 1961, 62.
[f]Loslier 1971, 100.
[g]Martinique, *DDA Service Statistique* 1975, 103, 115.
[h]DDA Douanes, cited in Martinique, Chambre de Commerce 1984, 15.

World War I. The western flanks of Mont Pelée had grown up into spontaneous bushlands and woodlands. Flocks of goats roamed wild, grazed on the sparse herbage, and browsed on the shoots of bushland species. In the 1930s dogs were used in rounding up these semiwild animals (Delawarde 1937a,104). A decade later Revert (1949a,336) wrote with reference to this area that "goats live outdoors all year round and go unfenced." Pigs in this area had become feral also in the 1930s and occasionally raided the gardens and provisions grounds of the small cultivator (Delawarde 1937a,178), although it had been the custom to tether pigs. In Diamant Commune, cattle and goats, and in Sainte-Anne and Marin communes, sheep and goats ran on large properties. Those in Diamant were in leguminous thorn scrub and regenerating dry woodland. In the eastern communes the animals were in savannas, in *Lantana* and *Croton* bushlands, and in occasional patches of littoral woodland and dry woodland. In 1964 and 1975 cattle were seen roaming free in the eastern end of the Caravelle Peninsula within the mosaic of savanna, seasonal forest, and littoral forest vegetation types. These were reported as parts of domestic flocks. Goats browsed in most of the small headlands of the southeastern littoral still in woodlands. However, nowhere have goats or sheep been allowed to degrade the landscape to the same extent as in Saint Kitts, Nevis, Antigua, and other parts of the Caribbean.

Although herds were building up in the mid-1970s, 60 percent of the cattle belonged to livestock holdings of fewer than nine head (France, DDA 1975,1). The more affluent peasants graze their cattle on their fallow grasslands. They may also permit cattle grazing by a neighbor who has no fallow land of his own. The manure is recognized as adequate payment (Horowitz 1967,29). In recent years this use of grassland fallow had proved profitable, and much of the landscape in the communes with high rural densities resembles pasture land. More than 12,000 ha of "natural" pasture are used in this way (France, DDA 1975,2; DDA 1981,n.p.).

Supplemental feed for horses, as for dairy cows, is generally cut and brought to the confined animals, which graze in fenced pastures. *Stenotaphrum secundatum, Andropogon* spp., *Panicum purpurascens, Paspalum dilatatum,* and *Panicum maximum* are all used as green hay. Other feed is derived from the mowings of the verge and from culling of the tops of garden *Ipomoea batatas* and *Dioscorea* spp. During the sugarcane harvest the small cultivator with a few head of livestock will try to get a supply of the "tops" of the sugarcane stalks to mix with molasses for feed.

Leaves of trees, especially during the dry season, are collected and given to the animals. *Acacias* spp., *Pithecellobium unguis-cati, P. dulce,* breadfruit *Artocarpus altilis,* and *Gliricidia sepium*[24] have been used as supplemental feed. During the dry season the grazed grasslands and savannas are generally burned, in part to furnish an ash dressing and to make room for new shoots, but also in part to destroy the larvae and eggs of numerous parasites and insects such as ticks and "*bêtes rouges*" (Delawarde 1937a,107).

For many years small cultivators butchered all but the reproducing stock during the dry season because of lack of water. With the bringing of water into the southeast, this practice is now less common, although according to informants, it was still standard in Diamant and Anses-d'Arlets communes in 1963. There were no reports of this practice in 1975, although between 1971 and 1973 some herds were cut because of the severe drought in the dry communes.

Clearing of forested or wooded land for the making of agricultural plots continued into the middle of the modern period. The Règlement Forestier of 1922 does not prohibit clearing; rather, it provides that clearings be authorized. Indeed, recognizing the impossibility of an immediate reversal of what had been appropriated as a right, officials made no effort to refuse all clearing, but directed attention toward intensifying "development." In 1925, according to a proclamation of the governor, forest guards were to "encourage among all small cultivators the development and intensification of provisions production."[25] The tendency was to encroach upon the forest borders, particularly those against state forests during the earlier part of the period, in spite of the drawing of boundaries and the employment of forest guards. The discipline and reorganization of agriculture under Admiral Robert during the blockade put an end to this casual tolerance.

Not all clearings went into permanent agricultural production. Even as late as the mid-1930s distinctions could be drawn between the *dégras* and the *bitué* (Delawarde 1935,95). The *dégras* were temporary clearings or swiddens "on the summits" to be tilled without benefit of manures or fertilizers; they were intended for three or four harvests before being allowed to revert to forest, that is, swiddens. The *bitués*[26] were particularly desirable agricultural clearings intended, because of either convenient location or superior soils, for permanent exploitation. The *bitué*, when cleared, was surrounded by fresh cuttings of *Gliricidia sepium* or gommier rouge *Bursera simaruba*, which eventually formed living fences. Alternatively, windbreaks were formed from trees left for the purpose.

Clearings in the seasonal forests and rainforests of Basse-Pointe and Macouba communes were made by the large landowners and in the south Atlantic communes in order to provide pastures for new cattle herds. The forest was bulldozed, the good timber was sawed into lumber, and the rest was left to dry and then was burned. The resulting cleared space was plowed and then kept free of shrubby species by periodic chaining[27] and by the use of herbicides. Clearing of leguminous thorn scrub and the developing dry woodland took place in the south in Diamant Commune, where citrus orchards and pastures, as well as coconut plantations, were made after World War II. Similar clearings were made in the Caribbean coast in Carbet and Prêcheur communes for pastures and coconut groves. Considerable expansion occurred in the 1970s and early 1980s.

It had become common practice during the modern period to burn the savannas and improved pastures annually in the seasonally dry areas of the island. Together with the practice of planting clones of imported grasses, this had effectively modified the species composition of the savannas.

Changes in planted vegetation have been by plant introductions and internal migration of vegetation types. The principal motives behind plant introductions in the modern period were the improvement of forage by exotic grasses and legumes and the improvement of commercial crop varieties. The agricultural research center at Duclos in Guadeloupe was responsible for some introductions, but well before it was formed, more species and varieties were imported by the estate owners and members of the research staffs of the gardens at Tivoli and Desclieux, which took over plant introductions when the botanical garden in St. Pierre was detroyed by the eruption of 1902.[28]

Natal grass, herbe à laine rose (*Rhynchelytrum repens*), and kikuyu grass (*Pennisetum clandestinum*) were introduced early in the modern period. Natal grass was introduced from Africa as pasture grass by the Station Agronomique in 1920 (Stehlé 1936,1:37). It has since naturalized in spontaneous grasslands, along roadsides, and in savannas in dry areas with montmorillonitic clay soils. Kikuyu grass, also from the South Africa, was imported to Martinique by Kervégant of the Service de l'Agriculture (Revert 1949a,335). Later in the period elephant grass (*Pennisetum purpureum*) was introduced in a number of varieties.[29] The elephant grass proper (*P. purpureum* var. *elephant*) was most successfully used for hay in the mountain-belt climates, where it was grown in pure populations, harvested periodically as hay. The variety called merker (*P. purpureum* var. *merkeri*) was introduced after World War II and proved very successful in a variety of climatic and edaphic situations. It was often used to improve already existing pastures. Pangola, also called digitaria by many cattle raisers (*Digitaria decumbens*), was an even later introduction by the Centre de Recherche des Antilles at Duclos (France, INRA 1950,16). This grass had revolutionized the livestock industry by converting farmers to planted pasture because four to five can be raised per ha as opposed to one head on natural pasture (Loslier 1971,100). Its persistent stolons had made it a poor species for pasture within a crop-rotation program, but it was used very successfully for permanent pasture development in low and midelevations. Guatemala, or Honduras grass (*Tripsacum laxum*), was introduced from Central America as forage grass, possibly also during this period (France, INRA 1950,17).

Among the leguminous introductions to the gardens at Tivoli was kudzu (*Pueraria phaseoloides*), a native of tropical Asia that as a green manure turned out to be a mixed blessing (Stehlé and Stehlé 1960,3: 86). In the dry-belt areas of the island it is useful in weed control and as a nitrogenous ground cover. In the middle and mountain belts it is an aggressive weedy climber, reaching to the tops of mature bananas, draping them destructively. The Centre de Recherche des Antilles introduced a number of tcha-tcha (*Crotalaria* spp.) and tested them for nitrogen fixation for use in coffee, cacao, and sugarcane lands. Some of these species are also good forage plants. Yellow pear or grandes sonnetts (*Crotalaria juncea*),[30] though already listed in Duss (1897,194), was reintroduced by the research station, and according to Stehlé and Stehlé (1960,3:44) "all the plants seen in the islands between 1934 and 1958 were in plant introduction centers, research installations, and private flower gardens." It has the greatest nitrogen content of all the species tested in 1953. Blue shak-shak or pistache violette (*C. verrucosa*), native to India, is found in sugarcane fields, pastures, and banana plantings.[31] Another leguminous plant useful for ground cover during the dry season that has become naturalized is pois jaune (*Vigna hosei*), a native of Borneo. It was introduced to the French Antilles in 1928 (Stehlé and Stehlé 1960,3:64).

Sisal (*Agave sisalana*) was introduced in 1924 as a possible commercial fiber-producing plant, but plans for developing this plant commercially were abandoned shortly after its introduction (Kervégant 1936a,58). However, it was still being recommended along with cotton as a good dry-region crop, and the fiber is used to make cloth for some rough clothing.

The plantation timber-tree introductions were big leaf or Honduran mahogany (*Swietenia macrophylla*), Caribbean pine (*Pinus caribaea*), and blue maho (*Hibiscus elatus*). Mahogany was an early introduction. Blue maho was introduced after 1960 (Kimber 1970,235). In 1979 the plant-introduction center nursery at La Donis had some thirty potential timber species under examination.

Hedge plants useful for their leaves were sought to supplement forage during the dry season. Pithecellobium (*Pithecellobium dulce*) was introduced from Mexico (Kervégant 1936a,28–90). This small tree, naturally reaching 5 to 10 m in height, makes a good living fence and will take pollarding. Its leaves were analyzed and found to be a useful feed supplement for cattle. Gliricidia (*Gliricidia sepium*) was reintroduced in the 1930s as a nurse tree for the coffee and cacao plantings, although originally it was brought in at the end of World War I, perhaps even just after the turn of the century. Its canopy characteristically passes more light than other nurse trees but gives adequate shade. It is said to be less susceptible to wind throw than are the native trees and earlier introductions, and it became the preferred nurse tree before the abandonment of these plantings. It was also found to be a fast-growing species for living fences. In the dry and littoral belts the tree is often used to enclose pastures because its leaves may be stripped during the dry season for use as fodder for cattle.[32]

New plant additions to dooryard gardens continued to be made, but the small ranges of tolerance of these species tended to limit most to only a part of the island. Gliricidia and kudzu were well distributed in gardens of the dry and middle belts. Sisal was introduced into gardens in the southeastern part of the island, and the yellow pea into some on the Caribbean north coast. There continued to be more fruit trees in these gardens than in monocrop orchards, and the greatest production came from these mixed plantings (France, DDA 1975). The areas dominated by these gardens remained where they had been in the previous period, although division of some former estate properties slightly increased the total surface in gardens, and land reform brought some prime agricultural land into the hands of small cultivators. Gardens continued to supply wood for charcoal and construction materials until after 1950. Fruit trees and shrubs having over a third of their numbers in dooryard gardens are the oranges—mandarin, sour, and sweet—breadfruit, nongrafted mangos and avocados, coffee, cacao, and the prunes (*Spondias* spp.) (Paquette 1978,19). The proportion of the population having access to or living on this kind of multicropping system was reduced during this period. There was no change in the bordering types and no significant regional movement of the borders.

Land in provisions gardens declined steadily, from the 12,300 ha of 1920 to a low of 4,200 ha in 1942 (Revert 1949a,319). The blockade caused a brief expansion of provisions lands to 7,000 ha in 1943, but the area declined to 3,500 ha in 1945 (Revert 1949a,338). In 1963 the land in provisions and low annuals was estimated at 4,000 ha, suggesting that the area in provisions gardens had remained relatively constant in those twenty years (Martinique, Préfecture 1964,48–50). By the early 1970s provisions were being censused with "cultures maraîchères" or "légume d'europe," which have been called truck gardens. There was a loss of provisions lands from the southern communes of the island, which had provided some 32 percent of the total agricultural area in 1935 (Martinique, *Bulletin Agricole* 1936,5(2):112).

There are several reasons for this decline. The first may be attributed to the increasing importance of imported foodstuffs made possible by the increased buying power of the majority of the population. The change in food habits associated with modernization is most pronounced in the area of starches; bread made of imported flour substantially replaced manioc flour products by World War II, thereby reducing the area in this crop (Revert 1949a,318). Canned goods and groceries imported from Europe sell in the grocery stores along with locally grown foodstuffs of the markets. Supermarkets, which opened for the first time in 1961, are stocked almost entirely with imported supplies. The buying habits of the urbanized population are changing, as more people are buying in one centralized location, rather than in the small specialty shops.

Some provision lands were replaced by bananas and a smaller proportion by truck gardens. The rest reverted to spontaneous vegetation or were put into pasture to support the increasing number of cattle. This vegetation type was on lands less favorable for cultivation of export crops because of steep slopes, poor soils, or droughty climates. By the mid 1970s the greatest concentration was in that part of those lands having good access to markets in the capital city and the other towns.

Very occasionally cacao and bananas are mixed in with these plants in provisions gardens. Bananas and tomatoes are frequently found together in Ajoupa-Bouillon, Grande Savane, while dasheen and bananas are found in Macouba. In Gros-Morne young plantings of Trinidad cacao were in the same mixture.

Local manufactures have been developed with varying success to stimulate the use of local crops and spontaneous plants. Chocolate factories were set up in the 1940s to use locally produced cacao and vanilla (Sylvestre and Lotaut 1947,1:305). A yam-processing plant was established in Sainte-Marie Commune to develop a powdered product that can be used to make instant mashed yams. But these manufactures have been less important in maintaining the cultivation of provisions than has the establishment of a large Martiniquan colony in Paris, which has provided a small export market for creole provisions in France. These products are airlifted to France, as are anthuriums, green beans, and eggplant.

After World War II, demands for European vegetables increased, and home gardens were not adequate for the demand in the population. The commune of Morne-Vert led in truck farming largely because of the influence of a metropolitan, Mme. Haumond. The field was carefully prepared; the individual rows of vegetables and flowers, which include carrots, leeks, celery, cabbage, string beans, parsley, onions, and cut flowers (South African daisies, gladioli, dahlias, and the like), were carefully weeded. However, tomatoes were field planted in such a garden. Fond-Saint-Denis Commune followed its neighbor in attention to this vegetation type. Most of these truck garden species are grown on small holdings, although some large fields of eggplant were grown by planters and subjected to wild fluctuations in the world market (Paquette 1977, 21–22). By the beginning of the 1970s these gardens were so generally dispersed that the collectors of agricultural statistics combined provisions and truck-garden produce in the last two censuses.

Canefice (*Cassia fistula*) orchards already established in Carbet, Case-Pilote, and Diamant were given desultory care throughout this period. A small quantity of exports, about 30 metric tons per year, during the first half of the period reflects not only the small European market but the remnant character of these orchards. Some of the canefice is collected from spontaneous trees colonizing the savannas of *côte-de-fer*. Kola (*Cola acuminata*) plantings initiated during the last period began to come into production during this period (Sylvestre and Lotaut 1947,2:304). Nuts are used locally to make a *vin de Kola*, and some nuts are exported to France. All trees are in small-sized orchards in Gros-Morne and Fond-Saint-Denis communes. They are situated in among the provisions gardens and cacao plantings or in secondary thickets regenerating in the previously cultivated lands.

Efforts begun in 1960 to develop markets in France for mangos (*Mangifera indica*) and avocados (*Persea gratissima*) have met with moderate success. Plantations were established in Prêcheur, Morne-Vert, and Carbet communes. In suitable sites mangos are interplanted with bananas to provide an income crop during the years the trees are coming into maturity. These plantings are at the expense of banana lands in Carbet, of second-growth seasonal forests in Prêcheur, and of savannas and provisions in Morne-Vert communes. Fruit trees have always been part of the dooryard gardens. What is new is the establishment of orchards for a local and an export market.

Coconut groves (*Cocos nucifera*) increased in importance as landscape features, especially after World War II. These plantings have always been on holdings of all sizes and tenure, but orchards are larger and more likely to be monocropped on large holdings. Plantings were made in virtually all the larger stream valleys of Prêcheur, reaching into the hills. The Prêcheur groves border banana plantings and provisions gardens as well as dry bushlands. The littoral on either side of Carbet for 5 km was covered with coconut groves. These were at the expense of second-growth seasonal forest and *Lantana* bushland, respectively. The Carbet plantings were between the road and the sea. Additional plantings of a substantial sort were made in the crowns of spurs in Carbet and Saint-Pierre communes, in Lamentin, on the plain between the sugarcane and littoral meadows, and in the southern peninsulas. Coconut groves, unlike the scattered coconut trees in dooryard gardens, are limited to the littoral and dry-belt zones. They are frequently bordered by mangrove swamps and littoral woodland as well as littoral meadow, sugarcane fields, and banana plantings. Labor costs have made the *coco nain*, a dwarf coconut variety, a favorite for new plantings, which is changing the lowland landscapes.

The copra oil was used in making margarine at a local factory, but some time before 1949 the use of coconuts began to be limited to fresh ripe coconuts for confections, and especially green coconuts for the coconut milk. The coir was used for caulking boats and stuffing mattresses, and the leaves in basketry.

The Chamber of Commerce and Industry is actively supporting the growth of agroindustries. The model example used in its report on possible agroindustries is the case of coconut punch, "the production of which was started in Martinique when there were no organized coconut plantations and coconuts even had to be imported." In 1984, because of the growth in sales, several coconut plantations were established, and there was talk of other manufactures, such as oils, shampoos, and creams, being developed to capitalize on the established enterprise (Martinique, Chambre de Commerce 1984,43).

Since 1924 the Service des Eaux et Forêts has planned to replace systematically as much of the spontaneous forest as possible with timber plantings. The first species planted was Honduran mahogany (*Swietenia macrophylla*); the next, Caribbean pine (*Pinus caribaea*) (Martinique 1961,54). When the departmental bureaucracy was reorganized, forests were put under the supervision of the Office National des Forêts (ONF), and a similar policy was followed. Timber plantings were made on private lands as well. The plantings were few and small in size between 1924 and 1947. Expansion increased more rapidly after 1948, so that in 1958 some 970 ha of timber plantings were reported. In 1970 the plantings amounted to about 1,000 ha. The total by 1982 was 1,300 ha in timber species, 1,200 ha in the Forêts Domaniales, and 100 ha in private holdings (Poupon 1982,2).

The best-established species of the timber plantations is the Honduran mahogany, also called the big-leaf mahogany. The second in importance is the small-leaf West Indian mahogany (*Swietenia mahagoni*). Plans called for at least 1,500 ha in these two species alone by the end of

1985. Considerably less important in areal extent is teak (*Tektona grandis*), although it has been planted almost as long as has big-leaf mahogany. Experimental plantations of tropical pines (*Pinus caribaea* and *P. occidentalis*) have proved successful. Simaruba or bois blanc (*Simaruba amara*) and blue maho (*Hibiscus elatus*) are still experimental.

Most of the first plantings of mahogany were made with the help of the small-plot farmer. These took place on land cleared by cultivators who had been given state lands to cultivate with the understanding that timber trees were to be planted before final abandonment of the plot. Government nurseries were set up to furnish mahogany seedlings started under local conditions. The strip or *en laôn* planting system in which lines of the native forest are left to form a forest community for the planted species was in favor in the mid 1960s but has been abandoned.[33] Teak plantings were made in hand-hewn forest clearings and thus form pure populations of canopy trees. The most recent clearing has been made with heavy machinery, but the nursery-grown seedling stock is hand planted. Since 1963 most timber plantings have been carried out by forestry personnel.

Almost all of the mahogany plantings suffer from infestations of shoot borer disease, although the damage is more extensive on the windward slopes (Lamb et al. 1960,51). The *en laôn* system cut down somewhat the rate of infestation by providing buffer areas between planted rows.[34] Sawmills were not to be set up by the Forest Service until 50 ha could be harvested at a time. This was not until 1985, when logging began in the Forêt de Montravail. However, small-scale cutting by local furniture manufacturers, coffin makers, and carpenters has been going on for some time.

Limes and lemons were grown in orchards for commercial supplies to process into juice and for the extracting of oil from the rind. Oranges of various species and varieties, including mandarin oranges and shaddock (*Citrus decumana*), appear in dooryard gardens. Most of the groves were in Sainte-Luce, Fort-de-France, and Saint-Joseph communes. No groves have been planted since World War II except in Vauclin Commune during the early 1980s. The latter were abandoned because praedial larceny caused the farmer to lose his crop. Oranges from the flanks of Mont Vauclin of Saint-Esprit Commune are well regarded locally (Sylvestre and Lotaut 1947,2:305). Colmet-Daage and his colleagues mapped very small groves in 1972, but most plantings are not recoverable at the scale of 1:50,000 (Martinique, *Carte des Cultures . . . 1976*).

During the modern period the principal expansion in bamboo brakes has been under the auspices of the highway maintenance staff. On the steeper downslopes of highways, bamboo makes an effective soil binder and retards erosion, and it is thought to reduce landslides significantly. The areally more significant plantings have been made along the roads straightened in the mountainous interior. Bamboo apparently is spontaneous in cliffy and steep slope sites in the middle-climate belt regions of the interior and east coast, bordering rainforest. The bamboo brakes in the cultivated zone are found in swales within sugarcane fields, banana fields, and pineapple plantings. These have become important habitats for animals previously occupying woodlands.

The cotton bush became less and less important in Martinique during this period. In 1935 cotton occupied 57 ha. In 1942 the total was 73 ha, the increase owing to the artificial stimulation of the blockade during the war (Revert 1949a,340). By 1963 the area had dwindled to a few remnant fields of less than 1 ha each in size, located in the hill country of Diamant and Anses-d'Arlets communes. These were located in the secondary leguminous scrub or next to provisions gardens and dooryard gardens. In 1960–61, in response to the cotton crisis in the British Antilles, new attempts to introduce Sea Island varieties occurred. Plantings subsidized by the Service Agricole were made in swidden clearings in the bushlands of Sainte Anne Peninsula. Efforts were not successful, and by the mid-1980s the only cotton plants were those in dooryard gardens and spontaneous individuals in Atlantic seashore sites.[35]

Cacao had been generally neglected since 1917, never having recovered from the destruction of the plantings by the eruption of Mont Pelée in 1902. Varieties grown in the 1940s, according to Sylvestre and Lotaut (1947,2:205), included Criollo, Venezuela, Forastero, and Calabacillo. In the late 1950s new clones, some hybrids, were introduced from Trinidad by the Service Agricole for plantings in different parts of the island in order to select the superior varieties for each locality. By 1959 more than 1,200 plants had been distributed to small cultivators in Lorrain and Rivière-Pilote communes.[36] Farm advisors then began an effort to renovate old plantings. Those old plantings not susceptible to upgrading by selective replacement because of heavy pathogen infestations were entirely replaced with the aid of government subsidies. The Institut Français du Café et du Cacao (IFCC) strongly supported the expansion of cacao areas in Martinique.[37]

The distribution of cacao plantings has not changed significantly during the modern period, but its importance has declined drastically. In 1936 Grand' Rivière, Prêcheur, Rivière-Pilote, and Vauclin communes had the largest areas in cacao (Martinique, *Bulletin Agricole* 1936 5[20]: 112). In 1947 Sylvestre and Lotaut listed Grand' Rivière and Macouba, Prêcheur, and Rivière-Pilote as the largest producers. In the uplands of Vauclin the cacao bushlands lost area to savannas. By 1973–74 cacao, like coffee, was censused by sampling because it was grown on small properties and in very small plantings (France, DDA 1975,115).

In 1935 the largest areas of coffee growing were in Saint-Joseph Commune, Gros-Morne, Lorrain, and Macouba communes. Coffee exports had declined to 30 metric tons in 1947. Parasitic diseases continued to plague the crop. Liberian coffee increased in area. In 1947 the largest plantings were in Macouba and Basse-Pointe (Sylvestre and Lotaut 1947,2:262). By the early 1960s four northern communes, Macouba, Basse-Pointe, Ajoupa-Bouillon, and Morne-Rouge formed one area of production. Saint-Joseph Commune was the other. Scattered bushlands in Morne-Vert, Lamentin, and Rivière-Salée were also located in the fields. They were surrounded by a variety of planted vegetation types. In contrast to its position of support for cacao, IFCC did not encourage coffee expansion (Coste 1959,15). Coffee persists only as a dooryard garden plant in those communes where it once flourished.

Tobacco areas fluctuated during the modern period, but no real change in distribution took place. Grand and Petit Martinique are the two varieties grown. Macouba plantings expanded during World War II, but the enterprises failed because of unskilled labor (Sylvestre and Lotaut 1947,2:262). Some tobacco continued to be grown in dooryard gardens for local use.

Peanuts declined in importance during the modern period. They are grown in pure populations or in fields often bordered by corn (*Zea mays*) plants. Requiring a loose, friable soil with good drainage, they are restricted to the shallow volcanics of the leeward slope. Fields of peanuts have been historically located in Carbet, Morne-Vert, and Prêcheur communes. All production is locally consumed. In 1935 there were 68 ha enumerated in the census. In 1942 that number had expanded to 396 ha but lost ground to provisions in 1943, with only 54 ha reported (Revert 1949a,319). In 1963 some 60 ha were estimated in nuts. The census of 1973–74 makes no separate entry for peanuts in the summary table. The peanuts are harvested in three to five months and may be planted continuously for a number of years in the same ground. Most of the land in Morne-Vert now in truck garden crops was formerly peanut lands. Peanuts are grown for home consumption and sale by street vendors. The decline has been substantial enough that one may say that peanuts are no longer a significant landscape feature in any region of Martinique.

The expansion of area in banana plantings is one of the principal developments in the plant cover history of modern times. The 2,141 ha reported for bananas in 1935 had expanded to 8,500 in 1963 and 9,094 ha in 1973 and then fell to 7,215 in 1981 (Kervégant 1936b,115, Beaulieu 1965,140; Linares 1984,36). This expansion was due primarily to the creation of a banana-boat fleet and to the protected market offered by France. Not only did the area in banana plantings increase, the location of banana lands was shifted during the modern period. Disease and water had been the two limiting factors; when these were relaxed, expansion into other environments proceeded as rapidly as the market expanded (Delawarde 1937a,38; Robequain 1958,411). Reduction of demand for West Indian bananas in the European Common Market is largely responsible for declines in production area. However, Hurricanes David (1979) and Allen (1980) resulted in great monetary losses, and some farmers decided against the investments required for replanting.

For most of the modern period before World War II and for a decade after, bananas for export were grown with family labor in those parts of the island with high rural population densities. After the war small planters gathered together to form the Société Fruitière Antillaise, which developed cooling facilities and packing sheds in the port. The difficulties besetting the sugarcane industry moved some planters to diversify.[38] The conversion of banana production from a family-style operation to a large-scale plantation crop and the formation of the Société d'Intérêt Collectif Agricole Bananière de la Martinique (SICABAM) for cooperative marketing helped rationalize production and lower costs. World War II put most of the small banana farmers out of export production. It was not until the formation of SATEC and SICA in the 1960s that some of the benefits of the cooperative movement were extended to small cultivators. However, by then the momentum of the large planter groups made it impossible for the small farmer to compete in the export market.[39] By 1973 bananas exceeded sugarcane as the large plantation crop. At the beginning (1921) 70 percent of the area in bananas was in small- and medium- sized holdings, but between 1940 and 1963 less than 38 percent of the land in bananas was in holdings 10 ha or smaller.[40] Bananas expanded into the lands taken out of sugarcane in the Atlantic north coast and much of the Lamentin Plain, and in savanna and sugarcane lands of Vauclin and Marin. Large commercial export banana production was missing from the lower tropical subhumid and lower tropical dry zones of the *côte-de-fer* and the southwestern communes.[41]

Two varieties of bananas have been grown for export on a large scale. These are *Poyo* or robusta and *Grande nain,* both cultivators of the Senensis group. The *Poyo* variety is water demanding but relatively wind resistant and is grown on the Atlantic north coast, where plantings lie directly in the path of the Trade Winds. The *Grande nain* is slightly more robust, more drought tolerant, but with a less well-developed root system and a more fragile fruit.[42] It is the variety most often grown on the Atlantic south coast.

Up until 1931 almost all banana imports to France were from the Canary Islands. After that date the French colonies were given a progressively larger part of the market. Although briefly interrupted by the war, importations increased until 1958, when the French market became saturated.[43] After considerable fluctuation in the market price, which was responsible for the elimination of some small-scale agriculturalists as producers, a division of the market was devised and local marketing procedures were developed. Two-thirds of the market was assigned to the French Antilles and one-third to African producers. In Martinique, in consequence of the crisis of 1958, producers joined to form SICABAM. Among the responsibilities assumed by the organization were the building of cooling rooms at the docks, the designing of packaging sheds, provision for transport, and the self-imposition of market quotas.

Commercial export production of pineapples in homogeneous populations had begun in the previous period, but it was not until the mid-1950s that pineapples became a really important landscape feature. In 1930 there were 130 ha in Gros-Morne Commune associated with a canning factory, in addition to the early pioneer planting of 2.2 ha in Saint-Joseph Commune (Revert 1949a,340). In 1940 there were some 300 ha (Sylvestre and Lotaut 1947,2:302). The principal varieties then grown were smooth Cayenne and Red Spanish.

After World War II, pineapple areas expanded, slowly increasing with the establishment of canneries in Gros-Morne, Saint-Joseph, Trinité, and Sainte-Marie communes (Sylvestre and Lotaut 1947,2:348).

Expansion was rapid until 1957, when a saturation of the fresh fruit market caused financial disaster to the smaller, less efficient producers. The industry recovered by 1960, and expansion recommenced. Most of the fields were planted in the smooth Cayenne "sets" imported from Santo Domingo. By 1963 there were 800 ha in pineapple planting, and by 1976, 1,000 ha. As with the other two grand export commodities, pineapple production declined in the early 1981 to 652 ha (see table 25).

Most of the production came from medium and large holdings (Beaulieu 1965,159). In 1961, seven producers out of forty-four produced 60 percent of the crop area. It is a labor-consuming crop, demanding some 500 days per ha per year for canning fruit and 200 to 300 days for fresh fruit production (Martinique, Conseil Général de la Martinique 1962). In the north Atlantic coast pineapple fields are an important landscape feature. Over 25 percent of the arable land in Basse-Pointe in 1981 was planted in pineapple, having gradually replaced some banana fields which had almost completely replaced sugarcane.

Land in sugarcane dropped by three-quarters during the modern period. Beginning in the 1920s there were 28,000 ha; it dropped to 14,000 in 1935, and then to 7,000 during blockade. After the war was over, production rose and averaged about 14,000 until the mid-1960s, when it started to drop again. By 1975 only 7,600 ha were in sugarcane production, and by the 1981 census, only 4,500 ha (see table 25). Between 1935 and 1963, when the regional area fluctuated but ended up with the same area, the sugar production actually doubled, indicating that a real intensification of production had taken place because of fertilization, improved varieties, and insect control. Sugarcane production areas migrated into areas with drier climates, gentler slopes, and closer to the few remaining sugar factories. Three-fourths of the sugarcane in 1963 was produced on holdings of 12 ha or more, and 50 percent in holdings of 50 ha or more (Beaulieu 1965,122). By 1973–74 very few small holdings produced any cane for exportable sugar. In 1963 only three communes had more than a third of the arable land in sugarcane (Linares 1984,31).

The two most important sugarcane pests are sugarcane mosaic and *Stries chlorotiques.* The only other pest of consequence is the cane borer (*Diatraea saccharalis*), which is also under biological control. The retreat of sugarcane lands has been paralleled by the reduction of numbers of central factories, or *usines.* In 1939 there were sixteen, in 1963 there were seven, in 1975 there were only two, and in 1981 only one, Usine Galion in Trinité Commune. Very good roads connect the estates in Rivière-Salée and Lamentin with the factory. The concentration of processing also affected the production of agricultural rum. Of the 162 distilleries enumerated in 1939, there were only about 20 in 1963 and 10 in 1981 (Linares 1984,12). The closing of the central sugar factories meant that those fields far from these factories were noncompetitive because of travel distances. In addition to this travel cost, the rising cost of labor, even the lack of it, caused additional difficulties for planters. Mechanization has been perceived as the answer, and there has been a drive to complete the mechanization of agricultural production and to automate the sugar factories. Mechanization puts an additional strain on the smaller landholders and has removed sugarcane from the lands in medium slopes.

Between 1963 and 1964 sugarcane land in Morne-Rouge Commune was reduced from 250 to 22 ha (Beaulieu 1965,132), and by 1973–74 it had dropped to 2 ha (France, DDA 1975,19). In the drier regions such as Diamant the area in sugarcane rose from 24 to 153 ha in 1963, but by 1975 there was none (France, DDA 1975,19). The lower tropical humid areas with flat or land in gentle slope had not suffered a decline in 1963 but had by 1975, illustrating that neither mechanization nor climate but distance from the central sugar factories combined with low world market price was responsible for this last decline in sugarcane areas.

During the modern period the use of spontaneous vegetation as pasture and browse rose and fell, declining in importance in comparison with planted and improved pasture. Most of the increase in livestock was accommodated on planted pastures and savannas enclosed in fences, although some herds and flocks established earlier continued to graze in relative freedom in the savannas and other spontaneous vegetation. The hand planting of bunch-grass clones in already pastured fields is the oldest form of improving pastures, and it continued up to the 1970s. Alternatively, Antigua hay grass or herbe à foin (*Dichanthium aristatum*) and, earlier, guinea grass (*Panicum maximum*) were sown in fields after the woody growth had been dragged off. Seeds of legumes petit trèfle (*Vigna hosei*) and indigo (*Indigofera*) are broadcast just before the rains begin. Recently guinea grass had been replaced by merkère (*Pennisetum purpureum* var. *merkeri*). This form of pasture improvement has been particularly important in the Atlantic south-coast littoral, where over three-fifths of the arable land is in some kind of pasture.

More recently, pastures along the whole of the Atlantic coast were improved with cuttings of pangola (*Digitaria decumbens*). Carpet grass (*Axanopus compressus*), a spontaneous stoloniferous grass, had been used with success to improve pastures in the middle-climate and mountain-climate belts. Saint Augustine grass or chiendent blanc (*Stenotaphrum secundatum*), another spontaneous forage grass, has been occasionally used to improve pasture mixes. The carrying capacity of pastures thus has been improved. Livestock ranchers have been converting open pasture to paddocks to permit rotation grazing. The larger fields in the savannas are also rotated, but with longer rotation periods.

The pastures cleared out of rainforest or second-growth seasonal forest or from the croplands in the lower and upper tropical humid zones are carefully prepared by plowing, seeding, or plugging and, during the period of establishment, by weeding out the spontaneous shrubby plants. The most successful of the combinations have been pangola and petit trèfle, Saint Augustine grass with indigo rampant (*Indigofera spicata*), and elephant grass with kudzu (*Pueraria phaseoloides*), trèfle rampant (*Desmodium axillare*), and cousin (*Desmodium cinereum*). Other grasses used include guinea grass, bahia grass (*Paspalum notatum*), and jaragua grass (*Hyparrhenia rufa*). Many other species can be found in these pastures, although the pangola and gazon have been particularly good competitors in these zones. In the dry and littoral belts (lower tropical subhumid and dry zones) the pangola has not been very successful. Some planters expressed disappointment with pangola because once established, it has proved difficult to remove. Pastures dominated by this grass cannot be used in a rotation scheme with other crops.

Hayfields continue to be small in scale and might properly be called hay plots. Guinea grass continues to be the species most often used, although elephant grasses are also grown to be cut for green hay and brought to the animals. Sweet sudan grass (*Sorghum sudanense*) and pangola are also grown for hay.

By the test of average carrying capacity, pasture improvement has been most successful during the modern period. In 1931 Rose-Rosette (1931,105) wrote that improvements were to be hoped for but that one head per hectare was the maximum carrying capacity to consider on local pastures. Thirty years later, the goal set by the agricultural service was five head per hectare (Martinique, Préfecture 1964,52). Animals fed on unimproved savannas generally have some supplemental feeding.

A considerable portion of the fallow and abandoned land taken out of planted vegetation went into urban use, including streets and roads. The forested lands no longer in swidden agricultural production of bananas and provisions or devoted to timber plantations have been regenerating to rainforest at different rates. In a few locations in Diamant, Sainte-Luce, and Sainte-Anne communes, old fields are passing to seasonal forest and dry woodland. Former sugarcane lands not in bananas or pineapples became savannas maintained by grazing and burning or were evolving to dry woodland or seasonal forest, as in Vauclin, François, and Carbet communes. The lands abandoned after destruction by Mont Pelée in 1929 developed into lightly grazed savannas and bushlands. Indeed, from Prêcheur to Schoelcher on the coast and Ajoupa-Bouillon to Morne Vert, between two-fifths and three-fifths of the arable land is in pastures. Much land formerly in provisions in Morne-Vert became long-term grass fallow grazed by increasing numbers of dairy and beef cattle.

The abandoned cacao lands of the northwest coast have been reclaimed by seasonal forest in which remnant cacao plants can still be seen. Cutting of experimental plots of timber trees is followed by replanting. There has been some regeneration of riverine facies of different forest types after provisions grounds have been abandoned, especially in the basin of the Lorrain River. Indeed, a considerable amount of the seasonal forest and rainforest has grown into their distinctive physiognomies.

As in previous periods, the vegetation on the island suffered the effects of natural disasters of volcanic eruptions, hurricanes, and tidal waves during the last period. Mont Pelée erupted again during 1929–30, spewing ashes over the whole of the island, especially in Ajoupa-

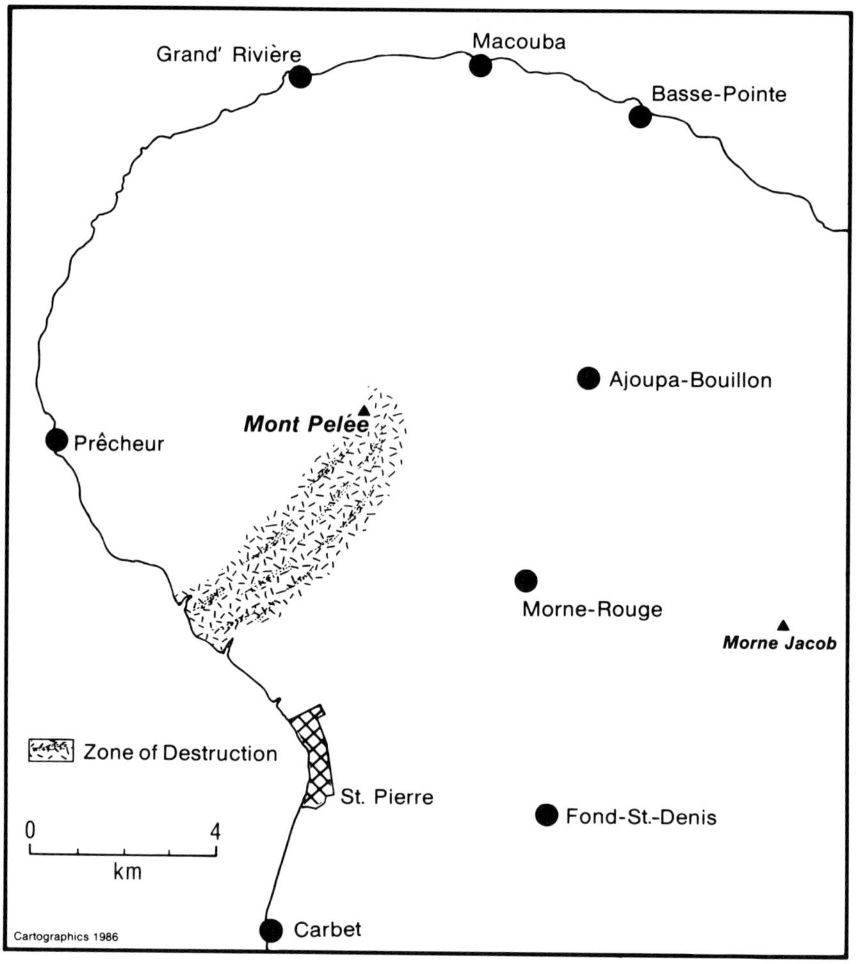

Map 34. The Mont Pelée eruptions of 1929–32. Source: Traineau 1982,69.

Bouillon (see map 34). However, the lava flows were all confined to the 150 ha summit area, and mudflows were confined to a sector in Prêcheur between the Sèche River and Morne Coffin. Destruction of the herbaceous vegetation and damage to the woody plants followed capricious patterns determined by the debris and blast patterns. The resettlement of persons after the explosion took place in the *hauteurs* of Lorrain, a pioneer settlement in rainforest (Delawarde 1937a,137). Champflore Basin residents were brought south to Camp Balata and other small settlements along La Trace highway to Fort-de-France.

Ajoupa-Bouillon was burned by hot gases, and parts of the Grand Savane *quartier* west of the mountain were covered with ash. Within ten years after the explosion, the inhabitants had returned to their old lands, for the most part reestablishing the same cadastral pattern. The principal lasting effect of the eruption in Ajoupa-Bouillon Commune was the turning of the inhabitants to provisions gardening and truck farming for the Fort-de-France market. The regeneration of vegetation on Mont

Pelée has been slow, and little of the destroyed elfin woodland, montane forest, or rainforest has regenerated. Recovery of the rest of the vegetation destroyed in the 1902 eruption proceeds slowly but with better success (Stehlé 1937,1944; Sastre 1978,92–93).

The tidal wave of the 1930s that hit Sainte-Marie is said to have torn away parts of the littoral hedge and caused considerable damage to properties in the town itself, but there were no effects on the vegetation still observable in 1963, and little change had taken place in the following twenty years.

A number of severe hurricanes occurred in the twentieth century; their paths across the island chain into the Caribbean Sea have been recorded (see map 35). Hurricane Edith hit Martinique in October, 1963. The uprooting of trees standing in isolated positions and some battering of the strand flora were the consequences in the spontaneous vegetation. Bananas, the most susceptible of the commercial crops to wind damage, were severely damaged, and a loss of 7,000 to 8,000 francs was reported (Martinique, Caisse Régionale 1964,8). Numerous coconut trees standing in isolated situations or in gardens were toppled. Losses of sugarcane and other commercial crops were minimal.

In 1967 Beulah caused damage to banana fields and exposed forest trees. Dorothy hit in 1970, devastating much of the southeastern part of the island, but it was Hurricanes David and Allen that caused the most severe damage. In 1979 David's high winds did much damage in banana fields on the windward side and felled many canopy trees in the forested areas. Damage to housing caused considerable distress to the population. Landslides tore out highways and impeded traffic. Hurricane David was the most severe of all, destroying hundreds of hectares of banana and sugarcane. The destruction in towns caused a significant drain on the economy, creating demands for relief monies. Hurricane Frederick caused

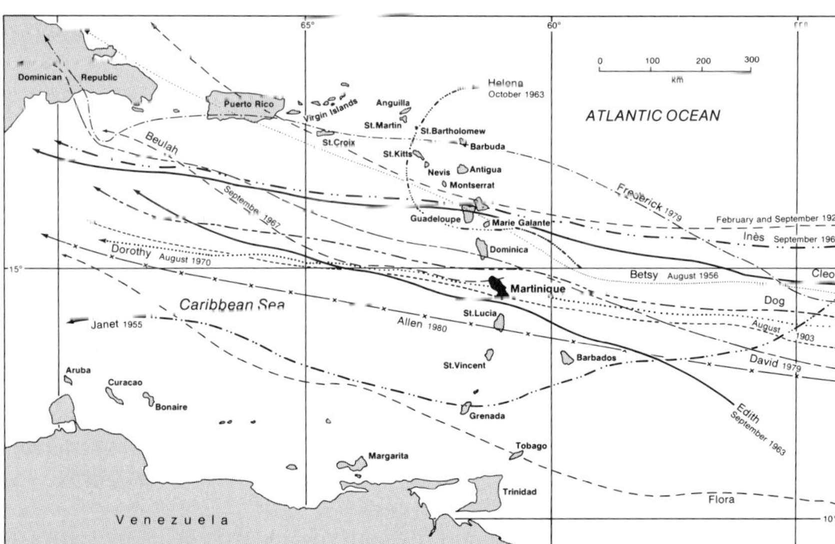

Map 35. Trajectories of the principal hurricanes of the twentieth century. Source: France, Centre National de la Recherche Scientifique 1977, U.S. Department of Commerce, NOAA 1985, special run.

little damage but brought heavy rains. Allen hit in 1980 just as people were recovering from the economic setbacks of David.

A number of hurricanes whose paths can be reconstructed caused severe damage in the vegetation and much economic distress. Hurricane damage in mature rainforests as well as pockets of seasonal forests was substantial, even in relatively protected areas between Mont Pelée and the northern pitons and mornes. Windfalls in all kinds of forests were still observable in the summer of 1985 in the northernmost part of the island. *Relevées* on ridges conducted in the spring of 1985 recorded broken stems of mature trees varying in diameter from 42 to 30 cm. Trees of comparable diameters not broken were in the highest canopy.

Severe drought hit Martinique in 1967, 1970 and 1971–73. During these dry periods some large estates went out of crop production entirely, turning to livestock grazing. The last drought was caused by the same mechanisms responsible for the great Sahelian drought of the same period. Recurrent drought must be thought of as a normal part of the local environment in the lower tropical semihumid and lower tropical dry zones. Many trees in dry woodlands and semideciduous forest are facultatively deciduous, an obvious functional adaptation to recurrent dry spells.

Notes

1. The civil service was directed from Paris, and service and *département* heads and higher-echelon personnel were rotated to and from metropolitan France. This provided for a steady stream of European-born functionaries who temporarily joined the middle class of local society (Horowitz 1967,15). There is no doubt that the practice led to a greater level of professionalism as well as the creation of a European standard for amenities and for making local administrative decisions (Burton 1978).

2. Between 1960 and 1970 the annual emigration averaged 2,000 persons per year. In 1970, 4,000 emigrated, and each year the number rose until 1974, when over 11,000 persons left. In 1975 the number was reduced to 6,600. By 1984 only 6,000 left for Europe or other islands (Martinique, Chambre de Commerce 1974,7).

3. France, INSEE 1975 and France, INSEE 1983. The definition of *urban* used in the French Antilles is different from that used in France. For Martinique the urban population is defined as "the total population of the commune of Fort-de-France and the nucleated populations of those other communes whose populations are over 2,000 inhabitants" (France, INSEE 1956–57,40).

4. Information provided by the Chambre Départemental d'Agriculture de la Martinique during a field trip organized in conjunction with the colloquium Farming Systems and Alternatives for Development, May 9–11, 1985, on the campus of the Université Antilles-Guyane. See also two useful works on the evolution of the land-tenure system in Martinique: Desruisseaux (1975) and Raymond Massé (1979).

5. For a discussion of the social and political reasoning behind these activities see Anonymous 1983, "Les Occupations de Terre en Guadeloupe," written by an activist in the Sainte-Rose area of Guadeloupe, a center of aggressive squatter activities.

6. Discussions with forest guard Vaillant and the deputy administrator of the local Office National des Forêts, Anecet, over the years 1963–85.

7. In 1955 there were 8,800 cars; in 1962, 20,000. Since 1970 more than 5,000 new cars and trucks a year have been imported to Martinique (Martinique, Préfecture 1977,8). In the last Plan Régional an item offered as part of the solution to the traffic problem was a change in attitude toward the use of public and private transportation (Martinique, Conseil Régional de la Martinique 1985,44). It will be hard to change the people's attitudes toward owning and using cars.

8. On July, 1960, a law was enacted introducing conservation into the French legal system, which applied to the Antilles. Its purpose was to preserve a portion of the country's natural beauty and to conserve its biological riches by placing areas under national protection for the pleasure, recreation, and improvement of present and future generations (France, Office National des Forêts, Parc Naturel Régional de la Martinique 1981,101). National parks were to be established by the Conseil d'Etat in consultation with local political units. A much earlier law of 1930, modified in 1967, established natural areas and controlled their use for scientific study. The Caravelle Nature Reserve was established in 1976 under this statute. In addition, natural reserves and the national park were authorized under a decree of March, 1967 (France, Office National des Forêts 1981,102).

9. France, Ministre de la Qualité de Vie 1977.

10. Whatever may have been French official policy with respect to diversification, it was always in terms of exportable commodities. The economic viability of the island vis-à-vis Europe is founded on its exporting capability and subsidies. The ratio of taxes to expenditures for Martinique in 1962 was 54:100, and the trend was downward. By 1975 imports exceeded exports fourfold. The budgetary deficits have been made up by the French treasury (Andic and Andic 1965,56). The export duties paid on sugar, molasses, rum, and other exports provided the balance to counter imports. Any diversification of the economy tending to develop internal markets at the expense of exports could not be tolerated. Since the duties provide the monies for public expenditures, an export-oriented economy must be preserved. Despite the Seventh Plan and those that followed, no effective stimulus to small-farm production of food had materialized by the mid-1980s.

11. Weed counts in estate fields conducted in 1962 and 1963 and again in 1975 revealed a greater decrease in the number of both of weedy species and of individuals in fields treated by chemical weeding practices than in fields not treated so. This fact is well known, and some advise widespread use of pesticides and herbicides. However, others are concerned about the long-range effects of their use and are reluctant to use chemicals.

12. The majority of the imported fertilizers still go to the large landholders, but the use of fertilizers by small cultivators showed a marked increase in the 1970s and 1980s.

13. The use of sprays is more efficient in dry climates. Fewer sprayings meant lower labor costs. However, the severe droughts of the first half of the 1970s reversed the trend temporarily. Former banana lands went into improved pasture, and some were abandoned entirely. After the drought few of these lands were replanted in bananas unless part of a land rotation scheme with sugarcane.

14. Martinique, SATEC 1963,23. The matter of neglected fishing opportunities has been the subject of a number of international conferences. It is generally agreed among social scientists and nutritionists that more fish can and should go into the Caribbean diet. The reasons for the reluctance to engage in fishing or to use fish more extensively is not really well understood. Part of the problem, not acknowledged by governments, is that there is a mutual disrespect of long standing between farmers and fisher folk on Martinique, as well as on the other islands of the Caribbean.

15. Martinique, *Bulletin Agricole* 1936 5(2):95–112. Although the census was conducted by the professional staff, the enumeration accounts for only some 80,000

ha. Kervégant (1936b,95) noted that some 7,000 ha in state-owned forest, urbanized areas, and roads were not counted in the census. The addition of these lands still leaves something less than 24,000 ha unaccounted for, or about one-fourth of the island's surface. Undoubtedly, the forested land was underestimated, as was the land in towns and roads. For the purpose of establishing the proportional representation of plant-cover types for 1935, the 7,000 ha in state forests was added to the *bois* figures. Not properly represented are bush fallow lands, which may well have occupied the larger portion of the unreported area, but which also included the *cinquante pas* lands, the rock desert pavement, and the mixed cactus desert. The area in coconuts was calculated by counting the trees and converting a number of trees to a surface area.

16. This last type was compiled from data for all sweet potatoes, yautias, dasheen, yams, manioc, grown either in pure populations or in mixed populations. The areal figures probably well below true values because of the difficulties in enumerating produce in dooryard gardens and polycultural provisions gardens, and the lack of interest in nonexport crops by the agricultural service at this time.

17. These are some high figures for interior communes that cannot be explained in terms of montane wet meadows and are most probably the forested lands on very humid andosols where the drainage is impeded, as in Saint-Joseph and Fort-de-France communes. Prêcheur Commune, with 813 ha censused in this vegetation type, poses a problem in interpretation. No mangroves have ever existed during human history, and the soils are dry shallow volcanics that drain easily. I have interpreted this figure as a correct enumeration attributed to the wrong commune, so it is included in the total but not in draft commune vegetation distribution maps.

18. The planted vegetation types are bananas, sugarcane, pineapples, tobacco, flowers, European-style vegetables called "legumes" (truck gardens) and provisions, fruit trees, mixed fruit trees, planted pasture, tree crops, and gardens. The spontaneous vegetation types included natural pastures (kept clear of spiny shrubs), fallow spontaneous grass and bushlands, woodlands and forests, and nonagricultural land (France, DDA 1975). These statistics include only areas considered within the agricultural area of the island and total 75,860 ha. The surface area in nonplanted vegetation is more nearly accurate than in any other census.

19. The breakdown included only banana plantings, sugarcane fields, truck gardens and provisions, tree crops, planted grasslands, spontaneous grasslands, other (*diverse*), and utilizable land (France, DDA 1975,19). The census is not only not so detailed by census tract as the previous one, it was also done by sampling. One in ten of the cultivators of landholdings between 3 and 50 ha were interviewed, as were one in fifty of those owning 0.5 ha or less. The totals for the island found on p. 12 of this manuscript census did not agree with the figures derived from summing the commune figures. There seems to be no reconciliation of the figures. When the island as a whole is discussed, figures from p. 12 of the manuscript are used; when the communes are discussed, the figures from p. 19 of the manuscript are used since proportional results are used.

20. This is considerably less than the 75,860 ha said to be agricultural land in the provisional census, which works out to 68.8 percent of the total surface of the island (France, DDA 1975,12).

21. Charcoal making was, until kerosene began to compete effectively, a remunerative occupation in comparison with farming (Delawarde 1937a,119). A charcoal furnace 2.5 m × 1.5 m takes about two weeks to complete: a week to cut the wood, two days to make the kiln, three days to fire it, a day and a half for cooling and sacking the charcoal. The charcoal pile is constructed by piling small roundwood and then large pieces on a bed of stones. This pile is covered with herbaceous plants and then earth, leaving four air holes at the corners. The smoke is

watched to see that it remains white (Delawarde 1937a,119). Many charcoal pits like these were observed by me in 1963, none in 1975, and then several in 1985, when charcoal was competitive again.

22. Sylvestre and Lotaut (1947,2:240) list herbalists for various communes. In the central markets of Fort-de-France, Ste. Marie, and St. Pierre women sell herbs as medicines: 1963, 1979, and 1985.

23. Sylvestre and Lotaut (1947,2:244,252) list three men in a tanning establishment in Lamentin, making it the largest of the five tanneries listed. Most of the leather locally tanned in the 1980s went into hand-made leather goods marketed as craft items or quality clothing and shoes.

24. A farmer in the southwestern peninsula had the leaves of gliricidia analyzed. According to Laboratoire Vizerne and Guillot in Marseilles, protein content was 24 percent; fatty compounds amounted to 4.55 percent. In Diamant Commune, where the dry season may exceed five months in some years, pastures are fenced in gliricidia, which sprouts easily from cuttings to make living fences. One planter has 15 km of fencing in gliricidia. It is pollarded in November toward the end of the rainy season, and the new leaves are stripped from the new branches in early February and April.

25. Martinique, *Bulletin Officiel*, "arrêté du gouverneur, 12 mars, 1925."

26. The creole term *bitué* is said to be derived from *habitué* (Delawarde 1937a,95).

27. This was observed especially in Morne-Vert and Sainte-Marie communes. Now in extensive pastures and fallow grasslands, scattered rows of trees can be understood only in terms of the local dominant wind direction and the previous cadastral pattern. These show up nicely on the modern topographic maps of 1:20,000, published by the Institut Géographie National.

28. The gardens at Desclieux had been inaugurated in 1918 at Fort-de-France to replace those of St. Pierre (Delawarde 1937a,80).

29. France, INRA, 1950,16. This vegetatively reproduced species is represented by five clonal varieties: var. *elephant*, var. *florida*, var. *merkeri*, var. *napier*, and var. *uganda*.

30. This is the sunn hemp of Asia, so called for its cultivation as a fiber plant (E.G.B. Gooding et al. 1965,203).

31. Other *Crotalaria* species introduced and distributed are: sonnette (or rattlebox, a generic common name) (*C. breviflora* and *C. intermedia*), introduced by exchange with Puerto Rico in 1950; *C. lanceolata* from east Africa; *C. vitellina* introduced in 1950 into testing plots; *C. paulina* introduced to Martinique via Puerto Rico in 1950; and *C. spectabilis* introduced from Puerto Rico to the French Antilles in 1939. The last is an ornamental and useful nitrogen-fixing plant of the cultivated vegetation naturalized in the middle-belt regions at the end of the 1950s (Stehlé and Stehlé 1960,3:22–56).

32. Mme. J. Roseau, Hon. Saint Charles of Diamant Commune, in the 1950s put in 15 km of living fence of gliricidia as insurance against severe drought damage to pastures. Emil Hayot claims that the planting of gliricidia has transformed the landscape he knew as a young man.

33. The timber trees were planted in strips 4 m wide, alternated with strips of undisturbed forest 10 m wide. The strips were along the contour of the slope or normal to it, depending largely on the direction of the prevailing wind and the exposure of the valley slopes. They may extend one-half kilometer or more. This strip is cleared by a forester and forest guides. Small seedlings are planted in the relatively undisturbed soil. Maintenance consists of cutting any regrowth in the sward and the selective cutting of very large trees from the intervening strips of

previously undisturbed forest. The timber trees soon overtop all but the tallest of the forest trees, and maintenance is minimal. No trees of marketable size were produced under this system, but the lower costs relative to clear-cutting and planting had made it seem an attractive form of forestry.

34. This is a practical application of sound ecological principle. Species in diverse plant communities are more resistant to species-specific pathogens than pure populations because of the relative isolation of trees from like members of the same species.

35. Spontaneous cottons tend to hybridize with the introduced cottons, thus making the continued importation of seed a necessity. The cost is perceived as prohibitive to the small farmers. This lack of interest is in marked contrast to the expansion of cotton in Montserrat during 1978–80.

36. The young cacao plants were made available at roughly a quarter of the real price, which indicates a strong governmental interest.

37. According to Coste (1959,19) all of Martinique between 200 m and 400 m that receives over 1,500 mm precipitation, does not suffer from a severe dry season, and is sheltered from the Trade Winds can be used for cacao. Though the cacao bushes will thrive up to 500 m elevation, the production declines markedly because of lower temperatures above 400 m.

38. According to the *Annuaire de la vie martiniquaise* (Martinique, *Annuaire* 1947) the largest holding of the twenty-six enumerated planters was 25 ha, the smallest 4 ha, but there were only 299 ha among them all. Ten other names were listed as banana planters, but the sizes of the holdings were not listed. However, some names were those of large planters, better known as sugarcane growers, so their plantings may have been larger than those eight of the other group.

39. In the beginning 70 percent of the area in bananas was in small and medium-sized holdings; by 1963 the largest portion (53 percent) was in larger holdings on former sugarcane lands.

40. In the 1 to 5 ha class were 2,360 ha, being 28 percent; in the 5 to 10 ha class were 765 ha, or 9 percent of the total area (Beaulieu 1965,140).

41. According to SICABAM a few hectares were found in Anses-d'Arlets, Sainte-Luce, and Sainte-Anne communes.

42. The *Gros Michel* variety, though one of a dozen fresh fruit bananas grown for local markets, is not satisfactory for plantation culture in Martinique. It is susceptible to Panama disease to which the other two are not, and its flavor is not particularly appreciated in metropolitan France.

43. Imports from the Canaries stopped completely just before the outbreak of World War II (Beaulieu 1965,156).

9 The Present

Those small patches of relatively undisturbed vegetation left in the plant cover of Martinique are artifacts, protected from exploitation by legislation or inaccessibility. Selective utilization, plant introductions, modification of habitats by settlement processes, clearing and burning, the introduction of grazing livestock, and changing crop patterns and agricultural technologies and industrialization are the major activities by which people have exploited the island. Volcanic explosions and hurricane strikes have continued to rearrange the plant mosaic and modify the species complement within it. During the period in which these assaults on the island ecology were being made, evolutionary processes continued to operate, although at rates modified by human interference. Degradation and aggradation processes continued in the physical landscape, as did long-range dispersal of plants by nonhuman agents. Regenerative sequences occurred in the spontaneous vegetation, leading toward putative elaborate, complex climax types or to some vegetation type in tenuous equilibrium with the processes extant.

Processes and Activities Affecting the Plant Geography

A summary of the kinds of impact on the different vegetation types is presented in table 30. The intensity of these effects can be generalized into four classes: insignificant, notable, substantial, and major. A discussion of the effects, by activity or process, on different vegetation types follows. Included in the discussion are references to the locations of the most intense impact.

The Effect of Settlement and Town Building

According to the 1982 census Martinique had 326,717 inhabitants (France, INSEE 1983,51). The island's gross density was 298 persons per square kilometer. The population was almost half urban (46 percent), with more than a third of it in the capital city. The countryside was losing people; in fact, two-thirds of the communes had lost population between the last two census enumerations (France, INSEE 1983,216). Those that had gained population were those having towns with major regional importance or those near Fort-de-France and the scenic Sainte-Anne and Diamant communes. The largest urban area sprawls over what was once coastal swamp, mangrove swamp woodland, littoral woodland, or seasonal forest, and its suburbs encroach on the rainforest. Coastal

Table 30. Activities and processes modifying the plant geography of Martinique, the native vegetation types most affected, and intensity of effect

	Activities and Processes										
Vegetation Types	Tourism, Recreation	Industrial Activities	Natural Disasters	Fallow, Abandonment	Tree Plantations	Planting, Ag. Tech.	Plant Introductions	Clearing, Burning	Grazing	Selective Utilization	Settlement (Ag., Towns)
Montane bushlands and meadows	I		S							I	
Elfin woodland			S							I	I
Montane forest			I	I					I	I	I
Rainforest	I	I		M	M	S	S	S	N	N	S
Seasonal forest	N	N	N	M	N	M	M	M	N	S	M
Dry woodland and bushlands	N	N	N	S	N	S	N	S	M	S	M
Mixed cactus scrub	I		N	I		I	N	I	S	N	I
Littoral woodland and hedge and meadows	M	S	I				I	S	M	N	M
Mangrove	N	S	I			M		I	I	N	N
Desert rock pavement	I						I		I		

NOTE: Intensity of effect: I = insignificant, N = notable, S = substantial, M = major.

vegetation types have been substantially and importantly affected by housing and mercantile establishments from the time of the first colonial period. The late Amerindian settlements were all coastal and the earlier such settlements largely so. Settlement activities have extended notably or substantially into most vegetation types. The exceptions are the montane types, which have experienced only marginal disturbance; the desert rock pavement, with little disturbance; and the high-elevation types, none at all, except for Mont Pelée.

Town buildings and streets replace native vegetation, and few plants exist except in parks, esplanades, and private gardens. Exotic plants are brought into the gardens of fanciers. In parks, the exotic is frequently featured over the native species. Although residential areas have substantial tree cover and some have dooryard gardens, much of the central town area is without vegetation. The paved streets increase runoff and reduce infiltration. Stream channels are confined and discharge debris into the sea, reducing alluvial deposition in the lower valley bot-

14. Population of Martinique, 1640–1982. Sources: Schnakenbourg 1973; France, Centre National de la Recherche Scientifique 1977; France, INSEE 1983.

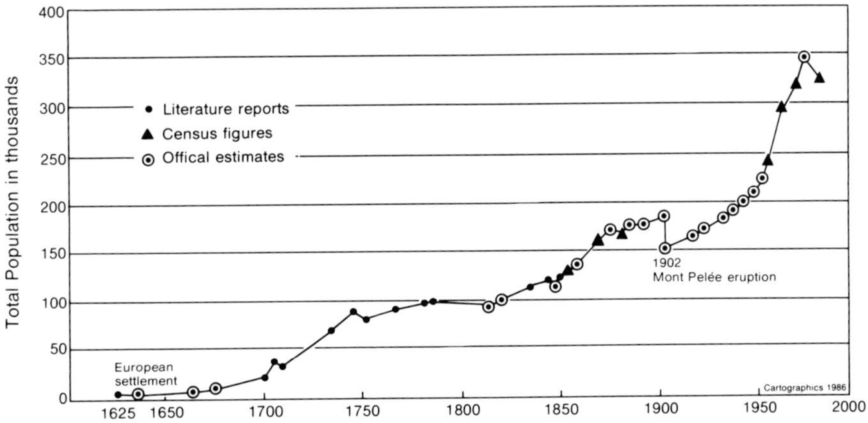

toms. Within the towns, micro-habitats available for spontaneous plants are more xeric and are colonized by pioneer, generally exotic weed populations.

Historically, the settlement pattern was dispersed and coastal, moving to the interior and occupying the lands below 500 m. The population remained largely dispersed, with many towns in the eighteenth and early nineteenth centuries having less population than the larger sugar estates, despite their central functions. The population reached a density of two hundred per sq km relatively quickly (David 1973a). After that, it increased in density slowly, only to fall again in the period between the last two censuses by seven persons per square kilometer (see figure 14).

This dispersed settlement pattern interrupted the fabric of spontaneous vegetation, creating more boundaries and increasing the edge effect. After the move to the towns began, the countryside regained larger unit areas of spontaneous vegetation. The rural population today is largely in the hilly portions of the seasonal forest and lower rainforest zones, on the mornes. There, the small cultivator still reigns, producing the provisions, truck produce, and small quantities of secondary export crops. The map of high rural densities corresponds to the seasonal forest and lower rainforest regions of the windward and central portions of the island up to 600 m, which is considered well beyond the limit for human settlement by the local population (see map 36).

It is in these areas of high rural density (150 per square kilometer and over) that the dooryard garden is a really significant landscape feature, where spontaneous vegetation is so intermingled with planted that it is difficult to segregate on small-scale maps. Small-plot farming puts its pattern on these landscapes as well. Old-field succession may pass to orchard savanna, only to be recultivated. Erosion is severe on steep slopes, and the natural fertility is lost or at least much reduced. The dry woodland, the thorn shrub, and the other xeric vegetation types are recovering their former size and returning to a classic physiognomy as the small-plot farmer moves onto lands once devoted almost exclusively to larger scale agricultural enterprises.

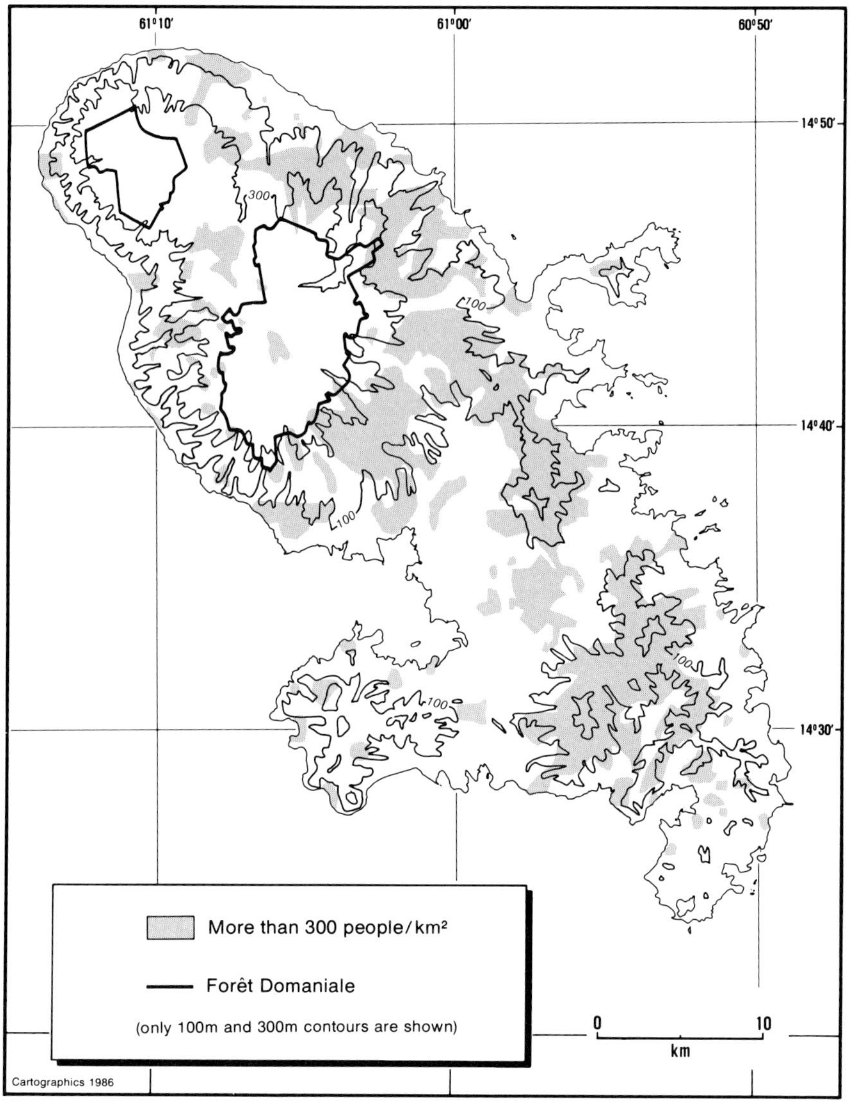

Map 36. Areas with high rural densities. Source: Compiled from France, INSEE 1983

Roads provide access to different parts of the island. Disturbance effects follow the roads. The road network gave an excellent clue, until recently, to the landscapes dominated by cultivated as opposed to spontaneous vegetation (see map 37).

The Effect of Selective Utilization

Culling of useful species from the vegetation modified the composition of the vegetation most substantially in the dry and littoral woodlands and the seasonal forest, but also notably in the rainforest mixed-cactus scrub and the littoral hedge. The increased importance of the pioneer *Tabebuia pallida, Bursera simaruba, Calophyllum calaba, Ceiba pentandra,*

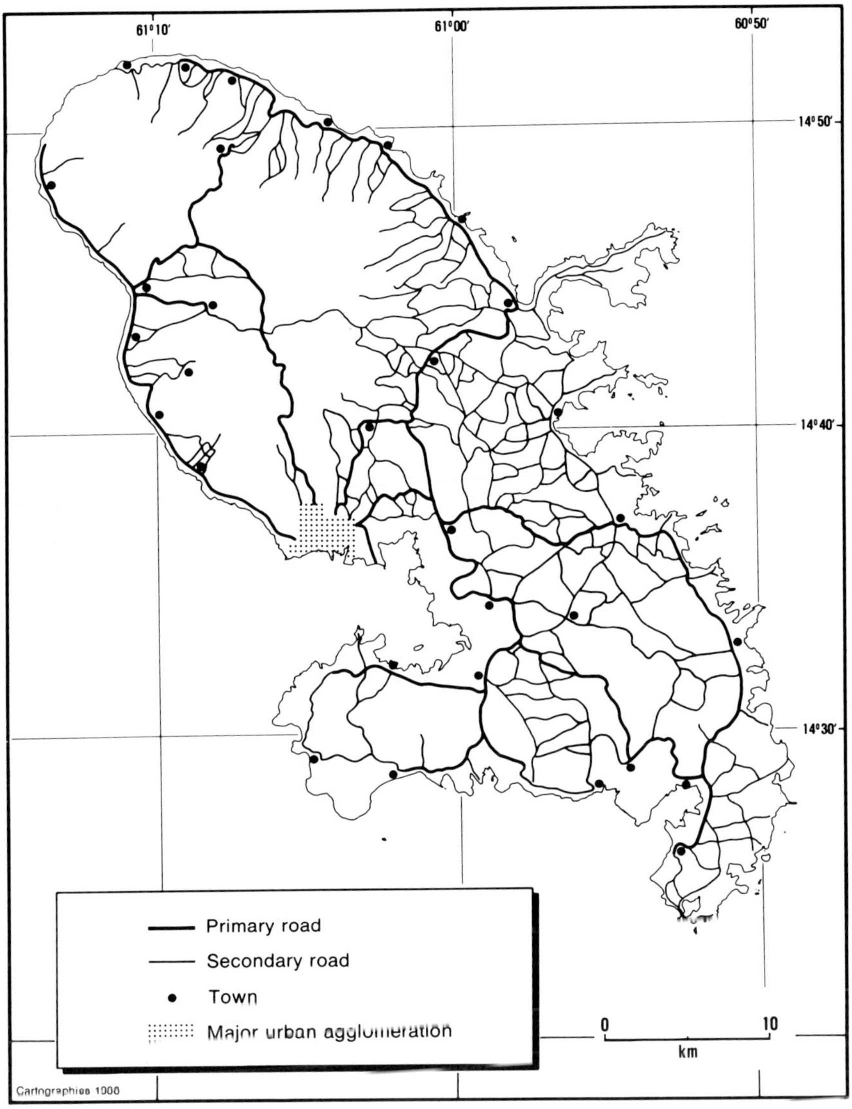

Map 37. Principal and secondary road net. Tertiary roads are not represented on the map. Source: France, IGN 1984.

and the decreased importance of *Rhyticocos amara, Mastichodendron foetidissimum,* and *Dacryodes excelsa* are examples of this effect. Reduction of some useful and attractive species has caused official conservation measures to be initiated. For example, in the eighteenth century *Guaiacum officinale* was protected, and in the twentieth century, *Prestoea montana.*

In other cases cutting has simplified the physiognomy. In the littoral hedges and woodlands *Hippomane mancinella* and *Coccoloba uvifera* have become much more important and evenly distributed, approaching consociations as other more useful trees have been eliminated.

The cutting down of large trees in the forest types created openings, similar to those associated with windfalls, increasing the availability of habitats for pioneer species. Selective cutting for lumber and charcoal making was occasionally so intense in areas around settlements that the forest was destroyed. These cleared areas did not always go into agriculture. Sometimes secondary successional bushlands resulted. Charcoal making and fires resulted in coppicing in the drier vegetation types and accentuated the trend toward more xeric forms of dry bushlands, particularly on the Caribbean coast.

Severe hunting pressure on larger animals such as the agouti and iguana and the extinction of the rice rat eliminated such animals as important dispersal agents of heavy or large-seeded plants. The extinction of parrots and other tasty game birds had much the same effect. This disruption of dispersal activities within selected vegetation types (the habitats) caused by such extinctions is evident. Just what the effects and the degree of disruptions that these losses of dispersal had on the vegetation composition can only be surmised. I suspect that in areas being protected for reestablishment of primary vegetation types, this loss will delay and perhaps confound attempts to reconstitute the original plant composition. For the present, the loss represents one of the changes in processes within the exosystem.

Selective culling resulted in some loss of complexity within native vegetation types. Disturbances that completely eliminated a species have greater importance on islands than in continental situations because individual species often have wider niches and, hence, play a greater role in the plant cover on the islands. It would seem that selective culling acts in the manner of a tree fall, and gap-phase regeneration strategies are set in train. An increase in such a process will differentially select for r- as opposed to K-reproductive strategies within the species assembly.

The Effect of Grazing Activities

Grazing by introduced herd animals affects the floristic constitution by elimination and by dispersal. The evidence is inconclusive with respect to total elimination of species by grazing; however, both sheep and cattle have tended to suppress the regeneration of some grasses and other plants to the point that local agronomists have sought palatable grasses that can withstand intense grazing; hence, the introduction of African grasses. Martinique has participated in the general Africanization of New World tropical savannas (Parsons 1979).

The increase of some species is clear and well established. Plants such as *Cassia emarginata* from xeric savannas and *Tabernaemontana citrifolia* from more mesic savannas have increased in importance because they are not palatable to stock and are left ungrazed. Others, such as *Leucaena leucocephala, Prosopis chilensis, Cassia fistula,* and *Citrus aurantifolia,* are actively dispersed by cattle. These animals find fruits palatable, and the seeds remain viable after being passed through the digestive tract. The physical destruction by browsing and grazing has retarded growth when plants have not been destroyed outright.

Grazing and browsing have modified the habitat by trampling and compacting the soil. Overgrazing has increased the exposure of the ground to heating and erosion. These effects have been particularly important in the vegetation of the dry and littoral belts, especially in the southeast and north Caribbean coast. They have been substantial in the mesic regions of midelevations as well, and notable in the more humid areas having raw ash soils.

The Effect of Clearing and Burning

In Martinique the clearing of woody growth, closed canopy forest has been the most important factor in modifying habitats. Clearing for permanent settlement and agricultural fields was supplemented by the clearing of swidden plots. These clearing activities affected most vegetation except the desert rock pavement, littoral meadow, and the high-altitude types. Clearing modified the environmental conditions at the ground. It introduced direct sunlight to forest soils and exposed them to impact by raindrops, thereby increasing the tendencies toward leaching, erosion, and modification of soil profiles. The effects of direct sunlight and higher temperature was very important in areas under continuous estate management, and resource depletion led to abandonment of much of that abused land.

Clearing also had an indirect effect on the species composition of some vegetation types. As was pointed out in the chapter on prehuman vegetation, the floristic composition tended generally toward homogeneity within vegetation types; that is, one plant association was represented in the environments occupied by each vegetation type. This was a function of the smaller number of species on the island and of its small area. Clearing has been so extensive in the last 400 years at the low and midelevations that formerly continuous vegetation has been broken up. As a result, subtypes dominated by a single species or a few species have replaced the more complex prototype, as in facies of the seasonal forest, the derived dry bushlands, and in the rainforest. These facies are examples of the breaking up of previously continuous, relatively homogeneous plant associations.[1] Except in the southern peninsulas, the island is sufficiently endowed with large enough spontaneous plant areas that most species of the flora have survived in one or another of these "refugia." These plants have, therefore, been available for later recolonization. When given the opportunity to reenter the vegetation, they have done so, but in a very episodic pattern.[2] Thus, while a number of species have become rare or been eliminated, the destruction has not been so complete as on islands with fewer spatial buffers (see Melville 1979).

Clearing for agriculture in aboriginal times was accomplished with the aid of fire. The burning was in part to clear the plots and in part to put an ash dressing on the soil as fertilizer; both were modifications of the habitat. In the dry bushlands the leaf litter is dry and brittle, and it decomposes slowly. Burning breaks down the fibers, accelerating the nutrient cycling, but much is consumed in the process. The early colonists continued the practice of the Indians, but in doing so they tended to destroy the woody vegetation more completely. During the dry season fires occasionally ran beyond the plots into the uncut vegetation. This accidental firing took place infrequently except in the west-coast lands.

Fire has, until the twentieth century, been important only as it was associated with clearing for agriculture. It was not important for pasture management because of the historical lack of interest in livestock ranching by the Martiniquans. When in fact the pasturing of livestock herds became important, as in Vauclin and Marin, burning became a regular practice.

Where regular burning or repeated burning took place, it modified the species composition by stimulating stump sprouting and increasing the germination of fire-adapted seeds. Firing has been moderately successful in controlling the invasion of grasslands and savannas by young woody growth; however, in the pasture lands dragging is still necessary. Although no studies have been made, local agronomists concur that some directional trend in species composition has been taking place. They do not think that the species composition in present pastures may be defined as "mature." The fire-adapted, intensively grazed grassland has not yet fully evolved in Martinique.

The Effect of Plant Introductions

Undoubtedly during the last 2,000 years disseminules of many species entered the Martiniquan ecosystem without the aid of humans. The vast majority were duplications of previously established plants. The ancient source areas were enriched by introductions that, in turn, became available then for natural dispersal to Martinique (see map 38 for the principal source areas for all known exotics).

Long-range dispersal was occurring before the arrival of people and continued afterward. The natural dispersal of logwood has already been suggested. *Terminalia catappa,* introduced to the Caribbean at the end of the eighteenth century, is part of the drift on most beaches in Martinique. It has become an important member of the littoral wood-

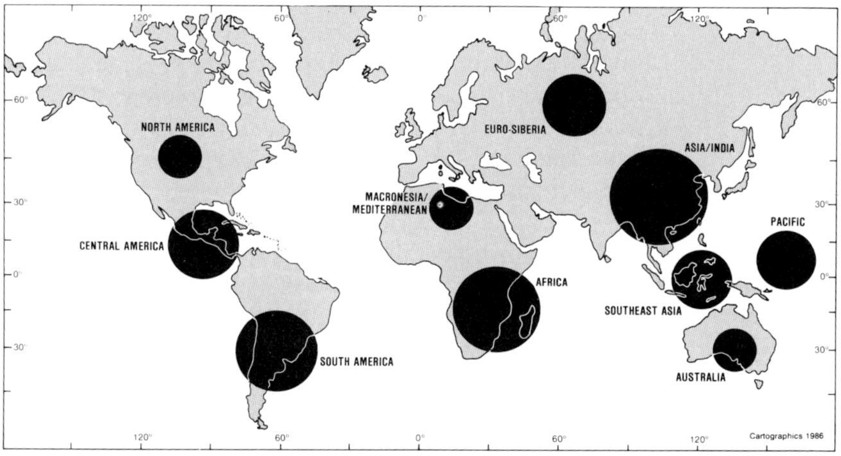

Map 38. Source areas for exotic plants of Martinique. Source: Compiled by author.

land. It has also been planted as a park and garden tree in lowland sites. Whether or not the first introduction into Martinique of this decorative species was by people or by currents within the Caribbean Sea, natural dispersal has certainly accounted for some of the inoculations. *Jatropha multifida* may well have been distributed by birds from South America, as well as introduced by the Indians. It is distributed by birds on the island now. *Capsicum frutescens* is also bird distributed and may likewise have entered the flora as an avian introduction. Since the arrival of humans, a different rate of introduction occurred because the dispersal of plants was more effective and the creation of new, disturbed habitats was involved in the settlement strategy. These new plants we can refer to as aliens.

Alien species are abundant in the modern flora of Martinique. Roughly one-fourth of the flora have been introduced from continental America, Africa, Europe, and Asia.[3] In the crop plant associations and the dooryard gardens, the exotics are most in evidence. Although a bare majority of planted vegetation types are dominated by Old World plants, the rest are introductions from Central and South America. Much of the low-elevation spontaneous vegetation is dominated by plants introduced by humans intentionally or as weeds. The leguminous thorn scrub is dominated by exotic plants, although native species are found within it. Rainforest clearings are dominated by exotics, and at lower elevations some exotics remain in spontaneous vegetation after the secondary succession has begun. The largest number of trail and roadside ruderals are introduced, weedy plants.

These introduced species dominate artificially prepared ground in which most native species, not being adapted to these open habitats, are noncompetitive, do not invade, or are forcibly eliminated by cultivation. Some of these introductions like *Swietenia mahagoni, Leucaena leucocephala, Psidium guajava, Syzygium jambos,* and *Mammea americana* have become important in the spontaneous vegetation and may represent the filling of a niche made available by selective culling. They may be better competitors for a niche previously occupied by plants with wide tolerance ranges, which now have retreated to more circumscribed niches. *Delonix regia* in the dry woodlands and *Tamarindus indica* and *Mangifera indica* in savannas probably represent remnants from previous plantings, even though the mango colonizes easily from pits tossed by the wayside. The flame tree is spontaneous in the southeastern mornes. A very few species have become aggressive, as have *Pueraria phaseoloides* and *Portulaca oleracea.*

Littoral vegetation types have proved relatively resistant to invasion by exotics except where they have been planted or have colonized into cleared-off areas. Coconuts are found on most beaches, but in every case, with Diamant the possible exception, the palms are clearly the result of planting. *Dodonaea viscosa* has entered the littoral hedge in protected segments of the south and southeastern coasts, *Thespesia populnea* on the windward sand ridges.[4] The East Indian almond colonizes beaches and littoral woodlands. One reason for the general lack of success by invaders in these vegetation types may be that more competitive species had already arrived by long-range dispersal prior to the advent of humans.

The result of the invasion by exotics has been to increase the diversity of the total flora and also to increase competition for open habitats. These colonizers of open ground have become dominant in much more area than pioneers were prior to the arrival of people.

The Effect of Agricultural Change and Land Reform

Although the clearing and burning of vegetation were the most important factors in modifying habitats, the management decisions with respect to which crops were planted and which techniques were used were most important in determining the actual plant cover in most lowland and midelevation environments. Crops have been important in the seasonal and lower rainforest regions, whereas tree plantings are significant in the rainforest and notable in the seasonal forests and dry woodlands.

Nineteen new planted vegetation types were introduced, all of which have simpler floristic composition and structural morphologies than the native types they replaced. Generally, there has been an increase in area in herbaceous types at the expense of area in woody types. Migratory fields, sugarcane, and fallow virtually replaced the seasonal forest during the colonial periods. Bananas and pineapple fields and tree farms took over much land previously in rainforest, and recently bananas have become important in the lowland seasonal vegetation as well. Bushland cultivated types were inserted into the native dry woodland, seasonal forest, rainforest, and dry bushlands, creating a mosaic of planted and spontaneous cover. Provisions gardens and truck flower gardens have been importantly associated with areas of dense settlement in a number of different vegetation types. The first were the vegetation types managed by swidden agricultural practices. The traditional attention to export crops on estate lands continues, but the dominance of sugarcane has been broken. Even bananas are losing ground to provisions grounds and vegetable gardens. Large agricultural units are being made larger to take advantage of mechanized production techniques and modern credit financing. This has led to an abandonment of much of the land in slope so that bush fallow rims the flatter, lower lands or the ridges of spurs. These trends in the post-estate agricultural dominance (Tibault 1985,663) era should continue unless the political status of Martinique should change to one of independence from France. Under those conditions I expect this trend to be reversed.

The increase of area devoted to silviculture is an objective in all modern land-use plans. Most of those areas have been taken from rainforest lands. Though care has been taken to keep stands isolated so that disease outbreaks can be contained, real isolation is effectively established by slope. The relatively level areas suitable for such tree plantings are small.

Cropping and tree planting create new habitat conditions. Erosion associated with crop agriculture modified soil profiles, and plowing reduced infiltration rates by compaction and by breaking down structure. The opening of forest land, particularly by bulldozer, encourages erosion. The desertification of the coastal portions of Martinique and Guadeloupe (Bassières 1923–24,4) was not caused, as Bassières thought, by a gradual reduction in rainfall, but by soil destruction.[5] Edaphic conditions were changed sufficiently so that the water-retentive capacities were reduced. Soil fertility loss, represented by the erosion of whole soil horizons, has so modified the water balance that more xeric conditions do now, in fact, exist in the southern regions, though not for climatic reasons. The extreme drought of the mid-1970s, which paralleled that in Africa, was a true climatic event that only exacerbated the problem created by poor soil management.

The Effect of Fallowing and Abandonment

Bush fallow was part of the management practice of the pre-European inhabitants. Savanna areas were presumably proportional to land in cultivation as long as the population remained constant. The European exploitation increased the area abandoned or in fallow well into the late nineteenth century.

Pioneering elements from bordering vegetation types entered the clearings, soon converting them from spontaneous grassy types to a variety of savannas and bushlands. These develop by secondary succession into woodlands or forest, depending on local habitat conditions. These early-stage, spontaneous vegetation types were dominated by rapidly colonizing native and introduced pioneers, which otherwise are not common in the vegetation.

In lowland areas heliophytic and xerophytic species have had a selective advantage over more shade-tolerant, less hardy elements in the flora. A major consequence of short-term fallow has been a gradual shift in floristic composition for the lowland types toward species having a variety of drought adaptations. In the more mesic environments plant succession following field abandonment has been in the direction of increased species complexity, leading to regeneration of rainforest and seasonal forest. Only some of the regeneration of spontaneous vegetation has been allowed a return to the original associations. A different complexity, presumably reduced with respect to the original types, has developed. It is, however, more complex than the planted types they replaced, especially those weeded by herbicides.

Under most conditions of fallow in colonial times the secondary succession was interrupted before succession had been completed, and the presumed complex types reestablished. Thus were developed the bois canon–balsa stands (*Cecropia peltata–Ochroma pyramidale*) that are larger in areas than those created by prehistoric windfalls, the crécré bushlands (*Miconia laevigata, M. prasina, M. striata*), and the tree fern colonies (*Cyathea arborea* and *C. imrayana*). The poirier-galba (*Tabebuia pallida* and *Calophyllum calaba*) forests and the pois doux forests (*Inga* spp.) of the seasonal forest zones were developed in the same fashion.

The Effects of Natural Catastrophes and Accelerated Natural Processes

The effects of volcanic eruptions were felt throughout the period of human occupation. Hot, incandescent gases burn and dry vegetation. *Nueés ardentes* carry ash, pumice blocks, and heavier extrusives, which mechanically break off and bury vegetation in their paths. The intense heat sterilizes the top soil layers, releasing ammonia, which acts as an ephemeral fertilizer. Paths of absolute destruction, as the experience of 1902 exemplified, do not completely eliminate any vegetation type. Montane bushland and elfin forest were destroyed on the southern slopes of Mont Pelée during the last two eruptions, but bushland and forest remained intact on the Pitons. Other types were fragmented and damaged, but patches were available to supply propagules as the habitats and communities slowly regenerated. Pioneer plants that are particularly tolerant of the raw volcanic debris soils actively colonized from other vegetation types where they were present as biological nomads or as minor elements in the vegetation.

High winds and heavy rains felled emergent trees, to create openings in forests, destroying hillsides of up to a few hundred meters square. At low elevations, where land had been cleared and put into planted vegetation, the damage from high waves was severe, even complete. Sugarcane is less susceptible to hurricane damage than bananas. Until crop insurance became available on bananas, the cautious planter and small cultivator left the more exposed fields in sugarcane. With the lack of any competitive advantage on the world market today, even that option no longer exists.

Although the direct destruction of vegetation by hurricanes and squalls is important, the indirect effects on plant cover have been in some cases more profound and longer lasting. The economic disaster following the destruction of crops sometimes caused estates to be abandoned when there was not enough capital to rebuild. This abandonment had a more lasting effect on the vegetation than did the hurricane damage proper.

Both landslides and high waters continued as important sculpturing agents in the terrain and disturbance factors in the vegetation. Unstable slopes were liable to slumping whenever saturated soil was overloaded. As roads were cut into unstable slopes, they increased the landslide hazard. During road building in the nineteenth and twentieth centuries, the vulnerable slopes have been planted with bamboo (*Bambusa vulgaris*) and vetiver grass (*Vetiveria zizanioides*). The soil-binding character of these root systems aids in stabilizing the slopes.

As the forests were cut down and infiltration rates reduced, the already fluctuating river regimes were amplified. Erosive action increased in severity in high flood stage, and the low stages tended to have less water. The denudation of the southern hill country is directly related to the lengthening of the season during which streambeds are dry.

The Effects of Industrial Activities

The first agricultural industry with significant impact on the vegetation was sugar making. Wood was needed to fire boilers in the sugarhouses of the seventeenth century. As sugarhouses were replaced by small sugar factories, the pressure held on the woodland and forest, largely the seasonal forest and dry woodland. Pressure declined when the larger sugar factory, and later the *usine* of the nineteenth century, began using coal and then bagasse, especially with the consolidations of these factories in the twentieth century.

Charcoal making was important and varied with respect to the size of the population. Charcoal was replaced as the dominant cooking fuel only after World War II, when kerosene began to be generally available in the towns and later in rural areas. The principal sources for charcoal were the thickets and woodlands that grew up in cleared forest lands. Today, charcoal is increasingly used for particular firing needs and for the cooking fires of the rural poor.

Pineapple-canning factories and banana-packing houses use much water, which is discharged into streams. This water is generally laden with herbicides and pesticides. The chemical pollution affects the downstream environments, as does residue from rum factories. These stress factors have acted for the most part at sublethal levels. The biochemical tests have not been made, but over time the effects must have been significant. Blue polyethylene bags, used extensively to protect banana stalks from insects and to facilitate uniform ripening of the hands, have become a visual blight in all banana-growing areas.

Industrialization in the more restricted sense is limited to a petroleum refinery, electric power plant, and a cement works, but considerable activity in manufacturing and warehousing has risen in the eastern margins of Fort-de-France and about Lamentin. The industrial manufacturing impact has been locally important on the mangrove swamp woodlands, on littoral hedge and woodland, and on meadows and savannas rimming mangrove in the Lamentin Plain.

The Effect of Tourism

The impact of tourists has been important largely in the coastal environments, especially on beaches, where the shrubby undergrowth has been cleared or trampled out of the hedges and woodlands. Clearing for trailer pads affects the trees, especially in the Sainte-Anne and Diamant communes. The international quality hotels have been built in the dry coastal vegetation types or near the mouths of small streams, virtually wiping out the riverine forests of the valley mouths. Trampling and picnic fires have further damaged the littoral by soil compacting and heating. Automobile pollutants contaminate the soils.

The tourism industry has been less noticeable in the more mesic and wet regions of the island. The widening of roads in the scenic interior (and coasts) has steepened slopes, increasing the likelihood of landslides. Trampling on Mont Pelée and plant collecting there and along the central tourist road, La Trace, has caused selective pressure on plants with showy flowers or leaves.

The Geography of Change and the Stability of the Island Plant Cover

The plant geographies of the different parts of Martinique have varied with disturbance, timing, and the regenerative capabilities of the several vegetation types. Areas of relative stability in vegetative cover contrast with those in which the plant cover has changed repeatedly. The reasons for stability differ for planted as opposed to spontaneous vegetation. Stability in spontaneous vegetation has depended upon extreme environmental conditions or inaccessibility. The extreme environments make the areas unattractive for settlement, agriculture, or industry, and inaccessibility makes areas too costly for exploitation. Stability in planted vegetation is dependent upon the factors influencing management decisions of the Martiniquans, natural disasters, and the market.

Regions of Comparative Stability Dominated by Planted Vegetation

Five areas may be considered particularly stable. They have little history of vegetation change since they became dominated by planted vegetation, other than the secular changes involved in fallow and field rotation (see map 39). These are the sugarcane lands of Lamentin and Trinité-Caravelle, the pastures and market gardens of Fond-Saint-Denis and Morne-Vert, and the northeastern lands of the Capesterre. In four of these areas the principal spontaneous vegetation to be victimized was the seasonal mesophytic forest, rich in useful species and occupying land topographically suitable for large-scale enterprises and climatically and edaphically suitable for sugarcane and bananas. The fifth area was in the upper rainforest belt farther from the coast and higher and cooler than the others.

The Lamentin Plain posed technical drainage problems so that the beginning of its exploitation was delayed until the last third of the seventeenth century and began with large properties that could support the expense. The rolling lands of Trinité Commune went into sugarcane soon after the port of Trinité became a *bourg* in 1658. Although provisions grounds, dooryard gardens, and some spontaneous vegetation was and is found in these areas, sugarcane has given character to the landscape. Enlargement of holdings accelerated in the nineteenth century and continues to this day, but the pattern of the landscape remains in standard-sized fields divided by grassed lanes. The recent significant changes have been a reduction in the number of holdings and the closing of sugar factories.

The intensive exploitation of Fond-Saint-Denis and of Morne-Vert had to wait until the emancipation of slavery released a land-hungry black population to colonize the middle slopes of the interior mountains. Charcoal makers who previously had invaded these interior highlands expanded their trade. Pasture animals grazed the grasses of the fallow fields so that bush fallow, as practiced in swidden agriculture, was

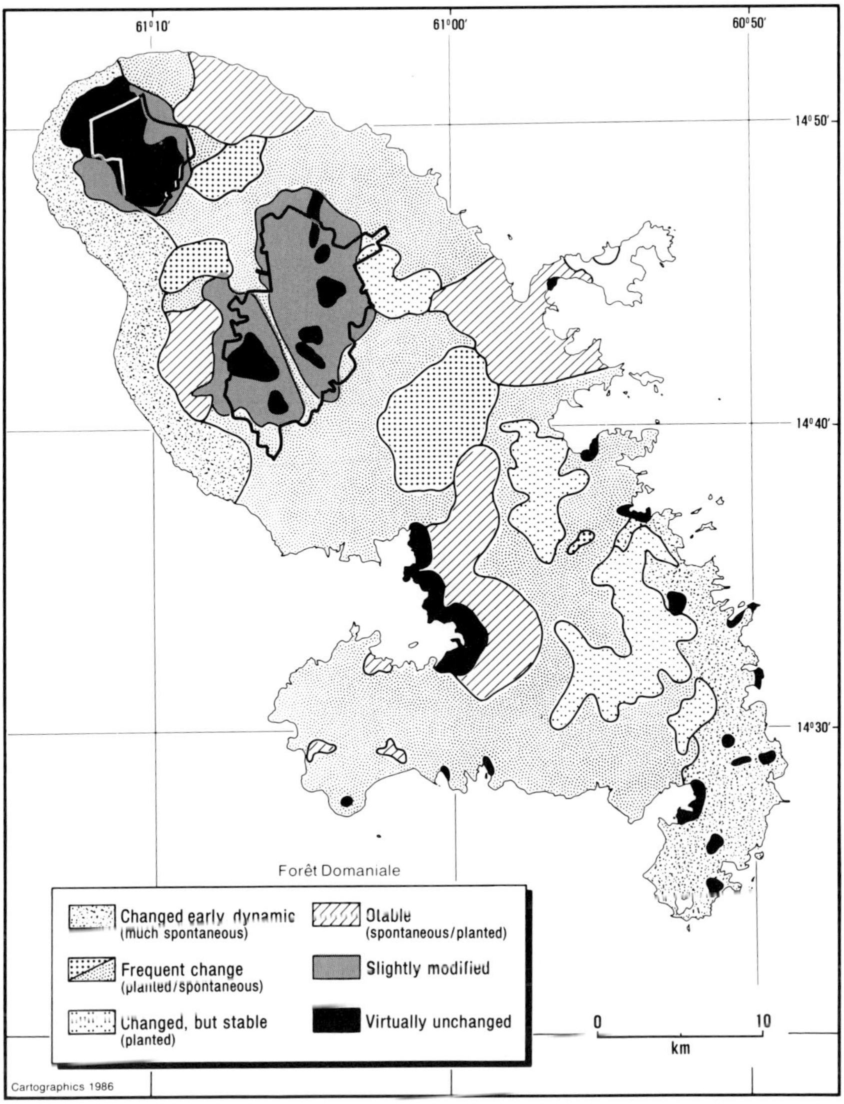

Map 39. Spatial pattern of change and stability in vegetation cover. Source: Compiled by author.

replaced by field rotation; thus, the term *en jachère* is more commonly used than *en frische* as the name for land in fallow.

The lower slopes of the Capesterre in the Basse-Pointe region had been dominated by sugarcane plantings since just after the expulsion of the Carib in 1658. By the end of the seventeenth century the original small holdings had been enlarged, and large holdings further expanded to provide suitable units for a semi-industrial agricultural system. The land was in sugarcane, provisions, and spontaneous vegetation for most of the period of European occupation, but the plant cover began to change in the early 1960s. It had changed over to banana plantings by the mid-1970s, and to bananas with pineapples and livestock grazing in the 1980s. The fertile lands immediately adjacent to rivers and streams have remained in the polyculture of mixed provisions.

Within the areas dominated by planted vegetation there is considerable ground in spontaneous cover that has not been modified substantially through the years. Those areas that have been only marginally disturbed include narrow stream courses or canyon walls still in mesophytic seasonal forest. Undrained low ground remains in spontaneous cover, as do the demonstrably poorer soils, when land use and population densities permit discrimination by landholders within the pattern of the soil distributions.

Regions Repeatedly Modified and Dominated by Planted Vegetation

Change in vegetation type has repeatedly modified the plant cover characteristics in the Carbet-Prêcheur Caribbean strip and in the eastern headlands and borders from the Vauclin region into the Sainte Anne Peninsula. The western littoral was the birthplace of the colony, originally conceived as a settlement of militia-serving small farmers producing exotic tropical crops. Crises in the tobacco market led to experimentation with other exportable commodities—indigo, canafistula, ginger, and sugarcane. Rapidly increasing population densities fractionated the holdings, increased the intensity of exploitation, and accelerated field rotation and the loss of soil fertility.

The collapse, under saturation, of market demand for canafistula, tobacco, and indigo led to the abandonment of those orchards and fields. Much of those lands were diverted into sugarcane production. Concentration so developed in sugarcane that the light, easily worked, ash-derived soils were soon exhausted of fertility. Pasture lands increased to feed the additional oxen required in the sugar estates. Provisions grounds expanded to meet the needs of the sugar estate field hands as well as the enlarged town populations. The maltreated steeper slopes reverted to dry *Acacia* bushland, which was periodically cleared and replanted. This practice led to the development of dry *Lantana* and thorn scrub bushlands containing a relatively large proportion of pioneer tree species of the dry woodland that today covers the uncultivated steep lands of the region.

With mechanization, the need for pastures was less, so some lands were reconverted to sugarcane. Also, with the loss of responsibility for feeding slaves after emancipation came the replacement of many provisions gardens by sugarcane acreage. However, north of Saint-Pierre provisions continued to be important. Peanuts began to be important and were grown in monocrop fields, as were provisions. A distinctive land-use pattern developed in the gentle hill slopes around Prêcheur. Small fields were and are referred to as *taches,* named for the daily work unit they represented. In the late nineteenth century flat lands of the crowns, spurs, and low coastal terraces, as well as wider stretches of beach, were put into coconut plantations. Mango and avocado orchards were interplanted with bananas and provisions on the gentler valley slopes. Chemical fertilizers and irrigation devices have facilitated a more continuous and intensive use of land. Thus, in spite of increased production values for these regions, much topographically inferior land is still in spontaneous second growth.

The coastal areas in the southeast, adjacent to Mont Vauclin, have had a similarly complex history of plant-cover replacement. Tobacco in the south was important only in the Sainte-Anne to Marin peninsula region, and it was soon replaced with sugarcane, which was the principal reason for clearing the native dry woodland of the region. By the mid-eighteenth century much of Sainte-Anne was in cotton, including the hills not used as sites for estate headquarters. There were, as well, scattered bushlands of coffee and cacao. Fields of cotton in the Vauclin region were interspersed with coffee and cacao lands and gardens. Sugarcane returned to importance in this area during the nineteenth century. Impoved pastures for quality livestock, mostly horses, were locally conspicuous in Vauclin Commune. The eruption of Mont Pelée, which devastated the planter class when St. Pierre was destroyed and reduced soil fertility under constant cropping, caused a retreat of sugarcane before spontaneous mixed cactus scrub and dry bushlands. On some of the former sugarcane land, planted pastures were interspersed with the spontaneous scrublands. Coffee, which once dominated the slopes of Mont Vauclin, has disappeared as a planted bushland, although individual bushes are found in the now-dominant dooryard gardens of this densely settled, conservative hill country.

Regions with Less Change Dominated by Planted Vegetation

Other regions had a less complete replacement of vegetation type in the early years, but have had much change in the past century. The Champflore Basin, which was colonized effectively in the 1860s and which had been in provisions, sugarcane, and bananas, is now dominated by pineapple fields and pasture. The Macouba region, renowned for its tobacco even into the midnineteenth century, has at present provisions gardens, dooryard gardens, planted pastures,and pineapple fields in the midslope areas and has timber plantations replacing the rainforest in the upper slopes. The Ajoupa-Bouillon region of peasant proprietors, with its pastures and provision gardens, was better known in the last century for its cacao and coffee. The eruptions of 1932 forced the evacuation of the majority of the population and destroyed much of the herbaceous vegetation. Sugarcane, pineapples, and bananas are now virtually restricted to fields in rotation cropping on the few estates reactivated after the abandonment in the 1930s. Though Saint Joseph still has coffee grown under trees, it has become more important as a center producing pineapple for canning and a provisions supply area for the expanding Fort-de-France market.

Stable Areas Dominated by Spontaneous Vegetation

The parts of the island dominated today by spontaneous vegetation are mostly those covered in second growth either regenerated after clearing and agricultural activities or modified by selective gathering. Some few areas have been relatively undisturbed by humans because of their unattractiveness for exploitation, their relative isolation from centers of population, or because they are above the culturally accepted altitudinal limit of settlement.

The areas of spontaneous vegetation occupied by climax vegetation are few and small. The elfin woodland found on the Carbet Pitons and Morne Jacob, as well as the montane bushland above it, has been crossed and recrossed by forest guards and amateur mountain climbers. Areas around resting places have had the vegetation trampled, but not frequently enough to modify that type significantly. The peak area of Mont Pelée has a cover of seral vegetation communities that have been more severely damaged by tourism. Collection of hearts of palm has reduced the number of mature palms. In spite of these impacts, the vegetation may be considered native, and no change in borders has resulted from human interference. The valley of the Lorrain River has the largest area of relatively undisturbed rainforest. However, the best examples of primary forest of *Sloanea-Dacryodes* as well as *Homalium racemosum* and *Lonchocarpus benthamianus* are found between the northern flanks of Mont Pelée and Morne Sibérie, Morne Platine, and Piton Mont Conil.

On the Atlantic coast the promontories in desert rock pavement are much as they are presumed to have been in pre-aboriginal times. There is little evidence of human habitation on them except at Macabou, where a tourist hotel was built that is now mostly condominiums for an expatriate enclave. The vegetation is not dense enough to prove attractive to cattle and sheep, which run loose in the adjacent bushland and woodlands. Mangrove swamps are stable areas except as they have been cleared and drained for industrial purposes around Lamentin and for agriculture in the Robert Haven.

Similarly native but distinctly seral in character are those plant communities established after natural disasters such as volcanic explosions and landslides. The *Pitcairnia-Guzmania* association of Mont Pelée, the scrubby populations of the fern *Dicranopteris bifida* on the high, steep, and rocky cliffs of the Pitons, and tree fern colonies of *Cyathea* on landslides, erosion-exposed slopes, or recent volcanic ejecta, all may occupy extensive areas for a long time because of the slow regeneration period and the frequency of disturbance.

Much larger areas are covered by regenerated growth. Most of the highlands of the interior mountains are in rainforest and montane forest, which were disturbed by swidden agricultural practices in the eighteenth and nineteenth centuries. The gross characteristics of two forest types have been regained under the conservation practices of the twentieth century. Small populations of *Miconia* thickets, *Syzygium jambos*, and bamboo patches witness the disturbed character of areas within these forests.

The southwestern peninsula, which was the source of the choice coffees of Martinique during the eighteenth century and had much of its surface in gardens, sugarcane, and savanna at various other times, is now mostly in second-growth dry woodland and in dry bushlands of acacias, logwood, or *Croton-Lantana-Cordia* types. Small patches of mature seasonal forest are found in steep ravines and small plateaus of Morne Gardier and Morne Larcher. In among this vegetation are numerous country homes. The peninsula of Sainte-Anne is an area with a sequence of vegetation change much like that of the Carbet-Prêcheur region except that it is today mostly in spontaneous vegetation. During the eighteenth century both were dominated by sugarcane and cotton. Later periods saw land redistribution, especially under the Crédit Foncier Colonial. Later abandonment led to secondary plant succession, culminating in the thorny scrub found today. Clearing associated with the building of resort hotels and vacation homes has opened some areas of scrub. This kind of clearing will probably increase.

Planted Vegetation with Many Conspicuous Spontaneous Elements

Three areas on the island are dominated by dooryard gardens and small-plot farms. These are Morne Pitault-Roches Carrés, Saint-Esprit–Mont Vauclin, and the Morne Crèvecoeur–Marin areas. These gardens are so fitted into remnants of the seasonal forest and dry woodland that the landscape is a mosaic of planted and spontaneous types. The dooryard garden itself is a somewhat ambiguous vegetation type since spontaneous elements play such a conspicuous part in both its morphology and its flora. Remnant populations of coffee can be found in the Vauclin hill country, left from the days of coffee dominance. Viewed from the air, the three areas appear wooded with a curiously patterned, discontinuous forest.

All three areas have high rural population densities. The Saint-Esprit–Morne Pitault and Mont Vauclin areas were colonized intensively by former slaves. The Crévecoeur area was divided up in the nineteenth century after the depression of 1880. Settlement preceded emancipation but was accelerated afterward. During the period when coffee was an important export, small properties were economic units. With the loss of this cash crop and the lack of any effective substitute, the inhabitants were forced into a subsistence orientation. This led to a more intense use of the dooryard garden–provisions garden complex. Today, part of the population is hired as seasonal labor, but there is often not enough work to go around, and others have immigrated to the capital.

Other areas in mixed planted and spontaneous vegetation are the savanna derivatives of the seasonal forest, rainforest, and dry woodland. These are scattered and not sufficiently large to be shown on a small-scale map.[6] They exist in the stretch of coast in the northwestern corner of Sainte-Luce Commune, and in the mid-elevations of the communes Sainte-Marie, Macouba, Lorrain, and in Ajoupa-Bouillon and Morne-Rouge. Although these tend to be maintained by grazing, their origins must be sought in clearings for agriculture.

Predictions Concerning Change and the Plant Cover Consequences

Historical evidence establishes that the plant cover of Martinique has been modified by repeated changes among the members of the ecosystem. About 1,500 species of plants have been introduced. The best estimate of losses has been made by Jacques Fournet (1978,11–16). He lists 97 species of phanerogams that have been lost from the flora. This makes a net gain of about 1,400 species. The flora has become richer. The indications from the history of past introductions are that agricultural and horticultural species, not weedy species, will be introduced, and these will go into plant assemblages that are carefully managed by humans. Some exceptions could be super species designed by science that could be competitive with the native vegetation. Grasses with the capability of fixing nitrogen or polyploids with greater seedling growth potentials are examples of such super species. I suggest that these will be few.

Associated with the changes in the flora are changes in the fauna. Numerous birds and several rodents have been eliminated, while congeries of household and barnyard animals have been introduced during the time of human occupation of the island. Some introductions have become feral: for example, some domesticated cats today, and cattle and pigs earlier. The mongoose has had some impact on the serpents it was introduced to kill, and even more impact on the bird and spider populations. The direction toward more livestock production will cause more land to be in grasslands, that is, pastures and hayfields. The loss of larger dispersal agents may well account for the slow recolonization of large seeded trees, especially in the rainforest.

The steep rise in human population growth has been abated by the acceptance of birth control and by immigration to Europe. Should attitudes toward birth control change or immigration be cut off, as seems to be happening, the growing population could be supported only by a massive industrialization. This could be accomplished only by the erection of strong trade barriers to protect infant industries, not likely under the present political situation. A more important opportunity for employment has been the development of the tertiary services. In 1982, 74 percent of the active population was in the tertiary sector. Less than 16 percent is in the primary sector. Education, tourism, and the arts are discussed with keen interest, but so far these areas have not produced the occupations hoped for. If the experiences in the rest of the Caribbean can be used for comparison, I think these sorts of outlets will be difficult to implement and may not be very important for some time.

Changes in the interaction processes have been numerous. New relationships between biota and the physical environment have resulted in the export of considerable energy and nutrients. The producers of sugarcane, bananas, and pineapples have exported many calories from the island to be consumed elsewhere. Although fertilizers have been imported, I believe there has been a net loss of energy. It would be interesting to examine this question more fully. If more land is devoted to livestock, we may find this export of energy somewhat reduced, since livestock (as consumers) are much less efficient converters of energy than plants.

Grazing by livestock has subjected the native flora to considerable pressure, decreasing the importance of some species in the native vegetation types and holding plant succession to herbaceous species generally.

Browsing by goats has changed the natural physiognomy of some woody species up to the browse line. Grazing and browsing animals accelerate the disperal of animal waste, which fertilizes fields.

Industrialization has served the needs of only a part of the redundant agricultural labor. It has served as an additional outlet for capital and energy, thus limiting the agricultural activities and so strengthening the ecosystem through diversification. Because fossil fuels and raw materials are lacking, most industry will need to be of the high-labor, low-material-cost types. With developments in fuels from sugarcane or other agricultural products, the energy shortage may not be so acute as today.

Perhaps the most important immediate factor affecting processes of vegetation change was the development of a conservation ethic on the part of a small but vocal and influential group of citizens. Partly as a result of tourism, which demonstrated that tropical vegetation in many forms is an attraction for tourists, but also as a result of a greater ecological awareness in France where most educated Martiniquans spend some time either in school or on holiday, the conservation movement and especially the preservationist philosophy is on the rise. It is behind the Association des Amis du Parc Naturel Régional and the establishment of the biological/geological reserves on the island. Four areas were set aside as reserves by 1979: (1) the mangrove swamps of the Lamentin Plain, (2) the natural and geological reserve of the southeast, (3) the integrated natural reserve of the Caravelle, and (4) the humid forest region of the northwest, centered on Piton Mont Conil. Numerous patches of forest and other fine vegetation types have been suggested for protection. The north coast from Anse Couleuvre to Grand' Rivière is now under protection as a preserve. As areas now recovering from agricultural exploitation in parts of the island are put under protection, there may well develop more good examples of mature vegetation in twenty-five years than have survived to the present time.

If Martinique should participate as one unit in a Windward Island Biosphere Reserve, as Putney (1983,9–11) proposed, protection could be given to a variety of landscapes. Not only would the fine examples of spontaneous vegetation be protected, but cultivated areas, where traditional farming systems dominate, could be preserved. The lands where resources have been depleted and that are in need of restoration might be set aside, permitting the regeneration of those resources (Riddle 1985, 252). If these areas evolved into forested areas and if the farming areas could be protected from noncompatible activities such as urbanization and tourism, the stability of the geographic pattern of the plant cover might be preserved.

Changes in habitat have been profound. Soils are much changed by erosion and compaction. The felling of forests has modified the climate at the ground. Tillage has also influenced the microclimates of the island. The surfacing of streets and roads and the building of concrete and manufactured-material housing also have modified local habitats beyond recognition. Irrigation modified the water balance locally and made possible the growing of plants unsuited to the natural environments. In other cases, the creation of artificial shade made possible the growing of tropical understory species on horticultural plots.

The dimensions of the system or systems on the island are variable and change with time, as agricultural production is tied to the world markets or the European Common Market, and with the possibility that overseas subsidies will be terminated. The political concerns of independence that have affected most of the Caribbean have had limited impact in Martinique. If such a movement were to become important and succeed, the island would be cut off from the monies that make it possible today to import food, construction materials, and manufactures and to enjoy one of the highest standards of living in the Caribbean. This might introduce a trend toward more sufficiency orientations in agriculture and the rise in importance of subsistence crops at the expense of other land uses. Tourism brings a different kind of impact, but also a different special character to the operative systems. The Club Mediterrané habitat appeals to one kind of tourist, the international quality hotel complex to another. If emigration were prohibited or severely limited, the high birth rate, despite its decline, would nonetheless still exert pressure as more people needed to be fed from the same resources of the small island. The processes and controls in operation today in Martinique are in precarious balance. Virtually any change in the system would be reflected in the plant geography of the island.

Notes

1. For example, in the rainforest, several associations developed out of the gommier, châtaignier, and bois blanc (*Dacryodes excelsa, Sloanea dentata, Simaruba amara*) vegetation types gave way to combinations of gommier châtaignier or châtaignier bois blanc. This interpretation of the species composition of modern stands contrasts that of other investigators.

2. I interpret the differences in *relevées* in different parts of the rainforest reported by Stehlé (1937a) and Portecôp (1979,51–53) not to environmental or altitudinal differences but to the chances of recolonization. For a discussion of the rare event see Kenneth Boulding 1978, and Gartner and McGuirk 1979, 1272–76.

3. This estimate is based upon the assumption of 4,000 species more or less in the flora. Fournet lists 2,779 phanerogams for Martinique and Guadeloupe. If one uses his figures, the proportion is more nearly 37 percent (Fournet 1978,11–16). New species records are being documented rather regularly by the team at the Gallerie Botanique, Fort-de-France, so I favor the higher figure.

4. These two species are considered by Fournet (1978) and Howard (1973) to be native, but I maintain that they are very early introductions to the Caribbean (Fournet 1978,1102–1103, 1516–17; Howard 1973). Howard is unsure about *Quassia* since it is both an ornamental and a medicinal (1979,36).

5. No definite indication can be found in the precipitation records that progressive desiccation has been under way since the seventeenth century.

6. See Jacques Portecôp's map of Martinique (1975).

Bibliography

Archival and Manuscript Materials
Archives Nationales, Paris, France (AN)

Archives du Ministère de la France d'Outre-Mer (FOM), formerly Ministère des Colonies.
Administrative Correspondence.
Series B1-B38. Letters sent by the minister on behalf of the king to the governors, and high officials of the American islands.
Series C4, box 4. Vegetation and scientific papers sent to the colonies.
C8 correspondence with Martinique
C8 A10 (1697–98). General correspondence of the Marquis d'Amblimont, governor, and Mr. Robert, intendant.
C8 A21 (1716). General correspondence of the Marquis Duquesne (quoted extensively by May 1930 and J. Petitjean-Roget 1980)
Series D. Collection of reports and letters concerning the introduction of vegetables, spices, cacao, citrus plants, forage plants, grain, and other plants.
See esp. D123, D124, D125, D212, D215, D216, D219, and D223. (Photocopies made available by B. Petitjean-Roget.)
Series F. Diverse services
F2 A Commercial companies
Box 11g, Dossier 1057. (Photocopy made available by B. Petitjean-Roget.)
Box 14. Memoirs and treaties concerning the commerce of the West Indies.
Box 15. Compagnie des Iles d'Amérique et des Indes Occidentales.
Box 17. Compagnie des Indes Occidentales (1664–1716).
Box 19. Compagnie Iles de l'Amérique (1635–1648).
F3. Collection of Moreau de Saint-Méry, Manuscript No. 26
"Historique de la Martinique (1635–1722). (Selected photocopies made available by E. Hayot and B. Petitjean-Roget.)
Series G. Généralités, including many important reports and much correspondence concerning plant introductions and agricultural practices.
See esp. G4 (50), G11 (121), G24 (184, 212, 222), G26 (243, 238, 240, 241, 244), G35 (280), G48 (370, 372, 373), G52 (434), G53 (502, 520), G56 (581, 582), G57 (587, 590), G87 (800), G236, G246, G252, G253, G259, G288. (Photocopies of these materials made available by B. Petitjean-Roget.)

Archives de la Marine (AM). Instructions to administrators.
Series B.

Bibliothèque Nationale, Paris, France (BN)

French manuscripts
No. 8027. Minutes of the Colbert letters (1669–71).
New French acquisitions
No. 9327. Notes on the population of Martinique, 1690–1700.
No. 9256–9510. Collection of documents by Pierre Margry.
No. 125, fol. 597. Colbert miscellaneous works.

Archives Départementales de la Guadeloupe, Basse-Terre Guadeloupe (ADG)

Materials made available by J.-P. Hervé, archivist.
Ballet Manuscript Collection. Compiled by Jules Ballet. 19 vols.

Archives Départementales de la Martinique, Fort-de-France, Martinique (ADM)

Materials made available by L. Chauleau, archivist.
Terrier of 1671. "Estat du procèz-verbal des terres de l'Isle de la Martinique." Prepared for the Compagnie des Indes Occidentales under the direction of Sieur Pélissier, one of the directors of that Company and resident of Martinique at the time of the survey. Copy in the Archives Départementales de la Martinique, Fort-de-France, Martinique, F.W.I. A photocopy of the Archives copy is in the possession of the author.

Maps

1645 *Carte de l'Isle de la Martinique.* By Sanson d'Abbeville. Paris: Pierre Mariette. Reprinted in J. Petitjean-Roget 1980,106–107.

1667 *L'isle de la Martinique.* By F. Blondel, sieur des Croisettes, cartographer. Bibliothèque Nationale, Cartes du Service Hydrographique, Portfolio 156, division 2, no. 5. Cited and reproduced in J. Petitjean-Roget 1980, 2: facing 1198.

1704 *Isle de la Martinique. A Pain des l'isle du Palais chex le sieur De Fer à la sphère royale avec privilège du Roi 1704.* (ADM) Reproduced by the Société d'Histoire de la Martinique.

1733 Houel. *Carte de l'Isle de la Martinique, colonie française de l'une des Isles Antilles de l'Amérique.* (ADM) Drafted from the manuscript plans of Mr. Houel, Engineer to the King. According to the astronomical observation and consistent with those *particuliers* of the Mr. Guillaume Delisle, Premier Geographer to His Majesty, of the Academy of Sciences and presented to Monseigneur, the Count de Maurepas, Secretary of State for the Départemental de la Marine. Amsterdam: Chez Jean Coveurs et Corneille Mortier. Archives Départementales de la Martinique, Fort-de-France, Martinique, F.W.I.

1734 Romain. *Plan de l'Isle de la Martinique.* (ADM) (Together within the cartouche) a map of the town of Fort St. Pierre (and the) plan of the Fort Saint Pierre, by J. Romain. Photographic reproduction. 91 cm × 92 cm.

1758 de Moras. *Carte réduite de l'Isle de la Martinique.* (ADM) Drafted at the Dépot des Cartes, Plans, et Journaux de la Marine. For the Service des Varsseau by order of Mr. de Moras, Minister and Secretary.

1762 Bellin. Carte de l'Isle de la Martinique. Drafted by Mr. Bellin . . . published by the heirs of Homan à Nuremburg in 1762. Vignaud. 64 × 54.6 cm.

1770 Moreau du Temple, *Carte géométrique et topographique de l'isle de la Martinique.* (ADM) Compiled and drafted by Moreau du Temple, Engineer-geographer to the King. *Carte géométrique et topographique.*

18th century *Carte de l'Isle de la Martinique, colonie française dans les Isles Antilles.* (ADM) Engraved. 40 cm × 31 cm.

18th century Sonne. *Carte de l'Isle de la Martinique, colonie française dans les Isles des Antilles.* (ADM) By M. Sonne, Hydrographic engineer of the Marine. 35.5 cm × 22 cm.

1962 Colmet-Daage, F. "Provisional Soils Map of Martinique, 1962." 1:50,000. In possession of the author.

1976 Martinique.
Cartes des Cultures

Miscellaneous

France, Service des Eaux et Forêts.
1964 Unpublished figures on lands in bois and forest. Courtesy of the director.

Hon. Paquemar, Vauclin, Martinique, F.W.I.
Estate records including nineteenth-century and twentieth-century field management plans made available by André Asselin.

Published Sources
Books, Articles and Pamphlets

Abbéville, C. d'
1614 *Histoire de la mission des pères Capucins en l'isle de Maraguan et terres circonvoisines.* Paris.

Acosta, J. de
1880 *The natural and moral history of the Indians.* Edited by C. R. Markam. London: The Hakluyt Society.

Adams, C. D.
1972 *Flowering plants of Jamaica.* With contributions by G. R. Proctor, R. W. Read et al. Mona, Jamaica: University of the West Indies.

Adelaide-Merlande, J.
1969 *Recherches sur l'économie de la Martinique à la fin du XXe siècle.* Etudes et documents du Centre d'Etudes Régionales Antilles-Guyane (CERAG), No. 1. Pointe-à-Pitre, Guadeloupe.

1977 "Problèmes de périodisation de l'histoire des Antilles." In *Quelques aspects de patrimoine culturel des Antilles,* pp. 53–64. Fort-de-France, Martinique: Centre Départemental de Documentation Pédagogique (CDDP).

1979 *Documents d'histoire antillaise et guyanaise 1814–1914.* Noyon, France: Imprimerie Finet.

1981a "Le régime du travail: Coercition-moralisation-immigration." *Historial Antillais* 4:125–47.

1981b "Les administrations abolitionistes–juin–oct. 1948." *Historial Antillais* 4:47–78.

1981c "Une conséquence de la crise sucrière: Les mutations de propriétés (habitations sucreries) à la Martinique en 1886–1887." *Historial Antillais* 4:319–27.

Albion, R. G.

1961 *The rise of New York Port, 1815–1860.* Hamden, Conn.: Archon Books.

Alegria, R.

1965 "On Puerto Rican archaeology." *American Antiquity* 3(2):246–49.

Alegria, R., H. B. Nicholson, and G. R. Willey

1983 "The archaic tradition in Puerto Rico." *American Antiquity* 21(2):113–21.

Allaire, L.

1973 *Vers une préhistoire des Petites Antilles.* Montreal: Centre de Recherches Caraïbes, Université de Montréal.

1977 "Late prehistory in Martinique and the Island Caribs: Problems in ethnic identification." Ph.D. diss., Yale University.

Allaire, L., and M. Mattioni

1983 "Boutbois et Le Godinot: Deux gisements acéramiques de la Martinique." *Proceedings of the 9th International Congress for the Study of the Pre–Columbian Cultures of the Lesser Antilles,* pp. 27–33. Montreal: Centre de Recherches Caraïbes, Université de Montréal.

Allan, J. A.

1957 *Grasses of Barbados.* Colonial Research Publication No. 23. London: Her Majesty's Stationery Office.

Amshoff, G. J. H.

1942 *The Grasses of the Dutch West Indies.* Kolonial Instituit Amsterdam, Afduling Handelsmuseum 225.

Andic, F.

1966 "Caribbean trade patterns." *Caribbean Studies* 6(2):46–58.

Andic, F., and S. Andic

1965 *Fiscal survey of the French Caribbean.* Rio Piedras: Institute of Caribbean Studies, University of Puerto Rico.

Andreieff, P., H. Bellon, and D. Westercamp

1976 "Cronométrie et stratigraphie comparée des édifices volcaniques et formations sédimentaires de la Martinique (Antilles françaises)." In *Publication-transactions de la VIIe Conférence Géologique des Caraïbes, 30 juin–12 juillet, 1974,* pp. 345–56. La Ravoire: Imprimeries Réunies de Chambéry.

Andrews, C. M.
[1896–98], 1914 *Guides to the materials for American history 1783, in the Public Record Office of Great Britain,* vol. 5. Washington, D.C.: Carnegie Institution. Reprint, New York: G. P. Putnam's Sons.
1915 "Anglo French commercial rivalry 1700–1750: The western phase." *American Historical Review* 20:339–556.
1934 *The colonial period of American history.* 4 vols. New Haven, Conn.: Yale University Press.

Andrews, K. R., ed.
1959 *English privateering voyages to the West Indies, 1508–1559.* Hakluyt Society, ser. 2, vol. 111. Cambridge: The University Press.
1964 *Elizabethan privateering during the Spanish War, 1585–1603.* Cambridge: The University Press.

Andrews, W. S.
1861 *Andrews' illustrations of the West Indies (1455–1526).* London: Day and Son.

Anghieria, P. M. d' [Peter Martyr]
1907 "De orbe novo. Les huit décades traduites du latin, avec notes et commentaires par P. Gaffard." In *Recueil de voyages et documents pour servir à l'histoire de la géographie depuis le XIIIe jusqu'à la fin du XVIe,* vol. 31. Paris.
1944 *Decades del Nuevo Mundo.* Translated by Joaquin Asensio. Buenos Aires: Editoriale Bajel.
1965 *The decades of the New Worlde, or West Indies.* Translated by Richard Eden. London: Hakluyt Society.

Anonymous [le P. Benin Bresson?]
[1649–59] 1975 *Histoire de l'isle de Grenade en Amérique.* Edited by Jacques Petitjean-Roget, translated by Elisabeth Crosnier. Montreal: Presses de l'Université de Montréal

Anonymous
1954 *Véritable relation de tout de qui s'est passé au voyage que de Bretigny fit à Amérique Occidentale.* Paris.

Anonymous [under the auspices of the Union des Travailleurs Agricoles, Union des Paysans Pauvres de la Guadeloupe]
1983 "Les occupations de terre en Guadeloupe." *Les Temps Modernes* 39 (441–42): 1974–87.

Anthiaume, Abbé
1916 *Cartes marines et constructions navales, voyages de découvertes chez les Normands.* 2 vols. Paris: F. Dumont.

Arciniegas, G.
1964 *Biografía del Caraïbe.* Buenos Aires: Editoriale Sudamericana.

Armytage, F.
1953 *The free port system in the British West Indies: A study in commercial policy, 1766–1822.* London: Published for the Royal Empire Society by Longmans, Green and Company.

Aspinall, A.
1928 *A wayfarer in the West Indies.* London: Methuen.

Asprey, G. F.
1959 "Vegetation in the Caribbean area." *Caribbean Quarterly* 5 (4):245–63.
Asprey, G. F., and A. R. Loveless
1958 "The dry evergreen formations of Jamaica. II. The raised coral beaches of the north coast." *Journal of Ecology* 46:547–70.
Asprey, G. F., and R. G. Robbins
1953 "The vegetation of Jamaica." *Ecological Monographs* 23:359–412.
Atkins, J.
[1735],1970 *A voyage to Guinea, Brasil, and the West Indies in His Majesty's ships, the "Swallow" and "Weymouth."* London: C. Ward and R. Chandler.
Atwood, T.
1791 *The history of the island of Dominica.* London: Privately printed for J. Johnson.
Aube, H. L. T.
1882 *La Martinique, son présent et son avenir.* Paris: Bergu-Lenraiult et Cie.
Auchinleck, G. G.
1956 "The rainfall of Antigua and Barbuda." N.p.: The Antigua Sugar Association.
Augelli, J. P.
1953 "Patterns and problems of land tenure in the Lesser Antilles: Antigua, B.W.I." *Economic Geography* 29:362–67.
Ault, W. O.
1965 *Open-field husbandry and the village community: A study of agrarian by-laws in medieval England.* Philadelphia: American Philosophical Society.
Ayensu, E. S.
1981 *Medicinal plants of the West Indies.* Algonac, Mich.: Reference Publications.
Baguenard, J. La.
1980 *Décentralisation territoriale.* Paris: Presses universitaires de France.
Baguet, H.
1905 *Droit français. Du régime des terres et de la condition des personnes aux Antilles françaises avant 1789.* Paris: Thèse de droit, Le Panvert.
Baker, H. G.
1965 "The evolution of the cultivated kapok tree: A probable west African product." In *Ecology and economic development in tropical Africa,* edited by David Prokensha. Berkeley: University of California Press.
Baker, J. N. L.
[1931],1937 *A history of geographical discovery and exploration.* London: G. G. Harrap.
Ballet, J.
1875 "Les Caraïbes." In *Congrès International des Américanistes. Compte-rendu de la première session,* pp. 394–438. Nancy, France: Chez Gustave Crépin-Leblond.
1890–99 *La Guadeloupe. Renseignements sur l'histoire, la flore, la faune, la géologie, la minéralogie, l'agriculture, le commerce, l'industrie, la législation, l'administration.* 5 vols. Basse-Terre, Guadeloupe: Imprimerie du Gouvernement.

Bambuck, C. A.
1935 *Histoire politique, économique et sociale de la Martinique sous l'ancien régime 1635–1789.* Paris: Librairie de Sciences Politiques et Sociales, M. Rivière.

Bamford, P. W.
1956 *Forests and French sea power, 1660–1789.* Toronto: Trinity.

Bandoian, G. A., and R. C. Murray
1974 "Pliocene-Pleistocene carbonate rocks of Bonaire, Netherlands Antilles." *Geological Society of America Bulletin* 85:1243–52.

Bangou, H.
1969 "L'abolition de l'esclavage dans la Caraïbe du point de vue de ses lois économiques et des initiatives humaines." In *Le passage de la société esclavagiste a la société post–esclavagiste aux Antilles au XIXe siècle,* pp. 7–8. Point-à-Pitre, Guadeloupe: Groupe Universitaire de Recherches Inter-Caraïbes.

Banquet, H.
1905 *Le régime des terres et la condition des personnes aux Antilles françaises avant 1789.* Paris: N.p.

Barbour, T.
1930 "Some faunistic change in the Lesser Antilles." *Proceedings of the New England Zoological Club* 11:73–85.

Barbour, W. R.
1942 "Forest types of tropical America." *Caribbean Forester* 3(4):137–50.

Barère de Vieuzac, B.
1794 *Rapport fait au nom du comité de salut public sur les colonies françaises isles du Vent, dans la séance du 19 thermidor, l'an 2 de la République Française et un Indivisible.* Paris: Published by order of the National Convention.

Barham, H.
1794 *Hortus Americanus: Containing an account of trees, shrubs, and other vegetable productions of South-America and West-India island and particularly of the island of Jamaica.* Kingston, Jamaica: Alexander-Aikman.

Barrabé, L.
1936 "La constitution géologique des Antilles." *Chroniques des Mines Coloniales,* pp. 214–27.
1939 "Geological structure of Lesser Antilles." *Pan American Geologist* 72:238–39.
1955 "Contribution à l'étude stratigraphique et tectonique des formations sédimentaires de la Martinique." *Revue, Institut Français du Pétrole et Annales des Combustibles Liquides* 10(5):295–308.

Barray, J., ed.
1963 *Plants and the mountains of Pacific peoples: A symposium.* Tenth Pacific Congress, Honolulu, Hawaii, 1961. Honolulu: Bishop Museum Press.

Barré, J.-M.
1963 *Grains d'or: Prose et poésie des Antilles et de Guyane.* Paris: Charles-Lavouzelle.

Barre, R.
1980 *Economie politique.* 2 vols. Paris: Presses universitaires de France.

Barrere, A.
1972 *Economie et institutions françaises.* 2 vols. 2nd ed. Paris: Dalloz.
Barrett, G. W., and R. Rosenberg
1981 *Stress effects on natural ecosystems.* Chichester: John Wiley and Sons.
Barrett, O. W.
1910 "Taotias, taros and dasheens." In *Promising root crops for the South,* pp. 7–29. Bureau of Plant Industry Bulletin 164. Washington, D.C.: U.S. Department of Agriculture.
1928 *The tropical crops.* New York: Macmillan.
Barrett, W.
1966 "Caribbean sugar production standards in the seventeenth and eighteenth centuries." In *Merchants and scholars,* edited by John Parker, pp. 147–70. Minneapolis: University of Minnesota Press.
Barrey, P.
1917 *Le Havre transatlantique du 1571 à 1610. Mémoires pour servir à l'histoire du commerce et de l'industrie en France.* Edited by M. Hayem. 5 vols. Paris: Hachette.
1918 *Les origines de la colonisation française aux Antilles: La Compagnie des Indes Occidentales.* Le Havre: Micaux.
Bartlett, H. H.
1953 "English names of some East-Indian plants and plant products." *Asa Gray Bulletin* 2(2):149–76.
Bassières, E.
1920–21 "Les irrigations à la Martinique." *Bulletin Agricole de la Martinique,* Oct.:19–22; Nov.:10–14.
1922a "Nos forêts." *Bulletin Agricole,* July:1–7.
1922b "Pluies et forêts." *Bulletin Agricole,* October:1–2.
1923–24 "Reboisement." *Bulletin Agricole de la Martinique,* Oct., 1923–24 Jan.–Apr., 1924.
Bates, H. W.
1878 *Central America in the West Indies and South America.* London: E. Stanford.
Bates, M.
1961 *The forest and the sea.* New York: Mentor.
Baude, T. H.
1918 *Etude sur les cinquante pas géométriques.* Fort-de-France, Martinique: Imprimerie Deslandes.
1931 "La Martinique." In *Guadeloupe, Guyane, Martinique, St. Pierre et Miquelon.* Exposition Coloniale Internationale. Paris: Libraire de Plas et G. Alexandre.
Beard, J. S.
1942 "Montane vegetation in the Antilles." *Caribbean Forester* 3(2):61–74.
1944a "Climax vegetation in tropical America." *Ecology* 25:127–58.
1944b "The natural vegetation of the island of Tobago, BWI." *Ecological Monographs* 74:135–63.
1944c "Provisional list of trees and shrubs of the Lesser Antilles." *Caribbean Forester* 5(2):48–66.

1946 *The natural vegetation of Trinidad.* Oxford Forestry Memoirs 20. Oxford: Clarendon Press.

1949a *The natural vegetation of the Windward and Leeward Islands.* Oxford Forestry Memoirs 21. Oxford: Clarendon Press.

1949b "Trees of the Windward and Leeward Islands." *Caribbean Forester* 12(1):38–50.

1953 "The savanna vegetation of northern tropical America." *Ecological Monographs* 23:149–215.

1955 "The classification of tropical American vegetation types." *Ecology* 36:89–100.

Beaulieu, A.

1965 "The changing plant associations of Martinique from pre–Columbian times to the present day." M.S. thesis, McGill University.

Bégon, M.

[1689],1925 "Lettres de Michel Bégon annotées par feu Louis Delavaud et Charles Dangibeard." *Annales historiques de la Sántonge et de l'Aunis,* vol. 47. Paris: Picard.

[1685],1930 "Michel Bégon et la commerce des îles en 1685." *Revue Historique des Antilles* 6 Apr.–Oct., 1930.

Bell, H. C.

1916 "British commercial policy in the West Indies, 1783–93." *English Historical Review* 31:429–41.

1917 "The West India trade before the American revolution." *American Historical Review* 22:272–87.

Bellin, J. N.

1758 *Description géographique des isles Antilles possedées par les Anglois.* Paris: Imprimerie de Didot.

Benjamins, H. D., and J. F. Snellmann

1917 *Encyclopaedie van Netherlandsch West Indies.* La Haye, Netherlands: Nijhoff.

Benoist, J.

1968 "Types de plantation et groupes sociaux a la Martinique." *Cahiers des Amériques Latines* 2:130–59.

Benzoni, G.

[1537],1857 *La historia del mundo nuovo.* Venice: N.p.

[1572],1857 *History of the New World, by Girolano Benzoni, of Milan showing his travels in America, from A.D. 1541 to 1556, with some particulars of the island of Canary.* Translated and edited by W. H. Smyth. First ser. 21. 2nd ed. London: Hakluyt Society.

Berenger, H., and C. Grahen

1935 *Tricentenaire du rattachement des Antilles.* Paris: Société d'Éditions Géographiques.

Bernissant, P.

1916 "Étude sur le régime agricole des Antilles françaises." Thèse de doctorat. Paris.

Berthe, J. P.

1959 "Sur l'histoire sucrière américaine." *Annales, Economies, Sociétés, Civilisations* Jan.–Mar., pp. 135–41.

Besson, M.
1928 *Les "frères de la Coste": Filibustier et corsaires d'après les textes et les estampes de l'époque.* Paris: Duchartre et Van Buggenhoudt.
1929 *The scourge of the Indies buccaneers, corsaires, and filibusters, from original texts and contemporary engravings.* New York: Random House. London: G. Routledge.
1931 *Histoire des colonies françaises.* Paris: Boivin.
1936 "De l'exposé colonial à la Conference 'Impériale'." *Grande Revue* 146:277–94, 458–70, 623–31; 147:110–18.

Biet, Father A.
1664 *Voyages de la France équinoxale en l'isle de Cayenne enterpris par les français en l'année 1652.* 3 vols. Paris: Clouzier.
[1664],1896 "Les Galibis, tableau veritable de leurs moeurs avec un vocabulaire de leur langue." Edited by A. Marre. *Revue de Linguistique,* July–Oct.

Birket-Smith, K.
1943 "The origin of maize cultivation." *Det Kgl Dansk Videnskabernes Selskav Historisk-Filologiske Meddelelser* (Copenhagen) 29(3).

Blackevelder, B. W., O. H. Pilkey, and J. D. Howard
1979 "Late Wisconsinean sea levels on the southeast U.S. Atlantic Shelf, based on in-place shoreline indicators." *Science* 204:618–20.

Blaut, J. M.
1959 "A study of cultural determinants of soil erosion and conservation in the Blue Mountains of Jamaica: Progress report." *Social and Economic Studies* 8(4):403–20.

Blérald, E. D.
1898 *La culture du tabac à la Martinique.* Paris: A. Challamel.

Blet, H.
1946–50 *Histoire de la colonisation française: Naissance et déclin d'un empire, des origines à 1789.* Paris: B. Arthaud.

Bloch, M.
1966 *French rural history: An essay on its basic characteristics.* Berkeley: University of California Press.

Blome, K.
1684 "Description de l'isle de la Jamaïque et de toutes celles que possèdent les Anglois dans l'Amérique avec des observations faites par le sieur Thomas Gouverneur de la Jamaïque, et autres personnes du puis." In *Le recueil de Justel.* Paris: Veuve Antoine Cellier.
1687 *The present state of His Majesty's isles and territories in America.* London: H. Clark for D. Newman.

Bloom, A. L.
1984 "Sea level and coastal changes." In *Late-Quaternary environments of the United States.* Vol. 2, *The Holocene,* pp. 42–51. Minneapolis: University of Minnesota Press.

Bond, J.
1948 "Origin of the bird fauna of the West Indies." *Wilson Bulletin* 30:207–99.
1960 *Birds of the West Indies.* 4th ed. London: Collins.

Bonne, R.
1780 *Les isles Antilles et le Golfe du Méxique.* Geneva: J. L. Pellet.
Bonnot, L.
1974 *Déficit pluviométrique en Martinique année 1973.* Fort-de-France, Martinique: Service Météorologique du Groupe Antilles-Guyane. (Mimeographed).
Bormann, F. H., and G. E. Likens
1979 *Pattern and process in a forested ecosystem.* New York: Springer-Verlag.
Boston *Gazette*
1790 July 5, 1790.
Boughey, A. S.
1957 "Ecological studies of tropical coastlines. I. The Gold Coast, West Africa." *Journal of Ecology* 45:665–87.
Boulding, K.
1978 *Ecodynamics.* Beverly Hills, Calif.: Sage Publications.
Bourde, A. J.
1953 *The influence of England on the French agronomes, 1750–1789.* Cambridge: The University Press.
Bourne, R.
1939 *Queen Anne's Navy in the West Indies.* New Haven, Conn.: Yale University Press.
Bousefield, E. L., and M. L. H. Thomas
1975 "Postglacial changes in the distributions of littoral marine invertebrates in the Canadian Atlantic region." In *Environmental Changes in the Maritimes,* pp. 47–60. Halifax, Nova Scotia: Institute of Science.
Bouton, Jacques, S.J.
1640 *Relation de l'éstablissement des François depuis l'an 1635, en l'isle de la Martinique, l'une des Antilles de l'Amérique, des mœurs des Sauvages de la situations et des autres singularitez de l'isle.* Paris: Sebastien Cramoisy.
[1640],1935 *Relation de l'éstablissement des François depuis l'an 1635, en l'isle de la Martinique, l'une des Antilles de l'Amérique.* Paris: S. Cramoisy. Quoted extensively in *Tercentenaire des Antilles,* edited by J. Rennard, pp. 35–82. Fort-de-France, Martinique: René Cottrel.
1963 *Relation de l'établissement des français en l'isle Martinique,* chap. 9–10. *Annales des Antilles* 11:97–103.
Bowin, C.
1976 *Caribbean gravity field and plate tectonics.* Special Paper 169. Denver: Geological Society of America.
Box, H. E.
1939 "Observations on the landslides in St. Lucia, B.W.I. in November, 1938." *Empire Forestry Journal* 18:119–21.
Boyce, S. G.
1954 "The salt spray community." *Ecological Monographs* 24:29–67.
Boyer de Peyreleau, E. E. baron de
1823 *Les Antilles françaises, particulièrement la Guadeloupe, depuis leur découverte jusqu'au 1er janvier 1823.* 3 vols. Paris: Béchet-Ainé.
1823 *Les Antilles françaises (surtout la Guadeloupe) depuis sa découverte jusqu'au 1er nov. 1823.* 3 vols. 2nd ed. Paris: L'Advocat.

1825 *Les Antilles françaises, particulièrement la Guadeloupe, depuis leur decouverte jusqu'au 1 janvier 1825.* 3 vols. 2nd ed. Paris: L'Advocat.

Boyer du Petit Puy, J. *See* Petit Puy, P. Boyer du

Bradshaw, A. D.

1971 "Plant evolution in extreme environments." In R. Geed, *Ecological genetics and evolution,* pp. 20–50. London: Blackwell Scientific Publications.

Bréard, Ch., and P. Bréard

1889 *Documents relatifs à la marine normande et à ses armements aux XVIe et XVIIe siècles pour le Canada, l'Afrique, les Antilles, le Brésil et les Indes.* Rouen, France: A. Lestringant.

Breen, H. H.

[1884],1970 *St. Lucia: Historical, statistical and descriptive.* London: Frank Cass.

Breton, Father R.

[1635], 1947 *Relatio gestorum a primis ordinis praedicatorum missionariis in insulis Americanis ditionis Gallicae praesertim apud Indios indigenas quos Caraïbes vulgo dicunt ab anno 1635 in annum 1643.*

1664 *Petit catéchisme ou sommaire des trois premières parties de la doctrine chrétienne, traduite du françois en langue des Caraïbes insulaires.* Auxerre, France: Gilles Bouquet.

[1665],1892 *Dictionnaire Caraïbe-François.* Auxerre, France: Gilles Bouquet. 1665 facsimile edition. Leipzig: Jules Platzman.

[1647],1929 "Relations du R. P. Breton." In *Les Caraïbes, la Guadeloupe, 1635–1656,* edited by J. Rennard. Paris: G. Fischer.

1667 *Grammaire caraïbe.* Auxerre, France: Gilles Bouquet.

1892 *. . . Caraïbe-François meslé de quantité de remarques historiques pour l'esclaircissement de la langue.*

1963 "De l'origine, moeurs, religion et autres façons de faire des Caraïbes appelés communément sauvages anciens habitants de la Guadeloupe." *Annales des Antilles* 11:105–28. (from an undated manuscript in Bibliothèque Nationale, manuscripts français 24574, fol. 23–50).

1978 *Relations de l'île de la Guadeloupe,* vol. 1. Vol. 3 of *Bibliothèque d'histoire antillaise.* Basse-Terre, Guadeloupe: Société d'Histoire de la Guadeloupe.

Breton, Father R., and Father Armand de La Paix

[1635–56], 1929 "Relation de l'île de la Guadeloupe." In *Les Caraïbes, la Guadeloupe, 1635–1656,* edited by J. Rennard, and in *Histoire Coloniale* 1:45–74.

Brieyer, F. G., J. T. A. Gurgeli, and E. Paterniani

1958 *Races of maize in Brazil and other eastern South American countries.* National Research Council Publication 593. Washington, D.C.: National Academy of Sciences.

Briton-Jones, H. R.

1927 *Report on visit to Dominica.* West Indies Pamphlet No. 473. Port-of-Spain, Trinidad: N.p.

Britton, N. L.

[1918],1965 *Flora of Bermuda.* New York: Hafner Publishing.

Britton, N. L., and P. Wilson

1923–30 *Botany of Puerto Rico and the Virgin Islands,* vols. 5–6. Scientific Survey of Puerto Rico and the Virgin Islands. New York: New York Academy of Sciences.

Broughton, A.
1794 *Hortus Eastensis: Or a catalogue of exotic plants cultivated in the botanic garden, in the mountains of Leguaner in the island of Jamaica.* Jamaica.

Brown, J. J.
1937 "William Bingham, agent of the Continental Congress in Martinique." *Pennsylvania Magazine of History* 61:54–87.

Brown, W. L.
1953 "Maize of the West Indies." *Tropical Agriculture* 20(7–9):141–70.

Browne, P.
1756 *The civil and natural history of Jamaica.* London: Privately printed.
1789 *The civil and natural history of Jamaica.* 2nd ed. London: Privately printed.

Bryson, R. A., and W. M. Wendland
1967 "Tentative climatic patterns for some late glacial and post glacial episodes in central North America." In *Life, land, and water,* edited by W. Mayer-Oakes, pp. 271–98. Winnipeg: University of Manitoba Press.

Bucher, W. H.
1947 "Problems of earth deformation illustrated by the Caribbean Sea basin." *Transactions of New York Academy of Science,* ser. 2. 10(3):98–116.

Buffin, A.
1970 *Monnaie et crédit en économie coloniale: Contribution à l'histoire économique de la Guadeloupe 1635–1919.* 2 vols. Paris: Université de Paris.
1973a "L'industrie, l'artisanat, le tourisme." In *Encyclopédie antillaise (Economie antillaise),* pp. 209–29. Pointe-à-Pitre, Guadeloupe: Editions Desormeaux.
1973b "L'agriculture." In *Encyclopédie antillaise (Economie antillaise),* pp. 231–51. Pointe-à-Pitre, Guadeloupe: Editions Desormeaux.
1979 *Monnaie et crédit en économie coloniale: Contribution à l'histoire économique de la Guadeloupe 1635–1919.* Vol. 8 of *Bibliothèque d'histoire antillaise.* Basse-Terre, Guadeloupe: Société d'histoire de la Guadeloupe.
1981 "La mise en place des usines centrales à la Martinique." *Historial Antillais* 4:185–213.

Buffon, A.
1920 "Rapport sur le recensement agricole." *Bulletin Agricole de la Martinique* 10:19–20.
1932 "La petite propriété rurale et la polyculture." *Bulletin Agricole de la Martinique* 42:1–3.

Bullen, R. P.
1965 "Caribbean symposium: Archeological chronology of Grenada." *American Antiquity* 31(2):237–41.

Bullen, R. P., and A. K. Bullen
1972 *Archaeological investigations on St. Vincent and the Grenadines.* Orlando, Fla.: William L. Bryant Foundation.

Bunce, E. T., J. D. Phillips, R. L. Chase, and C. O. Borvin
1970 "The Lesser Antilles arc and the eastern margin of the Caribbean Sea."

In vol. 4, pt. 2 of *The sea,* edited by A. E. Maxwell, pp. 359–85. New York: Interscience.

Burac, M.

1970 *Les plantation d'ananas dans la région du Gros-Morne, Martinique.* Bordeaux, France: Travail d'études et de Recherches-Lille.

1975 "L'économie rurale dans le Centre-Nord de la Martinique." *Contribution à l'étude de la crise d'une économie de plantation.* Montreal: Centre d'Etude de Géographie Tropicale.

1978 *Antilles d'hier et d'aujourd'hui.* Paris: Desormeaux.

1980 "L'aménagement de territoire à la Martinique." *Bulletin d'Information du CENADDOM* 54:17–27.

Bureau de Recherches de Géologie et des Mines (BRGM)

1974 *Livret-guide d'excursions dans les Antilles françaises.* Fort-de-France, Martinique: BRGM.

Burkill, F. H.

1954 "Aji and batata as group-names within the species *Ipomoea batatas.*" Ceiba (Tegucigalpa) 4(4):227.

Burton, R. D. E.

1978 *Assimilation or independence? Prospects for Martinique.* Montreal: McGill University Press.

Butel, P.

1982 *Les Caraïbes au temps des flibustiers: XVI*[e]*–XVII*[e] *siècles.* Paris: Aubier Montaigne.

Butterlin, J.

1956 *La constitution géologique et la structure des Antilles.* Bordeaux: Centre National de la Recherche Scientifique.

1977 *Géologie structurale de la région des Caraïbes (Mexique-Amérique Centrale-Antilles Cordillères Caraïbes).* Paris: Masson.

Butzer, K.

1972 *Environment and archaeology: An ecological approach to pre–history.* London: Methuen.

Byrne, A.

1980 "Man and the variable vulnerability of island life." *Atoll Research Bulletin* 240:1–200.

Byrne, A. R., and J. H. McAndren

1955 "Precolumbian purslane (*Portulaca oleraceae*) in the New World." *Nature* 253:726–727.

Cahingt, H.

1930 "Les registres du tabellionage de Dieppe et les Antilles." *Revue Historique des Antilles* 5(Jan.–Mar.): n.p.

Caines, C.

1801 *Letters on the cultivation of the Otaheite cane.* London: Privately printed.

Camp, H. W.

1954 "A possible source for American pre–Columbian gourds." *American Journal of Botany* 41:700–701.

Campan, G.

1959 *Notes sur la climatologie des Antilles et de la Guyane française.*

Monographies de la Météorologie Nationale No. 15. Paris: Service Météorologique du Groupe Antilles-Guyane.

Carbone, V. A.
1980 "Some problems in central paleoecology in the Caribbean area." In *Proceedings of the 8th International Congress for the Study of Pre–Columbian Cultures of the Lesser Antilles, 1979*, pp. 98–110. Tucson: Arizona State University.

Caribbean Commission, Central Secretarial
1950 *Population movements in the Caribbean.* Port-of-Spain, Trinidad: Malcolm J. Proudfoot.

Carlquist, S.
1906 "The biota of long distance dispersal. I. Principles of dispersal and evolution." *Quarterly Review of Biology* 41:247–70.
1965 *A natural history of the islands of the world.* Garden City, N.Y.: Natural History Press.
1974 *Island biology.* New York: Columbia University Press.

Carneiro, R. L.
1961 "Slash-and-burn cultivation among the Kuikuru and its implications for cultural development in the Amazon basin." In *The evolution of horticultural systems in native South America: Causes and consequences*, pp. 47–67. Caracas: Sociedad de Ciencias Naturales LaSalle.

Catesby, M.
1731–43 *Natural history of Carolina, Florida, and the Bahama Islands.* 2 vols. London: Privately printed.

Cattelíno, P. J., I. R. Nobel, R. O. Slayter, and S. R. Kessel
1979 "Predicting the multiple pathways of plant succession." *Environmental Management* 3:41–50.

[Chambon, A.]
1777 *Le guide du commerce de l'Amérique, principalement par le port de Marseille.* Avignon, France: J. Mossy.

Chambon, A.
1783 *"Traité général du commerce de l'Amérique par Marseille.* 2 vols. Amsterdam: Chez Marc-Michel Rey.

Champion, J.
1954 "La culture bananière aux Antilles et en Amérique centrale." *Fruits* no. 10(Nov.):433–35.
1963 *Le bananier, techniques agricoles et productions tropicales.* Paris: G. P. Maisonneuve and Larose.

Chanlatte-Baik, L. A.
1983 "Sorcé-Vieques: Climax cultural del Igneri y su participacion en los processos socio-culturales Antillanes." In *Comptes-rendus des communications du IXe congrès international d'étude des civilisations précolombiennes des Petites Antilles*, pp. 73–95. Montreal: Centre de Recherches Caraïbes, Université de Montréal.

Chanlatte-Baik, L. A., and Y. M. Narganes
1980 "La Hueca, Vieques: Nuevo complejo cultural agroalfarero en la arqueologia antillana." *Proceedings of the 8th International Congress for the*

Study of the Pre–Columbian Cultures of the Lesser Antilles, St. Kitts, 1980, pp. 501–17. Tempe: Arizona State University.

Chanvalon, T. de

1763 *Voyage à la Martinique.* Paris: P. Angrand.

Chapman, A., and V. W. Chapman

1911 *Encyclopaedia Britannica,* 11th ed., s.v. "Sugar."

Chauleau, L.

1973 *Histoire antillaise. La Martinique et la Guadeloupe du XVIIe à la fin du XIXe: Elements d'histoire antillaise.* Fort-de-France, Martinique: Désormeaux.

1978 *Guide des archives de la Martinique.* Fort-de-France, Martinique: Archives Départementales.

Chemillier-Gendreau, M.

1960 "La Réserve des 50 pas géométriques en Martinique." *Annales des Antilles* 8:3–47.

Chemin-Dupontes, P.

1909 *Les Petites Antilles. Etude sur leur évolution économique.* Paris: E. Guilmoto.

Cherbonnier, J.

1963 "Essai de classification des différents climats de la Martinique d'après les méthodes de Thornthwaite." Direction de la Météorologie Nationale, Service Météorologique du Groupe Antilles Guyane. Fort-de-France. (Mimeographed).

Chesters, G. J.

1869 *Transatlantic sketches in the West Indies, South America and the United States.* London: Smith Elder.

Chevalier, A.

1936 "Ce que l'Amérique a donné à l'ancien monde." *Revue de Botanique Appliquée et d'Agriculture Tropicale* 16:419.

1938 "Liaisons floristiques entre l'Afrique tropicale et l'Amérique equinoxiale." *Revue de Botanique Appliquée et d'Agriculture Tropicale* 208:855–62.

Chevillard, P. A.

[1659],1963 *Les desseins de son Éminence de Richelieu pour l'Amérique, ce qui s'y est passé de plus remarquable depuis l'établissement de colonies, et un ample traité du naturel, religion et moeurs des indiens insulaires et de la Terre Ferme.* Rennes, France: J. Durant. Excerpts reprinted in "Les Caraïbes vus par les premiers chroniqueurs français." *Annales des Antilles* 11:59–73.

Churchill, A., and J. Churchill

1752 *A collection of voyages and travels,* vols. 2 and 5. 4th ed. London: Privately printed.

Clark, A. H.

1959 *Three centuries and the island.* Toronto: University of Toronto Press.

1960 "Geographical change: A theme for economic history." *Journal of Economic History* 20:607–13.

Clarke, W.
1968 "The Ndwomba Basin, Bismarck Mountains, New Guinea: Place and people." Ph.D. diss., University of California, Berkeley.

Cléach, A.
1974 "Les progrès de l'élevage moderne en Guadeloupe." *Cahiers d'Outre Mer* 27:22–76.

Clerc, E.
1968 "Sites précolumbiens de la Grande-Terre de Guadeloupe." *Proceedings of the 2nd International Congress for the Study of the Pre–Columbian Cultures of the Lesser Antilles,* pp. 47–60. Barbados.
1971 "Les trois-pointes des sites précolombiens de la côte nordest de la Grande-Terre de la Guadeloupe." *Comptes-rendus du IVe congrès international pour l'étude des civilisations précolombiennes des Petites Antilles,* pp. 73–81. Saint Lucia.
1975 "Remarques sur quelques pierres à trois pointes provenant des gisements précolombiens de la côte nordest de Grande-Terre de Guadeloupe." *Compte-rendu du IVe Congrés International d'Études des Civilisation Précolombienne Petites Antilles,* pp. 42–43. Gainesville: Florida State Museum.

Cobo, B.
1890–95 "Historia del nuevo mundo." 4 vols. Seville: Impresor E. Rasco, for Sociedad de Bibliofilos Andaluces.

Coke, T.
1808–11 *A history of the West Indies containing the natural, civil, and ecclesiastical history of each island. . . .* Liverpool: Huttall, Fisher and Dixon.

Colmeiro, M.
1892 *Primeras noticias acerca de la vegetacion americana.* Madrid: Sucesores de Rivadeneyra.

Colmet-Daage, F.
1964 *Charactéristiques de quelques groupes de sols devenu des roches volcaniques aux Antilles françaises.* Bureau des Sols des Antilles. Fort-de-France, Martinique: Bureau de Recherche de Géologie et des Mines (BRGM).

Comité Technique de Fruit de la Martinique.
1959 *Fruits Antillais,* no. 12. Fort-de-France: Comité Technique de Fruit de la Martinique (CTFM).

Cook, O. F.
1893 "Food plants of ancient America." *Annual Report, Smithsonian Institution,* pp. 481–97. Washington, D.C.: Smithsonian Institution.

Cook, O. F., and G. N. Collins
1903 *Economic plants of Puerto Rico.* Washington, D.C.: Government Printing Office.

Coppier, G.
[1644],1963 *Histoire et voyage des Indes Occidentales,* chap. 8–13. Excerpts printed in *Annales des Antilles* 11:37–45.
1645 *Histoire et voyage des Indes Occidentales et de plusieurs autres régions maritimes et éloignées.* Lyon: d'Etain Cress for Jean Huguetan.

Coste, R.
1959 *Le caféier et le cacaoyer dans les départements antillais et en Guyane.* Fort-de-France, Martinique: Institut Français de Café et Cacao.

Crouse, N. M.
1943 *The French struggle for the West Indies, 1665–1713.* New York: Columbia University Press.

Cruxient, J., and S. Rouse
1969 "Early man in the West Indies." *Scientific American* 221(5):42–52.

Cutler, H. C., and T. W. Whitaker
1961 History and distribution of the cultivated cucurbits in the Americas." *American Antiquity* 26(4):469–85.

Dajani, T., J. Sabatier, and D. Rouchs, eds.
1981 *Historial Antillaise.* 6 vols. Pointe à Pitre; Guadeloupe: Dajani.

Dampière, J. de
1904 *Essai sur les sources de l'histoire des Antilles françaises (1492–1664).* Paris: A. Picart et Fils.

Dancer, T.
1792 *Catalogue of plants . . . in the Botanic Garden, Jamaica.* St. Jago de la Vega, Jamaica.
1793 "Catalogue of exotic plants, in the botanical garden of Jamaica, 1792." In B. Edwards, *The history, civil and commercial, of the British West Indies,* vol. 1, bk. 2, pp. 198–211. London: Privately printed.

Daney, M. S.
1846 *Histoire de la Martinique depuis la colonisation jusqu'en 1815.* 6 vols. Fort-Royal, Martinique: E. Ruelle.

Dansereau, P.
1951 "Description and recording of vegetation." *Ecology* 32:172–229.
1958 "A universal system for recording vegetation." *Contributions de l'Institut Botanique de l'Université de Montréal* 72:1–58.
1961 "The origin and growth of plant communities." In *Growth in living systems,* edited by M. X. Zarrow, pp. 567–603. New York: Basic Books.
1966 "Description and integrations of the plant communities." In *Studies on the vegetation of Puerto Rico,* pp. 3–45. Special Publication No. 1, Institute of Caribbean Science. Mayagüez, Puerto Rico: University of Puerto Rico.
1968 "Continuum concept of vegetation, forward." *Botanical Review* 34:253–54.

Darby, H. C.
1934 "The Fenland frontier in Anglo-Saxon England." *Antiquity* 8:185–201.
1956 "The clearing of the woodlands in Europe." In *Man's role in changing the face of the Earth,* edited by W. L. Thomas et al., pp. 183–216. Chicago: University of Chicago Press.

D'Arcy, F.
1929 "L'industrie sucrière aux isles au XVII siècle." *Revue Historique des Antilles* 2:3–16.

Darmezin de Garlande, F., and J. Poupon
1982 *L'art mobilier de la Martinique aux 18th and 19th siécles.* Département de

la Martinique, Office National des Forêts. Fort-de-France, Martinique: Imprimerie Hemmerlé.

Daubenmire, R.

1960 "Some major problems in vegetation classification." *Silva Fennenie* 105:22–25.

1966 "Vegetation: Identification of Tyrol communities." *Science* 151:291–98.

Dauden, J., and H. Guyot

1951 "Etude de l'attaque des bananeries antillaises par *Cosmopolites sordidus germ." Annales de l'Institut Français de Recherches Frutières Outre-Mer, Antilles Françaises* 3:135–40.

Dauvergne, R.

1948 "Les anciens plans ruraux des colonies françaises." *Revue d'Histoire des Colonies* 35:231–69.

David, Abbot B.

1973a *Les origines de la population martiniquaise du fil des ans (1635–1902).* Special issue no. 3 of *Annales des Antilles,* Mémoire de la Société d'Histoire de la Martinique.

1973b "La population d'un quartier de la Martinique au début du XIXe siècle d'après les registres paroisseaux Rivière-Pilote, 1802–1829." *Revue Française d'Histoire d'Outre-mer* 60(202):303–63.

Davies, J. of Kidwelly

1666 *The history of the Caribby Islands.* London: Privately printed.

Davis, D. D.

1974 "Some notes concerning the archaic occupation of Antigua." *Proceedings of the 5th International Congress for the Study of Pre-Columbian Cultures of the Lesser Antilles Antigua, July 22–28, 1973.* St. Michael's Mount, Antigua: Antigua Archaeological Society.

Davis, J. H.

1942 "The ecology of the vegetation and topography of the sand keys of Florida." *Papers from the Tortugas Laboratory, Carnegie Institution of Washington* 33:113–95.

Davis, M. B.

1981 "Quaternary history and stability of forest communities." In *Forest succession: Concepts and application,* edited by D. C. West, H. H. Shugart, and D. B. Bolkin. New York: Springer-Verlag.

Davis, T. A. W., and P. W. Richards

1933–34 "The vegetation of Moraballi Creek, British Guiana." *Journal of Ecology* 21:350–84; 22:106–55.

Davis, W. M.

1924 "The formation of the Lesser Antilles." *National Academy of Science Proceedings* 10:205–11.

Davy, J.

1854 *The West Indies, before and since slave emancipation, comprising the Windward and Leeward Islands military command.* London: W. and F. G. Cash.

de Baas, M.

[1674], 1960 "Rapport de M. de Baas, gouverneur des Iles d'Amérique à son ministre, Feb. 8, 1674." *Annales des Antilles* 8:10.

Debien, G.

1951 "Les engagés pour les Antilles: 1634–1715." *Revue d'Histoire des Colonies* 38:1–277.

1960 "Destinées d'esclaves à la Martinique." *Bulletin de l'Institut Français de Recherches Fruitières Outre-Mer, Antilles Françaises* 22:1–91.

1961 "Les travaux d'histoire sur les Antilles françaises, chronique bibliographique, 1959 et. 1960." *Revue Française d'Histoire d'Outre-Mer* 48:267–308.

1965 "Les origines des esclaves des Antilles." *Bulletin de l'Institut Français d'Afrique Noire XXII,* ser. B, 3–4:755–99.

1974 *Les esclaves aux Antilles françaises XVII–XVIIIe siècles.* Basse-Terre, Guadeloupe: Société d'Histoire de la Guadeloupe et de la Martinique; Fort-de-France: Société à Histoire de la Martinique.

1975 "Les origines des esclaves aux Antilles." *Bulletin de l'Institut Français d'Afrique Noire XXIX,* ser. B, 3–4:536–58.

Deerr, N.

1949 *The history of sugar.* 2 vols. London: Chapman and Hall.

1950 *Sugar and sugar cane: An elementary treatise on the agriculture of the sugar cane and the manufacture of the cane supply.* Aliricham, England: Norman Rodger, 1905.

de Goeje, C. H.

1939 "Nouvel examen des langues des Antilles." *Journal de la Société des Américanistes,* n.s. 31:1–120.

Delafosse, M.

1949 "La Rochelle et les Iles au XVIIe siècle." *Revue d'Histoire des Colonies* 36:238–81.

de Lanessan, J. L.

1886a *Les plantes utiles des colonies françaises.* Paris: Exposition Universelle d'Anvers, 1885.

1886b *L'expansion coloniale de la France, étude économique, politique et géographique sur les établissements français d'Outre-Mer.* Paris: Ancienne Librairie Germer Baillière et Cie.

Delawarde, J.-B.

1935 *Les défricheurs et les petits colons de la Martinique au XVIIe siècle.* Paris: Buffault.

1936 *Le Prêcheur. Histoire d'un établissement humain sur les pentes de la Pelée.* Paris: Buffault.

1937a *La vie paysanne à la Martinique.* Fort-de-France, Martinique: Imprimerie Officielle.

1937b *Préhistoire martiniquaise: Les gisements du Prêcheur et du Marigot.* Fort-de-France, Martinique: Imprimerie Officielle.

1938 "Les dernières Caraïbes: Leur vie dans une réserve de la Dominique." *Journal de la Société des Américanistes* n.s. 30(1):169–204.

Demangeon, A.

1905 *Les sources de la géographie de la France aux Archives Nationales.* Paris: Société Nouvelle de Librairie et d'Edition (G. Bellais).

Denis, S., ed.
1935 *Trois siècles de vie française: Ouvrage publié sous la direction de Serge Denis avec une introduction de L. Gallouedec.* New Orleans: G. Luzeray.

de Neufchateau, M. François
1811 *L'Art de multiplier les grains.* Paris: Chez Madame Huzard. Extracts by M. Tessier, *Annales de l'Agriculture Françoise* 46:59–90.

de Pilj, L. van. *See* van der Pilj, L.

de Poincy, Longvillier
[1640],1935 "Mémoires envoyés aux seigneurs de la compagnie des îles de l'Amérique" (extracted from B. N. Affairs Etrangères, Amerique, vol. 4, fol. 183ff.). In *Tricentenaire des Antilles,* edited by J. Rennard. Fort-de-France, Martinique: Réné Cottrel.

de Poras, D., and C. Columbo
[1503],1899 *"Littera rarissima" de Christophe Colomb sur la découverte de la terre Angers: Ferme accompagnée de l'itineraire de Diego de Poras.* Paris: Imprimé Camis et Cie., Oriental section, A. Burden.

de Provin, Pacifique. *See* Pacifique de Provin

de Rochefort, Charles
1658 *Histoire naturelle et morale des Isles Antilles de l'Amérique, enrichie des plusiers belles figures en taille douce, des placis et des routes les plus d'écrites, avec un vocabulaire Caraïbe.* Rotterdam: A. Leers.
[1665],1667 *Histoire naturelle des Antilles de l'Amérique.* 2 vols. Lyon: Chez Christophe Fourney.
1666 *The history of the Caribby-islands, viz., Barbados, St. Christophers, St. Vincents, Martinico, Dominco, Barbouthods, Monserrat, . . .* translated by John Davies. 2 vols. London: Printed for T. Dring and J. Starkley.

Deschamps, L.
1898 *Les colonies pendant la revolution, la constitution et la réforme coloniale.* Paris: Librairie Académique Didier, Perrin et Cie.

Deschamps, P., and M. Besson
1933 *Les colonies et la vie française pendant huit siècles.* Paris: Firmin-Didot

Descourtilz, M. E.
1833 *Flore pittoresque et médicale des Antilles.* 8 vols. Paris: Chappron.

Desportes, G.
1972 "Le régime des terres aux Iles du Vent: La Martinique." *Annales des Antilles* 16:1–106.

Desportes, M.
1954 *La régime des terres aux isles du vent: La Martinique.* Paris: Faculté de droit.

Desruisseaux, J.
1975 *La structure foncière de la Martinique.* Ste. Marie, Fond-Saint-Jacques: Centre de Recherches Caraïbes.

Dessalles, A.
1847–48 *Histoire générale des Antilles.* 5 vols. Paris: Libraire-éditeur.

Dessalles, P. R.
1786 *Annales de conseil souverain de la Martinique, ou tableau historique du gouvernement de cette colonie.* Paris: A. Bergerac, Chez J. B. Puynesge.

1847 *Histoire législative des Antilles; ou annales du conseil souverain de la Martinique.* Ser. 1, vol. 3 of *Histoire générale des Antilles,* edited and with notes by Adrien Dessalles. Paris: Libraire-éditeur.

1982 *Historique des troubles survenus à la Martinique pendant la révolution.* Presented by Dr. Henri de Frémont, Honorary Professor at the National Conservatory of Arts and Trades. (Pamphlet).

d'Harcourt, R.

1942 "Archéologie de la province d'Esmeraldas, Equateur: Ceramique, objects en pierre." *Journal de Société des Americanistes* 34:61–200.

Doyle, T. W.

1981 "The role of disturbance in gap dynamics of montane rain forest." In *Forest succession concepts and application,* edited by D. C. West, H. H. Shugart, and D. S. Botkin. New York: Springer-Verlag.

Drake, Sir Francis

1981 *Sir Francis Drake's West Indian voyage, 1585–1586.* Edited by Mary Frear Keeler. London: Hakluyt Society.

Dubois, M., and A. Terrier

1902 *Les colonies françaises: I un siècle d'expansion coloniale.* Exposition universelle de 1900, publication de la Commission. Paris: August Challamel. Fort-de-France: Société d'Histoire de la Martinique.

Duby, G.

1954 "La révolution agricole médiévale." *Revue de Géografie Lyon* 29:361–66.

1968 *Rural economy and country life in the medieval West.* Translated by Cynthia Potsan. London: Edward Arnold.

Dumoret, M.

1902 *Au pays du sucre.* Paris: H. E. Martin.

Dunn, R.

1972 *Sugar and slaves: The rise of the planter class in the English West Indies 1624–1713.* Chapel Hill: Published for the Institute of Early American History and Culture at Williamsburg, Va., by the University of North Carolina Press.

Dupontés, P. C.

1979 *Les Petites Antilles: Etude sur leur évolution économique.* Paris: Union Parisienne d'Imprimeries.

Dupuget, M.

1976 "Coup d'oeil rapide sur la physique générale et la minéralogie des Antilles. Annotations: Sur la soufrière de la Martinique." *Journal des Mines* 4:43–60.

Du Puis, Father M.

[1652],1972 *Relation de l'établissement d'une colonie françoise dans la Guadeloupe.* Caen, France: Marin Yvon. Reprint, Basse-Terre, Guadeloupe: Société d'Histoire de la Guadeloupe.

[1652],1963 "Relation de l'établissement d'une colonie française dans la Guadeloupe, Isle de l'Amérique et des moeurs des sauvages." *Annales des Antilles* 11:75–95.

Durand-Molard, E.

1807–11 *Code de la Martinique, 1807–1811.* 4 vols. St. Pierre, Martinique: Imprimerie du Gouvernement.

Duss, Father A.
1897 *Flore phanérogamique des Antilles françaises (Martinique et Guadeloupe).* 2 vols. Mâcon, France: Protat Frères, Imprimeurs. New edition, Fort-de-France, Martinique: Société de Distribution et de Culture, 1972.

Du Tertre, Father J. B.
1654 *Histoire générale des isles de S. Christophe, de la Guadeloupe, de la Martinique, et autres dans l'Amérique.* Paris: J. Langlois et E. Langlois.
[1667–71]
1973 *Histoire générale des Antilles habités par les Français.* 4 vols. in 3 books. New edition, Fort-de-France: Société d'Histoire de la Martinique.

Dutrone, La Couture
1790 *Précis sur la canne et d'en extraire le sucre.* Paris.

Eden, M. J.
1964 "The savanna ecosystem—Northern Rupununi, British Guiana." *McGill University, Savanna Research Series, I.* Montreal: Department of Geography. McGill University.

Edmond, J. B.
1963 "Analyse sommaire de l'économie Martiniquaise pour 1963." Fort-de-France, Martinique: "Préfecture de la Mártinique (Mimeographed).

Edwards, B.
1797 *An historical survey of the French colony in the island of St. Domingo.* London: Privately printed.
[1801],1959 *The history, civil and commercial, of the British Colonies in the West Indies.* 3 vols. London: Privately printed.
1801 *An historical survey of the island of St. Domingo, together with the account of the maroon Negroes in the island of Jamaica and a history of the war in the West Indies in 1793 and 1794.* London: Privately printed.
[1818],1948 *The history, civil and commercial, of the British West Indies.* 5 vols. London: Privately printed. Reprint, London: T. Miller.

Egler, F. E.
1942 "Indigene versus alien in the development of arid Hawaiian vegetation." *Ecology* 23:14–23.
1954 "Vegetation science concepts. 1. Initial floristic composition, a factor in old-field vegetation development." *Vegetatio* 14:412–17.
1958 *The ecology of invasions by animals and plants.* London: Methuen.
1968 "The contumacious continuum." In "The continuum concept of vegetation responses." *Botanical Review* 34:259–67.

Elson, J. A.
1969 "Late Quaternary marine submergence of Quebec." *Revue Géographique de Montréal* 23:247–58.

Ervan, J.
1952 "Frederich Pursh 1774–1820 and his botanical associates." *Proceedings, American Philosophical Society* 96:599–621.

Esquemelin, A. O.
[1684],1893 *The buccaneers of America; A true account of the most remarkable assaults committed of late years upon the coasts of the West Indies. . . .* Translated from Dutch and edited by W. S. Stally. London: Sonnenschein. Reprint, New York: Charles Scribner's Sons.

Etifier-Chalono, M. E.
1985 *Etude descriptive des jardin traditionnels des compagnes de Sainte Marie, Martinique.* 2 vols. Thèse de 3e siècle, Académie de Montpellier, Université des Sciences et Techniques du Languedoc.

Evans, F. C.
1956 "Ecosystem as the basic unit in ecology." *Science* 123:1127–28.

Evans, J. G., S. Limbrey, and H. Cleere, eds.
1975 *The effect of man on the landscape: The highland zone.* [London]: Council for British Archaeology Research Report 11.

Ewing, M., and J. L. Worzel
1954 "Gravity anomalies and structure of the West Indies, Part I." *Bulletin of the Geological Society of America* 65:165–74.

Fauquet, G.
1911 "Statistique agricole et industrielle de la Martinique." *Journal Officiel de la Martinique.*
1912 "Note sur la population de la Martinique." *Bulletins et Mémoires de la Société d'Anthropologie de Paris* 6 ser. 3:154–60.

Fawcett, W.
1904 *Guide to the botanic gardens, Castleton, Jamaica; with a plan and illustrations.* Kingston, Jamaica: Government Printing Office.

Fawcett, W., comp.
1891 *Economic plants. An index to economic products of the vegetable kingdom in Jamaica.* Kingston, Jamaica: Government Printing Establishment.

Fawcett, W., and A. B. Rendel
1910–36 *Flora of Jamaica.* Vols. 1, 3, 4 (1910–20), 5 (1926), and 8 (1936). London: Printed by order of the Trustees of the British Museum.
1921 *The banana: Its cultivation, distribution, and commercial uses.* London: Duckworth.

Ferriére, C. de
1770 *Nouveau commentaire sur la coutume de la Prévote et Vicomte de Paris.* 2 vols. 9th ed. Paris: Librairies Associées.

Fiard, J.-P. ed.
1979 *La forêt Martiniquaise: Présentation et proposition de mesures de protection.* Fort-de-France, Martinique: Office National des Forêts.

Fiedler, R. H., M. J. Labelle, and C. R. Lucas
1942 "The fisheries and fishery resources of the Caribbean area." Fishery Leaflet 259. Washington, D.C.: U.S. Fish and Wildlife Service, Department of Interior.

Fink, L. K., Jr.
1972 "Bathymetric and geologic studies of the Guadeloupe region, Lesser Antilles island arc." *Marine Geology of the Netherlands* 12:267–88.

Fink, L. K., Jr., C. T. Harper, J. J. Stipp, and F. Nagle
1972 "The tectonic significance of La Desirade. Possible relic sea floor crust." In *Transactions of the VI Caribbean Geological Conference,* edited by C. Pelzall, p. 302. Caracas: n.p.

Fisher, R. V., and H. H. Hess
1963 "Trenches." In vol. 3 of *The sea,* edited by M. N. Hill, pp. 411–36. New York: Interscience.

Fisher, R. V., A. L. Smith, and M. J. Roobol
1980 "Destruction of Saint Pierre, Martinique by ash-cloud surges, May 8 and 20, 1902." *Transactions of the VIII Caribbean Geological Conference.* Santo Domingo, Dominican Republic: n.p.
Flannigan, Mrs.
1844 *Antigua and the Antiguans.* 2 vols. London: Saunders and Otley.
Floren, L. L.
1948 *Bibliografía de la bibliografía dominicana* 8:68. Ciudad Trujillo, Dominican Republic: Roques Roman.
Folan, W. J., J. Gunn, J. D. Eaton, and R. W. Patch
1983 "Paleoclimatological patterning in southern Mesoamerica." *Journal of Field Archaeology* 10:443–68.
Forbes, J. D.
1937 *Port of Boston, 1783–1815.* Cambridge, Mass.: Harvard University Press.
Fosberg, F. R.
1960 "Vegetation of Micronesia, general description, the vegetation of the Marianas Islands and a detailed description of the vegetation of Guam." *Bulletin American Museum of Natural History* 119:1–75.
1961 *Qualitative description of the coral atoll ecosystem.* Atoll Research Bulletin No. 81, pp. 1–11. Washington, D.C.: National Research Council, Pacific Science Board.
1963 "The island ecosystem." In *Man's place in the island ecosystem,* edited by F. R. Fosberg, pp. 1–6. Honolulu: Bishop Museum Press.
1966 "The oceanic volcanic island ecosystem." In *The Galapagos,* edited by Robert I. Bowman, pp. 55–61. Berkeley: University of California Press.
1983 "The human factor in the biogeography of oceanic islands." *Compte-rendu Société Biogéographique* 59(2):147–90.
Fournet, J.
1978 *Flore illustrée des phanérogames de Guadeloupe et de Martinique.* Paris: Institut National de la Recherche Agronomique.
France
1883 *Notices statistiques sur les colonies françaises.* Paris.
1885 *Statistiques coloniales pour l'année 1883.* Paris.
1902 *Annuaire colonial: Agricole, commercial et industriel des colonies de la République française.* Paris.
1982 *Journal Officiel de la République Française.* Mar. 3, 1982.
France, Centre Technique de la Canne et de Sucre de la Guadeloupe et de la Martinique
1956–58 *Revue agricole sucrière et rhumière des Antilles françaises,* vols. 1–3.
France, Centre National de la Recherche Scientifique (CNRS), Institut Géographique National (IGN).
1977 *Atlas des départements d'Outre-Mer.* Vol. 2, *La Martinique.* Bordeaux-Tallance: CNRS-IGN.
France, Comité Technique des Fruit de la Martinique
1959 *Fruits Antillais,* No. 12. Fort-de-France: Comité Technique de Fruits de la Martinique.

France, Département d'Outre-Mer, Direction Départementale de l'Agriculture, Martinique (DDA)

1975 *Recensement général de l'agriculture, 1973. Résultats provisoires.* Département d'Outre-Mer, DOM. Paris: Ministère de l'Agriculture.

1976 *Recensement général de l'agriculture, 1973. Lots de réforme foncière.* Paris: Service Central Etudes Economique et Statistique (SCEES).

1978 *Visage de l'agriculture de demain dans le sud-est. Station d'essai de Sainte-Anne.* Dillon, Martinique: Imprimerie Bellepage.

1981 *Bilan de l'agriculture martiniquaise.* Fort-de-France, Martinique.

1983 *Recensement général de l'agriculture, 1981.* Fort-de-France, Martinique.

France, Direction de la Météorologie Nationale, Service Météorologique du Groupe Antilles-Guyane

1962 "Quelques éléments du climat de la Martinique." Fort-de-France, Martinique. (Mimeographed).

1973 "Le climat de la Martinique. Fascicule 11: Normales et statistiques." Fort-de-France, Martinique. (Mimeographed).

France, Institut Français de Recherches Frutières d'Outre-Mer

1960 *Fruits.* Fort-de-France, Martinique: Centre Technique des Fruits.

France, Institut National de la Recherche Agronomique (INRA)

1973 *Les nouvelles maraîchères et voirières de l'INRA aux Antilles.* Petit-Bourg, Guadeloupe: INRA-Centre de Recherches Agronomiques des Antilles Guyane (CRAAG).

France, Institut National de la Recherche Agronomique (INRA)

1950 *Index seminum antillarum gallicarum.* Duclos, Guadeloupe: Centre de Recherche des Antilles.

France, Institut National de la Statistique et des Etudes Economiques (INSEE)

ca. 1947 *Recensement de la population de la Martinique en 1946.* Paris: Service Colonial des Statistiques.

1955–60 *L'espace économique française.* Paris: Imprimerie Nationale, Presses universitaires de France.

1956–57 *Résultats statistiques du recensement général de la population des Départements d'Outre-Mer, effectué le 1er juillet 1954.* Paris: Imprimerie Nationale.

1961 *Situation démographique dans les départements de la Martinique, de la Guadeloupe, et de la Réunion en 1957 et 1958.* Paris.

1975 "Recensement général de la population." Thermafax copy of commune summary manuscript data at the INSEE offices, Fort-de-France, made available to the author May 2, 1975.

1983 *Résultats du recensement de la population des départements d'Outre-Mer., 9 mars 1982, Martinique.* Aix-en-Provence: Division des DOM-TOM, Direction Général de l'INSEE.

France, Ministère de l'Agriculture, Direction Départementale d'Agriculture (DDA)

1976 *Recensement général de l'agriculture, 1973. Lots de réforme foncière.* Fort-de-France, Martinique.

France, Ministère de l'Agriculture: Enquêtes et études statistiques (Service Central, Martinique)

1976 *Les exploitations agricoles cultivant des lots de la réforme financière, lots du recensement général agricole, 1973.* Paris: Service Central d'Etudes Economiques Statistiques (SCEES).

France, Ministère de la France d'Outre-Mer: Service Coloniale des Statistiques

1946 *Bulletin mensuel de statistiques. Supplément série etudes no. 13 recensement de la population de la Martinique en 1946.* Paris: Imprimerie Jemmapes.

France, Ministère de la France d'Outre-Mer, Service Contrôle de la Météorologie de la France d'Outre-Mer.

1958 *Extrait des annales des services météorologiques de la France d'Outre-Mer territoires français de l'Amérique, année 1954.* Paris: Ministère de la France. Service Météorologique du Groupe Antilles-Guyane.

France, Ministre de la Marine et des Colonies

1836 *Etats de population de culture et de commerce des colonies françaises pour 1836 avec le complément des états de 1835.* Paris.

France, Ministère de la Qualité de la Vie

1977 *Parc naturel de la Martinique: Charte constructive.* Neuilly sur Seine, France: Ministère de la Qualité de Vie.

1976 *Parc naturel régional de la Martinique: Charte constructive.* Aurillac, France: Imprimerie Moderne.

France, Office National des Forêts (ONF)

1981 *Parc naturel régional de la Martinique.* Edited by J.-P. Fiard. Fort-de-France, Martinique: ONF

François-Haugren, A.

1982 *Guides bibliographique de préhistoire de la Martinique de la formation géologiques à l'occupation française: 1635.* Fort-de-France, Martinique: N.p.

Fraser, H.

1956 "Forest preservation in the Windward Islands." *Caribbean Forester* 17 (1 and 2):25–28.

Freeman, W. G., and R. O. Williams

1928 *The useful and ornamental plants of Trinidad and Tobago.* 2nd ed. Port of Spain, Trinidad: Government Printing Office.

Friederici, G.

1947 *Amerikanistisches und Hilfswörterbuch, Woerterbuch, Universitaet Hamburg. Abhandlungen aus dem Gebiet der Auslandskunde,* vol. 53 (B); *Voelkerkunde, Kulturgeschichte und Spracheni,* vol. 29. Hamburg: Cram, de Gruyter.

Froger, F.

1698 *Relation d'un voyage fait en 1695, 1696, et 1697 aux côtes d'Afrique, Détroit de Magellan, Brésil, Cayenne et isles Antilles.* Paris: Dans l'îsle du Palais . . . et G. Saugrain.

Fryxell, Paul

1980 *The natural history of the cotton tribe.* College Station: Texas A&M University Press.

Gage, T.

1648 *The English-American, his travail by sea and land.* London: Printed by E. Cotes and John Sweeting.

1928 "A new survey of the West Indies, 1648." In *The English American,* edited by A. P. Newton. New York: G. Routledge & Sons.

Gallant, M. N.
1959 *Report to the Caribbean Commission on forestry and the timber trade in the Caribbean area.* Rome: Food and Agricultural Organization of the United Nations.

Garaud, L.
1895 *Trois ans à la Martinique: Etudes de moeurs, paysages et profils et portrait.* 2nd ed. Paris: Editions Alcide Picard et Kaan. Reprinted in 1978.

Garay, L. A., and H. R. Sweet
1974 *Orchidaceae.* Vol. 1 of *Flora of the Lesser Antilles.* Edited by R. A. Howard. Jamaica Plain, Mass.: Arnold Arboretum, Harvard University.

Garnier, A.
1969 *Journal du Conseil Garnier à la Martinique et à la Guadeloupe 1848–1855.* Edited by G. Dubien. Fort-de-France: Société d'Histoire de la Martinique.

Garnier, J.
1976 *Météorologie tropicale et ses aspects aux Antilles-Guyane.* Fort-de-France, Martinique: Centre Départemental de Documentation Pédagogue.

Gartner, S., and J. E. McQuirk
1979 "Terminal Cretaceous extinction scenarios for a catastrophe." *Science* 206:1272–76.

Gause, G. F.
1934 *The struggle for existence.* Baltimore: Williams and Wilkins.

Gauthier, J.
1974 "Etude des pâtes céramiques de la Martinique précolombienne." In *Compte-rendu du Ve Congrès International d'Études de Civilisation Précolombienne des Petites Antilles,* pp. 133–39. Saint Lucia.

Gautier du Tronchoy
1709 *Journal de la campagne des îsles l'Amérique, qu'a à fait par Monsieur D.* Troyes, France: J. le Febvre.

Gazin, G. J.
1926 *Eléments de bibliographie générale méthodique et historique de la Martinique (Antilles françaises).* Fort-de-France, Martinique: Imprimerie Antilles.

Geertz, C.
1963 *Agricultural involution: The processes of ecological change in Indonesia.* Berkeley: University of California Press.

Geoffroy, E.
1897 *Rapport de mission à la Martinique et à la Guyane.* Mâcon, France: Imprimerie de Protat Frères.

Gilbert, F. S.
1980 "The equilibrium theory of island biogeography: Fact or fiction." *Journal of Biogeography* 7:209–35.

Gillin, J.
1936 "The Barima River-Caribs of British Guiana." *Papers of Peabody Museum of American Archaeology and Ethnology* 14(2):1–274.

Gilmore, R. M.
1950 "Fauna and ethnozoology of South America." *Handbook of South American Indians* 6:345–464. Bureau of American Ethnology Bulletin 143. Washington, D.C.

Giraud, M.
1983 "Formation technique et développement économique en Martinique: Une relation problématique." *Les Temps Modernes* 39(441–442): 1924–39.

Girault, A.
1921 *Principes de colonisation et de législation coloniale.* 4 vols. Paris: Librairie du Recueil Sirey.

Glacken, C. J.
1967 *Traces on the Rhodian shore.* Berkeley: University of California Press.

Goebel, D. B.
1963 "The 'New England trade' and the French West Indies, 1763–1774: A study in trade policies." *William and Mary Quarterly* ser. 3, 20: 331–71.

Goeje, C. H. de. *See* de Goeje, C. H.

Good, R. D'O.
1974 *The geography of flowering plants.* 4th ed. London: Longman.

Gooding, E. G. B., A. R. Loveless, and G. R. Proctor
1965 *Flora of Barbados.* London: Her Majesty's Stationery Office.

Goodwin, R. C.
1979 "The prehistoric cultural ecology of St. Kitts, West Indies. A case study in island archaeology." Ph.D. diss., Arizona State University.

Gordon, B. L.
1957 *Human geography and ecology in the Sinu country of Columbia.* Ibero-Americana No. 39. Berkeley: University of California Press.

Goslinga, C.
1971 *The Dutch in the Caribbean and on the Wild Coast, 1580–1680.* Gainesville: University of Florida Press.

Gosselin, E.
1876 *Documents authentiques et inédits pour servir à l'histoire de la marine normande et du commerce rouennais pendant les XVI*[e] *et XVII*[e] *siecles.* Rouen, France: H. Boissel.

Goudie, A.
1982 *The human impact: Man's role in environmental change.* Cambridge, Mass.: MIT Press.

Goveia, V.
1965 *Slave society in the British Leeward Islands at the end of the 18th century.* New Haven, Conn.: Yale University Press.

Graham, A.
1973 *Vegetation and vegetational history of northern Latin America.* Amsterdam: Elsevier.

Graham, E. H.
1934 "Flora of the Kartabo region, British Guiana." *Annals of American Carnegie Museum* 22(1):17–292.

Graham, S.
1969 "La réforme foncière à la Martinique. Rapport de Terrain, été 1968." Montreal: Département d'Anthropologie Université de Montréal.
1970 "Le Larenty: Un case de reforme foncière à la Martinique." Master's thesis, University of Montreal.

Grand, R., and R. Delatouche
1950 *L'agriculture au moyen âge de la fin de l'Empire Romain au XVI siècle.* Vol. 3 of *L'agriculture à travers les âges.* Paris: E. de Bocard.

Gratiant, G.
1961 *Ile féderée française de la Martinique.* Paris: L. Soulange.

Grébert, R.
1934 "Les forêts de la Guadeloupe." *Bulletin d'Agence Générale des Colonies,* Mar., 1934.

Grillet, Father J.
[1668],1857 "Lettre du P. Grillet." In *Voyages et traveau des missionnaires de la Compagnie de Jésus.* Published by the Society of Jesus. Paris: Julien, Lanier, Cosnard.

Crime, J. P.
1979 *Plant strategies and vegetation processes.* Chichester: John Wiley & Sons.

Grisebach, A. H. R.
1864 *Flora of the British West Indian Islands.* London: Lovell, Reeve.
1963 *Flora of the British West Indian Islands.* Reprint, New York: Hafner Publishing Co.

Grünewald, H.
1965 "Geologie de la Martinique." Mémoire pour servir a l'explication de la carte géologique detaillée de la France. Paris: Imprimerie Nationale.

Guenant, J. P.
1976 *Données sur l'emploi en Martinique.* Fort-de-France: Institut National de la Statistique et des Etudes Economiques Service Interregional Antilles-Guyane, Service Départemental de la Martinique.

Guenin, E.
1901 *Ango et ses pilotes d'après des documents inédits, tirés des archives de France, de Portugal et d'Espagne.* Paris: Imprimerie Nationale.

Guérin, D.
1956 *Les Antilles décolonisées.* Introduction by Aimé Césaire. Paris: Présence Africaine.

Guérin, L.
1857–58 *Histoire maritime de France.* 6 vols. Paris: Dufour, Boulanger et Legrand.

Guet, J.
1893 *Les origines de la Martinique. Le colonel François Collart et la Martinique de son temps, colonisations, sièges, révoltes et combats de 1625 à 1720.* Vannes, France: Lafolye.

Gunn, C. R., and J. V. Dennis
1976 *World guide to tropical drift seeds and fruits.* New York: Demeter.

Guppy, H. B.
1917 *Plants, seeds, and currents in the West Indies and Azores.* London: Williams and Norgate.

Guy, C.
1900 *Les colonies françaises. III. La mise en valeur de notre domaine coloniale.* Exposition Universelle de 1900. Publication de la Commission. Paris: Augustin Challamel.

Guyot, A. D.
1946 *Les origines des plantes cultivées.* Paris: Presses universitaires de France.

Guyot, H.
1949 "L'avocatier dans les territoires français d'Outre-Mer." *Fruits* 4:141–45.
Haffer, Jr.
1969 "Speciation in Amazon forest birds." *Science* 165:131–37.
Hakluyt, R.
1903–1905 *The principal navigations, voyages, traffiques and discoveries of the English nation.* Glasgow: James MacLehose and Sons.
Halley, Father J., S.J.
1965 "Relation des îles de la Martinique et de Saint Christophe." Collection of Moreau, 841, fol. 158–63. Paris: Bibliothèque Nationale.
Hanotaux, G., and A. Martineau, eds.
1929a *Histoire des colonies françaises et de l'expansion de la France dans le monde.* Vol. 1, *L'Amérique.* Paris: Plon.
1929b *Histoire des colonies françaises et de l'expansion de la France dans le monde.* 5 vols. Paris: Plon for Société de l'Histoire Nationale.
Harcombe, P. A.
1977 "Nutrient accumulation by vegetation during the first year of recovery of a tropical forest ecosystem." In *Recovery and restoration of damaged ecosystems,* edited by J. J. Cairns and K. L. Dickson. Charlottesville: University of Virginia Press.
Hardy, F.
1931–36 "Studies in West Indian soils." *Tropical Agriculture.* Supplements 3–10.
1932 "Some aspects of the flora of Barbados." *Agricultural Journal* (Barbados) 1(1).
1942 "The need for soil surveys in the Caribbean region." *Tropical Agriculture* 23:197–292.
1973 *Soils of the British Commonwealth Caribbean region.* Saint Augustine, Trinidad: Department of Soil Science, University of the West Indies.
Hardy, F., and J. S. Beard
1949 *Soil formation in the British Caribbean volcanic islands.* Saint Augustine, Trinidad: Imperial College of Tropical Agriculture.
Hardy, F., and G. Rodriguez
1941 "Soil genesis from fragmental volcanic rocks in the Lesser Antilles." *Proceedings of Soil Science Society of America* 6:47–51.
Harlow, V. T.
1926 *History of Barbados, 1625–1685.* Oxford: Clarendon Press.
Harlow, V. T., ed.
1925 *Colonizing expeditions to the West Indies and Guiana, 1623–1667.* London: Hakluyt Society.
1932 *Raleigh's last voyage . . . to Guiana in the year 1617 and the fatal consequences of the same.* London: Argonaut Press.
Harris, D.
1962 "The invasion of oceanic island by alien plants: An example from the leeward islands, West Indies." *Transactions and Papers of the Institute of British Geographers* 31:67–82.
1965 *Plants, animals, and man in the outer Leeward Islands, West Indies: An ecological study of Antigua, Barbuda, and Anguilla.* Vol. 18, University of

California Publications in Geography. Berkeley: University of California Press.

1970 "Origins of agriculture in the tropics." *American Scientist* 60:180–93.

1978 "The environmental impact of traditional and modern agricultural systems." In *Conservation and agriculture,* edited by J. G. Hawkes. London: Duckworth.

Harris, R. C.

1966 *The seigneurial system in early Canada.* Madison: University of Wisconsin Press.

Harrisse, Henry

[1525?],1866 "Bartholomew Columbus account as compiled by Alessandro Zorzi or Strezzi." In *Bibliotheca americana vertustissima.* New York: G. P. Philes.

Hayot, E.

1971 Les gens de couleur libres du Fort Royal, 1679–1823. Paris: Société d'Histoire Outre Mer.

Hayot, O.

1881 *Matériaux pour l'histoire de la Martinique.* Fort-de-France, Martinique: n.p.

[1881],1938 *Matériaux pour l'histoire de la Martinique agricole.* Introduction by M. Charles Lapavra. Reprint, Fort-de-France, Martinique: Imprimerie du Gouvernement.

Hearn, L.

1890 *Two years in the French West Indies.* New York and London: Harper & Brothers.

Heathwole, H., and R. Levins

1972 "Trophic structure stability and faunal change during recolonization." *Ecology* 53:531–34.

1973 "Biogeography of the Puerto Rican Bank: Species-turnover on a small cay, Payo Ahogado." *Ecology* 54:1042–55.

Heilprin, A.

1903 *Mont Pelée and the tragedy of Martinique.* Philadelphia and London: Lippincott.

Heiser, C. B., Jr., and P. G. Smith

1953 "The cultivated capsicum peppers." *Economic Botany* 7:214–27.

Helms, S. R.

1938 "A Yankee trader views the French revolution in Martinique." *New England Quarterly* 2:802–807.

Hendrick, U. P., ed.

1919 *Sturtevant's notes on edible plants.* Albany, N.Y.: J. B. Lyon and Company.

Henriguez, U. P.

1938 "Para la historia de los indigenismos: Papa y batata, el enigma del aje, bonitato, caribe, palabras antillanas." *Biblioteca de Dialectología Hispanoamericana* 3. Buenos Aires: Instituto de Filologia, Facultad de Filosofía y Letras de la Universidad de Buenos Aires.

Herrera y Tordesillas, A.

[1601],1944 *Historia general de los hechos de los Castillanos en las islas tierra firme del Mar Oceano,* Decade I. Reprint. Asunción del Paraguay: Editorial Guarnia.

Herskovits, M. J.
1945 *The backgrounds of African art.* Denver: The Denver Art Museum.
1958 The myth of the Negro past. Boston: Beacon Press.

Hertwiz, S. R.
1981 *Regeneration of selected tropical tree species in Carcovado National Park, Costa Rica.* Vol. 24, University of California Publications in Geography. Berkeley: University of California Press.

Hill, A. R.
1975 "Ecosystems stability in relation to stresses caused by human activities." *Canadian Geography* 19:206–20.

Hill, M. N., ed.
1962–63 *The sea: Ideas and observations on progress in the study of the seas.* New York: Interscience.

Hill, R. T.
1905 "Pelée and the evolution of the windward archipelago." *Geological Society of America Bulletin* 16:243–88.

Hilliard d'Auberteuil, M. R.
1776–77 *Considérations sur l'état présent de la colonie française de Saint Domingue.* 2 vols. Paris: Grangé.

Hitchcock, A. S.
1936 *Manual of the grasses of the West Indies.* U.S. Department of Agriculture Miscellaneous Publication No. 243. Washington, D.C.

Hitchcock, A. S., and L. Chase
1950 *Manual of the grasses of the United States.* 2nd ed. Washington, D.C.: U.S. Department of Agriculture.

Hodge, W. H.
1954 "Flora of Dominica, B.W.I., Part I." *Lloydiana* 17 (1–3).

Hodge, W. H., and D. Taylor
1957 "The ethnobotany of the Island Caribs of Dominica." *da Webbia* 12:513–644.

Hoheukerk, L. S.
1924 "The forests of British Guiana." *Empire Forestry Journal* 3(2):169–79.

Holdridge, L. R.
1947 "Determination of world plant formations from simple climatic data." *Science* 105(2727):367–68.
1957 "The vegetation of mainland middle America." *Proceedings of the Eighth Pacific Science Congress* 4:148–61.
1967 *Life zone ecology.* San José, Costa Rica: Tropical Science Center.

Holdridge, L. R., W. C. Grenke, W. H. Hatheway, T. Liang, and J. A. Tosi.
1971 *Forest environments in tropical life zones: A pilot study.* Oxford: Pergamon Press.

Homans, J., Jr.
1857 *An historical and statistical account of the foreign commerce of the United States.* Compiled by J. Smith, Jr. New York: G. P. Putnam and Co.

Horowitz, M. M.
1961 "The Martiniquan East Indian cult of Maldevian." *Social and Economic Studies* 10(1):1–18.

1967 *Morne-Paysan, peasant village in Martinique.* New York: Holt, Rinehart, and Winston.

Hovey, E. O.
1902 "Martinique and St. Vincent: A preliminary report." *Bulletin of the American Museum of Natural History* 16:347.

Howard, R. A.
1952 *The vegetation of the Grenadines Windward Islands, B.W.I.* Contributions from the Gray Herbarium of Harvard University, No. 175 Gray Herbarium. Cambridge, Mass.: Harvard University Press.
1953 "Captain Bligh and the breadfruit." *Scientific American* 188 (March): 88–94.
1973 "The vegetation of the Antilles." In *Vegetation and vegetational history of northern Latin America,* edited by A. Graham, pp. 1–38. Amsterdam: Elsevier.
1979 *Monocotyledonae.* Vol. 3 of *Flora of the Lesser Antilles.* Jamaica Plain, Mass.: Arnold Arboretum, Harvard University.

Howard, L. A., ed.
1974–79 *Flora of the Lesser Antilles.* 3 vols. Jamaica Plain, Mass.: Arnold Arboretum, Harvard University.

Hughes, Rev. G.
1750 *The natural history of Barbados.* London: Privately printed.

Husnot, T.
1870 "Catalogue des cryptogames recueillies aux Antilles françaises en 1868 et essai sur leur distribution géographique dans ces îles." *Bulletin de la Société Linnéene de Normandie,* ser. 2, 4:19–74.

Hutchinson, J. B., R. A. Silow, and S. G. Stephens
1947 *The evolution of Gossypium and the differentiation of the cultivated cotton.* London: Oxford University Press.

Iltis, H.
1981 "The tropical forests—so rich, so fragile, so irreplaceable—is extermination or preservation to be their fate?" *Environmental Education Report* 9:2–3.

Imbart de la Tour J., F. Dorvault, and H. Lecomte
1900 *Les colonies françaises. V. Régime de la propriété, régime de la main-d'oeuvre d'agriculture aux colonies.* Exposition Universelle de 1900, Publications de la Commission. Paris: Editions Augustin Challamel.

Isambert, F. A., et al.
1821–33 *Laws and statutes, etc. Recueil général des anciennes lois françaises.* Collected by F. A. Isambert et al. 29 vols. Paris.

Isambert, M.
1850 *Lettre de M. Isambert contenant le rétablissement des faits importants relatifs aux événements des Antilles françaises en 1824 en 1829 et la réponse des colonies.* Paris: Imprimerie L. Martinet.

Jackson, M. H.
1958 "The economy of the French Caribbean." In *The Caribbean: British, Dutch, French, United States,* edited by A. Curtis Wilgus. Gainesville: University of Florida Press.

Jardel, J.-P., M. Nicolas, and C. Relouzat
1969 *Bibliographie de la Martinique.* Fort-de-France, Martinique: Imprimerie Antillaise St. Paul.

Jeffreys, T.
1762 *A description of the Spanish islands and settlements on the coasts of the West Indies.* London: T. Jeffreys.
1775 *The West India atlas or a general description of the West Indies.* London: R. Sayer and J. Bennett.

Jolly, A. L.
1956 "Small-scale farming in the West Indies." *World Crop* 8(5):173–76.

Josa, G.
1931 *Les industries du sucre et du rhum à la Martinique, 1639–1931.* Paris: Les Presses Modernes.

Josselin de Jong, J. P. B.
1947 "Archeological material from Saba and St. Eustatius, Lesser Antilles." In *Mededelingen Van Het.* Rijksmuseum voor Volkenkunde, Leyden No. 1. Leyden, Netherlands: E. J. Brill.

Journal d'Agriculture Tropical
1910,1911

Julius, C., and J.-C. Pons
1972 "Contribution à l'étude d'une mangrove et de la région des Vauclin." *Bulletin Institut Géologique Bassin d'Aquitaine* 12:181–82.

Kelman, M., J. Hudson, and K. Sanmugadao
1982 "Temporal variability in atmospheric nutrient influx to a tropical ecosystem." *Biotropica* 14(1):1–9.

Kervégant, D.
1927 "Notes de statistiques et d'Agricoles." *Bulletin Agricole de la Martinique* (2).
1936a "Les plantes utiles et ornementales." *Bulletin Agricole de la Martinique* n.s. 5(2):1–94.
1936b "Notes de statistiques et d'économie agricole." *Bulletin Agricole de la Martinique* n.s. 5(2):95–112.
1937 "Les plantes utiles et ornementales." *Bulletin Agricole de la Martinique* n.s. 6(2):1–154.

Kimber, C.
1966 "Dooryard gardens of Martinique." *Yearbook of the Association of Pacific Coast Geographers* 28:97–118. Corvallis: Oregon State University Press.
1969 "Recent historical plant geography of Martinique." Ph.D. diss., University of Wisconsin.
1970 "Blue mahoe." *Economic Botany* 24:233–40.

Knauss, J. A.
1963 "Equatorial current systems." In vol. 1 of *The sea,* edited by M. N. Hill, pp. 235–52. New York: Interscience.

Kirby, I. A. E.
1976 "The pre-Hispanic peopling of the Antilles." In *Compte-rendu des communications du VIe congrès international d'études des civilisations pré-Colombiennes des Petites Antilles,* pp. 14–19. Gainesville: Florida State Museum.

Kopp, M. A.
1930 "La colonisation rurale aux Antilles françaises." *Congrès de la Colonisation Rurale Algère, 1930, Comptes-Rendus* 3–4:209–26.
La Barre, A.-J. *See* Le Febvre de la Barre
Labat, Father J.-B.
1722 *Nouveau voyage aux isles de l'Amérique.* 6 vols. Paris: P. F. Giffart.
1724 *Nouveau voyage aux isles de l'Amérique.* 2 vols. La Haye, France: P. Husson.
1742 *Nouveau voyage aux isles de l'Amérique.* New ed. 8 vols. Paris: Theodore Le Gras.
1931 *Nouveau voyage aux isles de l'Amérique.* Extracts from the original work, first published in 1722. Paris: Les Oeuvres Représentatives.
La Borde, Father de
[1674],1684 *Relation de l'origine, moeurs, coustumes, religion, guerres et voyages des Caraïbes, sauvages des isles Antilles de l'Amérique.* The Caribs serving with Reverend Father Simon, S.J., taken from the office of Monsieur Blondel. In *Recueil de diverse voyages faites en Afrique,* edited by H. Justel. Paris: Veuve Antoine Cellier.
[1684],1886 *History of the origin, customs, religion, wars, and travels of the Caribs.* Translated by G. J. A. Bosche Reitz, *Timehre* (Georgetown Md.) 5:224–54.
Lack, D.
1976 *Island biology,* vol. 3 of *Studies in ecology.* Berkeley: University of California Press.
Lacroix, A.
1903 "Sur les principaux résultats de la mission de la Martinique." *Compte-rendu Académie des Sciences de Paris* 135:871–76.
1904 *La Montagne Pelée et ses éruptions.* Paris: Masson et Cie.
Lacroix, A., M. Rollet de l'Isle, and M. Giraud
1902 "Sur l'éruption de la Martinique." *Comptes Rendus* (Académie des Sciences, Paris) 135:382–489.
Laet, J. de
[1625],1640 *L'histoire du nouveau monde, ou description des Indes Occidentales.* Leyden, Netherlands: Chez B. and A. Elseviers.
Lamb, F. B., C. B. Briscoe, and G. H. Englerth
1960 "Recent observations on forestry in tropical America." *Caribbean Forester* 21:46–59.
Landes, M.
1879 "Rapport sur la situation de l'agriculture dans quelques colonies voisines." *Bulletin Agricole de la Martinique.* 2(3).
La Roncière, C. de
1899–32 *Histoire de la marine française.* 6 vols. Paris: Plon-Nourrit et Cie.
Lasaulce, M.
1901 *Rapport de mission sur l'étude et l'organisation des bases d'un régime forestier à la Martinique.* Fort-de-France, Martinique: Ministre des Colonies.
Las Casas, B. de
[1520], "Historia de las Indies." Vols. 62–66 of *Colección de documentos inéditos.*

1875–76 Madrid: Imprimir de M. Ginista.

[1527–61], 1875–76 *Historia general de las Indias.* Madrid: Biblioteca Autores Españoles.

1909 *Apologética historia de las Indias.* Madrid: Marcelino Ménédez y Pelayo.

1957 *Historia de las Indias.* Text photoreproduction. Perez de Tudela y Emilio Lopez Oto. Estudio Critico preliminar y edition Juan Perez de Tudela Buesco. Vols. 95 and 96, *Biblioteca de Autores Españoles.* Madrid.

1965 *Historia de las Indes.* 2nd ed. Edited by Augustin Millares Carlo, with an introductory essay by Lewis Hanke. Mexico City: Fondo de Cultura Económica.

[1552],1974 *In defense of the Indians: The defense of the Most Reverend Lord . . . against the persecutors and slanderers of the peoples of the New World discovered across the seas.* Translated and edited by Stafford Poole. DeKalb: Northern Illinois University Press.

1974 *The devastation of the Indies: A brief account.* Translated by Herma Briffault. New York: Seabury Press.

Lasserre, G.

1961 *La Guadeloupe, étude géographique.* 2 vols. Bordeaux: Union Française d'Impression.

Lathrap, D. W.

1970 *The upper Amazon.* London: Thames and Hudson.

1973 "The antiquity and importance of long-distance trade relationships in the moist tropics of Pre-Colombian South America." *World Archeology* 2(5)170–86.

Laudonnière, R. G. de

[1564],1954 *L'histoire notable de la Floride.* Paris: Guillaume Auray.

[1586],1958 *L'histoire notable de la Florida située en Indes Occidentales.* Reprinted in *Les français en Amérique pendant la deuxième moitié du XVIe siécle.* Paris: Suzanne Lussaynet, Presses universitaires de France.

La Vaissière, R. de

1910 *La Martinique en 1910.* Paris: A. Challamel.

Lavigne-Sainti, Suzanne, Marquise de

1935 *La Martinique au premier siècle de la colonisation, 1635–1742.* Nantes: Chantreau.

Lavollée, M. P.

1841 *Notes sur les cultures et la production de la Martinique et de la Guadeloupe.* Paris: Imprimerie Royale.

Leake, H. M.

1932 "Studies in tropical land tenure, the West Indies." *Tropical Agriculture* 9(9):272–76.

Le Blant, R.

1948 "Les études historiques sur la Martinique pour la période française jusqu'en 1789." *Revue d'Histoire des Colonies* 35:270–83.

Leblond, J. B.

1813 *Voyage aux Antilles et à l'Amérique méridionale, commencé en 1767 et fini en 1802.* Paris: Arthus-Bertrand, Librairie.

Lebrun, J.
1967 "L'écologie végétale, ses concepts et ses méthodes." *Revue des Questions Scientifiques* 17:321–51.

Leeds, A., ed.
1965 *Man, culture, and animals: The role of animals in human ecological adjustments.* Washington, D.C.: American Association for the Advancement of Science.

Le Febvre de la Barre, J.-A.
1666 *Description de la France eqvinoctiale, cy-devant appelée Guyane et par les Espagnols, El Dorado. Nouvellement remissons l'obeïssance du roy.* Paris: I. Ribov.
1671 *Relations de ce qui s'est passé dans les îles et terre-firme de l'Amérique pendant la dernière guerre avec l'Angleterre.* Paris: Jean de Clodoré.

Lefroy, J. H., ed.
[1877–97], 1981 *Memorials of the discovery and early settlement of the Bermudas or Somers Islands 1515–1685.* Reprint in 2 vols. Bermuda: Bermuda Historical Society.

Légier, E.
1901 "Sucrerie indigène et coloniale asiatique et africaine; la production, la propriété." *Revue des Deux Mondes* Ser. 2, 1:18–58.
1905 *La Martinique, la Guadeloupe: Considérations économiques sur l'avenir et la culture de la canne, la production du sucre et du rhum et les cultures secondaires dans les Antilles françaises. Notes de voyage.* Paris: Bureau de la Sucrerie Indigène et Coloniale.

Leiris, M.
1955 *Contacts de civilisation en Martinique et en Guadeloupe.* Paris: UNESCO/ Gallimard.

Leméry, H.
1936 *La révolution française à la Martinique.* Paris: Larose.

Léry, J. de
[1578],1880 *Histoire d'un voyage facit en la terre du Brésil.* A new edition with an introduction and notes by Paul Gafforel. 2 vols. Paris: A. Lemerre.
[1578],1957 *Journal de bord de Jean de Léry en la terre de Brésil, 1557.* Edited by M. R. Mayeux. Reprint, Paris: Editions de Paris.

Levin, S. A., ed.
1976 *Ecological theory and ecosystem models.* Indianapolis, Ind.: The Institute of Ecology.

Levins, R.
1968 *Evolution in changing environments.* Princeton, N.J.: Princeton University Press.

Lewis, J. F., and E. Robinson
1976 "A revised stratigraphy and geologic history of the Lesser Antilles." In *Publication, Transactions, VIIe Conference Géologique des Caraïbes à Saint-François, Guadeloupe,* pp. 340–44. Pointe-à-Pitre, Guadeloupe: Imprimeries Réunies de Chambéry.

Ligon, R.
[1657],1970 *A true and exact history of the island of Barbados.* London; Frank Cass.

Linares, C.
1984 *Documents pour l'étude de l'agriculture Martiniquaise, 1983.* Fort-de-France, Martinique: Ministre de l'Education Nationale, Académie Antilles-Guyane. Centre Régional de Documentation Pédagogique.

Lincoln, F. C.
1952 *Migration of birds.* Illustrated by Bob Hines. Garden City, N.Y.: Doubleday.

Lokke, C. L.
1932 *France and the colonial question, a study of contemporary French opinion, 1763–1801.* New York: Columbia University Press.

Lombard, P. A.
[1723–90] 1857 "Lettre du P. Lombard sur La Mission de Kourou." In *Voyages et traveaux des missionaires de la Compagnie Jésus, publié par des pères de la même compagnie,* pp. 260–342. Paris: Julian, Lanier, Cosnard.

Long, E.
[1774],1970 *The history of Jamaica.* 3 vols. London: Frank Cass.

Lonn, E.
1919 "The French Council of Commerce in relation to American trade." *Mississippi Valley Historical Review* 6:192–219.

Loslier, L.
1971 "La Martinique: Essai de détermination des régions naturelles, des régions de service, et des regíons agricoles et analyse de leur intégration." M.S. thesis, University of Sherbrooke.

Lotaut, E., and E. Sylvestre
1936 *Annuaire de la Martinique.* Fort-de-France, Martinique: Imprimerie de Gouvernement.
1947 *Annuaire de la Martinique.* 2 vols. Fort-de-France, Martinique: Imprimerie de Gouvernement.

Lothrop, S. K.
1964 *Les trésors de l'Amérique précolombienne.* Geneva: Albert Skira.

Loveless, A. R.
1960 "The vegetation of Antigua, West Indies." *Journal of Ecology* 48:495–527.

Loveless, A. R., and G. F. Asprey
1957 "The dry evergreen formations of Jamaica. The limestone hills of the South Coast." *Journal of Ecology* 45:799–822.

Loven, S.
1935 *Origins of the Tainan culture, West Indies.* Göteborg, Sweden: Elanders Boktryckeri Aktiebolag.

Lowe, R.
1951 *The Codrington correspondence, 1743–1851.* London: Robson Lowe.

Lowenthal, D.
1972 *West Indian societies.* Published for the Institute of Race Relations in collaboration with American Geographical Society. Oxford: Oxford University Press.

Lucas, C. P.
1905 *A historical geography of the British Colonies, West Indies.* Oxford: Clarendon Press.

Lugo, A. E.

1978 "Stress and ecosystem." In *Energy and environmental stress in aquatic ecosystems*, edited by J. H. Thorp and J. W. Gibbons, pp. 62–107. Springfield, Va.: Technical Information Center, U.S. Department of Energy.

Lugo, A. E., M. Applefield, D. J. Pool, and R. B. McDonald

1983 "The impact of Hurricane David on the forests of Dominica." *Canadian Journal of Forest Research* 13(2):201–11.

Lundell, C. L.

1937 *The vegetation of the Petén.* Carnegie Institution of Washington Publication No. 478. Washington, D.C.

MacArthur, R. A.

1972 *Geographical ecology: Pattern in the distribution of species.* New York: Harper & Row.

MacArthur, R. A., and E. O. Wilson

1967 *Theory of island biogeography.* Princeton, N.J.: Princeton University Press.

McCloy, S. T.

1966 *The Negro in the French West Indies.* Lexington: University of Kentucky Press.

Magellan, J. H. de

1786 "Description of the French *toise*." *Annals of agriculture and Other Useful Arts* 7:318–31.

Maggiolo, M. V.

1971 "El rito de la cahoba entre los aborigenes antillanos." *Revista Dominicana de arqueología y anthropología* 1(1). Santo Domingo: Universidad Autonoma de Santo Domingo.

Malik, M.

ca. 1960 *L'etude agronomique des climats du Gabon.* Libreville, Gabon: Services Météorologiques de la France d'Outre-Mer. (Mimeographed).

Mann, P., and K. Burke

1984 "Neotectonics of the Caribbean." *Reviews of Geophysics and Space Physics* 22(4):309–62.

Margalef, R.

1957 "La teoría de la información en ecología." *Memoria, Real Academia de Ciencias y Artes* (Barcelona) 33:373–79.

1968 *Perspectives in ecological theory.* Chicago: University of Chicago Press.

1981 "Stress in ecosystems: A future approach." In *Stress effects on natural ecosystems*, edited by G. W. Barrett and R. Rosenberg. Chichester: John Wiley and Sons.

Margry, P.

1867 *Relations et mémoires inédits pour servir à l'histoire de la France dans les pays d'outre-mer.* Drawn from the archives of the Ministère de la Marine et des Colonies by Pierre Margry. Paris: Challamel, elder.

1878 "Origines françaises des pays d'outre-mer: Les seigneurs de la Martinique." *Revue Maritime et Colonial* Supplement (July–Sept.): 1–61. Paris: Berger-Levrault.

1897 "La Martinique en 1696. Mémoire de Fr. Robert, intendant des îles françaises de l'Amérique sur l'établissement de la Martinique." In *Relations pour servir à l'histoire de la France d'Outre-Mer*, pp. 1–225. Paris: Achille Fauré.

Marshall, R. C.
1934 *Physiography and vegetation of Trinidad and Tobago*. Oxford Forestry Memoirs No. 17. Oxford: Clarendon Press.

Martin, R. M.
1839 *Statistics of the colonies of the British Empire*. London: Whitaker.

Martin, S.
1773 *An essay upon plantership*. 5th ed. London: Whitaker.
1792 "An essay on plantership." *Annals of Agriculture and Other Useful Arts* 18:236–308.

Martineau, A., and L. P. May
1935 *Trois siècles d'histoire antillaise. Martinique et Guadeloupe, de 1635 à nos jours*. Paris: Société de l'histoire des colonies françaises et Librairie Leroux.

Martinique
1827 *Lois et statutes*. Vol. 43, no. 14.
1841–1925 *Bulletin officiel de la Martinique*. Fort-de-France, Martinique. 1841, 1842 1845, 1856, 1869, 1925.
1842 *Courrier de la Martinique*. Nov. 15, 1842.
1900 *Annuaire de la Martinique . . . 1900*. Fort-de-France, Martinique.
1908,1910 *L'Almanach Martinique Illustré*. Fort-de-France.
1911 "Statistiques agricoles et industrielles de la Martinique, août 1911." *Journal Officiel de la Martinique*.
1912 *Annuaire de la Martinique pour l'année 1912*. Fort-de-France.
1918,1936 *Bulletin Aqricole*.
1922 *Règlement forestier du 30 septembre 1922*. Fort-de-France: Caisse Générale.
1947 *Annuaire de la vie martiniquaise*. 2 vols. Fort-de-France: Imprimerie Officielle.
1961 *Annuaire de la Martinique, 1956–1959*. Fort-de-France.

Martinique, Archives Départementales de la Martinique, Service Educatif
1977 *La Martinique pendant la révolution de 1789. Edition des textes*. Fort-de-France: Centre Départemental de Documentation Pedagogique.
1984 *L'habitation, domaine terrien des Antilles avant et aprés la révolution industrielle*. Fort-de-France: Centre Régional de Documentation Pedagogique.

Martinique, Caisse Régionale de Crédit Agricole Mutuel de la Martinique
1964 *Assemblée Générale Ordinaire du 23 juin 1964*. Fort-de-France

Martinique, Chambre de Commerce et d'Industrie de la Martinique
1984 *Martinique: Your business partner*. Fort-de-France: Chambre de Commerce et d'Industrie de la Martinique.
1985 "Réforme foncière-Land Reforms." Memorandum for participants of Symposium on Caribbean Farming Systems and Alternatives for Development, May, 1985.

Martinique, Colonie de la
1918 *Inauguration du Jardin Desclieux.* Fort-de-France, Martinique: Imprimerie du Gouvernement.
Martinique, Commission Météorologique Départementale
1975 *Bulletin climatologique mensuel de la Martinique janv.–dec.* Fort-de-France: Service Météorologique de la Martinique.
Martinique, Conseil Général de la Martinique
1868,1884,1888 *Mémoires.*
1902 *Comptes-rendus.* Fort-de-France: Imprimerie du Gouvernement.
1938 *Décret du 9 mars 1938 fixant les conditions d'exportation des bananes fraîches.* Fort-de-France: Imprimerie du Gouvernement.
1962 *Proposition pour le IVe Plan, 1962–1965.* Fort-de-France: Imprimerie du Gouvernement.
Martinique, Conseil Régional de la Martinique
1976 *Déliberation du Conseil Régional de la Martinique, sept. 1976.* Fort-de-France: Caisse Régionale.
1985 *Plan régional, 1984.* Fort-de-France: Caisse Régionale. (Mimeographed).
Martinique, Direction Départementale de l'Agriculture, Service Statistique
1975 *Le cheptel bovin en 1975: Enquête sur la structure du cheptel bovin en Martinique, fin 1974 debut 1975.* Fort-de-France: Service Regional de Statistique Agricole, Département d'Outre-Mer.
Martinique, Préfecture
1963,1977, 1984 *La Martinique.* Fort-de-France: Editions Berger (1977,1984).
Martinique, Service des Eaux et Forêts
1947 *Règlements relatifs au régime forestier à la police des propriétés rurales de la chasse et des cours d'eau.* Fort-de-France: Imprimerie du Gouvernement.
Martinique, Service Météorologique, Direction de la Météorologie National
n.d. *Note sur le climat de la Martinique.* Fort-de-France, Martinique. (Mimeographed).
Martinique, Société d'Assistance de la Technique et du Crédit Social d'Outre-Mer (SATEC)
1963 Bulletin d'information des cadres et administrateurs SATEC-SICA. Fort-de-France, Martinique. (Mimeographed).
Martin-Kaye, P. H. A.
1959 *The geology of the Leeward and British Virgin Islands.* Castries, Saint Lucia: Voice Publishing.
1969 "A summary of the geology of the Lesser Antilles." *Overseas Geological Mineral Resources* 10(2):172–206. Geological Bulletin.
1977 "A summary of the geology of the Lesser Antilles." In *West Indian Island Arcs,* edited by Peter H. Marttson. Stroudsburg, Pa.: Dowden, Hutchinson, and Russ.
Masefield, G. B.
1949 *A handbook of tropical agriculture.* Oxford: Clarendon Press.
1950 *A short history of agriculture in the British Colonies.* Oxford: Clarendon Press.

Massé, R.
1979 *Rapport sur la petite exploitation agricole aux Antilles françaises.* Montreal: Centre de Recherches Caraïbes, Université de Montréal.

Mathias du Puis, P. F.
[1652],1972 *Relation de l'éstablissement d'une colonie française dans la Guadeloupe, isle de l'Amérique, et des moeurs des sauvages.* Caen, France: Marin Yvon. Reprint in *Bibliothèque d'Histoire Antillaise,* vol. 1. Basse-Terre, Guadeloupe: Société d'Histoire de la Guadeloupe.

Mattioni, M.
1972 *Que sait le Caraïbe?* Fort-de-France: Musée Départemental de la Martinique.
1973 *Troisième compte-rendu de fouilles au lieu-dit Vivé-Canidés.* Fort-de-France: Musée Départemental de la Martinique.
1980 "Salvage excavation at the Fond-Brulé site." In *Proceedings of the 8th International Congress for the Study of Pre–Columbian Cultures of the Lesser Antilles,* pp. 553–66. Anthropological Research Paper No. 22. Tucson: Arizona State University.
1984 *Le site archéologique de Vivé.* Fort-de-France: Direction des Antiquités de la Martinique.

Mattioni, M., and R. P. Bullen
1970 "A chronological chart for the Lesser Antilles: Site dated by ceramic typology." *Proceedings of the 3rd International Congress for the Study of Pre–Columbian Civilizations of the Lesser Antilles.* St. Georges, Grenada: Grenada National Museum.
1973 "Some ceramic variations at Vivé, Martinique." *Actes du XLe Congrès International des Américanistes, 1972.* Genoa: Tilger.

Maurice, J.
1963 "L'expansion de l'économie agricole de la Martinique." *Revue Economique Française* 85(1):23–28.

Maurile de Saint-Michel, Father
1652 *Voyage des isles camericanes en l'Amérique qui font partie des Indes Occidentales avec l'establissement des R. P. Carmes.* Le Mans: Hiérome Olivier.
[1652],1963 "Voyage des isles camericanes en l'Amérique," chap. 20–21. *Annales des Antilles* 11:47–95.

May, L. P.
1929 "Plan parcellaire du Quartier du Gros-Morne de la Martinique en 1705." *Revue Historique des Antilles* 3:31–39.
1930 *Histoire économique de la Martinique (1635–1763).* Paris: Marcel Rivière, Les Presses Modernes.

Maycock, J. D.
1830 *Flora Barbadensis.* London: J. Ridgeway.

Mazé, H.
1892 *Contribution à la flore de la Guadeloupe.* Basse-Terre, Guadeloupe: Imprimerie du Gouvernement.

Melville, R.
1979 "Endangered island floras." In *Plants and islands,* edited by D. Bramwell, pp. 361–77. London: Academic Press.

Merrens, H. R.
1965 "Historical geography and early American history." *William and Mary Quarterly* ser. 3, 22:529–42.
Merriam, C. H.
1896 "Laws of temperature control of the geographic distribution of terrestrial animals and plants." *National Geographic* 6:229–38.
1899 "Life zones and crop zones." *Bulletin of the Biological Survey* 10:9–79.
Merrill, G. C.
1954 *The historical geography of St. Kitts and Nevis, the West Indies.* Mexico City: Instituto Pan Americano de Geografia e Historia and Editorial Fournier S.A.
Mervoyer, B., and D. Westercamp
1974 "Introduction à la géologie des Petites Antilles Françaises." In *Livret-guide d'excursions dans les Antilles Françaises. VIIe conference géologique des Caraïbes, 30 juin–12 juillet, 1974.* Pointe-à-Pitre, Guadeloupe: Bureau de Recherches Géologiques et Minières (BRGM).
Mestres, R.
1956 "Les problémes des salaires." *Marches Coloniaux* 12(557):1947–48.
Metraux, A.
1948 "Étude sur l'agriculture paysanne dans une valée haïtienne." *Acta Americana* 4:173–91.
Mims, S. L.
1912 *Colbert's West India policy.* New Haven, Conn.: Yale University Press.
Mintz, S. W.
1967 "Petits cultivateurs et propriétaires ruraux dans la région des Caraïbes." In *Les problèmes agraires des Amériques Latines.* Paris: Editions du Centre National de la Recherche Scientifique.
Mismer, C.
1890 *Souvenirs de la Martinique et du Méxique pendant l'intervention française.* Paris: Hachette.
Mitchell, II.
1963 *Europe in the Caribbean: The policies of Great Britain, France and the Netherlands toward their West Indian territories in the twentieth century.* Stanford, Calif.: Hispanic American Society.
Mohr, E. C. J., and F. A. Van Baren
1954 *Tropical soils.* The Hague: W. Van Hoeve.
Monet, H.
1892 *La Martinique, par Henri Monet.* Paris: A. Savine.
Monkhouse, A. M. A.
1910 *Encyclopaedia Britannica* 11th ed., s.v. "cotton."
Moreau de Jonnes, A.
1822 *Histoire physique des Antilles françaises, savoir la Martinique et les îles de la Guadeloupe.* Paris: Imprimerie Migeret.
Moreau de Saint-Méry, M. L. E., ed. (or Moreau de Saint-Rémy).
1784–86 *Lois et constitutions des colonies françaises de l'Amérique sous-le-vent de 1790.* Vols. 1–3. Paris: Privately printed.
[1797],1958 *Description topographique, physique, civile, politique et historique de la par-*

tie française de Saint-Domingue. 3 vols. Republished by Société de l'Histoire des Colonies Françaises. Paris: Librairie Larose.

Morton, J. F.
1981 *Atlas of medicinal plants of middle America.* Springfield, Ill.: Chas. C. Thomas.

Moseley, H. N.
1872 *Notes by a naturalist on the "Challenger."* London: J. Murray.

Muller, F.
1872 *Catalogue of books, maps, plates on America and of a remarkable collection of early voyages . . . with bibliographical and historical notes and presenting an essay towards a Dutch American bibliography.* Amsterdam: F. Muller.

Nagel, F., J. J. Stippe, and D. E. Fisher
1976 "K-Ar geochronology of the Limestone Caribbees and Martinique, Lesser Antilles, W.I." *Earth Planet Science Letter* 29:401–12.

Navarreti, M. F. de
1825–29 *Colección de los viages y descubrimientos que hicieron por mar los Españoles.* 3 vols. Madrid: n.p.

Newton, A. P.
1914 *The colonizing activities of the English Puritans: The last phase of the Elizabethan struggle with Spain.* New Haven, Conn.: Yale University Press.
1929 *The British Empire since 1783: Its political and economic development.* London: Methuen.
1933 *The European nations in the West Indies, 1493–1688.* London: Adams and Charles Black.

Nicholson, D. V.
1976a "Artifacts of preceramic Antigua." In *Proceedings of the 6th International Congress for the Study of Pre–Columbian Cultures of the Lesser Antilles,* pp. 264–68. Gainesville: Florida State Museum.
1976b "Precolumbian seafaring capabilities in the Lesser Antilles." *Compte-rendu des communications du VIe Congrés International d'Etudes des Civilisation Précolumbiennes des Petites Antilles,* edited by R. P. Bullin, pp. 98–105. Gainesville: Florida State Museum, University of Florida.

Nielsen, M. S.
1965 *Introduction to the flowering plants of West Africa.* London: University of London Press.

Nietschmann, B.
1979 *Caribbean edge: The coming of modern times to isolated people and wildlife.* Indianapolis, Ind.: Bobbs-Merrill.

Nitz, H.-J.
1963 "The church as colonist: The Benedictine Abbey of Lorsch and planned Woldhufen colonization in the Odenwald." *Journal of Historical Geography* 9:105–26.

Nordenskiold, E.
1922 *Deductions suggested by the geographical distribution of some post–Columbian words used by the Indians of South America.* Comparative Ethnographical Studies No. 5. Göteborg, Sweden: Elanders Boktryckeri Aktiebolag.

Noter, R. de
1930 *Le jardin potager colonial. Légumes éxotiques et d'Europe, cultures familiales et industrielles.* Paris: Société d'Éditions Géographiques, Maritimes et Coloniales.
Nousanne, H. de
1927 *Nouveau petit Larousse illustré.* Paris: Larousse.
1936 *La France missionaire aux Antilles: Guadeloupe, Martinique, Trinidad.* Paris: Lethielleux.
Nussbaum, F. L.
1923 *Commercial policy in the French revolution: A study of the career of G. J. A. Ducher.* Washington, D.C.: American Historical Association.
Oaks, H., and J. T. Thorpe
1951 "Characteristics of dark-clay soils of warm regions variously called rendzina, black cotton soils, regurs, and tirs." *Proceedings Soil Science Society of America* 15:347–54.
Ober, F. A.
1880 *Camps in the Caribbees: The adventures of a naturalist in the Lesser Antilles.* Boston: Lee and Shepard.
1895 "Aborigines of the West Indies." *Proceedings of the American Antiquarian Society* n.s. 9:270–313.
Odum, E. P.
1969 "The strategy of ecosystem development." *Science* 164:262–70.
1981 *Fundamentals of ecology.* 3rd ed. Philadelphia: W. B. Sanders.
Oexmelin, A. O. *See* Esquemelin, A. O.
Ogilby, J.
1671 *America.* London: Privately printed.
Otto, J. S., and N. E. Anderson
1982 "Slash-and-burn in the Highlands South: A problem in comparative agriculture." *Agricultural History* 24:131–47.
Oviedo y Valdes, G. F. de
[1526],1959 *Natural history of the West Indies (1526).* Translated and edited S. A. Stoudemire. Chapel Hill: University of North Carolina Press.
[1535], 1851–55 *Historia general y natural de las Indias y tierra ferme de Mar Oceano.* Edited by Jose Amador de los Rios. 4 vols. Madrid: Biblioteca de Autores Españoles.
[1535],1959 *Historia general y natural de las Indias,* vol. 1. Edited by Juan Pérez de Tudela Bueso. In *Biblioteca de Autores Españoles.* Madrid: Ediciones Atlas.
Pacifique de Provin
[1646],1939 *Brève relation du voyage des Iles de l'Amérique.* Paris: Chez Nicolas et Jean de la Coste. Reprint, Paris: Godefroy.
Pago, G.
1981 "L'insurrection du sud de la Martinique." *Historial Antillais* 4:219–58.
Paquette, R.
1978 "La petite exploitation agricole en Martinique." *Environnement Africain* n.s. 3:77–93.
1982 *Désengagement paysan et sous-production alimentaire: Martinique, Marie*

Galante, Barbade. Montreal: Presses de l'Université de Montréal, Les Editions de l'Université de Sherbrooke.

Pardon, N.
1878 *La Martinique depuis sa découverte jusqu'à nos jours.* Paris: Challamel.

Pares, R.
1932 "War, trade in the West Indies." *Revue d'Histoire des Colonies* 20:41–76.
1936 *War and trade in the West Indies, 1739–1763.* Oxford: Clarendon Press.
1956 *Yankees and Creoles: The trade between North America and the West Indies before the American Revolution.* London: Longmans, Green and Co.
1960 *Merchants and planters.* Economic History Review Supplement No. 4. Cambridge: The University Press.

Parker, F.
1958 "Political development in the French Caribbean." In *The Caribbean British, Dutch, French, United States,* edited by A. Curtis Wilgus. Gainesville: University of Florida Press.

Parry, J. H., and Sherlock, P. M.
1956 *A short history of the West Indies.* 2nd ed. New York: St. Martin's Press.

Parsons, J.
1969 "Ridged fields in the Rio Guayas Valley, Ecuador." *American Antiquity* 34:76–80.
1979 "Spread of African pasture grasses to the American tropics." *Journal of Range Management* 25(1):12–17.

Parsons, R. F.
1969 "Distribution and paleogeography of two mallee species of *Eucalyptus* in southern Australia." *Australian Journal of Botany* 17:323–330.

Peeters, A.
1976 "Le petit paysannat martiniquais et son environnement végétal, recherches en cours." *Journal d'Agriculture Tropicale et de Botanique Appliquée* 23:49–56.

Pelleprat, Father P.-I., S.J.
[1665],1857 *Relation des Missions des P.P. de la Compagnie de Jésus dans les isles et dans la terre ferme de l'Amérique Méridionale.* 2 vols. Paris: Sébastien Cramoisy. Republished in *Voyages et traveaux des missionaires de la Compagnie de Jésus publié par les pères de la même compagnie,* pp. 1–180. Paris: Julien, Lanier, Cosnard, et Ce,
[1665],1963 "Relation des Missions des P.P. de la Compagnie de Jésus par le Père Pierre Pelleprat." *Annales des Antilles* 11:28–36.

Pendleton, R. L.
1954 "The place of tropical soils in feeding the world." *Ceiba* 4(4):201–22.

Perret, F. A.
1937 *The eruption of Mt. Pelée 1929–1932.* Carnegie Institution of Washington Publication No. 458. Washington, D.C.

Perrodin, B.
1958 "Rapport de Mission en Jamaïque et au Guatemala." Basse-Terre, Guadeloupe: Bureau de Génie Rural. (Mimeographed).

Perronnette, H.
1963 *Comment se nourir aux Antilles.* Fort-de-France, Martinique: N.p.

Perruset, M.
1974 "Le climat de la Martinique." Fort-de-France, Martinique: Service Météorologique du Groupe Antilles-Guyane. (Mimeographed).

Peter, G.
1966 *Industrie et surpeuplement aux Antilles.* Fort-de-France, Martinique: Centre d'Études Régionales Antilles-Guyane (CERAG).

Peterkin, J.
1790 *A treatise on planting, from the origin of the semen to ebullition, with a correct mode of distillation, and a melioration on the whole process progressively.* Basse-Terre, Saint Christopher: Printed by E. L. Low.

Petitjean-Roget, B.
1981 "La mise en place des usines à sucre." *Historial Antillais* 4:185–218.
1983 "Pour comprendre la situation économique des Antilles." *Les Temps Modernes* 39 (441–42):1853–71.

Petitjean-Roget, H.
1974 "La Savane des Pétrifications (Martinique): Un gisement de l'âge lithique." *Proceedings of the 5th International Congress for the Study of Pre-Columbian Cultures of the Lesser Antilles,* pp. 82–93. Michael's Mount, Antigua: The Antiguan Archaeological Society, The Antigua National Trust.
1975 *Contributions à l'étude de la préhistoire des Petites Antilles.* 2 vols. Fort-de-France, Martinique: Privately printed.
1978 "Reconnaissance archéologique à l'île de la Dominique (West Indies)." In *VIIe Congrès International d'Études des Civilisations Précolombiennes des Petites Antilles 1977,* pp. 81–98. Montreal: Centre de Recherches Caraïbes.
1980 "Les populations amérindiennes: Aspects de la préhistoire antillaise." *Historial Antillais* 1:75–152.

Petitjean-Roget, J.
1955a "Le peuplement de la Martinique sous l'ancien régime." *La Revue Française,* July.
1955b "Les protestants à la Martinique sous l'ancien régime." *Revue d'Histoire des Colonies* 42:220–65.
1957 "Ses enroyés du roi à la Martinique." *Annales des Antilles* 4:52–83.
1968 "Étude d'un horizon Arawak et Proto Arawak à la Martinique à partir du niveau II du Diamant." *Compte-rendu IIe Congrès International d'Études des Civilisations Précolombiennes des Petites Antilles.* Barbados.
1970a "Historique de quelques cultures aux Antilles." *Parallèles* 34:1–10.
1970b "L'archéologie Martiniquaise. L'archéologie précolombienne aux Antilles Françaises." Parallèles (Fort-de-France) 36–37:1–67.
1970c "Le Fonds Saint-Jacques." *Annales des Antilles* 15:41–71.
1980 *La société d'habitation à la Martinique: Un demi-siècle de formation 1635–1685.* Paris: Librairie Honoré.

Petit-Puy, P. Boyer du
1654 *Véritable relation de tout ce qui s'est fait et passé au voyage que M. de Brétigny (fit) à l'Amérique Occidentale.* Paris: N.p.

Peuchet, M.
1821 *État des colonies et du commerce des Européens dans les deux Indes depuis 1783 jusqu'en 1821.* Paris: A. Costes.

Peytrad, L.
1897 *L'esclavage aux Antilles françaises avant 1789.* Paris: Librairie Hachettes.

Pianka, E. R.
1975 "Niche relations of desert lizards." In *Ecology and evolution,* edited by M. L. Cody and J. R. Diamond, pp. 292–314. Cambridge, Mass.: Harvard University Press.

Pinchon, Father R.
1935 "Les peuples pré-colombiens et leurs migrations dans les Petits-Antilles." *Bulletin de la Société Française d'Histoire Naturelle des Antilles* 3:19–36.
1952 "Introduction à l'archéologie martiniquaise." *Journal de la Société des Américanistes* n.s. 41:305–52.
1953 "L'avifaune ancienne des Antilles françaises." *Bulletin de la Société Française d'Histoire Naturelle des Antilles* 2:30–41.
1967 *Quelques aspects de la nature aux Antilles.* 3 vols. Caen, France: M. Ozanna.
1971 "Les mammifères domestiques." *D'autres aspects de la nature aux Antilles.* Fort-de-France.
1973 *Nature Antillaise. Collection Encyclopédie Antillaise,* no. 6. Pointe-à-Pitre, Guadeloupe: Emile Désormeaux.

Pitkin, T.
1816 *A statistical view of the commerce of the United States of America.* Hartford, Conn.: Charles Hosmer.

Pitman, F. W.
1917 *The development of the British West Indies, 1700–1763.* New Haven, Conn.: Yale University Press

Pittman, E. D.
1974 "Porosity and permeability changes during diagenesis of Pleistocene corals, Barbados, West Indies." *Geological Society of American Bulletin* 85:1811–20.

Plumier, C.
1693 *Description des plantes de l'Amérique avec leurs figures.* Paris: Imprimerie Royale.
1755 *Plantarum americanum fasciculus primus (—decimus), continens plantas, quas olim Carolus Plumierus, botanicorum princeps.* Amsterdam: Sumtibus Auctoris Protant in Horto Medico, at que Apud Viduam et Fillum S. Schouten.

Poey, A.
1855 "A chronological table, comprising 400 cyclonic hurricanes which have occurred in the West Indies and in the North Atlantic within 362 years, from 1493–1855." *Journal of the Royal Geographical Society* 25:291–328.

Pons, J.-C., and C. Julius
1973 *La Savane des pétrifications de la Martinique.* Fort-de-France, Martinique: Imprimerie Berger-Bellepage.

Portecôp. J.
1978 "Documents, proposition pour la protection des forêts domaniales du centre de la Martinique." *Environnement Africain,* supplement to n.s. 3:203–13.
1979 "Phytogéographie, cartographie écologique et eménagement dans une île tropicale: Le cas de la Martinique." *Extrait de documents de cartographie écologique.* Grenoble, France: Laboratoire de Biologie Végétale de l'Universitié de Grenoble.

Poupon, J.
1983a *La protection des pas géométriques de caractère naturel et forestier par l'Office National des Forêts.* Fort-de-France, Martinique: Ministère de l'Education National, Académie des Antilles et de la Guyane. Centre Régional de Documentation Pédagogique. (CRDP)
1983b *La protection des pas géométriques de caractère naturel et forestier par l'Office Nationale des Forêts.* Fort-de-France, Martinique: Centre Régional de Documentation Pédagogique.

Pouquet, J.
1952 *Les Antilles françaises.* Paris: Presses universitaires de France.

Poyer, J.
[1808],1971 *History of Barbados, from the first discovery of the island in the year 1605, till the accession of Lord Seaforth, 1801.* London: Printed for J. Mawman. Reprint ed., London: Frank Cass.

Prato-Perelli, A. da
1983 "Relations existant au début de la colonisation espagnole entre les populations caraïbes des Petites Antilles et celles due Venezuela." In *Comptes-rendus des communications du IXe Congrés International d'Etudes des Civilisations Précolombiennes des Petites Antilles,* pp. 459–83. Montreal: Centre de Recherches Caraïbes, Université de Montréal.

Priestley, H.
1939 *France overseas through the old regime: A study of European expansion.* New York and London: Appleton-Century.

Proctor, G. R.
1977 *Pteridophyta.* Vol. 2 of *Flora of the Lesser Antilles.* Edited by R. A. Howard. Jamaica Plain, Mass.: Arnold Arboretum, Harvard University.

Puis, Reverend Father M. du
1652 *Relation de l'éstablissement d'une colonie française dans la Guadeloupe, isle de l'Amérique et des moeurs des sauvages.* Caen, France: Marin-Yvon.

Purchas, S.
1905–1907 *Hakluytus posthumus or purchas his pilgrims,* vol. 15. Glasgow: J. MacLehose and Sons.

Putney, A. D.
1983 "Basis for the selection of protected areas in the Lesser Antilles." In *Proceedings, Workshop on Biosphere Reserves,* pp. 1–11. St. Johns, American Virgin Islands.

Putz, F. E.
1983 "Treefall pits and mounds, buried seeds, and the importance of soil disturbances to pioneer tree species on Barro Colorado Island, Panama." *Ecology* 64:969–1074.

Questel, A.
1941 *The flora of the island of St. Bartholomew F. W.I. and its origin.* Basse-Terre, Guadeloupe: Imprimerie Catholique.
1951 *La flore de la Guadeloupe et dépendances.* Paris: L. LeCharles.
Ravaisson-Mollien, F. N. N., ed.
1866–1904 *Archives de la Bastille.* 19 vols Paris: A. Durand et Pedone-Lauriel.
Raynal, G. T. F.
1770–74 *Histoire philosophique et politique des établissements et du commerce des Europiens dans les Indes.* 7 vols. Amsterdam and Maestricht: Jean-Edme Dufour.
1777 *History of the East and West Indies.* Translated by J. Justamond. London: J. Mundell.
1780 *Histoire philosophique et politique des établissements et du commerce des Européens dans les deux Indes.* 5 vols. Geneva: J. L. Pellet.
1798 *A philosophical and political history of the settlements and trades of the Europeans in the East and West Indies.* Translated by J. Justamond. 6 vols. London: J. Mundell.
Reichlen, H., and P. Barret
1940 "Contribution à l'archéologie de la Martinique: Le gisement de l'Anse Belleville." *Journal de la Société des Americanistes* 32:227–59.
Reis Altschul, Siri von
1964 *A taxonomic study of the genus* Anadenanthera. In Contributions from the Gray Herbarium of Harvard University 1903. Cambridge, Mass.: Harvard University Press.
Renard, R. A. D.
1973 "La Martinique de 1848 à 1870." *Cahier du Groupe Universitaire de Recherches Inter-Caraïbes* 12. Point-à-Pitre, Guadeloupe: Université Antilles-Guyane.
1982 *A social history of Guadeloupe and Martinique in the post emancipation nineteenth century, 1848–1902.* Saint Augustine, Trinidad: N.p.
Renaudot, T.
1639 *Recueil des Gazettes. Nouvelles relations extraordinaires et autres récits des choses avenues toute l'année 1638.* Paris: N.p.
Rennard, Abbot J.
1927 *Origine des paroisses et quartiers de la Martinique.* Fort-de-France, Martinique: Imprimerie Antillais.
1929a *Précis d'histoire de la Martinique.* Annecy, France: Imprimerie Commerciale.
1929b "Histoire des vingt premières années de la colonisation de la Guadeloupe d'après les relations du R. P. Breton." In *Les Caraïbes, la Guadeloupe (1635–1656),* pp. 1–82. Paris: G. Ficker.
1931 *Essai bibliographique sur l'histoire religieuse des Antilles françaises.* Paris: Secrétariat Général S.J.
1935a "Arrivée des femmes aux Antilles." *Revue d'Histoire des Colonies* 23: 135–48.
1935b *Baas, Blénac ou des Antilles françaises au XVIIe siècle.* Thonon-les-Bains, Switzerland: Imprimerie le Messager.

1935c *Tricentenaire des Antilles. Guadeloupe-Martinique 1635–1935.* Fort-de-France, Martinique: R. Cottrel.

1946 *Le commerce aux Antilles 1635–1860.* Fort-de-France, Martinique: Imprimerie du Service de l'Information.

1954 *Histoire religieuse des Antilles françaises des origines à 1914 d'aprés des documents inédits.* Paris: Société de l'Histoire des Colonies Françaises et Librairie Larose.

Renouard de Sainte-Croix, F. *See* Saint-Croix, F. Renouard, Marquis de

Revert, E.

1931 "La Montagne Pelée et ses dernières éruptions." *Annales de Géographie* 225:275–91.

1938 "Note sur la population de la Martinique." *Congrès International de la Population 1937, VI Démographie de la France d'Outre-Mer* 5:28–97.

1947 *Géographie de la Martinique.* Fort-de-France, Martinique: Imprimerie Officielle.

1949a *La Martinique: Étude géographique.* Paris: Nouvelles Editions Latines.

1949b *St. Pierre et Miquelon.* Paris: Editions Maritimes et Coloniales.

1951 *De quelques aspects du folklore martiniquais: La magie antillaise.* Paris: Editions Bellenand.

1954 *Les Antilles.* Paris: A. Colin.

1958 *Entres les deux Amériques: Le monde caraïbe.* Paris: Editions Françaises.

Richards, P. W.

1962 *The tropical rainforest: An ecological study.* Cambridge: The University Press.

1963 "Ecological notes on West African vegetation." *Journal of Ecology* 51: 123–49, 529–54.

Riddle, J.

1985 "Farmland management and biosphere reserves in the Lesser Antilles." *Proceedings Small Farm Systems of the Caribbean.* 20th Annual Meeting Caribbean Food Crops Society, Saint Croix, U.S. Virgin Islands, 1984. Fredericksburg, Saint Croix: Published by the Eastern Caribbean Center, College of the Virgin Islands and the Caribbean Food Crops Society at Antilles Graphic Arts.

Ridley, H. N.

1930 *The dispersal of plants throughout the world.* Ashford, Kent: L. Reeve.

Robequain, C.

1958 *Madegascar et les bases dispersées de l'Union Française.* Paris: Presses universitaires de France.

Roberts, L. M., U. J. Grant, and R. E. Ramirez

1957 *Races of maize in Colombia.* National Academy of Science–National Research Council Publication 550. Washington, D.C.: National Academy of Sciences.

Roberts, S. H.

1929 *History of French colonial policy, 1870–1925.* London: P. S. King. Reprint ed., London: Archon Books, 1963.

Roberts, W. A.

1942 *The French in the West Indies.* Indianapolis, Ind.: Bobbs-Merrill.

Rochefort, C.-C. de. *See* de Rochefort, C.
Rochette, A.
1897 *Étude sur les rapports commerciaux de la France et de ses colonies.* Paris: A. Pedone.
Rollet, B.
1969 "La régeneration naturelle en forêt dense humide sempervirentide plaine de la Guyane Vénézuélienne." *Bois, Forêts Tropicales* 124:19–38.
Rolph, G. M.
1917 *Something about sugar: Its history, growth, manufacture and distribution.* San Francisco: J. J. Newbegin.
Romer, A.
1933 "Étude sur la climatologie de la colonie." Fort-de-France, Martinique: Service de Météorologie. (Mimeographed).
Roncière, C. B. de la
1899 *Histoire de la Marine Française.* Paris: E. Plon, Nourrit.
1929–33 *Histoire des colonies françaises et de l'expansion de la France dans le monde.* 6 vols. Paris: Plon.
Roobol, M. J., H. Petitjean-Roget, and A. L. Smith
1976 "Mt. Pelée and the island population of Martinique." In *Proceedings of the 6th Congress for the Study of Pre–Columbian Cultures of the Lesser Antilles, 1927,* pp. 46–53. Pointe-à-Pitre, Guadeloupe: Centre Universitaire Antilles-Guyane.
Roobol, M. J., and A. L. Smith
1976 "Mt. Pelée, Martinique: A pattern of alternating eruptive styles." *Geology* 5:277–83.
1980 "Pumice eruptions of the Lesser Antilles." Bulletin Volcanologique 43(2):277–86.
Ropa, D. L.
1959 "La société coloniale de Santo Domingo à la veille de l'occupation française." *Revue Française d'Histoire d'Outre-Mer* 46:155–98.
Rosenblat, A.
1954 *La poblacion indigena y el mestezaje en America,* vol. 1. Buenos Aires: Editorial Nova.
Rose-Rosette, F.
1931 "Notice sur l'agriculture." *Bulletin Agricole* 21:105.
1953 *La Martinique: Madenina, 1502–1932.* Paris: Editions A. and P. S. Jarach.
Roth, H. L.
1887 "The aborigines of Hispaniola." *Journal of the Royal Anthropological Institute of Great Britain and Ireland* 16(3):247–86.
Roumain, J.
1942 "Contribution à l'étude de l'ethnobotanique précolombienne des Grandes Antilles." *Bulletin du Bureau d'Ethnologie de la Republique d'Haïti* 1. Port-au-Prince: Haïti (Republic) Bureau d'Ethnologie.
Rouse, I.
1948 "The Arawak." In *Handbook of South American Indians,* vol. 4, edited by J. Steward. Washington, D.C.: Bureau of American Ethnology.

1951 "Areas and periods of culture in the Greater Antilles." *Southwestern Journal of Anthropology* 7(3):248–65.

1952 "Puerto Rican prehistory." *Scientific survey of Puerto Rico and the Virgin Islands,* vol. 18, pt. 3–4. New York: New York Academy of Sciences.

1956 "Settlement patterns in the Caribbean area." In *Prehistoric settlement patterns in the New World,* edited by Gordon Willey, pp. 165–72. Viking Fund Publication in Anthropology No. 23. New York: Viking Fund.

1962 "The intermediate area, Amazonia and the Caribbean area." In *Courses toward urban life,* edited by R. Braid and G. Willey. Chicago: Aldine.

1964a "The Caribbean area." In *Prehistoric man in the New World,* edited by Jesse D. Jennings and Edward Norbeck, pp. 389–417. Chicago: Published for William Marsh Rice University by the University of Chicago Press.

1964b "Prehistory of the West Indies." *Science* 144:507.

1976a "The Saladoid sequence on Antigua and its aftermath." *Compte-rendu des communications du VIe Congrès International d'Etudes des Civilisations Precolombiennes des Petites Antilles,* edited by R. B. Bullen, pp. 35–41. Gainesville: Florida State Museum, University of Florida.

1976b "Peopling of the Americas." *Quaternary Research* 6(4):597–612.

Rouse, I., and L. Allaire

1978 *Caribbean in chronologies in New World archeology,* edited by R. Taylor and W. C. Meighan, pp. 431–81. New York: Academic Press.

Rufz de Lavison, E.

1850 *Études historiques et statistiques sur la population de la Martinique.* 2 vols. St. Pierre, Martinique: Imprimerie du Gouvernement.

Ruppé, R. J.

1980 "Sea-level rise and Caribbean prehistory." In *Proceedings of the 8th International Congress for the Study of the Pre-Columbian Cultures of the Lesser Antilles,* pp. 331–39. Anthropological Research Papers No. 22. Tucson: Arizona State University.

Sable, V.

1955 *La transformation des îles d'Amérique en départements français.* Paris: Editions Larose.

Safford, W. E.

1915 "Food plants and textiles of ancient America." In *Proceedings of the 19th International Congress of Americanists,* pp. 12–30. Washington, D.C.: Congress of Americanists.

1916 "Identity of cohoba, the narcotic snuff of ancient Haiti." *Journal of the Washington Academy of Science* 6:547–62.

Saint-Croix, F. Renouard, Marquis de

1820 *La statistique de la Martinique.* Paris: Chaumerol, Librairie Palais Royal.

1822 *Statistiques de la Martinique, orñee d'une carte de cette île avec les documents authentiques de sa population, de son commerce et de sa consommation annuelle et de ses revenues.* 2 vols. Paris: Chaumerol.

Saintoyant, J. F.

1929 *La colonisation française sous l'Ancien Régime.* 2 vols. Paris: La Renaissance du Livre.

1931 *La colonisation française pendant la période Napoléonienne.* Paris: La Renaissance du Livre.
1947 *La colonisation européenne du XVe au XIXe siècle.* Paris: La Renaissance du Livre.

Sands, W. N.
1912 "An account of the return of vegetation and the revival of agriculture in the area of devastation by the Soufrière of St. Vincent in 1902–1903." *West Indian Bulletin* 12:22–33.

Sastre, C.
1978 "Plantes menacées de Guadeloupe et de Martinique. I: Éspèces altitudinales." *Bulletin, Musée National d'Histoire Naturelle.* ser. 3, no. 519, Sept.–Oct. Paris.

Sastre, C., A. Coute, P. Sastre, S. Tiller, and A. Tiller
1982 *Restauration et évolution des écosystémes après des catastrophes naturelles: Le volcanisme antillais.* Paris: Museum National d'Histoire Naturelle.

Satineau, M.
1928 *Histoire de la Guadeloupe sous l'Ancien Régime (1635–1789).* 2 vols. Paris: Payot.

Sauer, C. O.
1950 "Cultivated plants of South and Central America." In *Handbook of South American Indians,* edited by J. Steward, 5:487–533. Washington, D.C.: Bureau of American Ethnography, Government Printing Office.
1952 *Agricultural origins and dispersals.* New York: American Geographical Society.
1966 *The early Spanish main.* Berkeley: University of California Press.

Sauer, J. D.
1959 "Coastal pioneer plants of the Caribbean and Gulf of Mexico." Madison, Wis.: Departments of Botany and Geography, University of Wisconsin. (Mimeographed).
1961 *Coastal plant geography of Mauritius.* Louisiana State University Coastal Studies Series No. 5. Baton Rouge: Louisiana State University Press.
1965 "Geographical reconnaissance of Western Australian seashore vegetation." *Australian Journal of Botany* 13(1):36–69.
1967 "Geographic reconnaissance of seashore vegetation along the Mexican Gulf Coast." *Louisiana State University Coastal Studies Series* 21:1–25.
1969 "Oceanic islands and biogeographical theory: A review." *Geographical Review* 59:582–93.
1982 *Cayman Islands seashore vegetation: A study in comparative biogeography.* University of California Publications in Geography No. 25. Berkeley: University of California.

Savine, A.
1902 "La Martinique et les Antilles au VIIe siècle, Relation du P. Jean Hallay de la Compagnie de Jésus (1657)." *Nouvelle Revue Rétrospective* 98:73–95.

Schnakenbourg, C.
1973 "Les sucreries de la Guadeloupe dans la deuxième moitié du XVIIIe siècle 1760–1790: Contribution à l'étude de la crise de l'économie co-

loniale à la fin de l'ancien régime." Thèse 3éme cycle. Paris: Presses universitaires de France.

1977 "Statistiques pour l'histoire de l'économie de plantation en Guadeloupe et en Martinique (1635–1835)." *Annales des Antilles* 21:1–121.

1980 *La crise du système esclavagiste 1835–1847.* Paris: L'Harmattan.

Schomburgk, R. H.

1848 *History of Barbados.* London: Longman, Brown, Green and Longman.

Schumann, K.

1886 "Bombacaceae." In *Flora Brasiliensis,* edited by C. P. F. Martius, 12:3. Monaco.

See, H.

1925 "Le commerce de Saint Malô au XVIIIe siècle d'après les papiers des Magon." *Mémoires et documents pour servir à l'histoire du commerce et de l'industrie en France.* Edited by Julien Hanem. 9th ser. Paris: Maisonneuve.

Shephard, C. Y.

1947 "Peasant agriculture in the Leeward and Windward Islands, part I: The development of peasant agriculture." *Tropical Agriculture* 24:61–67.

Sheridan, R. B.

1957 "Letters from a sugar plantation in Antigua, 1739–1758." *Agricultural History* 31:3–23.

1961 "The rise of a colonial gentry: A case study of Antigua, 1730–1755." *Economic History Review* 20(3):342–57.

1970 *The development of the plantations to 1750: An era of Caribbean prosperity 1750–1775.* Bridgetown, Barbados: Caribbean Universities Press.

Shimwell, D. W.

1971 *The description and classification of vegetation.* London: Sidgwick and Jackson.

Shugart, H. J., Jr., and D. C. West

1981 "Long-term dynamics of forest ecosystems." *American Scientist* 69:647–52.

Shure, D. J., and E. J. Hunt

1981 "Ecological responses to enrichment perturbation in a pine forest." In *Stress effects on natural ecosystems,* edited by G. W. Barrett and R. Rosenberg, pp. 103–14. Chichester: John Wiley & Sons.

Sidney-Daney, M. *See* Daney, M. S.

Simberloff, D. S.

1974 "Equilibrium theory of island biogeography and ecology." *Annual Review of Ecological Systematics* 5:161–82.

Simmonds, N. W.

1960 "The growth of post-war West Indian banana trades." *Tropical Agriculture* 37(2):79–85.

Simmons, I. G.

1979 *Biogeography: Natural and cultural.* London: Arnold.

Simpson, G. G.

1952 "Probabilities of dispersal in geologic time." *Bulletin of the American Museum of Natural History* 99:163–76.

1956 "Zoogeography of West Indian land mammals." *American Museum Novitates* 1759:1–28.

1967 "The beginning of the age of mammals in South America, part 2, Systematics: Notoungulala, concluded." *Bulletin of the American Museum of Natural History* 137:1–260.

Sitterson, J. C.

1953 *Sugar country: The cane sugar industry in the South, 1753–1950.* Lexington: University of Kentucky Press.

Slater, M. K.

1977 *The Caribbean family: Legitimacy in Martinique.* New York: St. Martin's Press.

Sleight, F. W.

1965 "Certain environmental considerations in West Indian archaeology." *American Antiquity* 31:226–31.

Sloane, H.

1701–25 *A voyage to the Islands Madera, Barbados, Nievis, St. Christophers and Jamaica, with the natural history of the herbs and trees with some relations concerning the neighboring continent, and islands of America. . .* 2 vols. London: Privately printed.

Smith, A. L., and M. J. Roobol

1980 "Late prehistoric and historic activity of Mt. Pelée. Martinique." Abstract. *9th Caribbean Geology Conference Publications.* Santo Domingo, Dominican Republic: N.p.

Smith, J.

1630 *True travels, adventures, and observations of Captain John Smith.* In *Europe, Asia, Africa, and America from Anno Domini 1593–1629.* London: Printed by J. H. for T. Salter.

Smith, W.

1745 *The natural history of Nevis and the rest of the English Leeward Charibbee Islands in America.* Cambridge: J. Bentham.

Sobesky, A.

1940 "La location des terres domaniales pour cultures vivrières à la Martinique." *Bulletin Agricole* 30: 246–59.

Southey, C. T.

1827 *Chronological history of the West Indies.* London: Longman, Rees, Orme, Brown & Green.

Spate, O. H. K.

1963 "Islands and men." In *Man's place in the island ecosystem,* edited by F. R. Fosberg. Honolulu: Bishop Museum Press.

Standley, P. C.

1923–25 "Trees and shrubs of Mexico." *Contributions from the U.S. National Herbarium* 23:1–1721. Washington, D.C.: The Smithsonian Institution.

1930 "Flora of Yucatan." In *Botany* 3:157–492. Chicago: Field Museum of Natural History.

1937–38 "Flora of Costa Rica." *Field Museum of Natural History Botanical Series* 18:1–4.

Standley, P. C., and S. J. Record
1936 "The forests and flora of British Honduras." *Field Museum of Natural History Botanical Series* 12:1–450.

Stearns, C. D.
1976 "Estimates of the position of sea level between 140,000 and 75,000 years ago." *Quaternary Research* 6:445–49.

Stehlé, H.
1935 *I. Essai d'écologie et de géographie botanique.* Basse-Terre, Guadeloupe: Gouvernement de la Guadeloupe et Dépendances.
1936 *Essai d'écologie et de géographie botanique.* Vol. 1 of *Flore de la Guadeloupe et dépendances.* Basse-Terre, Guadeloupe: Imprimerie Catholique.
1937a "Esquisse des associations végétales de la Martinique." *Bulletin Agricole* 6(3–4):193–264.
1937b "Les associations végétales de la Guadeloupe et leur intérêt dans la valorisation rationnelle." *Revue Botanique Appliquée et Agriculture Tropicale* 12(186):98–109; 12(187):188–95.
1938 *Flore de la Guadeloupe et dépendances. I. Catalogue des phanérogames et fougères, avec contribution à la flore de la Martinique.* Basse-Terre, Guadeloupe: Gouvernement de la Guadeloupe et Dépendances. Imprimerie Catholique.
1939 *Flore descriptive des Antilles françaises. I. Orchidales.* Fort-de-France: Imprimerie Officielle de la Martinique.
1940 *Flore descriptive des Antilles françaises. II. Piperales.* Fort-de-France, Martinique: Imprimerie du Government.
1941 "Conditions éco-sociologiques et évolution des forêts des Antilles françaises." *The Caribbean Forester* 2(4):154–59.
1944 "Le sens de l'évolution du tapis végétal de la Madinina caraïbe à la Martinique actuelle." *Martinique: Revue Tremestrielle* 3:110–14.
1945–46 "Les types forestiers des iles caraïbes." *Caribbean Forester* 6(supplement):273–474; 7(supplement):337–709.
1946a "La première conférence internationale d'amélioration du sol et d'exploitation forestière." *Revue Botanique Appliquée et Agriculture Tropicale* 27(285–86):328–29.
1946b "Végétation d'origine asiatique ou africaine naturalisée dans l'archipel caraïbe." *Revue Internationale Botanique Appliquée et Agriculture Tropicale* 26 (281–82):100–106.
1947 *La végétation sylvatique de l'archipel Caraïbe.* Thèse d'Ingénieur-Docteur, Université de Montpellier.
1954 "La protection de la nature en Amérique tropicale." *Rapport Congrès International de Botanique de Paris,* pp. 146–153. Paris: Société de Botanique de Paris.
1955a "Les principaux types de pâtures et savanes paturelles et les mélanges fourragers dans l'archipel caraïbe." Etude No. 36. Basse-Terre, Guadeloupe: Centre des Recherches Agronomiques des Antilles et de la Guyane Française (CRAAG).
1955b "Les principales plantes ruderales de l'Archipel Antillais." *VIIIe Congrès International de Botanique de Paris 2–14 juillet 1954.* Etude No. 46.

Basse-Terre, Guadeloupe: Centre des Recherches Agronomiques des Antilles et de la Guyane Française (CRAAG).

1956a *Investigation sur les plantes fourragères dans les territoires caraïbes.* Port-of-Spain, Trinidad: Commission des Caraïbes.

1956b *Survey of forage crops in the Caribbean.* Port-of-Spain, Trinidad: Caribbean Commission.

1956c "Les variétés de canne à sucre aux Antilles françaises." *Revue d'agriculture: le sucre et rhum Antilles françaises* 1(Jan.–Mar.):11–13.

1957 "Le problème de l'érosion des sols agropédiques aux Antilles et les moyens de protection." *Revue agriculture sucrière et rhumière des Antilles françaises* 2(4):1–10.

1966 "Quelques mises au point historiques relatives à l'introduction de végétaux économiques aux Antilles françaises." *Bulletin de la Société d'Histoire de la Guadeloupe,* No. 526. Basse-Terre, Guadeloupe.

1976 "Les monocotylédones des Antilles françaises." *Bulletin de la Société Botanique de France* 123:441–54.

1979–80 "Premières observations sur la reconstitution du tapis végétal sur le volcan de la Soufrière de Guadeloupe après les éruptions de juillet-août 1976 et mars 1977." *Bulletin de la Société Botanique Française* 126(3): 349–59; 127(2):159–68.

Stehlé, H., and M. Stehlé

1947 "Supplementary list of the trees and shrubs of the Lesser Antilles." *Caribbean Forester* 8:91–111.

1950 "L'origine et introduction des végétaux économiques aux Antilles françaises." *Marchés Coloniaux du Monde.* Special issue for May 13.

1957 *Flore agronomique des Antilles françaises. I. Flore des champs de canne à sucre.* Pointe-à-Pitre, Guadeloupe: Imprimerie Parisienne.

1958 *Flore agronomique des Antilles françaises. X. Flore ornementale.* Basse-Terre, Guadeloupe: Imprimerie Officielle.

1960 *Flore agronomique des Antilles françaises. III. Flore des légumineuses et anti-érosions.* Basse-Terre, Guadeloupe: Imprimerie Officielle.

Stehlé, H., M. Stehlé, and L. Quentin

1937–49 *Flore de la Guadeloupe et dépendances. II. Catalogue des phanerogames et des fougères* pt. 1, (Basse-Terre, 1937); pt. 2 (Montpellier, 1948); pt. 3 (Montpellier, 1949).

Stephens, S. G.

1944 "Cotton growing in the West Indies during the 18th and 19th centuries." *Tropical Agriculture* 21(2):23–29.

Steward, J. H.

1953 "Evolution and process." In *Anthropology Today,* edited by A. L. Kroeber, pp. 313–26. Chicago: University of Chicago Press.

1948 "Introduction." *Handbook of South American Indians.* Vol. 4, *The circum-Caribbean* tribes, pp. 1–41. Washington, D.C.: Government Printing Office.

1955 *Theory of culture change: The methodology of multilinear evolution.* Urbana: University of Illinois Press.

Steward, J. H., and L. C. Faron

1959 *Native peoples of South America.* New York: McGraw-Hill.

Stipp, J. J., F. Nagel, and D. C. Fisher
1974 "Potassium-argon géochronology Martinique." *Compte-rendu VIIe conférence géologique des Caraïbes.* Saint-François, Guadeloupe: Imprimerie Réunies de Chambéry.

Stockdale, Sir F. A.
1915 *Guide to the botanic gardens, Dominica.* Roseau, Dominica: Department of Agriculture.

Stoddart, D. R.
1965 "British Honduras cays and the low wooded island problem." *Transactions of the Institute of British Geographers* 36:131–47.
1974 "Post–hurricane changes on the British Honduras reefs: Resurvey of 1972." *In Proceedings of the Second International Coral Reef Symposium,* edited by A. M. Cameron et al., 2:473–83. Brisbane, Australia: The Greater Barrier Reef Committee.
1980 "II. Geography and contemporary exploration." *Geographical Journal* 146(2):169–73.

Stoddard, D. R., and F. R. Fosberg
1982 "Species-area relationships on small islands: Floristic data from Belizean sand cays." In *The Atlantic barrier reef ecosystem at Carrie Bow Cay, Belize, 1:Structure and communities,* edited by K. Rüztzler and I. G. Macintyre, 12:539. Washington, D.C.: Smithsonian Contributions to the Marine Sciences, Smithsonian Institution.

Stoffers, A. L.
1956 *The vegetation of the Netherlands Antilles.* Utrecht: Natuur-wetenschappelijke Studiekring voor Suriname en de Nederlandse Antillen.

Stokes, J.
1792 "West Indian agriculture." *Annals of Agriculture and Other Useful Arts* 18:148–57.

Stover, J. F.
1958 "French-American trade during the confederation, 1781–1789." *North Carolina Historical Review* 35(4):399–414.

Street, J.
1960 "Historical and economic geography of the southwest peninsula of Haiti." Ph.D. diss., University of California, Berkeley.

Stuart, V.
1891 *Adventures amidst the equatorial forests and rivers of South America; also in the West Indies and the wilds of Florida.* London: J. Murray.

Sturtevant, W. C.
1961 "Taino agriculture." In *The evolution of horticultural systems in native South America, Anthropologica, Supplement Public,* edited by J. Wilbert, vol. 2. Caracas: Sociedad de Ciencias Naturales La Salle.
1969 "History and ethnography of some West Indian starches." In *The domestication and exploitation of plants and animals,* edited by P. S. Ucko and G. W. Dembleby, pp. 177–99. Chicago: Aldine.

Suarez, C.
1921 *Vocabulario Cubano: Supplemento a la 14[a] edicion del diccionario de R. A. de la lengua.* Havana: R. Velsos.

Sugar
1888–92 Vols. 2–5.
Sutty, L.
1983 "Liaison Arawak-Caliviny-Carib between Grenada and St. Vincent, Lesser Antilles." In *Comptes-rendus des communications du IXe Congrés International d'Etudes des Civilisations Précolombiennes des Petites Antilles, Santo Domingo, 1981*, pp. 145–53. Montreal: Centre de Recherches Caraïbes, Université de Montréal.
Sylvestre, E., and E. Lotaut
1947 *L'annuaire de la vie martiniquaise.* 2 vols. Fort-de-France, Martinique. Imprimerie du Gouvernement.
Tansley, A. G.
1935 "The use and abuse of vegetational concepts and terms." *Ecology* 16:284–307.
1939 *The British Islands and their vegetation.* Cambridge: The University Press.
Tansley, A. G., and T. F. Chipp
1926 *Aims and methods in the study of vegetation.* London: British Empire Vegetation Committee.
Tardieu, P. A.
1817 *Atlas géographique des colonies.* Paris: A. Eymery.
ca. 1832 *Atlas géographique, statistique et progressif des départements de la France et de ses colonies.* Edited by Pierre Tardieu. Paris: E. Dieulouard.
Taylor, D. M.
1938 *The Caribs of Dominica.* Bureau of American Ethnology Bulletin 119. 3:103–59. Washington, D.C.: Government Printing Office.
1949 "The interpretation of some documentary evidence on Carïb culture." *Southwestern Journal of Anthropology* 5:379–92.
1951 *The Black Carib of British Honduras.* Viking Fund Publications in Anthropology No. 17. New York; Wenner-Gren Foundation for Anthropological Research.
1957 "Aji and Batata." *American Anthropologist* 59(4):704–705.
1959a "Historical implications of linguistic data on the floods of the Island Carib and Black Carib." *Actas del XXXIII Congreso Internacional de Americanistas* 1:295–308. San José, Costa Rica: International de Americanistas.
1959b "Morpheme mergers in Island Carib." *International Journal of American Linguistics* 25(3):190–95.
1968 "New language for old in the West Indies." In *Readings in the Sociology of Language*, edited by J. A. Fishman, pp. 607–19. Paris: Mouton.
TeBrake, W. H.
1985 *Medieval frontier: Culture and ecology in Rijnland.* College Station: Texas A&M University Press.
Ternaux, H.
1837 *Bibliothèque américaine, ou catalogue des ouvrages relatifs à l'Amérique, qui ont paru depuis sa découverte jusqu'à l'an 1700.* Paris: Arthus-Bertrand.
Theveneau, A., R. Martin, and G. Sinthe
1964 *Types de lemas aux Antilles françaises.* Monographie de la Météorologie Nationale No. 31. Paris: Météorologie Nationale.

Thevet, A.
1571 *La cosmographie universelle d'André Thevet, cosmographe du roy.* Paris: P. L'Huiller.
[1575],1953 "La cosmographie universelle." In *Les français en Amérique pendant la deuxième moitié du XVIe siècle. Le Brésil et les brésiliens,* edited by Suzanne Lursaguet. Paris: Presses universités de France.
Thibault de Chanvalon. *See* Chanvalon, T. de
Thom, B. G.
1967 "Mangrove ecology and deltaic geomorphology: Tabasco, Mexico." *Journal of Ecology* 55:301–43.
Thompson, L. G., E. Mosley-Thompson, J. F. Bolzan, and B. R. Koci
1985 "A 1500-year record of tropical precipitation in ice cores from the Quelecarya Ice Cap, Peru." *Science* 229:971–73.
Thornthwaite, C. W.
1948 "Approach towards a rational classification of climate." *Geographical Review* 38:55–96.
Thornton, A. P.
1956 *West-Indian policy under the restoration.* Oxford: Clarendon Press.
Tibault, H. L.
1985 "Etude des systèmes agraires et développement: l'agriculture d'aurès plantation à la Martinique." *Systèmes de production agricole Caribiens et alternatives de développement,* pp. 663–78. Fort-de-France, Martinique: Imprimerie Berger Bellepage.
Tillon, M. R.
1975 "La production forestière." *Bulletin d'information.* CENADDOM No. 25. Fort-de-France.
Tomblin, J. K.
1975 "The Lesser Antilles and Aves Ridge." In *Ocean basins and margins,* vol. 3 of *Gulf of Mexico and the Caribbean Sea,* edited by A. E. M. Nairn and F. G. Stehli, pp. 467–500. New York: Plenum Press.
Tosi, J. A., and R. Voertman
1964 "Some environmental factors in the economic development of the tropics." *Economic Geography* 40:189–205.
Traineau, H.
1982 "Contribution à l'étude géologique de la Montagne Pelée, Martinique: Evolution de l'activité éruptive au cours de la période récente." Thèse de doctorat 3e cycle, l'Université de Paris-Sud, Centre d'Orsay.
Traineau, H., D. Westercamp, and C. Coulon
1983 "Mélanges magmatiques à la Montagne Pelée (Martinique): Origine des éruptions de type Saint Vincent." *Bulletin Volcanologique* 46(3): 243–69.
Tromel, P.
1861 *Bibliothèque américaine: Catalogue, raisonné d'une collection de livres precieux sur l'Amérique, parus depuis sa découverte jusqu'à l'an 1700.* Leipzig: F. A. Brockhaus.
Twist, T. F.
1953 "Weather of the eastern Caribbean." *Weather* 303–304.

U.S. Department of Commerce, National Oceanographic and Aeronautical Administration (NOAA)
1985 *Atlas of tropical cyclones of the North Atlantic 1871–1980 (with storm track maps updated through 1984).* Revised. Asheville, N.C.: NOAA.

U.S. Department of Commerce, Weather Bureau
1938 *Atlas of climatic charts of the oceans.* Washington, D.C.

U.S. Department of the Navy
1981 *Surface currents: Southwest North Atlantic Ocean including the Gulf of Mexico and Caribbean Sea.* Naval Oceanographic Office Special Publication 1400–NA9. National Science Testing Laboratory Station (NSTL), Miss.

van der Hammen, T.
1974 "The Pleistocene changes of vegetation and climate in tropical South America." *Journal of Biogeography* 1:3–26.

van der Pijl, L.
1972 *Principles of dispersal in higher plants.* 2nd ed. Berlin: Springer-Verlag.

van Steenis, C. G. G. J.
1958 "Rejuvenation as a factor for judging the status of vegetation types: The biological nomad theory." In *Acts Colloquia of Kandy,* pp. 212–15. Paris: UNESCO.
1972 "*Northofagus,* key genus to plant geography." In *Taxonomy, phytogeography, and evolution,* edited by D. H. Valentine, pp. 257–88. London: Academic Press.

van Valen, L.
1965 "Morphological variation and width of ecological niche." *American Naturalist* 99:377–90.
1975 "Life, death and energy of a tree." *Biotropica* 7:259–69.

Vassière, R. de la
1910 *La Martinique en 1910.* Paris: N.p.

Vathaire, M. M., ed.
1959 *Essai de documentation économique, sociale, politique des Antilles françaises, 1946–1958, Guadeloupe-Martinique.* Paris: N.p.

Vaughn, T. W.
1918 "Geologic history of Central America and the West Indies during Cenozoic time." *Geological Society America Bulletin* 29:615–30.
1919 "Contributions to the geology and paleontology of the Canal Zone, Panama, and geologically related areas in Central America and the West Indies." *United States National Museum Bulletin* 103:189–524, 547–612.

Vazques de Espinosa, A.
[1620],1942 *Compendeum and description of the West Indies.* Translated by C. U. Clark. Washington, D.C.: Smithsonian Institution.

Veblen, T. T.
1981 "Forest dynamics in south-central Chile." *Journal of Biogeography* 8:211–47.

Veblen, T. T., and D. H. Ashton
1979 "Successional patterns above timberline in south-central Chile." *Vegetatio* 40(1):39–47.

Velez, I.
1957 *Herbaceous angiosperms of the Lesser Antilles.* Rio Piedras: Biology Department of the Interamerican University of Puerto Rico.

Vignols, L.
1928 "Les Antilles françaises sous l'ancien régime: Aspects économiques et sociaux. L'institution des engagés (1626–1774)." Supplement to *Revue d'Histoire Economique et Sociale.* Paris: Rivière.
1935 "Une question mal posée, le travail manuel des blancs et des esclaves aux Antilles." *Revue Historique* Mar.–Apr., 308–16.

Vuilleumier, B. S.
1971 "Pleistocene changes in the flora and fauna of South America." *Science* 1973:771–80.

Vuilleumier, F.
1971 "Qu'est-ce que la biogéographie?" *Compte-Rendu Lecuté Biographie* 475:41–66.

Wagner, P.
1962 "Natural and artificial zonation in a vegetation cover." *Geographical Review* 52:253–74.

Walker, J. B.
1980 "Analysis and replication of the lithic artifacts from the sugar factory pier site, St. Kitts, West Indies." M.A. thesis, Washington State University.

Warner, G. F., ed.
1901 *The voyages of Sir Robert Dudley to the West Indies and Guiana.* ser. 2, vol. 3. London: Hakluyt Society.

Wassen, S. H.
1965 "The use of some specific kinds of South American snuffs and related paraphernalia." *Etnologiska Studier* (Göteborg, Sweden), 28:1–116.

Wassen, S. H., and B. Holmstead
1963 "The use of *Paricá,* an ethnological and pharmacological review." *Ethnos* 1:5–45.

Watters, D. R.
1981 "Linking oceanography to prehistoric archaeology." *Oceanus* 24(2): 11–19.

Watters, D. W.
1983 "Assessing the ocean's roles in Antillean prehistory." *Comptes-rendus des communications du IX^a Congrès International Colombiennes des Petites Antilles,* pp. 531–41. Montreal: Centre de Recherches Caraïbes, Université de Montréal.

Watts, A. P.
1924 *Une histoire des colonies Anglaises aux Antilles de 1649 à 1660.* Paris: Presses universitaires de France.

Watts, A. S.
1947 "Pattern and process in the plant community." *Journal of Ecology* 35:1–22.

Watts, D.
1966 *Man's influence on the vegetation of Barbados 1627 to 1800.* Occasional

Papers in Geography No. 4. Hull, Yorkshire: University of Hull Publications.
1970 "Persistence and change in the vegetation of oceanic islands." *Canadian Geographer* 14:91–109.
1971 *Principles of biogeography: An introduction to the functional mechanisms of ecosystems.* London: McGraw-Hill.

Waugh, A.
1964 *A family of islands; a history of the West Indies from 1492 to 1898, with an epilogue: Sketchy events from the Spanish American War to the 1960's.* Garden City, N.Y.: Doubleday.

Weatherwax, R.
1954 "The taxonomy, genetics, production and uses of cultivated species of Cucurbita." *Economic Botany* 4(1):52–81.

Webb, L. J.
1958 "Cyclones as an ecological factor in tropical lowland rainforest, North Queensland." *Australian Journal of Botany* 6:220–28.

Welch, B. L.
1939 *The West Indies yearbook, including also the Bermudas, the Bahamas, British Guiana, and British Honduras.* Montreal.
1962 "Aspects of succession in shallow coastal waters of the Caribbean." Ph.D. diss., Duke University.

Wendland, W. M., and R. A. Bryson
1974 "Dating climatic episodes of the Holocene." *Quaternary Research* 4:9–24.

Westercamp, D.
1972 "Contribution à l'étude du volcanisme en Martinique." Thèse 3e cycle, Université Paris-Sud Orsay.
1975a "Petrology of the volcanic rocks of Martinique, West Indies." *Bulletin Volcanologique* 39(2):175–200.
1975b "Stratigraphie des roches volcaniques du massif du Carbet, Martinique, F.W.I." In *Publication, transactions de la VIIe Conférence Géologique des Caraïbes, Pointe-à-Pitre, Guadeloupe, 1974,* pp. 427–35. Paris: Imprimeries Réunies de Chambéry.

Westercamp, D., and B. Mervoyer
1975 "Les séries volcanique de la Martinique et de la Guadeloupe (F.W.I.)." In *Publication, transactions de la VIIe Conférence Géologique des Caraïbes, Pointe-à-Pitre, Guadeloupe, 1974,* pp. 437–50. Paris: Imprimeries Réunies de Chambéry.

Westercamp, D., and H. Tazieff
1980 *Martinique, Guadeloupe, Saint-Martin, La Désirade, Guides géologiques régionaux.* Paris: Masson.

Westercamp, D., B. Mervoyer, and C. Alsac
1974 "Les séries magmatiques de la Martinique et de la Guadeloupe, Petites Antilles Françaises: Rapports stratigraphiques et petrographiques." *Recueil des résumés de communication. VIIe Conférence Géologique des Caraïbes, Pointe-à-Pitre, Guadeloupe.* Pointe-à-Pitre, Guadeloupe: Service de l'Industrie et des Mines, Régions: Antilles-Guyane Française.

Westerman, J. C.
1935 *The Netherlands and the United States: Their relations in the beginning of the nineteenth century.* The Hague: M. Nijhoff.

Whitbeck, R. H.
1933 "The Lesser Antilles—past and present." *Annals of the Association of American Geographers* 28:21–26.

Whitmore, T. C.
1974 *Change with time and the role of cyclones in tropical rainforests on Kolombangana, Solomon Islands.* Oxford: Commonwealth Forestry Institute.
1983 "Secondary succession from seed in tropical rain-forests." *Forestry Abstracts* 44:767–79.

Whitson, A. M.
1930 "The outlook of the continental American Colonies on the British West Indies, 1760–1775." *Political Science Quarterly* 45:56–86.

Whittaker, R. H.
1967 "Gradient analysis of vegetation." *Biological Review* 49:207–67.
1970 *Communities and ecosystems.* London: Macmillan.
1975 *Communities and ecosystems.* 2nd ed. New York: Macmillan.

Whittaker, R. H., and S. A. Levin, eds.
1975 *Niche: Theory and application.* Stroudsburg, Pa.: Dowden, Hutchinson, and Ross.

Wiley, J.
1806 *Hortus Eastensis: Or, a catalogue of exotic plants cultivated in the botanic gardens . . . in the island of Jamaica.* Kingston: N.p.

Wilgus, A. C., ed.
1958 *The Caribbean: British, Dutch, French, United States.* Gainesville: University of Florida Press.

Wilkinson, H. C.
1950 *Bermuda in the Old Empire . . . 1684–1789.* New York: Oxford University Press.

Williams, R. O., and E. E. Cheesman
1928 *Flora of Trinidad and Tobago.* Port-of-Spain, Trinidad: Government Printing Office.

Williamson, J. A.
1926 *The Caribbee Islands under the proprietary patents.* London: Oxford University Press.

Wilson, A. M.
1936 "The logwood trade in the seventeenth and eighteenth centuries." In *Essays in the history of modern Europe,* edited by D. C. McKay, pp. 1–15. New York: Harper & Brothers.

Wright, H.E.
1976 "The dynamic nature of Holocene vegetation." *Quaternary Review* 6:581–96.

Wroth, L. C.
1930 *Acts of the French royal administration concerning Canada, Guiana, the West Indies and Louisiana prior to 1791.* New York: New York Public Library.

Ybarra, A. M. de
1907 "The letter of Dr. Diego Alvarez Chanca, dates 1494 relating to several voyages of Columbus." *Smithsonian Miscellaneous Collection* 48:428–57.

Young, A.
1788 "West Indian agriculture." *Annals of Agriculture and Other Useful Arts* 10:335–62.

Young, W.
1801 "A tour through the several islands of Barbados, St. Vincent, Antigua, Tobago, and Grenada, in the years 1791 and 1792." In B. Edwards, *An historical survey of the islands of Saint Domingo . . . in 1793 and 1794*, pp. 259–301. London: J. Stockdahl.

Maps

Bellin, J. N.
1758 *Carte de l'isle de la Martinique pour servir à l'histoire général des voyages.* Vignaud. 36 × 54.5 cm.

Benoit, J., and C. Crabot
1960 *Carte des exploitations et des industries agricoles de la Martinique.* Drafted and compiled by Jean Benoit and Christain Crabot. Published under the auspices of the Historical Society. 1:50,000. 120 × 92 cm.

Bonne, R.
ca. 1788 *Carte de l'isle de la Martinique, colonie françoise dans les Isles Antilles.* by R. Bonne, Engineer, Hydrographer of the Marine.

Case, J., and T. L. Holcombe
1980 "Geologic-tectonic map of the Caribbean." *U.S. Geological Survey Miscellaneous Investigation.* 1:1,100.

Desruisseaux, J.
1969 *La structure de la propriété en Martinique.* 1:100,000. Fond-St. Jacques, Martinique: Centre de Recherche Caraïbe.
1970 *Contribution cartographique à l' étude de la vue rurale en Martinique.* Bordelle, France; Institute de Géographie.
1973 *Carte de la structure fonciére de la Martinique.* Montreal: Centre de Recherches Caraïbes and Centre ORSTOM des Antilles 1:50,000.

France, Institut Géographique National (IGN)
1982 *Carte touristique de la Martinique* Paris: IGN. 1:100,000.
1984 *Carte touristique de la Martinique* Paris: IGN. 1:100,000.

France, Institut Géographique National (IGN)
1955 *Carte de la Martinique.* 1:20,000. Sheets 1–23.

Grünewald, H.
1958 *Carte géologique du département de la Martinique.* Printed by IGN. 1:50,000. North Sheet 71.5 × 105 cm; South Sheet 67 × 112 cm.

Portecôp, J.
1978 Carte écologique de la Martinique à 1:75,000. In *Documents de cartographie écologique 20.* Laboratoire de Biologie Végétale. Grenoble, France: Imprimerie Louis-Jean.

Société d'Editions Géographiques, Maritimes et Coloniales
1931 *Martinique, Guadeloupe, Guyane, Saint Pierre, Miquelin.* Paris: Société d'éditions géographiques, maritimes et coloniales.

Westercamp, D., and H. Traineau
1983 *Montagne Pelée, Martinique. Carte géologique.* 1:20,000. Service Géologique National BRGM.

Appendix A. Systematic List of Plant Species Mentioned in the Text

FERNS AND FERN ALLIES

Cyatheaceae

Cyathea arborea (L.) Smith
Cyathea muricata Willd. = *Hemitelia muricata* (Willd.) Fee*

Gleicheniaceae

Ctenitis excelsa (Desv.) Proctor = *Dryopteris excelsa* (Desv.) C. Chr. 1959**
Dicranopteris bifida (Willd.) Maxon 974, 1991
Dicranopteris flexuosa (Schrad.) L. M. Underw. 2074
Dicranopteris pectinata (Willd.) L. M. Underw.
Gleichenia bifida (Willd.) Spreng.
Gleichenia farinosa (Kaulf.) Hook.

Hymenophyllaceae

Hymenophyllum hirtellum Sw.
Trichomanes alatum Sw.
Trichomanes hymenophylloides van der Bosch.
Trichomanes trigonum Desv.

Lycopodiaceae

Lycopodium sp. 972
Lycopodium cernuum L. 2007
Lycopodium clavatum L.
Lycopodium linifolium L.
Lycopodium reflexum Lam.
Lycopodium sieberianum Spring 1154, 968
Lycopodium verticillatum L. f.

Ophioglossaceae

Ophioglossum reticulatum L.

*Older name or names replaced by synonymy. Nomenclature used in this work is that of Garay and Sweet 1974 for the Orchidaceae, Proctor 1977 for the Pteridophyta, Howard 1979 for the monocotyledoneae. For the other species, Fournet 1978 is followed. The older names are found in much of the ecological and botanical literature. Acknowledgment is made of the courteous help in identification given me by Henri Stehle, Hugh Iltis, Richard Howard, Jacques Portecôp, and Jean-Pierre Fiard.

**Collections made by the author are deposited in the Herbarium, University of Wisconsin at Madison, Wis. A smaller number are deposited in the Tracy Herbarium, Texas A&M University, College Station, Tex. A number of these specimens were identified by specialists in the group. Where these specimens carry names not recognized in the compiled floras, I have kept the designations of the expert who identified it, as in *Dicranopteris bifida* (Willd.) Maxon.

Pandanaceae

Pandanus hahnii Warb. 1687
Pandanus maximus Martelli 1689
Pandanus odoratissimus L. f. = *P. tectorius* Soland 1692
Pandanus utilis Bory
Pandanus veitchii Veitch ex Dillière

Polypodiaceae

Acrostichum aureum L. 1586
Adiantopsis radiata (L.) Fee 1253
Adiantum fragile Sw. 1593
Adiantum tenerum Sw.
Adiantum tetraphyllum H. & B.
Asplenium formosum Willd.
Asplenium salicifolium L.
Blechnum divergens (Kunze) Mittenius = *B. striatum* (Sw.) C. Chr. 970, 981, 1159
Blechnum occidentale L. 530, 531
Blechnum ryanii (Kaulf.) Hieron.
Cochilidium linearifolium (Desv.) Maxon ex C. Chr.
Elaphoglossum dussii Underw.
Elaphoglossum feei (Bory ex Fee) T. Moore
Elaphoglossum glabellum J. Smith
Elaphoglossum lingua Brack.
Elaphoglossum plumieri T. Moore
Grammitis asplenifolia (L.) Proctor
Grammitis harti (Jenman) Proctor
Grammitis serrulata (Sw.) Sw.
Hymenodium crinitum (L.) Fee
Nephrolepis biserrata (Sw.) Schott
Nephrolepis rivularis (Vahl) C. Chr. 730, 1252
Pityrogramma sp. 2158
Pityrogramma calomelanos (L.) Link 695, 876
Pityrogramma chrysophylla (Sw.) Link 878
Polybotrya cervina (L.) Kaulf.
Polypodium astrolepis Liebm.
Polypodium aureum L. var. *aureum* Howard
Polypodium glaucophyllum Kunze
Polypodium piloselloides Linn.
Polypodium polypodioides (L.) Watt.
Pteris biaurita Linn.
Salpichloena volubilis (Kaulf.) J. Smith
Stigmatopteris rotundata (Willd.) C. Chr.
Tectaria heracleifolia (Willd.) L. M. Underw.
Tectaria trifoliata (L.) Cav.
Thelypteris balbisii (Spreng.) Ching.
Thelypteris normalis (C. Chr.) Moxley
Thelypteris reticulata (L.) Proctor
Trichomanes alatum Sw.

Selaginellaceae

Selaginella flabellata (L.) Spring 2214, 1951
Selaginella substipitata Spring

Cycadaceae

Cycas circinalis L.

GYMNOSPERMS

Araucariaceae

Araucaria heterophylla (Salisb.) Franco

Pinaceae

Pinus caribaea Morelet

Podocarpaceae

Podocarpus coriaceus L. C. Rich.

ANGIOSPERMS

Acanthaceae

Acanthaceae spp. 807, 815, 828, 1504
Asystasia gangetica (L.) T. And.
Barleria lupulina Lindl. 1327
Blechum brownei Juss. = *B. pyramidatum* (Lam.) Urb. 1143, 1345, 1404, 1504
Justicia pectoralis Jacq. = *Dianthera pectoralis* Murr. 1250
Odontonema nitida (Jacq.) O. Kuntze
Oplonia microphylla (Lam.) Stearn = *Anthacanthus microphyllus* (Jacq.) Nees
Oplonia spinosa (Jacq.) Raf. = *Anthacanthus spinosus* (Jacq.) Nees = *Justicia spinosa* Jacq.
Ruellia tuberosa L. 1344, 1510
Teliostachya alopecuroidea (Vahl) Nees = *Ruellia alopecuroidea* Vahl
Thunbergia alata Bojer ex Sims. 781, 839, 1099, 1220, 1221
Thunbergia fragrans Roxb. 1091

Agavaceae

Agave caribaccola Trel.
Agave dussiana Trel.
Agave sisalana Perrine
Furcraea tuberosa (Mill.) Ait. f. = *Agave campanulata* Sesse & Moq.

Aizoaceae

Sesuvium portulacastrum L. 211

Amaranthaceae

Achyranthes indica (L.) Mill. 645, 666, 680, 1097, 1487
Alternanthera bettzickiana (Regel.) Standl.
Alternanthera brasiliana (L.) Ktze. = *Mogiphanes Jacquini* Schrad. 1130, 1212
Alternanthera sessilis (L.) R. Br. 491
Amaranthus dubius Mart. 593-B, 1176, 1254, 1456
Amaranthus spinosus L. 404, 1181, 1465, 2177
Amaranthus viridis L. = *A. gracilis* Desf. 1499
Gomphrena globosa L.
Iresine diffusa Willd. = *I. celosia* L. 1141
Philoxerus vermicularis (L.) Nutt. 439, 656, 1547, 1553, 2104

Amaryllidaceae

Pancratium caribaeum L. = *Hymenocallis expansa* (Herb.) Herb.
Zephyranthes grandiflora Lindl.

Ammiaceae = Umbelliferae

Apium graveolens L.
Apium petroselinum L.
Centella asiatica (L.) Urb.
Daucus carota L.
Eryngium foetidum L.
Pastinaca sativa L.

Anacardiaceae

Anacardium occidentale L. 613
Comocladia dodonaea (L.) Urb. = *Ilex dodonaea* L.
Comocladia undulata Urb.
Mangifera indica L. 528
Spondias dulcis Forst. = *S. cytherea* Tuss.
Spondias mombin L.
Spondias purpurea L.
Spondias purpurea forma *lutea* Fawc. and Rendle

Annonaceae

Annona glabra L.
Annona muricata L. 561, 616
Annona reticulata L.
Annona squamosa L.
Guatteria caribaea Ubr.
Oxandra laurifolia (Sw.) A. Rich.

Apocynaceae (contains Asclepiadaceae)

Allamanda cathartica L.
Asclepias curassavica L. 855
Calotropis procera (Ait.) Ait. f. 658
Catharanthus roseus L. = *Vinca rosea* L.
Plumieria alba L. = *P. hypoleuca* Gasp.
Plumieria rubra L.
Rauvolfia biauriculata J. Muell.
Rauvolfia viridis Roem. and Schult. = *Rauwolfia lamarkii* A.DC. = *R. latifolia* A.DC.
Rhabdadenia biflora (Jacq.) Muell-Arg. = *Echites biflora* Jacq.
Tabernaemontana citrifolia L. = *T. laurifolia* R. & S.
Tetraphylla L. 554, 1240, 1437
Urechites lutea L. Britt 1210

Aquifoliaceae

Ilex macfadyenii (Walp.) Rehder var. *caribaea* Stehlé & Quentin
Ilex nitida (Vahl) Maxim.
Ilex sideroxyloides (Sw.) Griseb.

Araceae

Anthurium andraeanum Linden (Hort)
Anthurium cordatum (L.) Schott
Anthurium martinicense Engler = *A. lanceolatum* (L.) Schott

Anthurium palmatum (L.) Schott = *Pothos pentaphylla* Sims
Anthurium willdenowii Kunth = *Pothos lanceolatus* Maycock
Colocasia esculenta (L.) Schott 1904, 1919, 2193, 2198
Dieffenbachia seguine (Jacq.) Schott
Monstera pertusa (L.) De Vriese = *M. adansonii* Schott
Montrichardia arborescens (L.) Schott
Philodendron dussii Engler = *P. oxycardium* Schott
Philodendron giganteum Schott
Philodendron lingulatum (L.) C. Koch
Xanthosoma brasiliense (Desf.) Engl.
Xanthosoma sagittifolium (L.) Schott = *X. atrovirens* C. Koch and Bouché 516
Xanthosoma undipes (C. Koch) C. Koch = *X. jacquinii* Schott

Araliaceae

Didymopanax attenuatum E. March
Didymopanax urbanianum E. March
Nothopanax cochleatum Merr., Miq.
Oreopanax capitatum (Jacq.) Dcne. and Planch.
Oreopanax dussii Kr. and Urb.
Polyscias balfouriana Bailey

Arecaceae = Palmae

Acrocomia aculeata (Jacq.) Lodd. = *A. sclerocarpa* Mart. 1272
Aiphanes minima (Gaertn.) Burr = *A. corrallina* (Mart.) Wendl.
Areca catechu L.
Coccothrinax barbadensis (Lodd. ex Mart.) Becc. = *C. dussiana* L. H. Bailey
Cocos nucifera L. 496
Geonome martinicensis Mart. = *G. pinnatifrons* Willd.
Phoenix dactylifera L.
Prestoea montana (R. Graham) Nichols = *Euterpe globosa* Gaertn.
Rhyticocos amara (Jacq.) Becc.
Roystonea oleracea (Jacq.) Cook = *Oreodoxa oleracea* Mart.
Roystonea regia (HBK) O. F. Cook

Asteraceae = Compositeae

Artemesia absinthium L.
Baccharis pedunculatum (Mill.) Cabera
Bidens cynapifolia HBK 745, 1127
Bidens pilosa L.
Borrichia arborescens (L.) DC. 1556
Clibadium erosum (Sw.) DC. 899
Conyza bonariensis (L.) Cronq.
Conyza canadensis (L.) Cronq. 678, 854, 1175
Elephantopus scaber L. 594, 829, 1258
Emilia coccinea (Sims.) Sweet 573, 635, 1133, 1134, 1515, 2166
Emilia fosbergii Nicholson
Emilia sonchifolia (L.) DC. 586, 1515-B, 2163, 2209
Eupatorium sp. 626, 788, 789
Eupatorium impetiolare Griseb. 983
Eupatorium magdalenae Stehlé = *Critonia macropoda* DC.
Eupatorium ovatum L. 1255, 1578

Lactuca sativa L.
Mikania hookeriana DC.
Mikania latifolia Sm. 818
Mikania micrantha HBK 514, 706, 1178, 1965, 2134, 2165
Mikania ovalis Griseb.
Neurolaena lobata (L.) R. Br. = *Conyza lobata* L.
Parthenium hysterophorus L. 406, 541, 668, 1362
Pectis elongata HBK
Pectis humifusa Sw. 1381, 1571, 2110
Pectis martinicensis Urb.
Pluchea symphytifolia (Mills) Gillis = *P. carolinensis* (Jacq.) G. Don
Rolandra fruticosa (L.) O. Kuntze 871
Senecio lucidus (Sw.) DC.
Senecioides cinerea (L.) O. Kuntze = *Vernonia cinerea* (L.) Lees 707, 1059, 1106
Solidago serotina Ait.
Spilanthes uliginosa Sw.
Syndrella nodiflora (L.) Gaertn. 702, 811, 1079
Tragopogon porrifolius L.
Vernonia arborescens Sw. = *V. icosantha* DC.
Wedelia calycina L. C. Rich. = *W. jacquini* L. C. Rich. = *W. parviflora* L. C. Rich. 908, 1387, 1447
Wedelia trilobata (L.) Hitchc. = *Wedelia carnosa* L. C. Rich. 426, 445, 511, 874, 1081, 2209

Batidaceae

Batis maritima L.

Begoniaceae

Begonia hirtella Link.
Begonia macrophylla Sesse & Moc. 1163
Begonia odorata Willd.

Bignoniaceae

Catalpa longissima (Jacq.) Sims
Crescentia cucurbitina L.
Crescentia cujete L. = *C. fasciculata* Miers.
Cydista aequinoctialis (L.) Miers. 1576
Macfadyena unguis-cati (L.) A. H. Gentry
Schlegelia axillaris Griseb.
Tabebuia leucoxyla DC.
Tabebuia pallida (Lindl.) Miers. = *T. pentaphylla* (L.) Hemsl. 651, 768, 884, 1077, 1192, 1385
Tecoma stans (L.) Juss. 653, 1202

Bixaceae

Bixa orellana L.

Bombacaceae

Ceiba pentandra (L.) Gaertn. 618
Ochroma pyramidale (Cav.) Urb.

Boraginaceae

Beurreria succulenta Jacq. = *Ehretia bourreria* L.
Cordia sp. 443, 571, 1189, 1256, 1489

Cordia alliodora (R. & P.) Oken = *C. gerascanthus* Jacq.
Cordia collococca L. = *C. glabra* Britt. and Wilson 1113
Cordia curassavica (Jacq.) Roem. and Schult. 1060
Cordia elliptica Sw. = *C. laevigata* auct. non Lam. 456
Cordia globosa (Jacq.) HBK
Cordia sulcata DC.
Heliotropium angiospermum Murr.
Heliotropium curassavicum L. 1997
Heliotropium ternatum Vahl
Mallotonia gnaphalodes (L.) Britton = *Tournefortia gnaphalodes* (L.) R. Br. = *Heliotropium gnaphalodes* L. 1560
Rochefortia cuneata Sw.
Tournefortia bicolor Sw.
Tournefortia maculata Jacq. 568
Tournefortia ternatum Vahl 932, 2114
Tournefortia volubilis L. 557, 1378

Brassicaceae = Cruciferae

Brassica oleracea L.
Brassica integrifolia (West) O. E. Schultz
Cakile lanceolata (Willd.) O. E. Schultz
Lepidium virginicum L.
Rorippa nasturtium-aquaticum (L.) Hayek = *Nasturtium officinale* R. Br.

Bromeliaceae

Unidentified bromeliad 1207
Aechmea lingulata (L.) Baker = *Wittmackia lingulata* (L.) Mez
Ananas comosus (L.) Merr. = *A. sativus* Schult.
Bromelia plumieri (E. Morr.) L. B. Smith = *Karatas plumieri* E. Morr.
Catopsis floribunda (Brongn.) L. B. Smith
Glomeropitcairnia penduliflora (Griseb.) Mez 460
Guzmania lingulata (L.) Mez
Guzmania plumierii (Griseb.) Mez 1156
Pitcairnia angustifolia Aiton = *P. gracilis* Mez 608-A, 1457-B
Pitcairnia bifrons (Lindl.) R. W. Read = *P. bracteata* Dryand
Pitcairnia spicata (Lam.) Mez
Tillandsia sp. 886
Tillandsia polystachya L.
Tillandsia utriculata L.

Burmanniaceae

Apteria aphylla (Nutt.) Barnh. = *A. hymenanthera* Miq.

Burseraceae

Bursera simaruba (L.) Sarg. = *Elaphrium simaruba* Rose 881
Dacryodes excelsa Vahl = *D. hexandra* Griseb. = *Amyris hexandra* Hamitt. 453
Protium atenuatum (Rose) Urb.

Cactaceae

Cephalocereus nobilis (Haw.) Britt. & Rose
Hylocereus trigonus (Haw.) Safford
Opuntia antillana Br. & R.
Opuntia dillenii (Ker-Gawl) Haw.

Opuntia triacantha (Willd.) Sweet
Opuntia tuna Mill.

Campanulaceae. *See* Lobeliaceae

Canellaceae

Canella winterana (L.) Gaertn.

Cannaceae

Canna indica L. = *C. edulis* Ker-Gawl

Capparidaceae

Capparis baducca Jacq. = *C. frondosa* L.
Capparis cynophallophora L. = *C. cynophallophora* L. var. *normalis* Eichl. 1541
Capparis flexuosa L. 1078, 1191, 1203
Capparis hasta Jacq. = *C. coccolobaefolia* Mart. = *C. ferruginea* auct., non L. 1474
Capparis indica (L.) Fawc. & Rendl. 1382, 1434, 1435-B
Capparis jamaicensis Jacq.
Cleome aculeata L. 1056, 1094, 1179, 1511
Cleome spinosa Jacq. 2136, 2240
Cleome viscosa L. 630, 1341, 1918

Caprifoliaceae

Lonicera japonica Thunb.
Sambucus simpsonii Rehd. 547

Caricaceae

Carica papaya L. 1107

Caryophyllaceae

Drymaria cordata (L.) Willd. 520, 734, 1139

Cassythaceae

Cassytha filiformis L.

Casuarinaceae

Casuarina equisetifolia L. 1414

Celastraceae

Cassine xylocarpa Vent. = *Elaeodendron xylocarpum* (Vent.) DC.
Crossopetalum rhacoma Crantz = *Rhacoma crossopetalon* L.
Gyminda latifolia (Sw.) Urb.
Maytenus elliptica (Lam.) Kr. and Urb.
Maytenus guianensis Klotzsch.
Schaefferia frutescens Jacq.

Chenopodiaceae

Beta vulgaris L.
Chenopodium ambrosioides L.
Salicornia ambigua Mich.

Chloranthaceae

Hedyosmum arborescens Sw. 2016

Clusiaceae = Guttiferae

Calophyllum calaba L. = *C. antillanum* Britton = *C. inophyllum* L. 846, 1248, 1257
Clusia alba L.
Clusia plukenetii Urb.
Clusia rosea Jacq. 721
Garcinia humilis (Vahl) Adams

Mammea americana L. 1896
Marila racemosa Sw.
Rheedia macrophylla Planch. and Triana
Symphonia globulifera L. f. = *Moronobea coccinea* Duss
Tovomita plumieri Griseb.

Combretaceae

Buchenavia capitata (Vahl) Eichl.
Bucida buceras L. = *Buceras bucida* Crantz
Conocarpus erecta L. 434, 1459, 1592
Laguncularia racemosa (L.) Gaertn. f. 1548, 1585
Terminalia catappa L.
Terminalia latifolia Sw. 1112

Commelinaceae

Commelina diffusa Burm. f. = *C. longicaulis* Jacq.
Commelina elegans HBK 665
Rhoeo spathacea (Sw.) Stearn
Tradescantia elongata G. F. W. Meyer = *T. pallida* (Rose) Hunt 621

Compositae. *See* Asteraceae

Convolvulaceae

Calonyction aculeatum (L.) House
Calonyction tuba (Schlet.) Colla 1565
Cuscuta americana L.
Evolvulus convolvuloides (Willd. ex Schultes) Stearn = *E. glaber* Spreng. 1562
Evolvulus nummularius (L.) L.
Evolvulus sericeus Sw. 1294, 1343, 2101
Ipomoea sp. 558, 564, 670, 1384, 1519, 1582
Ipomoea batatas L. 870
Ipomoea pes-caprea Roth. 423, 591
Ipomoea tiliacea (Willd.) Choisy 451, 783, 1165
Quamoclit hederifolia (L.) Choisy
Quamoclit pinnata (Desr.) Bojer

Crassulaceae

Kalanchoe pinnata (Lam.) Pers. = *Bryophyllum pinnatum* (Lam.) Oken

Cucurbitaceae

Cayaponia americana (Lam.) Cong.
Cayaponia racemosa (Sw.) Cong.
Cucumis anguria L.
Cucumis melo L.
Cucumis sativus L.
Cucurbita cucurbitina L.
Cucurbita moschata (Duch. ex Lam.) Duchesne ex Poir 2217
Cucurbita pepo L.
Lagenaria siceraria (Molina) Standl.
Luffa aegyptiaca Mill. = *L. cylindrica* (Laur.) Roem.
Momordica charantia L. 552, 602, 643, 663, 1186
Sechium edule (Jacq.) Sw.

Cunoniaceae

Weinmannia pinnata L.

Cyclanthaceae

Asplundia rigida (Aubl.) Harl. = *Carludovica plumieri* Kunth 468
Cyclanthus bipartitus Poit.

Cymodoceae

Halodule wrightii Aschers

Cyperaceae

Abilgaardia monostachya (L.) Vahl 1171, 1296, 1480, 1498-B, 2105
Bulbostylis barbata (Rottb.) C. B. Clarke
Bulbostylis capillaris (L.) Kunth = *B. antillana* (Britton) Fernald 925, 1051-A, 2152
Cyperus alopecuroides Rottb.
Cyperus digitatus Roxb. aff. 2157
Cyperus distictus L. f. 896
Cyperus elegans L. = *C. viscosus* Sw.
Cyperus laxus Lam. = *C. diffusus* Vahl 1946, 2182
Cyperus luzulae (L.) Retz. 449
Cyperus rotundus L. 490, 588, 656, 1055, 2149
Cyperus sphacelatus Rottb. 588, 1048, 1050, 1052
Cyperus surinamensis Rottb.
Dichromena nervosa Vahl = *D. ciliata* Vahl = *Rhynchospora nervosa* (Vahl) Boecklr. 907, 1307
Dichromena radicans Schlecht. & Cham.
Dichromena stellata (Lam.) Fournet
Eleocharis sp. 2005
Eleocharis flavescens (Poir.) Urban
Eleocharis geniculata (L.) Roem. & Schult. = *E. caribaea* (Rottb.) Blake = *E. diphylla* (Retz.) Vahl 724, 1053, 1498, 1958, 1959-B
Fimbristylis complanata (Retz.) Link.
Fimbristylis dichotoma (L.) Vahl = *F. annua* (All.) Roem. & Schult.
Fimbristylis ferruginea (L.) Vahl 415, 441, 1494
Fimbristylis spathacea Roth.
Fuirena umbellata Rottb. 2011-B, 2175
Kyllinga odorata Vahl
Kyllinga pumila Michx. = *Cyperus densicaepitosus* Mattf. & Kukenth.
Machaerina restioides (Sw.) Vahl = *Cladium restioides* (Sw.) Benth.
Mariscus alternifolia Vahl = *Cyperus subumbellatus* Kukenth. 590, 1049
Mariscus ligularis (L.) Urb. = *Cyperus ligularis* L.
Mariscus planifolius (L. C. Rich.) Urb. = *Cyperus planifolius* L. C. Rich. 424, 894
Mariscus tenuis (Sw.) C. B. Clarke = *Cyperus tenuis* Sw. 1054
Pycraeus polystachyos (Rottb.) P. Beauv. = *Cyperus polystachyos* Rottb. 2148
Rhynchospora corymbosa (L.) Britt.
Rhynchospora longiflora Presl. 980
Rhynchospora polyphylla Vahl 1981
Rhynchospora radicans (Schlecht. & Cham.) H. Pfeiffer = *Dichromena radicans* Schlecht. & Cham. 737
Scleria latifolia Sw.
Scleria lithosperma (L.) Sw. = *S. filiformis* Sw. 906, 928, 1473
Scleria pterota Presl. 861, 921

Scleria secans (L.) Urb. = *S. reflexa* HBK
Torulinium odoratum (L.) Hoop. = *Cyperus odoratus* L. 2156

Dichapatalaceae

Tapura latifolia Benth. = *T. antillana* Gleason

Dioscoreaceae

Dioscorea alata L. 808, 2186
Dioscorea cayennensis Lam. 867-A, 1924, 1933
Dioscorea esculenta (Lour.) Burkill
Dioscorea polygonoides H. & B. 716-B, 1955
Dioscorea trifida L. f. 2184

Ebanaceae

Diospyros revoluta Poir.

Elaeocarpaceae

Sloanea dussii Urb.
Sloanea massoni Sw. = *S. dentata* L. 1228
Sloanea truncata Urb. = *S. sinemarensis* Aubl. 2212

Ericaceae (contains Vacciniaceae)

Gaultheria swartzii R. A. Howard
Symphysia racemosa (Vahl) Stearn = *Hornemannia racemosa* Vahl

Erythroxylaceae

Erythroxylon sp. 1580
Erythroxylon brevipes DC. = *E. areolatum* L. 1397
Erythroxylon ovatum Cav. 1208
Erythroxylon squamatum Sw. 1392, 2030

Euphorbiaceae

Acalypha arevensis Poepp. & End. 492, 1147, 1234
Amanoa caribaea Kr. and Urb.
Argythamnia candicans Sw. = *Ditaxis polygama* (Jacq.) Wheeler 1380, 1575
Argythamnia lancifolia Muell.-Arg. 1434
Bernardia corensis (Jacq.) Muell.-Arg.
Croton astroites Dryand. in Ait. 888
Croton bixoides Geiseler 1391, 1448
Croton corylifolius Lam.
Croton flavens L. = *C. balsamifera* Jacq. 438, 885, 1214
Croton guildingii Griseb.
Croton hircinus Vent. = *C. populifolius* Lam. 1359, 1548
Croton niveus Jacq. 553, 918, 1235, 1440, 1546
Croton populifolius Sw. vel. aff. 1360, 1577
Drypetes alba Poit.
Drypetes glauca var. *macrocarpa* Kr. and Urb
Drypetes serrata Kr. and Urb.
Euphorbia sp. 1476
Euphorbia hirta L. 628, 1108, 2228
Euphorbia hypericifolia L. 617, 1310, 1435, 2153
Euphorbia hyssopifolia L. = *E. maculata* L. 1310, 1435, 1512, 1523
Euphorbia nutans Lag. 1129
Euphorbia pilulifera L. 548,725
Euphorbia serpens HBK 440

Euphorbia thymifolia L. = *E. maculate* var. *thymifolia* Griseb. 1445
Hippomane mancinella L.
Hura crepitans L.
Jatropha curcas L. 660
Jatropha gossypifolia L. 403, 1226
Jatropha multifida L.
Manihot esculenta Cranz. = *M. utillissima* Pohl.
Margaritaria nobilis L. f.
Phyllanthus acidus (L.) Skeels
Phyllanthus amarus Schum. & Thonn. 1509, 1514
Phyllanthus caroliniensis ssp. *guianensis* (Kl.) Webs. 1146
Phyllanthus niruri L. 842, 1144
Phyllanthus urinaria L. 486, 685, 804, 1971, 1979, 2202
Richeria grandis Vahl
Ricinus communis L. 601
Sapium caribaeum Urb.
Tragia volubilis L. 405, 1235, 1435

Flacourtiaceae

Casearia decandra Jacq. 500
Flacourtia indica (Burm. f.) Merr. = *F. ramontchi* L'Hérit.
Homalium racemosum Jacq.
Xylosma buxifolium A. Gray

Gentianaceae

Enicostema verticillatum (L.) Engl. = *Slevogetia occidentalis* Griseb. 1307
Leiphaimos aphylla (Jacq.) Gilg

Gesneriaceae

Alloplectus cristatus (L.) Mart.
Besleria coriacea Urb.
Besleria lutea L. var. *alpestris* Urb.
Besleria lutea L. var. *vulgaris* Urb.
Columnea scandens var. *typica* Stehlé
Gesneria ventricosa Sw. f. var. *montana* (Dcne.) Stehlé
Gesneria ventricosa Sw. f. var. *obovata* (Urb.) Stehlé
Isostoma longiflora (L.) Peterm. 719
Nautilocalyx melittifolius (L.) Wiehler var. *typicus* (Urb.) Fournet = *Episcia melittifolia* (L.) Mart.

Goodeniaceae

Scaevola plumieri Vahl

Hydrocharitaceae

Thalassia testudinum Banks and Solander

Heliconiaceae

Heliconia bihai (L.) L. 446
Heliconia caribaea Lam.

Iridaceae

Trimezia martinicensis (Jacq.) Kunth. 873, 1151, 1303

Juncaceae

Juncus bufonius L.

Lamiaceae = Labiatea

Hyptis sp. 1988
Hyptis atrorubens Poit. 864
Hyptis lanceolata Poir.
Hyptis pectinata (L.) Poir. 1148, 1413
Hyptis verticillata Jacq. 796
Ischaemum latifolium (Spreng.) Kunth
Leonotis nepetaefolia (L.) R. Br. 661, 1067, 1111
Marsypianthes chamaedrys (Vahl) Kuntze 663, 1271, 1940, 2005, 2174
Ocimum micranthum Willd. 542, 1324, 1529
Rosmarinus officinalis L.
Salvia serotina L. 880
Scutellaria havanensis Jacq. = *S. purpurascens* Griseb.
Thymus vulgaris L.

Lauraceae

Unidentified Lauraceae 45
Aniba bracteata (Nees) Mez
Aniba ramageana Mez
Beilschmiedia pendula (Sw.) Hemsl. = *Hufelandia pendula* Nees
Endlicheria sericea Nees
Nectandra antillana Meissn.
Nectandra coriacea (Sw.) Griseb.
Nectandra dominicana Mez
Nectandra membranacea (Sw.) Griseb.
Nectandra patens (Sw.) Griseb.
Ocotea cernua Mez
Ocotea eggersiana Mez
Ocotea floribunda (Sw.) Mez
Ocotea leucoxylon (Sw.) G. de la Maza
Ocotea martinicensis Mez
Ocotea rodioei Mez = *Nectandra rodioei* Hook.
Persea americana Mill. = *Persea gratissima* Geartn.
Persea drymifolia Mill.
Persea urbaniana Mez
Phoebe elongata (Vahl) Nees

Leguminosae (contains Fabaceae, Mimosaceae, and Caesalpiniaceae)

Abrus precatorius L. 1168, 1216
Acacia curassavica L.
Acacia farnesiana (L.) Willd. 649, 1348
Acacia guadalupensis DC.
Acacia macrantha H. & B.
Acacia muricata (L.) Willd.
Acacia nilotica (L.) Delile 1472
Acacia paniculata Willd.
Acacia parvifolia Duss
Acacia scleroxyla Tuss.
Acacia tamarindifolia (L.) Willd. 1213

Acacia tortuosa (L.) Willd. 1218, 1472
Adenanthera pavonia L.
Aeschynomene sensitiva Sw. 597, 634, 649, 1920
Agati grandiflora (L.) Desv. = *Sesban grandiflorus* Poir.
Albizia lebbeck (L.) Benth.
Alysicarpus vaginalis (L.) DC. 1326
Arachis hypogaea L.
Bauhinia excisa (Griseb.) Hemsl.
Bauhinia monandra Kurz.
Caesalpinia pulcherrima Sw. 601
Cajanus cajan (L.) Millsp. = *C. indicus* Spreng.
Calliandra latifolia Griseb.
Calliandra tergemina (L.) Benth. ex Griseb.
Calliandra tetragonia Benth. 1229, 1264
Calopogonium caeruleum (Benth.) Hemsl. 625, 1251
Calopogonium mucunoides Desv. 625-A, 739
Canavalia ensiformis L.
Canavalia gladiata (Jacq.) DC.
Canavalia maritima (Aubl.) Thou. 1572, 1575
Canavalia obtusifolia DC.
Cassia alata L. 1263
Cassia chamaecrista L.
Cassia emarginata L. 612, 1273, 1532
Cassia fistula L. 610, 614, 1228
Cassia obtusifolia L.
Cassia occidentalis L. 618-B
Cassia patellaria DC.
Centrosema plumieri (Turp. and Pers.) Benth.
Centrosema pubescens L. 1519
Centrosema virginianum (L.) Benth. 1317, 1357, 1438, 1517, 2117
Chaetocalyx scandens (L.) Urb. = *C. vincentinus* DC.
Clitoria rubiginosa Juss.
Clitoria ternatea L.
Cracca caribae (Jacq.) Benth.
Crotalaria alata (Buch.) Ham.
Crotalaria anagyroides HBK
Crotalaria breviflora DC.
Crotalaria grantiana Harvey
Crotalaria incana L. = *C. purpurascens* Lam. = *C. pubescens* Moench.
Crotalaria intermedia Klotszchy
Crotalaria juncea L.
Crotalaria lanceolata E. Mayer
Crotalaria mucronata Desv. = *C. striata* DC. 452, 546, 653, 1549, 2029
Crotalaria mysorensis Roth.
Crotalaria paulina Schrank
Crotalaria quinquifolia L.
Crotalaria retusa L. 541, 600, 639, 1085, 1304, 1455, 1496
Crotalaria sericea Burm. f.

Crotalaria spectabilis DC.
Crotalaria verrucosa L.
Crotalaria vitellina Ker.
Dalbergia monetaria L. f.
Delonix regia (Bojer) Raf. = *Poinciana regia* Bojer
Desmanthus depressus H. & B.
Desmanthus virgatus (L.) Willd. 1217, 2107
Desmodium axillare (Sw.) DC.
Desmodium canum (Gmel.) Schinz & Thell.
Desmodium cinereum Poepp.
Desmodium frutescens (Jacq.) Schindl. = *D. canum* (Gmel.) Schinz & Thell. 1295, 1346, 1441, 1525, 1956
Desmodium girans DC.
Desmodium gyroides DC.
Desmodium latifolium DC.
Desmodium molle (Vahl) DC.
Desmodium procumbens (Mill.) Hitch. 718
Desmodium scorpiurus (Sw.) Desv. 1441, 1956
Desmodium tortuosum (Sw.) DC. 1527, 2008
Desmodium triflorum DC. 1986
Desmodium uncinatum (Jacq.) DC.
Dichrostachys cinerea (L.) Wight and Arn. 2135
Dussia martinicensis Kr. and Urb.
Erythrina corallodendron L. 583
Erythrina variegata L.
Flemingia strobilifera R. Br.
Galactia dubia DC.
Galactia rubra (Jacq.) Urb.
Galactia striata (Jacq.) Urb. var. *caribae* Stehlé
Gliricidia sepium (Jacq.) Steud. 648, 1328
Glycine max (L.) Merr.
Geoffroea inermis W. Wright = *Andira inermis* HBK = *A. jamaicensis* Urb.
Guilandina crista (L.) Small
Haematoxylon campechianum L. = *H. brasiletto* Karst. 550, 552, 1417, 1433, 1479
Hymenaea courbaril L. 529
Indigofera disperma L.
Indigofera hirsuta L.
Indigofera spicata Forsk. = *I. hendecaphylla* Jacq.
Indigofera subulata L.
Indigofera suffruticosa Mill. = *I. anil* L. 1306
Indigofera sumatrana Gaertn.
Indigofera tinctoria L. 544, 677
Inga coruscans HBK
Inga ingoides L. 1269
Inga laurina (Sw.) Willd. 1266
Inga vera Willd.
Krameria ixina L.

Lablab niger Medic. = *Dolichos lablab* L.
Leucaena brachycarpa Urb.
Leucaena leucocephala (Lam.) DeWit = *L. glauca* (L.) Benth. 1454, 1500
Licaria salicifolia (Sw.) Kosterm. = *Acrodiclidium solicifolium* (Sw.) Griseb.
Lonchocarpus benthamianus Pittier = *L. violaceus* Benth. 1439
Lonchocarpus pentaphyllus (Poir.) DC. = *L. latifolius* (Willd.) HBK
Lonchocarpus sericeus (Poir.) HBK
Lourea vespertilionis Desv.
Machaerium lunatum (L. f.) Ducke = *Drepanocarpus lunatus* G. F. W. Meyer = *Pterocarpus lunatus* L. f.
Macroptilium lathyroides (L.) Urb. 674, 785
Medicago arborea L.
Medicago sativa L.
Melibomia supina (Sw.) Britton
Melilotus alba Desr.
Mimosa ceratonia L. 1495
Mimosa pigra L. 455, 631
Mimosa pudica L. 1265, 1322, 1349, 1521, 1548
Moringa oleifera Lam.
Mucuna pruriens (L.) DC. 1230
Mucuna sloanea Fawc. & Rendle
Mucuna urens (L.) DC.
Ormosia krugii Urb.
Ormosia monosperma (Sw.) Urb.
Pachyrrhizus erosus (L.) Urb. = *P. angulatus* L. C. Rich.
Parkinsonia aculeata L.
Phaseolus adenanthus G. F. W. Meyer
Phaseolus aureus Roxb.
Phaseolus calcaratus Roxb.
Phaseolus coccineus L.
Phaseolus lunatus L.
Phaseolus mungo L. = *Vigna mungo* (L.) Hepper
Phaseolus vulgaris L. 840, 1224
Piptadenia perigrina (L.) Benth.
Piscidia piscipula (L.) Sarg.
Pithecellobium jupunba (Willd.) Urb.
Pithecellobium tortum Mart. 1215, 1352
Pithecellobium unguis-cati (L.) Benth. 1450, 1468, 1587
Pterocarpus officinalis Jacq.
Pueraria phaseoloides (Roxb.) Benth. 596
Rhynchosia caribaea DC. 671
Rhynchosia minima (L.) DC.
Rhynchosia phaseoloides (Sw.) DC.
Rhynchosia reticulata (Sw.) DC.
Sesbania sesban (L.) Merrill = *S. aegyptiaca* Poir.
Sesbania emerus (Aubl.) Urb.
Sesbania sericea (Willd.) Link
Sophora tomentosa L.

Sorghum bicolor L. = *S. vulgare* L.
Stizolobium aterrimum Piper and Tracy
Stizolobium deeringianum Bert. = *Muncuna deeringiana* (Bort) Fournet
Stylosanthes hamata (L.) Taubert
Tamarindus indica L.
Tephrosia candida DC.
Tephrosia cinerea (L.) Pers. 647
Tephrosia noctiflora Bojer
Tephrosia senna HBK = *T. cathartica* (Sesse & Moc.) Urb.
Teramnus labialis (L. f.) Spreng.
Vicia faba L.
Vigna hosei (Craig.) Baker 683
Vigna luteola (Jacq.) Benth. 776, 788, 2025
Vigna marina (Burm. f.) Merr. 1985, 2036
Vigna mungo (L.) Hepper
Vigna repens (L.) Kuntze
Vigna sinensis (Torm.) Hassk. ex Engl.
Vigna unguiculata (L.) Walp. = *V. sesquipedalis* L.
Vigna vexillata (L.) A. Rich.
Voandzela subterranea DC.
Zornia gemella (Willd.) Vog. = *Z. diphylla* Pers.
Zornia microphylla Desr.
Zygia latifolia (L.) Fawc. and Rendl.

Lentibulariaceae

Utricularia jamesoniana Oliver

Liliaceae

Allium porrum L.
Allium sativum L.
Allium scheonoprasum L.
Aloe vera (L.) Burm. = *A. barbadensis* Mill. f.
Asparagus officinalis L.
Cordyline fructosa (L.) A. Chev.
Dracaena fragrans Ker-Gawl
Sansevieria cylindrica Boj. = *S. guineensis* (L.) Willd. = *S. metallica* Gerome and Labroy

Linaceae

Linum usitatissmum L.

Lobeliaceae = Campanulaceae

Centropogon berterianus (Spreng.) A. DC.
Isostoma longiflora (L.) Petern
Lobelia flavescens (Presl.) E. Wimm. = *L. stricta* Sw. 984
Lobelia guadelupensis Urb.

Loganiaceae

Spigelia anthelmia L. 563, 1239

Loranthaceae

Dendrophthora hexastichum (DC.) Griseb.
Dendrophthora martinicensis Kuijt.
Phoradendron sp. 2073

Phoradendron chrysocarpum Kr. and Urb. = *P. martinicense* Griseb.
Phoradendron dussii (Urb.) Trel.
Phoradendron trinervium (Lam.) Griseb. 917, 1236, 1319, 1453

Lythraceae

Cuphea carthagenesis (Jacq.) Macbr. = *C. balsamona* Cham. & Schlecht. 1152, 1309, 1330, 1945, 2003
Cuphea strigillosa subsp. *nitidens* Koehne 930

Magnoliaceae

Liriodendron tulipifera L.
Talauma dodecapetala (Lam.) Urb.

Malpighiaceae

Brachypteris ovata (Cav.) Small 1240
Bunchosia glandulifera (Jacq.) HBK
Bunchosia nitida (Jacq.) L. C. Rich.
Byrsonima coriacea var. *spicata* (Cav.) Ndz. = *B. spicata* (Cav.) DC.
Byrsonima coriaceae (Sw.) var. *picata* (Cav.) Nied.
Byrsonima lucida (Sw.) DC.
Byrsonima martinicensis Kr. and Urb.
Heteropterys platyptera DC.
Heteropterys purpurea (L.) HBK
Malpighia coccigera L. 1383
Malpighia glabra L.
Malpighia punicifolia L.
Malpighia urens L.
Stigmaphyllon cordifolium Ndz.
Stigmaphyllon convolvulifolium (Cav.) A. Juss.
Stigmaphyllon periplocifolium (Desf.) Juss. = *S. lingulatum* (Poir.) Small 911, 914, 933

Malvaceae

Gossypium arboreum L.
Gossypium barbadense L. 792, 1068, 1542
Gossypium herbaceum L.
Gossypium hirsutum L. var. *marie-galante* Hutch.
Herissantia crispa (L.) Brizicky
Hibiscus abelmoschus L.
Hibiscus elatus Sw.
Hibiscus esculentus L.
Hibiscus radiatus Cav. = *H. cannabinus* L. var. *unidens* Hochr.
Hibiscus sabdariffa L.
Hibiscus tiliaceus L.
Hibiscus tulipiflorus Hook.
Pavonia fruticosa (Mill.) Stehlé
Pavonia spicata Cav.
Pavonia spinifex (L.) Cav. = *P. scabra* (Presl.) Stehlé
Sida acuta Burm. 540-B
Sida ciliaris L.
Sida rhombifolia L. 540, 694
Sida spinosa L.

Thespesia populnea (L.) Soland. 770
Urena lobata L.
Urena trilobata Vell. 2071

Marantaceae

Calathea allouia (Aubl.) Lindl.
Calathea lutea (Aubl.) Meyer
Ischnosiphon arouma (Aubl.) Koern.
Maranta arundinacea L.

Marcgraviaceae

Marcgravia rectiflora Tr. and Plach. 1417

Melastomaceae

Charianthus corymbosus (L. C. Rich.) Cogn. = *Melastoma corymbosus* L. C. Rich. 955
Charianthus nodosus (Desr.) Triana
Charianthus purpureus var. *rugosus* Hodge 985
Clidemia sp. 1983, 2015
Clidemia hirta Cogn. 746, 1247, 2173
Clidemia umbrosa (Sw.) Cogn.
Conostegia calyptrata (Desv.) D. Don 949
Henriettea lateriflora (Vahl) Howard = *Henriettella lateriflora* Vahl
Miconia sp. 465, 939
Miconia coriacea (Sw.) DC.
Miconia furfuracea (Vahl) Griseb. = *Melastomata furfuraceum* Vahl
Miconia laevigata (L.) DC.
Miconia martinicensis Cogn. in DC.
Miconia mirabilis (Aubl.) L.O. Wms. = *Miconia guianensis* Cogn. 935, 937
Miconia prasina (Sw.) DC.
Miconia striata (Vahl) Cogn.
Miconia trichotoma (Desr.) DC.
Miconia virescens (Vahl) Triana 810, 1984
Miconia vulcanica Naud. 987
Nepsera aquatica (Aubl.) Naud.
Pterolepis sp. = *Arthrostemma* sp. 723
Pterolepis glomerata (Rottb.) Miq. 735, 1249, 1994, 2147
Tetrazygia discolor (L.) DC. var. *villosa* DC.
Tibouchina chamaecistus (Naud.) Cogn. 973

Meliaceae

Carapa guianensis Aubl.
Cedrela mexicana M. J. Roem. = *C. odorata* auct. mult. non L.
Guarea glabra Vahl
Guarea kunthianum A. Juss. var. *hahnii* DC.
Guarea perrottetiana A. Juss.
Guarea ramiflora Vent.
Khaya senegalensis (Desr.) A. Juss.
Swietenia macrophylla King
Swietenia mahagoni Jacq.
Trichilia simplicifolia Spreng.

Menispermaceae

Cissampelos pareira L.

Moraceae

Artocarpus altilis (Park.) Fosberg = *A. communis* J. R. and G. Forst.
Artocarpus heterophyllus Lam. = *A. integrifolia* L.
Cecropia peltata L.
Chlorophora tinctoria (L.) Gaudich. ex Benth.
Ficus insipida Willd. = *F. krugiara* Warb.
Ficus perforata L.
Ficus trigonata L. = *F. crassinervia* Desf. ex Willd. 843-A
Morus alba L.
Morus nigra L.

Moringaceae

Moringa oleifera Lam.

Musaceae (contains Sterlitziaceae)

Heliconia bihai L.
Heliconia caribaea Lam.
Musa paradisiaca L. = *M. sapientium* = *M. acuminata* Colla = *M. bulbisiana*
Musa textilis Née

Myoporaceae

Bontia daphnoides L. 1539

Myristicaceae

Virola surinamensis (Roland) Warh.

Myrsinaceae

Ardisia magdalenae Stehlé
Ardisia obovata Desv.
Ardisia solanacea Roxb. = *A. coriacea* Maycock
Rapanea coriacea (Sw.) Mez

Myrtaceae

Myrtaceae sp. 900
Calyptranthes elegans Kr. and Urb.
Calyptranthes fasciculata Berg. = *C. sericea* Griseb.
Eucalyptus sp.
Eugenia sp. 547, 910, 916
Eugenia axillaris (Sw.) Willd.
Eugenia brachystachus Berg.
Eugenia chrysobalanoides DC.
Eugenia ligustrina (Sw.) Willd.
Eugenia lambertiana DC.
Eugenia monticola (Sw.) DC.
Eugenia octopleura Kr. and Urb.
Eugenia pseudopsidium Jacq.
Eugenia tapacumensis Berg.
Eugenia uniflora L.
Myrcia spp.
Myrcia citrifolia (Aubl). Urb. = *M. coreacea* Vahl 934, 1377
Myrcia deflexa (Poir.) DC.

Myrcia fallax (Rich.) DC. var. *dussii* Kr. and Urb.
Myrcia leptoclada DC.
Myrcia martinicensis Kr. and Urb.
Myrcianthes fragrans (Sw.) McVaugh = *Eugenia fragrans* (Sw.) Willd. = *Anamomis fragrans* (Sw.) Griseb.
Myrciaria floribunda Berg = *Eugenia floribunda* West
Myrcia splendens (Sw.) DC.
Pimenta racemosa (Mill.) J. W. Moore 1268
Psidium guajava L. 566, 572, 1350, 1576
Syzygium jambos (L.) Alst. = *Eugenia jambos* L. = *Jambosa vulgaris* DC. 844

Nyctaginaceae

Boerhaavia diffusa L. 636
Boerhaavia erecta L. 598, 1237
Mirabilis jalapa L.
Pisonia sp. 442
Pisonia aculeata L. 1451
Pisonia fragrans Dumont-Cours
Pisonia helleri Standl.
Pisonia subcordata Sw. var. *typica* Heimrl.

Ochnaceae

Ouratea longifolia (DC.) Engl.
Sauvagesia erecta L. 865, 1284, 1949, 2167, 2178

Oenotheraceae = Onagraceae

Ludwigia erecta (L.) Hara = *Jussiaea erecta* L. 702, 2170
Ludwigia hissopifolia (G. Don) Exell = *Jussiaea linifolia* Vahl 690, 1929, 2003
Ludwigia octovalvis (Jacq.) Raven

Olacaceae

Heisteria coccinea Jacq.
Ximenia americana L.

Oleaceae

Chionanthus compactus Sw. = *Linociera caribaea* Knobl.
Forestiera rhamnifolia Griseb. var. *martinicensis* Kr. and Urb.

Orchidaceae

Elleanthus cephalotus Garay & Sweet = *E. capitatus* Rchb. f.
Elleanthus dussii Cogn.
Epidendrum sp.
Epidendrum ciliare L.
Epidendrum difforme Jacq.
Epidendrum dendrobioides Thunb. = *E. durum* Lindl. 977
Epidendrum dussii Stehlé
Epidendrum miserrimum Rchb. f.
Epidendrum nocturnum Jacq.
Epidendrum ramosum Jacq.
Epidendrum pallidiflorum Hook.
Habenaria monorrhiza (Sw.) Rchb. f. = *H. maculosa* Lindl.
Ionopsis satyrioides (Sw.) Rchb. f.
Isochilus linearis (Jacq.) R. Br.
Jacquiniella teretifolia (Sw.) Britt. and Wils. = *Epidendrum teretifolium* Sw.

Maxillaria coccinea (Jacq.) L. O. Wms. 459
Octomeria graminifolia (L.) R. Br.
Phyllogonium sp.
Pleurothallis sp. 462
Pleurothallis imraei Lindl.
Pleurothallis rucifolia (Jacq.) R. Br.
Polystachya concreta (Jacq.) Garay & Sweet
Ponthieva petiolata Lindl.
Spiranthes torta (Thunb.) Garay & Sweet = *S. apiculata* Lindl.
Stelis scabrida Lindl.
Vanilla planifolia G. Jackson in Andr.
Vanilla pompona Schiede

Oxalidaceae

Averrhoa bilimbi L.
Averrhoa carambola L.
Oxalis corniculata L.
Oxalis corymbosa DC. = *O. martiana* Zucc. 560, 1132, 2150
Oxalis frutescens L. 902, 1293

Pandanaceae

Pandanus hahnii Warb.
Pandanus maximus Martelli
Pandanus veitchii Veitch ex Dillière

Papaveraceae

Argemone alba L. 1149

Passifloraceae

Passiflora andersonii DC.
Passiflora foetida L. 1581
Passiflora laurifolia L. 1596
Passiflora rotundifolia L.
Passiflora rubra L.
Passiflora quadrangularis L.
Passiflora suberosa L.

Pedaliaceae

Sesamum indicum L. = *S. orientale* L.

Phytolaccaceae

Microtea debilis Sw. 1138
Petiveria alliacea L. 599
Rivina humilis L.

Piperaceae

Lepianthes peltata (L.) Raf. = *Piper pellatum* L. = *Pothomorphe peltata* Miq.
Peperomia spp.
Peperomia hernandifolia (Vahl) Dietr.
Peperomia hirtella Miq. in Hook. = *P. subbracteiflora* C. DC.
Peperomia nigropunctata Miq.
Peperomia questeliana Stehlé and Trel.
Peperomia stehleana Trel.
Peperomia trifolia (L.) A. Dietr.
Piper spp.

Piper aduncum L.
Piper aequale Vahl
Piper amalago L.
Piper dilatatum L. C. Rich.
Piper dussii C. DC.
Piper glabrescens (Miq.) C. DC.
Piper reticulatum L.
Pothomorphe peltata (L.) Miq.
Sarchorhachis incurva (Seib.) Trel.

Plantaginaceae

Plantago major L.

Poaceae = Graminae

Anatherum bicorne (L.) Beauv.
Andropogon bicornis L. 875
Andropogon glomeratus (Walt.) B. S. P.
Andropogon leucostachyus HBK = *A. pertusus* (L.) Willd. 1298
Andropogon sericatus Swallen 1300-B
Arundo donax L. 1076, 1941
Axonopus compressus (Sw.) Beauv. 537, 1526, 1550, 2041
Bacopa monnieri (L.) Pennell = *Bramia monnieri* L.
Bambusa arundinacea Retz.
Bambusa vulgaris Schrad. 495, 801, 820, 1486
Bothriochloa pertusa (L.) A. Camus
Bouteloua americana (L. f.) Scribn.
Brachiaria fasciculata (Sw.) St. Blake = *Panicum fasciculatum* Sw. 432
Brachiaria purpurascens (Ruddi) Henr. = *Panicum purpurascens* Raddi = *P. muticum* Forsk.
Brachiaria reptans (L.) Gard. & C. E. Hubb. = *Panicum reptans* L. 780
Brachiaria subquadripara (Trin.) Hitchc.
Cenchrus brownii Roem. & Schult. 2226
Cenchrus echinatus L. 646, 784, 1505
Chloris ciliata Sw. 1484
Chloris gayana Kunth
Chloris inflata Link. = *C. barbata* Sw. = *C. paraquarensis* Steud. 539, 776, 1073, 1484, 1599, 2232
Chloris radiata (L.) Sw.
Chloris virgata Sw. 1503
Coix lachryma-jobi L. 609
Cymbopogon citratus (DC. ex Ness) Stapf. = *C. nardus* (L.) Rendle = *Andropogon nardus* L. 1274
Cynodon dactylon (L.) Pers. 420, 669, 727
Dactyloctenium aegyptium (L.) Beauv. 619, 786, 1104, 1478, 1492
Dichanthium aristatum (Poir.) C. E. Hubb.
Dichanthium caricosum (L.) Stapf. = *Andropogon caricosus* L. 1536
Dichanthium ischaemum (L.) Roberty
Digitaria ciliaria (Ketz) Koel.
Digitaria decumbens Stent.
Digitaria horizontalis Willd. 644, 741, 1061

Digitaria longiflora (Ritz.) Pers. 1297
Digitaria sanguinalis (L.) Scop. 1502, 2162
Digitaria violascens Link. 822, 2052
Echinocloa colonum (L.) Link. 672, 1498, 2145, 2176
Echinocloa polystachya (HBK) Hitchc.
Eleusine indica (L.) Gaertn. 518, 620, 664, 764, 781, 856, 860, 1100, 1101, 1180, 2154
Eragrostis ciliaris (L.) R. Br. 629
Eragrostis pilosa (L.) Beauv.
Eragrostis tephrosanthos Schult.
Eriochloa polystachya HBK
Eriochloa punctata (L.) Desv.
Gynerium sagittatum (Aubl.) Beauv.
Heteropogon contortus (L.) Beauv.
Hyparrhenia hirta (L.) Stapf.
Hyparrhenia rufa (Nees) Stapf.
Ischaemum latifolium (Spreng.) Kunth 1952, 1992
Isachne disperma (Lam.) Doell
Isachne rigidifolia (Poir.) Urb.
Lasiacis divaricata (L.) Hitchc.
Lasiacis sloanei (Griseb.) Hitchc. 805
Lasiacis sorghoides (Desv.) Hitchc. & Chase 1185
Leptochloa filiformis (Lam.) Beauv.
Leptochloa virgata (L.) Beauv. 775
Melinis minutiflora Beauv.
Ochnanthus pallens (Sw.) Munro
Olyra latifolia L.
Oplismenus burmanni (Retz.) Beauv. 535
Oplismenus hirtellus (L.) Beauv. spp. *setarius* (Lam.) Mez
Oryza sativa L.
Panicum hirsutum Sw. = *P. purpurascens* Raddi
Panicum laxum Sw. = *P. boliviense* Hack. 1990, 2165
Panicum maximum Jacq. 1462, 2235, 2239
Panicum muticum Forsk.
Panicum pilosum Sw. 503, 2195
Panicum stoloniferum Poir.
Panicum trichoides Sw. 1174
Pappophorum pappiferum (Lam.) Kuntze 681, 879, 1058
Paspalum conjugatum Berg. 448, 731, 736, 821, 1957, 2152
Paspalum dilatatum Poir.
Paspalum distichum L.
Paspalum fimbriatum HBK 1507, 1528, 2231
Paspalum laxum Lam. 1103
Paspalum notatum Flugge
Paspalum paniculatum L. 744, 2237
Paspalum pleostachyum Doell = *P. secans* Hitchc. & Chase 2233
Paspalum saccharoides Nees

Paspalum vaginatum Sw. 821-B, 1169, 1325
Paspalum virgatum L. 412, 774, 1485, 2238
Pennisetum clandestinum Schumach.
Pennisetum purpureum Schumach.
Pennisetum setosum (Sw.) L. C. Rich.
Pharus glaber HBK
Pharus latifolius L. 1286
Phragmites australis (Cav.) Trin.
Rhynchelytrum repens (Willd.) C. E. Hubb. = *Tricolaena repens* (Willd.) Hitch.
Saccharum cayennense (Beauv.) Benth.
Saccharum officinarum L.
Saccharum sponteneum L.
Schizachyrium microstachyum (Desv.) Roseng. = *Andropogon condensatus* HBK 1259, 1300, 1376-B
Setaria geniculata (Lam.) Beauv. 733
Setaria glauca (L.) Beauv. = *S. lutescens* (Weigel) C. E. Hubb. 1993
Setaria indica Beauv.
Setaria paniculifera (Steud.) Fourn. 809
Setaria setosa (Sw.) Beauv. 2176
Sorghum bicolor (L.) Moench = *S. vulgare* (L.) Pers.
Sorghum halepense (L.) Pers.
Spartina patens (Ait.) Muhl. 722
Sporobolus indicus (L.) R. Br. = *S. poiretii* (R. & S.) Hitchc. 444, 682, 728, 1483
Sporobolus purpurascens (Sw.) Hamilt.
Sporobolus tenuissimus (Schrank) Kuntze
Sporobolus virginicus (L.) Kunth 762, 1555, 1558, 1574, 2106
Stenotaphrum secundatum (Walt.) Kuntze
Tragus berteronianus Schult.
Trichachne insularis (L.) Nees 622, 1075
Tripsacum laxum Nash
Vetiveria zizanioides (L.) Nash
Zea mays L. 869

Polygalaceae

Polygala paniculata L. 477, 1948, 2160, 2179, 2216
Securidaca diversifolia (L.) Blake

Polygonaceae

Antigonon leptopus Hook. & Arn.
Coccoloba ascendens Duss
Coccoloba pubescens L. = *C. grandifolia* Jacq. = *C. latifolia* Lam. 897
Coccoloba swartzii Meissn. = *C. diversifolia* Griseb. 1402
Coccoloba uvifera L. 416, 891-B, 903, 1569
Coccoloba venosa L. = *C. nivea* Jacq. 1563
Rheum rhaponticum L.
Rumex acetosa L.

Portulacaceae

Portulaca halimoides L.
Portulaca oleracea L. 508, 1071, 1238, 1570

Potamogetonaceae

Potamogeton nodosus Poir.

Punicaceae

Punica granatum L.

Ranunculaceae

Clematis dioica L.

Rhamnaceae

Colubrina arborescens (Mill.) Sarg.
Colubrina elliptica (Sw.) Brizicky & Stern = *C. reclinata* (L'Hérit.) Brongn.
Colubrina arborescens (Mill.) Sarg.
Colubrina elliptica (Sw.) Brizicky & Stern = *C. reclinata* (L'Hérit.) Brongn
Colubrina ferrugimora Brongn. = *C. arborescens* (Mill.) Sarg.
Gouania lupuloides (L.) Urb. 1110
Krugiodendron ferreum (Vahl) Urb.
Reynosia unoinata Urb.
Ziziphus mauritiana Lam. 1342

Rhizophoraceae

Cassipourea guianensis Aubl. = *C. elliptica* Poir.
Rhizophora sp.
Rhizophora mangle L. 1540, 1595

Rosaceae

Chrysobalanus cuspidatus Griseb.
Chrysobalanus icaco L. var. *genuina* Stehlé 1399, 1436
Chrysobalanus icaco L. var. *pellocarpa* (G. F. W. Meyer) DC.
Dalbergia monetaria L. f.
Eriobotrya japonica Lindl.
Hirtella triandra Sw.
Licania ternatensis Hook. f.
Prunus acutissima Urb.
Prunus dussii Kr. and Urb.

Rubiaceae

Rubiaceae spp. 509, 2019, 2050, 2197
Borreria laevis (Lam.) Griseb. 407, 545, 837, 1294-A, 1444, 1534, 1906
Borreria latifolia (Aubl.) Schum.
Borreria ocymoides (Burm. f.) DC. = *B. parviflora* Meyer
Borreria verticillata (L.) Meyer 927
Cephaëlis axillaris Sw. 982
Cephaëlis mucosa Sw.
Cephaëlis swartzii DC.
Chiococca alba (L.) Hitchc.
Chimarrhis cymosa Jacq. 1980
Coffea arabica L.
Coffea liberica Hiern. 851
Erithalis acuminata Kr. and Urb.
Erithalis fruticosa L. 1566
Exostemma caribaeum (Jacq.) R. and S.
Exostemma sanctae-lucia (Kentish) J. Britton
Faramea occidentalis (L.) Rich.

Genipa americana L.
Gonzalagunia hirsuta (Jacq.) Schum. 1950
Guettarda parviflora Vahl
Guettarda scabra (L.) Lam. 926
Hamelia patens Jacq. = *H. erecta* Jacq. 1408
Hemidiodia ocymifolia (Willd.) K. Schum. 534, 817, 843, 1995, 2203
Hillia parasitica Jacq. 461
Ixora ferrea (Jacq.) Benth.
Manettia domincensis Wernham
Morinda citrifolia L. 889
Oldenlandia herbacea (L.) DC. 687
Palicourea crocea (Sw.) R. & S.
Palicourea riparia Benth. in Hook.
Psychotria sp. 1429, 1999
Psychotria antillana Howard = *P. floribunda* H.B.K.
Psychotria berteriana DC.
Psychotria guadelupensis (DC.) Howard 971
Psychotria nervosa Sw.
Psychotria nitida Willd.
Psychotria tenuifolia Sw.
Randia sp. 1338, 1395
Randia aculeata L. = *Randia mitis* L. 901, 1339, 1340, 1453, 2000, 2033
Rondeletia martinicensis Kr. & Urb.
Rondeletia parviflora Poir.
Rondeletia stereocarpa Griseb.
Rudgea citrifolia (Sw.) Shum. = *R. caribaea* Benth.
Schradera vahlii Steyerm.
Spermacoce tenuior L. 1142, 1928, 2143
Strumpfia maritima Jacq. 1551

Ruppiaceae

Ruppia maritima L.

Rutaceae

Amyris elemifera L.
Amyris maritima Jacq.
Citrus aurantifolia (Christm.) Swingle
Citrus aurantium L.
Citrus grandis (L.) Osbeck = *C. decumana* L. = *C. maxima* (Burm.) Merrill
Citrus limon (L.) Burm. f.
Citrus medica L.
Citrus reticulata Blanco = *C. nobilis* Andrews
Citrus sinensis (L.) Osb.
Pilocarpus racemosus Vahl
Triphasia trifolia (Burm. f.) P. Wilson
Zanthoxylum sp. 1329, 1361
Zanthoxylum caribaeum Lam.
Zanthoxylum flavum Vahl
Zanthoxylum martinicense (Lam.) DC.
Zanthoxylum microcarpum Griseb.

Zanthoxylum monophyllum (Lam.) P. Wils. 1335, 1355, 1394
Zanthoxylum punctatum Vahl = *Z. trifoliatum* Wright = *Fagara trifoliata* Sw. 919
Zanthoxylum spinifex (Jacq.) DC. = *Fagara spinifex* Jacq.
Zanthoxylum vulgaris L. 494

Sabiaceae

Meliosma herbertii Rolfe var. *martinicensis* Kr. and Urb.
Meliosma pardonii Kr. and Urb.

Sapindaceae

Allophylus racemosus Sw.
Cardiospermum microcarpum HBK 501, 1358, 4444-B
Dodonaea viscosa Jacq. 413, 417, 1579
Litchi chinensis Sonner.
Melicoccus bijugatus Jacq. = *Melicocca bijuga* L.
Paullinia cururu L. 1588
Paullinia pinnata L. 1166
Paullinia respertilio Sw.
Sapindus saponaria L.

Sapotaceae

Achras sapota L. = *Sapota achras* Mill. = *Nispero achras* Aubreville
Chrysophyllum argenteum Jacq. = *C. glabrum* Jacq.
Chrysophyllum cainito L.
Chrysophyllum coeruleum Jacq.
Manilkara bidentata A. Chev. = *Mimusops riedleana* Pierre = *M. martinicensis* Pierre
Mastichodendron foetidissimum (Jacq.) Cronq. = *Bumelia mastichodendron* R. & S. = *Sideroxylon foetidissimum* Jacq.
Micropholis sp. 848
Pouteria sp.
Pouteria chrysophylloides (Pierre) Stehlé
Pouteria dussiana (Pierre) Stehlé = *Lucuma dussiana* Pierre
Pouteria hahniana (Pierre) Stehlé = *Ocythece hahniana* Pierre
Pouteria martinicensis (Pierre) Stehlé = *Lucuma martinicensis* Pierre
Pouteria semicarpifolia Pierre

Saxifragaceae

Hydrangea macrophylla Hort. 2013

Scrophulariaceae

Capraria biflora L. 1389, 1452, 1901
Lindernia rotundifolia (L.) Alst.
Scoparia dulcis L.
Stemodia verticillata (Mill.) Hassl.

Simarubaceae

Castelaria nicholsoni (Hook.) Small
Picramnia pentandra Sw. = *Tariri pentandra* Baill.
Picrasma antillana (Eggers) Urb.
Simaruba amara Aubl.

Smilacaceae

Smilax cariacea Spreng.
Smilax guianensis Vitm.

Solanaceae

Capsicum annuum L.
Capsicum frutescens L.
Capsicum sinense Jacq.
Cestrum alternifolium (Jacq.) O. E. Schultz var. *tipicum* Stehlé
Cestrum laurifolium L'Hérit.
Cestrum megalophyllum Dunal
Cestrum nocturnum L. var. *typicum* O. E. Schultz = *Chiococca nocturna* Jacq.
Datura metel L.
Datura stramonium L.
Datura suaveolens L.
Dunalia arborescens (L.) Sleumer
Lycopersicum esculentum Mill.
Nicotiana rustica L.
Nicotiana tabacum L. 1070
Solandra grandiflora Sw. = *Swartzia grandiflora* Gmelin
Solanum ficifolium Ortega 747, 1109
Solanum ignaeum L. 1193
Solanum lanceifolium Jacq. var. *volubile* (Sw.) Sendtn. 524, 890
Solanum melongena L.
Solanum nigrum L. 493, 1125
Solanum racemosum L. 401, 882, 1188
Solanum torvum Sw.
Solanum triste Jacq. 488

Staphyleaceae

Turpinia occidentalis (Sw.) G. Don

Sterculiaceae

Cola acuminata (P. Beauv.) Schott & Endl.
Guazuma ulmifolia Lam.
Melochia polystachya HBK 1548
Sterculia caribaea R. Br.
Theobroma cacao L. 502, 525, 536, 826
Waltheria indica L. = *W. americana* L. 1232, 1305

Styracaceae

Styrax glaber Sw.

Surianaceae

Suriana maritima L. 1552

Symplocaceae

Symplocos martinicensis Jacq.

Theophrastaceae

Jacquinia arborea Vahl

Thymeleaceae

Daphnopsis americana (Mill.) J. Johnst. spp. *caribaea* (Griseb.) Nevling
Lagetta lagetto (W. Wright) Nash = *L. lintearia* Lam.

Tiliaceae

Corchorus capsularis L.
Corchorus olitorius L.
Corchorus siliquosus L.

Sloanea dussii Urb.
Sloanea massoni (Sw.) = *S. plumierii* Aubl. = *S. dentata* L.
Sloanea truncata Urb. = *S. sinemariensis* Sw.
Triumfetta grandiflora Vahl
Triumfetta lappula L.
Triumfetta semitriloba Jacq.

Ulmaceae

Celtis iguanaea (Jacq.) Sarg.
Trema lamarckiana (Roem. & Schult.) Bl.

Umbelliferae = Ammiaceae

Urticaceae

Boehmeria nivea (L.) Gaud.
Laportea aestuans (L.) Chew.
Phenax sp. 2171
Phenax sonneratii (Poir.) Wedd. 487, 726-B, 827, 1126, 2190
Pilea duchassaingii Urb.
Pilea inequalis (Juss.) Wedd.
Pilea microphylla (L.) Liebm. 512
Pilea nummulariaefolia (Sw.) Wedd.
Pilea parietaria (L.) Bl. 1162

Verbenaceae

Aegiphila martinicensis Jacq. 1080
Avicennia germinans L. = *A. nitida* Jacq.
Citharexylum fruticosum L. 524, 849, 1895, 2035
Citharexylum spinosum L.
Clerodendrom philippinum Schau. = *C. fragans* Vent. var. *pleniflorum* Schau. 1066
Cornutia pyramidata L.
Holmskioldia sanguinea Ketz
Lantana sp. 1198, 1336
Lantana aculeata (L.) Moldenke 425
Lantana camara L. 567, 569, 598, 1064, 1136
Lantana involucrata L. 405, 429, 587, 913, 2116
Lippia nodiflora (L.) L. C. Rich. in Michx. 1550, 2108
Petitia domingensis Jacq. 857
Petrea kohautiana Presl.
Petrea volubilis L.
Priva lappulacea (L.) Pers. 841, 1497
Stachytarpheta sp. 2043
Stachytarpheta australis var. *neocaledonica* Moldenke 555
Stachytarpheta cayennensis Vahl
Stachytarpheta jamaicensis (L.) Vahl 475, 1439, 1996
Tektona grandis L. f. = *Tectonia grandis* L.
Vitex divaricata Sw.

Violaceae

Viola stipularis Sw. 640, 1157

Vitaceae

Cissus sicyoides L.

Zannichelliaceae

Cymodocea filliformis (Kütz) Correll

Zingiberaceae

Aframomum granum-paradisi (Hook.) K. Schum.
Curcuma amada Roxb.
Curcuma domestica Valeton = *C. longa* Koenig.
Hedychium coronarium Koenig.
Zingiber officinale Rosc.

Zygophyllaceae

Guaiacum officinale L.
Guaiacum sanctum L.

Appendix B. Vernacular Names and Their Scientific Equivalents

The plant names listed here were collected in 1962, 1963, and 1975, in conversations or discussions with the rural inhabitants and forest guards of Martinique and with Henri and Madelaine Stehlé. I gratefully acknowledge the help of the Stehlés and Lucienne Maurice in sorting out the common names used. Some of the names reported here may no longer be in use, since vernacular naming is a dynamic process and since the publication of Fournet's flora (1978) has codified usage. For the latter reason, no common names were collected in 1985. The scientific names have been rectified insofar as possible with synonymy.

abaca	*Musa textilis* Née
abajou rouge	*Cedrela mexicana* M. J. Roem = *C. odorata* L.
abricotier bâtard	*Byrsonima laevigata* (L. C. Rich) DC.
abricotier-pays	*Mammea americana* L.
absinthe	*Artemisia absinthium* L.
absinthe bord-de-mer	*Heliotropium curassavicum* L.
acacia blanc	*Acacia paniculata* Willd.
acacia blanc	*Desmanthus depressus* H. & B.
acacia courant	*Desmanthus virgatus* (L.) Willd.
acacia de Saint-Domingue	*Dichrostachys cinerea* (L.) Wight and Arn.
acacia fougère	*Pithecellobium jupunda* (Willd.) Urb.
acacia piquant	*Acacia macrantha* H. & B.
acajou	*Cedrela mexicana* M. J. Roem.
acajou	*Swietenia* spp.
acajou amer	*Cedrela mexicana* M. J. Roem.
acajou à meubles	*Cedrela mexicana* M. J. Roem.
acajou blanc	*Simaruba amara* Aubl.
acajou de Saint-Domingue	*Swietenia mahagoni* Jacq.
acajou rouge	*Carapa guianensis* Aubl.
acajou rouge	*Cedrela mexicana* M. J. Roem.
acajou senti	*Cedrela mexicana* M. J. Roem.
acomas	*Homalium racemosa* Jacq.
acomat	*Chrysophyllum argenteum* Jacq.
acomat	*Homalium racemosum* Jacq.
acomat	*Mastichodendron foetidissimum* (Jacq.) Cronq.
acomat bâtard	*Margaritaria nobilis* L. f.
acomat côtelette	*Beurreria succulenta* Jacq.
acomat franc	*Mastichodendron foetidissimum* (Jacq.) Cronq.

acomat grandes-feuilles	*Sloanea massoni* Sw.
acomat-hêtre	*Homalium racemosum* Jacq.
agave	*Agave dussiana* Trel.
agave	*Furcraea tuberosa* (Mill.) Ait. f.
aje	*Ipomoea batatas* L.
allamanda	*Allamanda cathartica* L.
amandier	*Terminalia catappa* L.
amarante-bord-de-mer	*Philoxerus vermicularis* (L.) Nutt.
amourette	*Acacia guadelupensis* DC.
amourette	*Oplonia microphylla* (Lam.) Stearn.
amourette	*Oplonia spinosa* (Jacq.) Raf.
ananas	*Musa paradisiaca* L.
ananas bois	*Guzmania lingulata* (L.) Mez
ananas bois	*Guzmania megastachya* Mez
ananas-grand-bois	*Glomeropitcairnia penduliflora* (Griseb.) Mez
ananas-grand bois	*Vriesea guadelupensis* (Bak.) Mez
ananas rouge	*Pitcairnia angustifolia* Soland. in Ait.
ananas sauvage	*Aechmea serrata* (L.) Mez
ananas sauvage	*Tillandsia utriculata* L.
anatto	*Bixa orellana* L.
angelin	*Geoffroea inermis* W. Wright
angelin-bois	*Ormosia monosperma* (Sw.) Urb.
angelin-olive	*Geoffroea inermis* W. Wright
angelin-palmiste	*Geoffroea inermis* W. Wright
angelin-palmiste	*Ormosia monosperma* (Sw.) Urb.
anthurium	*Anthurium willdenowii* Kunth
arachide	*Arachis hypogaea* L.
aralie	*Didymopanax urbanianum* E. March
aralie blanc	*Oreopanax capitatum* (Jacq.) Dcne. & Planch.
aralie des hauts	*Oreopanax capitatum* (Jacq.) Dcne. & Planch.
aralie-grandes-feuilles	*Clusia plukenetii* Urb.
aralie montagne	*Didymopanax attenuatum* E. Marsh
aralie-petities-feuilles	*Ficus perforata* L.
aralie rose	*Clusia plukenetii* Urb.
arbre à lait	*Rauvolfia biauriculata* J. Muell.
arbre à pain	*Artocarpus altilis* (Park.) Fosberg
arouma	*Ischnosiphon arouma* (Aubl.) Koern.
asperge	*Asparagus officinalis* L.
attrape-sot	*Xylosma buxifolium* A. Gray
ausu	*Pimenta racemosa* (Mill.) J. W. Moore
avocatier	*Persea americana* Mill.
balais onze heures	*Sida rhombifolia* L.
balais savane	*Corchorus siliquosus* L.
balata	*Manilkara bidentata* A. Chev.
balata rouge	*Pouteria hahniana* (Pierre) Stehlé
balisier	*Heliconia bihai* L.
balisier	*Heliconia caribaea* Lam.
bambou	*Bambusa vulgaris* Schrad.

banane	*Musa paradisiaca* L.
barac	*Pouteria hahniana* (Pierre) Stehlé
barbade	*Maranta* spp.
barbasco	*Canella winterana* (L.) Gaertn.
bâtard	*Forestiera rhamnifolia* Griseb.
bâtard	*Myrcianthes fragrans* (Sw.) McVaugh
bâtard de fougère	*Pithecellobium jupunba* (Willd.) Urb.
baume blanc	*Croton bixoides* Geiseler
baume grandes-feuilles	*Croton corylifolius* Lam.
bois à enivrer	*Piscidia piscipula* (L.) Sarg.
bois agouti	*Vitex divaricata* Sw.
bois à graines rouges	*Erythroxylon squamatum* Vahl
bois anoli	*Marila racemosa* Sw.
bois à rames	*Pouteria chrysophylloides* (Pierre) Stehlé
bois arcoquois	*Buchenavia capitata* (Vahl) Eichl.
bois baguette	*Myrcia fallax* (Rich.) DC. var. *dussii* Kr. and Urb.
bois balata	*Manilkara bidentata* A. Chev.
bois balata	*Pouteria hahniana* (Pierre) Stehlé
bois bander	*Richeria grandis* Vahl
bois blanc	*Simaruba amara* Aubl.
bois cabrit	*Aegiphila martinicensis* Jacq.
bois cachiman	*Marila racemosa* Sw.
bois cachiman	*Talauma dodecapetala* (Lam.) Urb.
bois caco	*Guarea perrottetiana* A. Juss.
bois café	*Drypetes glauca* var. *macrocarpa* Kr. and Urb.
bois café bois	*Bunchosia glandulifera* (Jacq.) HBK
bois canon	*Cecropia peltata* L.
bois carré	*Citharexylum fruticosum* L.
bois cassant	*Besleria lutea* L. forma *vulgaris* Urb.
bois cassant	*Rudgea caribaea* Benth.
bois casse rose	*Marila racemosa* Sw.
bois chandelle	*Amyris elemifera* L.
bois chandelle	*Zanthoxylum spinifex* (Jacq.) DC.
bois chandelle noir	*Erithalis fruticosa* L.
bois-charbon	*Byrsonima coriacea* (SW.) DC var. *spicata* (Cav.) Ndz.
bois-Chypre	*Cordia alliodora* (R. & P.) Oken
bois-Chypre	*Phoebe elongata* Vahl
bois ciceron	*Pithecellobium jupunba* (L.) Urb.
bois cierge	*Piper aequale* Vahl
bois cierge	*Piper glabrescens* (Miq.) C. DC.
bois citron	*Ilex sideroxyloides* (Sw.) Griseb.
bois citron	*Maytenus elliptica* Kr. and Urb.
bois-cochon	*Dacryodes excelsa* Vahl
bois colique	*Aniba bracteata* (Nees) Mez
bois côte	*Tapura latifolia* Benth.
bois côtelette	*Miconia laevigata* (L.) DC.

bois côtelette	*Tapura latifolia* Benth.
bois côtelette montagne	*Miconia coriacea* DC.
bois côtelette rouge	*Miconia trichotoma* (Desr.) DC.
bois côte noir	*Tapura latifolia* Benth.
bois créole	*Calyptranthes elegans* Kr. and Urb.
bois créole	*Eugenia* spp.
bois créole	*Myrcia berberis* DC.
bois créole	*Myrcianthes fragrans* (Sw.) McVaugh
bois d'ail	*Cassipourea guianensis* Aubl.
bois d'amadebois cachiman	*Marila racemosa* Sw.
bois d'ande	*Richeria grandis* Vahl
bois d'ébène	*Rochefortia cuneata* Sw.
bois de fer	*Aegiphila martinicensis* Jacq.
bois de fer	*Krugiodendron ferreum* (Vahl.) Urb.
bois de fer	*Licania ternatensis* Hook. f.
bois de fer bâtard	*Trichilia simplicifolia* Spreng.
bois de fer blanc	*Dussia martinicensis* Kr. and Urb.
bois de fer blanc	*Guettarda parviflora* Vahl
bois de fer rouge	*Ixora ferrea* (Jacq.) Benth.
bois de fer rouge	*Ormosia monosperma* (Sw.) Urb.
bois de l'an	*Oxandra laurifolia* (Sw.) A. Rich
bois de l'anglais	*Guatteria caribaea* Urb.
bois de l'eau	*Hedyosmum arborescens* Sw.
bois de masse	*Licania ternatensis* Hook. f.
bois de masse	*Tapura latifolia* Benth.
bois dentelle	*Lagetta lagetto* (W. Wright) Nash
bois de rose	*Cordia alliodora* (R. & P.) Oken
bois de rose	*Guarea kunthianum* (A. Juss) var. *hahnii* DC.
bois de rose	*Marila racemosa* Sw.
bois de sept-ans	*Meliosma herbertii* Rolfe var. *martinicensis* Kr. and Urb.
bois de soie	*Sapium caribaeum* Urb.
bois diable	*Margaritaria nobilis* L. f.
bois d'Inde	*Pimenta racemosa* (Mill.) J. W. Moore
bois d'Inde bâtard	*Myrcianthes fragrans* (Sw.) McVaugh
bois d'orme	*Guazuma ulmifolia* Lam.
bois doux	*Nectandra patens* (Sw.) Griseb.
bois doux	*Phoebe elongata* (Vahl) Nees
bois doux blanc	*Endlicheria sericea* Nees
bois doux-couronne	*Ocotea leucoxylon* (Sw.) G. de la Maza
boix doux Isabelle	*Ocotea cernua* Mez
bois doux jaune	*Ocotea leucoxylon* G. de la Maza
bois doux muscade	*Nectandra dominicana* Mez
bois doux noir	*Nectandra coriacea* (Sw.) Griseb.
bois doux pimenté	*Ocotea leucoxylon* Mez
bois du diable	*Hura crepitans* L.
bois envirant	*Clibadium erosum* (Sw.) DC.
bois flambeau	*Amyris elemifera* L.

bois flambeau	*Exostemma caribaeum* (Jacq.) R. & S.
bois flambeau	*Zanthoxylum* spp.
bois flambeau montagne	*Erithalis acuminata* Kr. and Urb.
bois fleurs jaunes	*Tecoma stans* (L.) Juss.
bois flot	*Hibiscus tiliaceus* L.
bois flot	*Ochroma pyramidale* (Cav.) Urb.
bois flot des hauts	*Hibiscus tulipiflorus* Hook
bois fragile	*Hedyosmum arborescens* Sw.
bois gamelle	*Dussia martinicensis* Kr. and Urb.
bois gli-gli	*Buchenavia capitata* (Vahl) Eichl.
bois gli-gli	*Bucida buceras* L.
bois gli-gli	*Chrysophyllum argenteum* Jacq.
bois gli-gli	*Guettarda parviflora* Vahl
bois glu	*Chrysophyllum argenteum* Jacq.
bois (à) graines rouges	*Erythroxylon squamatum* Vahl
bois grillé	*Eugenia tapacumensis* Berg
bois gris	*Licania ternatensis* Hook. f.
bois grosse peau	*Nectandra patens* (Sw.) Griseb.
bois guépois	*Myrcia leptoclada* DC.
bois Guillaume	*Baccharis pedunculata* (Mill.) Cabrera
bois jaune	*Aniba bracteata* (Nees) Mez
bois jaune	*Casearia decandra* Jacq.
bois jaune	*Ixora ferrea* (Jacq.) Benth.
bois jaune	*Zanthoxylum flavum* Vahl
bois kaki	*Chrysophyllum argenteum* Jacq.
bois la glu	*Sapium caribaeum* Urb.
bois lait	*Tabernaemontana citrifolia* L.
bois lance	*Randia aculeata* L.
bois laurier	Lauraceae
bois-l'encens	*Protium attenuatum* (Rose) Urb.
bois lépuni	*Amyris maritima* Jacq.
bois lézard	*Vitex divaricata* Sw.
bois liége	*Ochroma pyramidale* (Cav.) Urb.
bois mabi	*Colubrina elliptica* (Sw.) Brizicky & Stern
bois mabouya	*Capparis baducca* L.
bois-madame	*Guettarda scabra* (L.) Lam.
bois madame	*Picramnia pentandra* Sw.
bois marbré	*Richeria grandis* Vahl
bois-marguerite	*Cephaëlis axillaris* Sw.
bois-marguerite	*Cephaëlis swartzii* DC.
bois merle-jaune	*Rudgea citrifolia* (Sw.) Schum.
bois mille-branches	*Margaritaria nobilis* L.
bois montagnes	*Picramnia pentandra* Sw.
bois-mousseux	*Sapindus saponaria* L.
bois négresse	*Nectandra coriacea* (Sw.) Griseb.
bois noir	*Capparis cynophallophora* L. var. *normalis* Eichl.
bois noyau	*Prunus dussii* Kr. and Urb.
bois-noyer	*Zanthoxylum flavum* Vahl

bois noyer	*Picrasma antillana* (Eggers) Urb.
bois-olive	*Geoffroea inermis* W. Wright
bois olivier	*Buchenavia capitata* (Vahl) Eichl.
bois-pain	*Endlicheria sericea* Nees
bois pain	*Pouteria dussiana* (Pierre) Stehlé
bois-pays	*Croton corylifolius* Lam.
bois-petit-chique	*Ardisia obovata* Desv.
bois-petit Jean	*Ocotea cernua* Mez
bois-pilori	*Turpinia occidentalis* (Sw.) G. Don
bois-pin	*Talauma dodecapetala* (Lam.) Urb.
bois-piquet	*Ixora ferrea* (Jacq.) Benth.
bois-piquette	*Erythroxylon squamatum* Sw.
bois-pistolet	*Guarea* spp.
bois plié	*Eugenia pseudopsidium* Jacq.
bois-poisson	*Cestrum laurifolium* L' Hérit
bois-poisson	*Picramnia pentandra* Sw.
bois pripri	*Ochroma pyramidale* (Cav.) Urb.
bois-puant	*Capparis indica* (L.) Fawc. & Rendl.
bois puant	*Ximenia americana* L.
bois rada	*Byrsonima martinicensis* Kr. and Urb.
bois-raisin	*Ardisia magdalenae* Stehlé
bois résolu	*Chimarrhis cymosa* Jacq.
bois résolu	*Licania ternatensis* Hook. f.
bois rivière	*Chimarrhis cymosa* Jacq.
bois rouge	*Amanoa caribaea* Kr. and Urb.
bois rouge	*Coccoloba pubescens* L.
bois rouge	*Coccoloba swartzii* Meissn
bois rouge à balles	*Guarea perrottetiana* A. Juss.
bois rouge à grives	*Erythroxylon squamatum* Sw.
bois savon	*Cestrum laurifolium* L'Hérit
bois savonnette	*Lonchocarpus benthamianus* Pittier
bois sept ans	*Meliosma herbertii* Rolfe var. *martinicensis* Kr. and Urb.
bois tabac	*Exostemma sanctae luciae* (Kentish) J. Britten
bois-tan	*Byrsonima laevigata* (L. C. Rich.) DC.
bois tan	*Byrsonima martinicensis* Kr. and Urb.
bois-tan	*Byrsonima cariacea* (Sw.) DC.
bois tan-montage	*Byrsonima martinicensis* Kr. and Urb.
bois ti-feuilles	*Eugenia ligustrina* (Sw.) Willd.
bois trompette	*Cecropia peltata* L.
bois vert	*Rochefortia cuneata* Sw.
bois-vignot	*Turpinia occidentalis* (Sw.) G. Don
bois-vinette	*Erythroxylon brevipes* DC.
bois-vinette	*Erythroxylon ovatum* Cav.
bois violet	*Meliosma herbertii* Rolfe var. *martinicensis* Kr. and Urb.
bios violon	*Nectandra coriacea* (Sw.) Griseb.
bougainvillée	*Bougainvillea* spp.

bouton d'or	*Spilanthes uliginosa* Sw.
bouton d'or	*Wedelia trilobata* (L.) Hitch.
boyau chat	*Smilax* spp.
brésillette	*Erythroxylon brevipes* DC.
brésillette	*Myrcia splendens* (Sw.) DC.
buis	*Chrysophyllum argenteum* Jacq.
cabouya	*Sporobolus indicus* (L.) R. Br.
cacao	*Theobroma cacao* L.
caca-ravet	*Forestiera rhamnifolia* Griseb. var. *martinicensis* Kr. and Urb.
caca-ravet	*Henriettea lateriflora* (Vahl) Howard
caca-ravet	*Rapanea coriacea* (Sw.) Mez
caca-ravet	*Trema lamarckiana* (Roem. & Schult) Bl.
cachibou	*Asplundia rigida* (Aubl.) Harl.
cachiman	*Annona reticulata* L.
cachiman bois	*Guatteria caribaea* Urb.
cachiman-montagne	*Talauma dodecapetala* (Lam.) Urb.
caconnier blanc	*Dussia martinicensis* Kr. & Urb.
caconnier rouge	*Ormosia monosperma* (Sw.) Urb.
café	*Coffea arabica* L.
café bâtard montagne	*Rudgea citrifolia* (Sw.) Shum.
café-bois	*Byrsonima trinitensis* Juss.
café-bois	*Ixora ferrea* (Jacq.) Benth.
café-bois	*Maytenus guianensis* Klotzsch
café-bois	*Ouratea longifolia* (DC) Engl.
café-bois	*Psychotria berteriana* DC.
café grand bois	*Bunchosia nitida* (Jacq.) L. C. Rich.
café grand bois	*Drypetes glauca* Vahl var. *macrocarpa* Kr. and Urb.
café-liberica	*Coffea liberica* Hiern.
café-marron	*Drypetes alba* Poit.
café-marron	*Psychotria berteriana* DC.
café montagne	*Rudgea citrifolia* (Sw.) Schum.
cäimitier-bois	*Chrysophyllum argenteum* Jacq.
cäimitier-bois	*Chrysophyllum coeruleum* Jacq.
calaba	*Calophyllum calaba* L.
calalu	*Xanthosoma saggtifolium* (L.) Schott
calebassier	*Crescentia cujete* L.
calumet	*Dicranopteris bifida* (Willd.) Maxon
calumet	*Isachne disperma* (Lam.) Doell
calumet	*Lasiacis divaricata* (L.) Hitchc.
campêche	*Haematoxylon campechianum* L. H.
canamaze	*Axonopus compressus* (Sw.) Beauv.
canéfice	*Cassia fistula* L.
cannelle blanche	*Canella winterana* (L.) Gaertn.
canellier bâtard	*Canella winterana* (L.) Gaertn.
canne	*Saccharum officinarum* L.
canne d'eau	*Hedychium coronarium* Koenig

canne-rivière	*Hedychium coronarium* Koenig
caoba	*Swietenia mahagoni* (L.) Jacq.
caracoli	*Tektona grandis* L. f.
carambole	*Averrhoa carambola* L.
carapate	*Amanoa caribaea* Kr. and Urb.
carapate	*Carapa guianensis* Aubl.
carapate	*Ricinus communis* L.
caratas	*Furcraea tuberosa* Ait. f.
casse savane	*Cassia emarginata* L.
cassia alata	*Cassia alata* L.
cèdre argente	*Endlicheria sericea* Nees
cerise-du-pays	*Malpighia* spp.
cerise-du-pays	*Malpighia glabra* L.
chadèque	*Citrus grandis* (L.) Osbeck
chapeau chinois	*Holmskioldia sanguinea* Retz
châtaignier	*Artocarpus altilis* (Park.) Fosberg
châtaignier	*Sloanea truncata* Urb.
châtaignier-grandes-feuilles	*Sloanea massoni* Sw.
châtaignier-grandes-feuilles	*Sterculia caribaea* R. Br. in Benn.
châtaignier montagne	*Sloanea truncata* Urb.
châtaignier-petit-coco	*Sloanea dussii* Urb.
châtaignier-petites-feuilles	*Sloanea truncata* Urb.
Chêne d'Amerique	*Catalpa longissima* (Jacq.) Sims.
chêne noir-pays	*Catalpa longissima* (Jacq.) Sims.
chevalier-onze-heures	*Pectis humifusa* Sw.
cheveux-béké	*Juncus bufonius* L.
cheveux de Vénus	*Quamoclit pinnata* (Desr.) Bojer
chiendent	*Cynodon dactylon* (L.) Pers.
chiendent blanc	*Stenotaphrum secundatum* (Walt.) Kuntze
chou caräibe	*Xanthosoma sagittifolium* (L.) Schott.
chou montagne	*Prestoea montana* (R. Graham) Nichols
chou palmiste	*Prestoea montana* (R. Graham) Nichols
christophine	*Sechium edule* (Jacq.) Sw.
cierge	*Cephalocereus nobilis* (Hawe.) Britt. & Rose
citron	*Citrus limon* (L.) Burm. f.
citronelle	*Cymbopogon nardus* (L.) Rendl.
citronnier	*Cestrum laurifolium* L'Hérit.
citronnier	*Citrus medica* L.
citronnier	*Ilex nitida* (Vahl) Maxim.
citronnier blanc	*Ilex sideroxyloides* (Sw.) Griseb.
coco	*Cocos nucifera* L.
coco ravet	*Casearia decandra* Jacq.
cocotier	*Cocos nucifera* L.
cocoyer	*Sloanea dussii* Urb.
colibri	*Agati grandiflora* (L.) Desv.
concombre	*Cucumis sativus* L.
concombre marron	*Cayaponia americana* (Lam.) Cogn.
copahu	*Croton flavens* L.

corde à violon	*Galactica dubia* DC.
corossol	*Guatteria caribaea* Urb.
corossol-montagne	*Guatteria caribaea* Urb.
côtelette	*Citharexylum fruticosum* L.
côtelette	*Tetrazygia discolor* (L.) DC. var. *villosa* DC.
côtelette bois	*Charianthus corymbosus* (L. C. Rich.) Cogn.
côtelette-grand-bois	*Conostegia calyptrata* (Desr.) D. Don
coton-France	*Calotropis procera* (Ait.) R. Br.
coton soie	*Calotropis procera* (Ait.) R. Br.
cotonnier Marie-Galante	*Gossypium hirsutum* var. *marie-galante* Hutch.
cotonnier "sea island"	*Gossypium barbadense* L.
courbaril	*Hymenaea courbaril* L.
courbaril jaune	*Hymenaea courbaril* L.
cousin	*Desmodium* spp.
cousins-mahots	*Urena lobata* L.
cousses-couche	*Dioscorea trifidia* L.f.
cousse-couche douce	*Dioscorea esculenta* (Lour.) Prain.
cousse-couche Guinée	*Dioscorea cayennensis* Lam.
crécré blanc	*Miconia striata* (Vahl) Cogn.
crécré blanc	*Tetrazygia discolor* (L.) DC. var. *villosa* DC.
crécré-grand-bois	*Clidemia umbrosa* Cogn.
crécré-grand-bois	*Henrietta lateriflora* (Vahl) Howard
crécré-grande-feuille	*Clidemia umbrosa* (Sw.) Cogn.
crécré noir	*Charianthus corymbosus* (L. C. Rich) Cogn.
créosote	*Spilanthes uliginosa* Sw.
cresson	*Lepidium virginicum* L.
cresson d'eau	*Rorippa nasturtium-aquaticum* (L.) Hayek
cresson sauvage	*Lindernia rotundifolia* (L.) Alst.
croc-à-chien	*Celtis iguanaea* (Jacq.) Sarg.
croc-à-chien	*Pisonia helleri* Standl.
croc-chien	*Machaerium lunatum* (L. f.) Ducke
croton	*Croton niveus* Jacq.
cypre	*Phoebe elongata* (Vahl) Nees
cypre orange	*Styrax glaber* Sw.
dachine	*Colocasia esculenta* (L.) Schott.
dasheen	*Colocasia esculenta* (L.) Schott.
diaballe	*Pithecellobium unguis-cati* (L.) Benth.
doux blanc	*Inga laurina* (Sw.) Willd.
ébène	*Diospyros revoluta* Poir.
ébène jaune	*Homalium racemosum* Jacq.
ébène vert	*Tabebuia leucoxyla* DC.
enivrage	*Piscidia piscipula* (L.) Sarg.
envers	*Maranta arundinacea* L.
épinard	*Alternanthera bettzickiana* (Regel) Standl.
escambron	*Adelia ricinella* L.
espinillo	*Adelia ricinella* L.
eucalyptus	*Eucalyptus* spp.
faux ipéca	*Cephaëlis swartzii* DC.

feuilles	*Trema lamarckiana* (Roem. & Schult.) Bl.
figue-banane	*Musa paradisiaca* L.
figuier	*Ficus citrifolia* Mill.
figuier-à-agoutis	*Ficus insipida* Willd.
figuier-aralie	*Oreopanax capitatum* (Jacq.) Dcne & Planch.
figuier maudit	*Clusia alba* L.
figuier maudit	*Clusia rosea* Jacq.
figuier maudit	*Ficus insipida* Willd.
figuier maudit	*Ficus rosea* Jacq.
figuier maudit montagne	*Clusia plukenetii* Urb.
flambeau caraïbe	*Pilocarpus racemosus* Vahl
flambeau noir	*Pilocarpus racemosus* Vahl
flamboyant	*Delonix regia* Boj.
fleur blanche	*Utricularia jamesoniana* Oliver
fleur d'amour	*Acacia paniculata* Willd.
fleur-montagne	*Lobelia flavescens* (Presl.) E. Wimm.
fleur-trompette	*Datura metel* L.
fleur violette	*Apteria aphylla* (Nutt.) Barnh.
forte ventura	*Lonchocarpus pentaphyllus* (Poir.) DC.
fougère	*Blechum occidentale* L.
fougère	*Elaphoglossum dussii* Underw.
fougère	*Maxillaria coccinea* (Jacq.) L.O. Wms.
fougère	*Pithecellobium jupunba* (Willd.) Urb.
fromager	*Ceiba pentandra* (L.) Gaertn.
fromager-mapou	*Ochroma pyramidale* (Cav.) Urb.
fuschia-montagne	*Charianthus nodosus* (Desr.) Triana
gaïac	*Guaiacum officinale* L.
galba	*Calophyllum calaba* L.
galba odorant	*Calophyllum calaba* L.
gazon	*Axonopus compressus* (Sw.) Beauv.
génipa	*Genipa americana* L.
girasole	*Helianthus annus* L.
giraumon	*Cucurbita moschata* (Duch. ex Lam) Duch. ex Poir.
giraumon	*Erythroxylon squamatum* Sw.
gironflée	*Peperomia hirtella* Miq. in Hook.
giroflé-grande feuille	*Peperomia hernandifolia* (Vahl) Dietr.
gli-gli	*Buchenavia capitata* (Vahl) Eichl.
gombo	*Hibiscus esculentus* L.
gombo à fibre	*Hibiscus radiatus* Cav.
gombo grand bois	*Hibiscus tulipiflorus* Hook.
gommier	*Dacryodes excelsa* Vahl
gommier à canots	*Dacryodes excelsa* Vahl
gommier-barrière	*Bursera simaruba* (L.) Sarg.
gommier blanc	*Dacryodes excelsa* Vahl
gommier l'encens	*Protium attenuatum* (Rose) Urb.
gommier rouge	*Bursera simaruba* (L.) Jacq.
gongle hout	*Margaritaria nobilis* L. f.

gosco	*Cassine xylocarpa* Vent.
goyavier	*Psidium guajava* L.
goyavier bâtard	*Guettarda scabra* (L.) Lam.
goyavier bois	*Eugenia octopleura* Kr. and Urb.
goyavier bois	*Myrcianthes fragrans* (Sw.) McVaugh
goyavier-montagne	*Eugenia pseudopsidium* Jacq.
goyavier-montagne	*Myrcia deflexa* var. *dussii* Kr. and Urb.
goyavier-montagne	*Myrcianthes fragrans* (Sw.) McVaugh
goyavier-que-de-rat	*Myrcia deflexa* var. *dussii* Kr. and Urb.
goyavier rose	*Guarea kunthianum* A. Juss. var. *hahnii* DC.
graine bleue	*Cephaëlis swartzii* DC.
graine bleue des hauts	*Cephaëlis axillaris* Sw.
graine bleue montagne	*Symplocos martinicensis* Jacq.
graine-Job	*Coix lacryma-jobi* L.
graines bleues	*Forestiera rhamnifolia* Griseb. var. *martinicensis* Kr. and Urb.
graines bleues	*Symplocos martinicensis* Jacq.
graines vertes	*Ilex macfadyenii* (Walp.) Rehder
graines vertes	*Meliosma herbertii* Rolfe var. *martinicensis* Kr. and Urb.
graines vertes	*Meliosma pardonii* Kr. & Urb.
graines violettes	*Meliosma herbertii* var. *martinicensis* Kr. and Urb.
graine vert	*Picrasma antillana* (Eggers.) Urb.
grand baume	*Croton corylifolius* Lam.
grand cousin-bois	*Triumfetta grandiflora* Vahl
grande seguine	*Philodendron giganteum* Schott
grande vinette	*Erythroxylon squamatum* Vahl
grand gironflé	*Ludwigia octovalvis* (Jacq.) Raven.
grand mapou	*Cordia sulcata* DC.
grand merisier	*Eugenia chrysobalanoides* DC.
grand roseau	*Gynerium sagittatum* (Aubl.) Beauv.
grand'savane	*Melibomia supina* (Sw.) Britton
gratte-jambe	*Celtis iguanaea* (Jacq.) Sarg.
grigri	*Aiphanes minima* (Gaertn.) Burr.
groseille	*Hibiscus sabdariffa* L.
gros-fort	*Mikania latifolia* Sw.
gros pois-pois	*Centrosema plumieri* (Turp. and Pers.) Benth.
grosse herbe grasse	*Rhoeo spathecea* (Sw.) Stern.
grosse raquette	*Opuntia tuna* (L.) Mill.
guaba	*Inga vera* Willd.
guama	*Inga laurina* (Sw.) Willd.
guayabacon	*Myrcia leptoclada* DC.
guayacan	*Guaiacum officinale* L.
guele de loup montagne	*Gesneria ventricosa* Sw. f. var. *obovata* (Urb.) Stehlé
guépois	*Myrcia leptoclada* DC.
guépois-bâtard	*Eugenia octopleura* Kr. and Urb.

guéri-tout	*Cordia globosa* (Jacq.) HBK
gumbo	*Hibiscus esculentus* L.
haricot	*Phaseolus vulgaris* L.
herbe à boutons	*Hyptis atrorubens* Poit.
herbe à chevaux	*Abildgaardia monostachya* (L.) Vahl
herbe à cornette	*Borreria ocymoides* (Burm. f.) DC.
herbe à couteau	*Scleria latifolia* Sw.
herbe à envirer	*Tephrosia cinerea* (L.) Pers. var. *littoralis* (Jacq.) Benth.
herbe à foin	*Dichanthium aristatum* (Poir.) C. E. Hubb.
herbe à laine	*Rhynchelytrum repens* (Willd.) C. E. Hubb.
herbe à lapin	*Senecio lucidus* (Sw.) DC.
herbe à miel	*Hyptis lanceolata* Poir.
herbe à miel	*Rhynchelytrum repens* (Willd.) C. E. Hubb.
herbe à mouche	*Nepsera aquatica* (Aubl.) Naud.
herbe à pique	*Besleria coriacea* Urb.
herbe à pique	*Nuerolaena lobata* (L.) R. Br.
herbe à têtes	*Kyllinga odorata* Vahl
herbe-caprit	*Zornia microphylla* Desv.
herbe chapalet	*Sophora tomentosa* L.
herbe-collant	*Cenchrus echinatus* L.
herbe coq d'Inde	*Conyza canadensis* (L) Cronq.
herbe cotelette	*Clidemia umbrosa* (Sw.) Cogn.
herbe conteau	*Ischaemum latifolium* (Spreng.) Kunth
herbe conteau	*Scleria pterota* Presl.
herbe de guinée	*Panicum maximum* Jacq.
herbe de para	*Brachiaria purpurascens* (Raddi) Hems.
herbe de Para	*Eriochloa polystachya* HBK
herbe de Rhodes	*Chloris gayana* Kunth
herbe fine	*Abilgaardia monostachya* (L.) Vahl
herbe fine	*Axonopus compressus* (Sw.) Beauv.
herbe fine	*Leptochloa virgata* (L.) Beauv.
herbe grasse	*Commelina* spp.
herbe grasse	*Tradescantia* spp.
herbe kangourou	*Hyparrhenia rufa* (Nees) Stapf
herbe maconet	*Hemidiodia ocymifolia* (Willd.) K. Schum.
herbe melinis	*Melinis minutiflora* Beauv.
herbe mouton	*Paspalum notatum* Flugge
herbe pou-bois	*Centella asiatica* (L.) Urb.
herbe-rasoir	*Scleria secans* (L.) Urb.
herbe-scorpion	*Heliotropium angiospermum* Murr.
herbe sûre	*Axonopus compressus* (Sw.) Beauv.
herbe violette	*Alternanthera bettzickiana* (Regel.) Standl.
herbes éléphant	*Pennisetum purpureum* Schumach.
higuerillo	*Vitex divaricata* Sw.
higuillo	*Margaritaria nobilis* L. f.
honteuse mâle	*Aeschynomene sensitiva* Sw.
honteuse mâle	*Cassia patellaria* DC.

icaque	*Chrysobalanus icaco* L.
icaque à poils	*Hirtella triandra* Sw.
icaque à ramiers	*Hirtella triandra* Sw.
icaque grand bois	*Chrysobalanus cuspidatus* Griseb.
icaque-montagne	*Chrysobalanus cuspidatus* Griseb.
icaque noire	*Chrysobalanus icaco* L. var. *pellocarpa* (GFW Meyer) DC.
icaque pendant	*Hirtella triandra* Sw.
igname	*Dioscorea* spp.
igname grand bois	*Dioscorea polygonoides* H. & B.
igname sauvage	*Dioscorea polygonoides* H. & B.
ignames blanches	*Dioscorea alata* L.
ignames jaunes	*Dioscorea cayennensis* Lam.
immortelle	*Erythrina corallodendron* L.
indigo	*Indigofera* spp.
indigo faux	*Indigofera suffruticosa* Mill.
indigo marron	*Tephrosia senna* HBK
indigo poilu	*Indigofera hirsuta* L.
indigo rampant	*Indigofera spicata* Forsk.
indigo vrai	*Indigofera tinctoria* L.
ipéca bâtard	*Asclepias curassavica* L.
ipéca bâtard	*Cephaëlis axillaris* Sw.
ipéca bâtard	*Cephaëlis muscosa* Sw.
ixora	*Ixora coccinea* L.
jacquet brûlot	*Coccoloba ascendens* Duss
jacquier	*Artocarpus heterophyllus* Lam.
jak	*Artocarpus heterophyllus* Lam.
jasmin bâtard	*Cestrum alternifolium* (Jacq.) O. E. Schulz var. *typicum* Stehlé
jasmin bois	*Cestrum megalophyllum* Dunal
jasmin bois	*Hillia parasitica* Jacq.
jasmin bois	*Schradera vahlii* Steyerm.
javillo	*Hura crepitans* L.
josé vrai	*Symphysia racemosa* (Vahl) Stearn.
jujube (surette)	*Ziziphus mauritiana* Lam.
karatas	*Furcraea tuberosa* Ait. f.
kénaf	*Hibiscus sabdariffa* L.
kola	*Cola acuminata* (P. Beauv.) Schott. & Endl
kudzu	*Pueraria phaseoloides* (Roxb.) Benth.
kush-kush	*Dioscorea trifida* L.
langue à boeuf	*Ouratea longifolia* (DC.) Engl.
langue boeuf	*Asplundia rigida* (Aubl.) Harl.
larmes de Job	*Coix lacryma-jobi* L.
latanier à balai	*Coccothrinax barbadenis* (Lodd. ex Mart) Becc.
laurier-avocat	*Persea urbaniana* Mez
laurier bois	*Aniba bracteata* (Nees) Mez
laurier-bord-de-mer	*Ocotea martinicensis* Mez
laurier caillé	*Endlicheria sericea* Nees

laurier cannelle	*Phoebe elongata* (Vahl) Nees
laurier cerise	*Nectandra antillana* Miessn.
laurier cerise	*Prunus dussii* Kr. and Urb.
laurier-chypre	*Phoebe elongata* (Vahl) Nees
laurier-chypre	*Styrax glaber* Sw.
laurier doux	*Nectandra patens* (Sw.) Griseb.
laurier-falaise	*Aniba ramageana* Mez
laurier fine	*Ocotea leucoxylon* (Sw.) G. de la Maza
laurier fine	*Nectandra coriacea* (Sw.) Griseb.
laurier franc	*Nectandra antillana* Meissen
laurier Isabelle	*Ocotea cernua* Mez
laurier Isabelle rouge	*Nectandra patens* (Sw.) Griseb.
laurier jaune	*Aniba bracteata* (Nees) Mez
laurier Madame	*Ocotea leucoxylon* (Sw.) G. de la Maza
laurier noir	*Nectandra coriacea* (Sw.) Griseb.
laurier-orange	*Styrax glaber* Sw.
laurier-rose-montagne	*Podocarpus coriaceus* L.C. Rich.
l'envers blanc	*Maranta arundinacea* L.
lépiné blanc	*Zanthoxylum caribaeum* Lam.
lépiné jaune	*Zanthoxylum martinicense* (Lam.) DC.
lépiné rouge	*Zanthoxylum punctatum* Vahl
lépineux blanc	*Zanthoxylum spinifex* (Jacq.) DC.
liane à barrique	*Connarus grandifolius* Planch.
liane à barrique	*Dalbergia monetaria* L. f.
liane à chique	*Tournefortia bicolor* Sw.
liane à crabe	*Cydista aequinoctialis* (L.) Miers
liane à crabes	*Clematis dioica* L.
liane à eau	*Mikania latifolia* Sm.
liane à l'encre	*Passifolia rotundifolia* L.
liane à paniers	*Cydista aequinoctialis* (L.) Miers
liane bamboche	*Smilax guianensis* Vitm
liane blanche	*Schradera vahlii* Steyerm.
liane blanche des hauts	*Manettia dominicensis* Wernham
liane caraïbe	*Heteropteris purpurea* (L.) HBK
liane ciel-de-crabe	*Schlegelia axillaris* Griseb.
liane corde	*Chaetocalyx scandens* (L.) Urb
liane crabe	*Cydista aequinoctialis* (L.) Miers
liane guafre	*Mikania latifolia* Sm.
liane-jasmin	*Schradera vahlii* Steyerm.
liane-mibi	*Stigmaphyllon convolvulifolium* (Cav.) A. Juss
liane-percée	*Monstera pertusa* (L.) de Vriese
liane-persil	*Paullinia vespertilio* Sw.
liane-pomme	*Solandra grandiflora* Sw.
liane rouge	*Centropogon berterianus* (Spreng.) A. DC.
liane rude	*Petrea volubilis* L.
liane-savon	*Chaetocalyx scandens* (L.) Urb.
liane sirop	*Columnea scandens* L. var. *typica* Stehlé
liane trompette	*Solandra grandiflora* Sw.

liege	*Ochroma pyramidale* (Cav.) Urb.
lignum vitae	*Guaiacum officinale* L.
limon	*Citrus limon* (L.) Burm. f.
lin	*Isochilus linearis* (Jacq.) R. Br.
lin bâtard	*Isochilus linearis* (Jacq.) R. Br.
litchie	*Litchi chinensis* Sonn.
loquat	*Eriobotrya japonica* (Thumb.) Lindl.
luzerne arborescente	*Medicago arborea* L.
luzerne herbacée	*Medicago sativa* L.
mabouya	*Capparis baducca* L.
macata bourse	*Leucaena leucocephala* (Lam.) DeWit
macornet	*Borreria verticillata* Mey.
magnolia	*Talauma dodecapetala* (Lam.) Urb.
mahogani	*Swietenia mahagoni* (L.) Jacq.
mahogani du Honduras	*Swietenia macrophylla* King
mahogani du pays	*Swietenia mahagoni* (L.) Jacq.
mahot	*Hibiscus elatus* Sw.
mahot	*Hibiscus tiliaceus* L.
mahot	*Triumfetta lappula* L.
mahot	*Urena lobata* L.
mahot-bré	*Cordia collococca* L.
mahot-cochon	*Sterculia caribaea* R. Br. in Benn.
mahot-cousin	*Urena lobata* L.
mahot grand bois	*Hibiscus tulipiflorus* Hook.
mahot-grandes-feuilles	*Cordia sulcata* DC.
mahot l'anglais	*Guatteria caribaea* Urb.
mahot mare	*Pavonia spicata* Car.
mahot noir	*Guatteria caribaea* Urb.
maïs doux	*Zea mays* L.
malagueta	*Pimenta racemosa* (Mill.) J.W. Moore
malanga sauvage	*Montrichardia arborescens* (L.) Schott
malimbé	*Piper dilatatum* L. C. Rich
malimbé-grandes-feuilles	*Piper aduncum* L.
malnommée	*Euphorbia hirta* L.
mamain	*Annona glabra* L.
mambi	*Colubrina elliptica* (Sw.) Brizicky & Stern.
mamey	*Mammea americana* L.
mampou	*Pisonia fragrans* Dumont-Cours
mancenillier	*Hippomane mancinella* L.
mandarine	*Citrus reticulata* Blanco
manger-lapin	*Solidago serotina* Ait.
mangle blanc	*Laguncularia racemosa* (L.) Griseb.
mangle bois	*Tovomita plumieri* Griseb.
mangle rouge	*Rhizophora mangle* L.
mangle gris	*Conocarpus erecta* L.
mangle médaille	*Machaerium lunatum* (L. f.) Ducke
mangle montagne	*Clusia plukenetii* Urb.
mangue	*Mangifera indica* L.

manioc	*Manihot esculenta* Cranz
mapou	*Ochroma pyramidale* (Cav.) Urb.
mapou	*Pisonia fragrans* Dumont-Cours
mapou-baril	*Sterculia caribaea* R. Br. in Benn.
mapou-grand-bois	*Cordia elliptica* Sw.
mapou gris	*Pisonia fragrans* Dumont-Cours
mapou-lélé	*Cordia elliptica* Sw.
mapou puant	*Sterculia caribaea* R. Br. in Benn.
maran	*Croton astroites* Ait.
marbri	*Richeria grandis* Vahl
mari	*Calophyllum calaba* L.
maria	*Calophyllum calaba* L.
marie honte	*Mimosa pudica* L.
marocaine	*Hyparrhenia hirta* (L.) Stapf
marron	*Myrcianthes fragrans* (Sw.) McVaugh
maruba	*Simaruba amara* Aubl.
maurécif	*Byrsonima coriacea* (Sw.) DC. var. *spicata* (Cav.) Ndz.
mauricif	*Byrsonima laevigata* (L. C. Rich). DC.
médaille	*Dalbergia monetaria* L. f.
médecinier	*Jatropha curcas* L.
médecinier bâtard	*Jatropha gossypifolia* L.
melilot	*Melilotus alba* Desr.
melilot blanc	*Melilotus alba* Desr.
mélongène bois	*Solanum lanceifolium* Jacq. var. volubile (Sw.) Sendtn.
merisier	*Eugenia axillaris* (Sw.) Willd.
merisier bois	*Eugenia axillaris* (Sw.) Willd.
merisier bois	*Eugenia brachystachys* Berg.
merisier de Madegascar	*Flacourtia indica* (Burm. f) Merr.
merisier jaune	*Eugenia lambertiana* DC.
merisier noir	*Eugenia liqustrina* (Sw.) Willd.
merisier petites-feuilles	*Eugenia axillaris* (Sw.) Willd.
merisier ti-feuilles	*Eugenia lambertiana* DC.
merker	*Pennisetum purpureum* Schum.
mibi	*Stigmaphyllon puberum* (L. C. Rich.) A. Juss.
mibi	*Stigmaphyllon convolvulifolium* (Cav.) A. Juss.
mille fleur	*Lantana camara* L.
moabite	*Pithecellobium unguis cati* (L.) Benth.
moca	*Geoffroea inermis* W. Wright
mombin	*Spondias mombin* L.
montagne	*Sapindus saponaria* L.
moralon	*Coccoloba pubescens* L.
mourron	*Peperomia nigropunctata* Miq.
mourron-bois	*Peperomia hirtella* Miq. in Hook.
mourron gironflée	*Peperomia trifolia* (L.) A. Dietr.
mourron-grand-bois	*Peperomia hernandifolia* (Vahl) Dietr.
mourron-grand-bois	*Peperomia questeliana* Stehlé & Trel.

muguet-grand-bois	*Leiphaimos aphylla* (Jacq.) Gilg.
muguet jaune	*Leiphaimos aphylla* (Jacq.) Gilg.
mûrier-pays	*Chlorophora tinctoria* (L.) Gaud.
muscadier à grive	*Virola surinamensis* Warh.
nappe montagne	*Mikania ovalis* Griseb.
neflier bois	*Guarea perrottetiana* A. Juss.
negresse	*Nectandra coriacea* (Sw.) Griseb.
noyau France	*Prunus dussii* Kr. and Urb.
noyer	*Zanthoxylum flavum* Vahl
oeil de bourrique	*Mucuna sloanea* Fawc. & Rendl.
olivier	*Byrsonima lucida* (Sw.) DC.
olivier grand bois	*Buchenavia capitata* (Vahl) Eichl.
orange	*Citrus sinensis* (L.) Osbeck.
orange amère	*Citrus aurantium* L.
orange sûre	*Citrus aurantium* L.
orme-petites-feuilles	*Trema lamarckiana* (Roem. & Schult.) Bl.
ortie savane	*Croton hirtus* L'Hérit.
oseille-bois	*Begonia macrophylla* Lam.
oseille-grand-bois	*Begonia odorata* Willd.
pain d'épice	*Pouteria dussiana* (Pierre) Stehlé
palétuvier	*Amanoa caribaea* Kr. and Urb.
palétuvier-grand-bois	*Tovomita plumieri* Griseb.
palétuvier gris	*Amanoa caribaea* Kr. and Urb.
palétuvier-gris-montagne	*Amanoa caribaea* Kr. and Urb.
palétuvier jaune	*Symphonia globulifera* L. f.
palétuvier montagne	*Clusia plukenetii* Urb.
palétuvier montagne	*Tovomita plumieri* Griseb.
palmier à balai	*Coccothrinax dussiana* L. H. Bailey
palmiste amêr	*Rhyticocos amara* (Jacq.) Becc.
palmiste montagne	*Prestoea montana* (R. Graham) Nichols
pamplemousse	*Citrus paradisi* Macf.
pangola	*Digitaria decumbens* Stent.
papaya	*Carica papaya* L.
papillon végétal	*Agati grandiflora* (L.) Desv.
parasol agouti	*Selaginella flabellata* (L.) Spring
paspalum	*Paspalum dilatatum* Poir.
patate douce	*Ipomoea batatas* L.
patate macaque	*Ruellia tuberosa* L.
patte à canard	*Passiflora andersonni* DC.
patte à canard	*Passiflora rotundifolia* L.
patte lapin	*Ochroma pyramidale* (Cav.) Urb.
peste à pou	*Picrasma antillana* (Eggers) Urb.
petit ananas bois	*Catopsis floribunda* (Brogn.) L. B. Smith
petit-baume	*Croton guildingii* Griseb.
petit bouis	*Chrysophyllum argenteum* Jacq.
petit calumet	*Isachne rigidifolia* (Poir.) Urb.
petit citronnier	*Ilex macfadyenii* (Walp.) Rehder
petit coco	*Randia aculeata* L.

petit coco	*Rhyticocos amara* (Jacq.) Becc.
petite siguine	*Anthurium martinicense* Engler
petite verveine	*Heliotropium angiospermum* Murr.
petites-feuilles	*Erythroxylon ovatum* Cav.
petites-feuilles	*Eugenia monticola* (Sw.) DC.
petites-feuilles-des-hauts	*Myrcia fallax* (Rich.) DC. var. *dussii* Kr. and Urb.
petit foin	*Heteropogon contortus* (L.) Beauv.
petit ipéca bâtard	*Ruellia tuberosa* L.
petit Jean	*Nectandra patens* (Sw.) Griseb.
petit merisier	*Gyminda latifolia* (Sw.) Urb.
petit merisier	*Schaefferia frutescens* Jacq.
petit ortie	*Pilea inaequalis* (Juss.) Wedd.
petit pied poule	*Chloris gayana* Kunth
petit pied poule	*Chloris radiata* (L.) Sw.
petit pois jaune	*Vigna hosei* (Craig.) Baker
petit résolu	*Rondeletia stereocarpa* Griseb.
petit siguine	*Anthurium martinicense* Engler
petit trèfle	*Desmodium triflorum* (L.) DC.
pied poule	*Digitaria decumbens* Stent.
pied poule	*Digitaria ciliaris* (Ketz.) Koel.
pied poule	*Eleusine indica* (L.) Gaertn.
picanier	*Oplonia microphyllus* (Lam.) Stearn.
picanier	*Oplonia spinosus* (Jacq.) Raf.
piment	*Capsicum annuum* L.
piment	*Capsicum frutescens* L.
piment bâtard	*Erythroxylon squamatum* Sw.
pin caraïbe	*Pinus caribaea* Morlet
pistache	*Arachis hypogaea* L.
poinciana	*Delonix regia* (Bojer) Raf.
poirier	*Tabebuia pallida* (Lindl.) Miers
poirier blanc	*Tabebuia pallida* (Lindl.) Miers
poirier du nord	*Tabebuia pallida* (Lindl.) Miers
poirier gris	*Tabebuia pallida* (Lindl.) Miers
poirier rouge	*Tabebuia pallida* (Lindl.) Miers
pois à gratter	*Mucuna pruriens* (L.) DC.
pois aiguille-mare	*Sesbania sesban* (L.) Merrill
pois bâtard	*Phaseolus calcaratus* Roxb.
pois bois-de-mer	*Canavalia maritima* (Aubl.) Thou.
pois boucoussou	*Dolichos lablab* L.
pois cabrit	*Canavalia maritima* (Aube.) Thou.
pois d'Angole	*Cajanus cajan* (L.) Millsp.
pois de bois	*Cajanus cajan* (L.) Millsp.
pois de sept ans gros	*Phaseolus lunatus* L.
pois doré	*Phaseolus aureus* Roxb.
pois doux blanc	*Inga laurina* (Sw.) Willd.
pois doux gris	*Inga vera Willd.*
pois enivrant	*Teramnus labialis* (L. f.) Spreng.

pois ficelle	*Galactia striata* var. *caribaea* Urb.
pois-hallier	*Rhynchosia phaseoloides* (Sw.) DC.
pois indiens	*Dolichos lablab* L.
pois jaune	*Vigna marina* (Burm. f.) Merrill
pois liane	*Clitoria ternatea* L.
pois liane	*Galactia striata* (Jacq.) Urb. var. *caribaea* Urb.
pois marron	*Centrosema pubescens* Benth.
pois marron	*Phaseolus adenanthus* G. F. W. Meyer
pois mars	*Sesbania emerus* (Aubl.) Urb.
pois pigeon	*Vigna marina* (Burm. f.) Merrill
pois pigeon	*Vigna sinensis* (Torn.) Hassk.
pois poison	*Macroptilium lathyroides* (L.) Urb. ex Engl.
pois rampant	*Vigna luteola* (Jacq.) Benth.
pois rouge	*Galactia rubra* (Jacq.) Urb.
pois rouge	*Macroptilium lathyroides* (L.) Urb.
pois sabre blanc	*Canavalia gladiata* (Jacq.) DC.
pois sabre rose	*Canavalia ensiformis* L. DC.
pois-savon	*Phaseolus lunatus* L.
pois sucrier	*Rhynchosia minima* (L.) DC.
pois velu	*Calopogonium mucunoides* Desv.
pois yeux noirs	*Vigna unguiculata* (L.) Walp.
pois z'oiseaux	*Cracca caribaea* (Jacq.) Benth.
pois z'oiseaux	*Rhynchosia reticulata* (Sw.) DC.
pois z'oiseaux	*Teramnus labialis* (L. f.) Spreng.
pois zombi	*Macroptilium lathyroides* (L.) Urb.
pois zombi	*Vigna vexillata* (L.) A. Rich.
poix doux marron	*Catalpa longissima* (Jacq.) Sims
pomme-acajou	*Anacardium occidentale* L.
pomme-cajou	*Anacardium occidentale* L.
pomme-cythère	*Spondias dulcis* Forst.
pomme-liane-bâtard	*Passiflora suberosa* L.
pomme-liane bois	*Passiflora rubra* L.
pomme pain	*Pouteria dussiana* (Pierre) Stehlé
pomme pin	*Talauma dodecapetala* (Lam.) Urb.
pommier rose	*Syzygium jambos* (L.) Alst.
pourpier	*Portulaca halimoides* L.
pourpier	*Portulaca oleracea* L.
prune	*Chrysophyllum argenteum* Jacq.
pruneau noir	*Ilex nitida* (Vahl) Maxim.
prune bord-de-mer	*Cassine xylocarpa* Vent.
prune bord-de-mer	*Ximenia americana* L.
prune café	*Bunchosia glandulifera* (Jacq.) HBK
prune du Japon	*Eriobotrya japonica* (Thunb.) Lindl.
prune mombin	*Spondias mombin* L.
puerto-rico	*Zea mays* L.
queue à rat	*Piper dilataum* L. C. Rich
queue de lézard	*Piper dussii* C. DC.
queue de rat	*Pennisetum setosum* (Sw.) L. C. Rich

queue de renard	*Teliostachya alopecuroidea* (Vahl) Nees
queue la rat	*Piper dussii* C. DC.
queue la souris	*Peperomia stehleana* var. *ajoupana* Trel.
quinquina caraïbe	*Exostemma sanctae-luciae* (Kentish) J. Britton
quinquina-piton	*Exostemma sanctea-luciae* (Kentish) J. Britton
radegonde	*Catalpa longissima* (Jacq.) Sims
raisin coudre	*Coccoloba venosa* L.
raisinier	*Coccoloba uvifera* L.
raisinier bâtard	*Coccoloba venosa* L.
raisinier bois	*Coccoloba swartzii* Meissn.
raisinier-bord-de-mer	*Coccoloba uvifera* L.
raisinier-grandes-feuilles	*Coccoloba pubescens* L.
raisinier-gros-morne	*Richeria grandis* Vahl
ramie	*Boehmeria nivea* (L.) Gaud.
raquette	*Opuntia dillenii* (Ker-Gawl) Haw.
raquette douce	*Opuntia tuna* (L.) Mill.
raquette volante	*Opuntia triacantha* (Willd.) Sweet
résolu	*Chimarrhis cymosa* Jacq.
résolu	*Licania ternatensis* Hook. f.
résolu montagne	*Rondelitia parviflora* Poir.
résolu montagne	*Rondelitia stereocarpa* Griseb.
ricinella	*Adelia rincinella* L.
roble	*Tabebuia pallida* (Lindl.) Miers
romarin	*Nothopanax cochleatum* Merr., Miq.
romarin	*Polyscias balfouriana* Bailey
rose	*Rosa* spp.
roseau d'Inde	*Gynerium sagittatum* (Aubl.) Beauv.
roselle	*Hibiscus sabdariffa* L.
roucou	*Bixa orellana* L.
sablier	*Hura crepitans* L.
sainfoin	*Indigofera spicata* Forsk
sang-dragon	*Dracaena fragrans* Ker-Gawl
sapotillier	*Achras sapota* Mill.
savonette	*Lonchocarpus benthamianus* Pittier
savonette-grande-bois	*Lonchocarpus pentaphyllus* (Poir.) DC.
savonnette	*Sapindus saponaria* L.
siguine rouge	*Anthurium cordatum* (L.) Schott
sorghos	*Sorghum bicolor* (L.) Moench
sou marqué	*Cassia obtusifolia* L.
soulier zombi	*Scutellaria havanensis* Jacq.
sunabo	*Pavonia spicata* Cav.
sureau bâtard	*Cestrum alternifolium* (Jacq.) O.E. Schultz var. *typicum* Stehlé
sureau gros	*Aegiphila martinicensis* Jacq.
surelle	*Phyllanthus acidus* (L.) Skeels
surette	*Ziziphus mauritiana* Lam.
tabac à diable	*Neurolaena lobata* (L.) R. Br.
tabac diable bâtard	*Eupatorium magdalenae* Stehlé

tamarin	*Tamrindus indica* L.
tamarin bâtard	*Pithecellobium jupunba* (L.) Urb.
tamarin-bois	*Pithecellobium jupunba* (L.) Urb.
tannia	*Xanthosoma sagittifolium* L.
tcha-tcha	*Crotalaria* spp.
teck	*Tektona grandis* L. f.
teigne	*Stemodia verticillata* (Mill.) Hassl.
teigne	*Zornia microphylla* Desv.
tendre à caillou	*Acacia muricata* (L.) Willd.
tête à nègre	*Triumfetta lappula* L.
thé savane	*Sauvagesia erecta* L.
thé savane	*Scoparia dulcis* L.
thym	*Thymus vulgaris* L.
thym-montagne	*Tibouchina chamaecistus* (Naud.) Cogn.
ti-bambou	*Isachne disperma* (Lam.) Doell
ti-baume	*Croton astroites* Ait.
ti-baume	*Croton flavens* L.
ti-baume	*Croton guildingii* Griseb.
ti-bonbon	*Croton hircinus* Vent
ti-coco	*Randea aculeata* L.
ti-teigne	*Alysicarpus vaginalis* (L.) DC. var. *tipicus* King
ti-teigne	*Pilea duchassaingii* Urb.
toloman	*Canna indica* L.
topitambou	*Calathea allouia* (Aubl.) Lindl.
trèfle	*Desmodium gyroides* DC.
trèfle	*Desmodium uncinatum* (Jacq.) DC.
trèfle gazon	*Desmodium triflorum* (L.) DC.
trèfle jaune	*Stylosanthes hamata* (L.) Taubert
trèfle jaune	*Vigna hosei* (Craig) Backer
trèfle rampant	*Desmodium axillare* (Sw.) DC.
trinidad	*Phaseolus mungo* L.
trompette	*Datura suaveolens* L.
trompette du jugement	*Datura suaveolens* L.
tuluman	*Calathea allouia* (Aubl.) Lindl.
vacoua	*Pandanus utilis* Bory
vanille	*Vanilla planifolia* G. Jackson in Andr.
vanillon	*Vanilla pompona* (L.) Schiede
vert	*Rochefortia cuneata* Sw.
verveine bord de mer	*Heliotropium curassavicum* L.
violette	*Viola stipularis* Sw.
vois dande	*Richeria grandis* Vahl
wappe-montagne	*Mikania ovalis* Griseb.
zeb couteau	*Scleria* spp.
zeb coq d'Inde	*Conyza bonariensis* (L.) Cronq.

Appendix C. Quality Timbers from Martinique Forests*

MAGNOLIACEAE

*Talauma plumierie*** equals *T. dodecapetala* **(Lam.) Urb.*** (Bois pin, cachiman de montagne):**

Large tree from 25 to 35 m high and 1.5 m in diameter.

Hard wood. Construction and joinery wood.

Density: 0.556. In all the tall forests to 750 m elevation.

Lyriodendron tulysifera **equals** *Liriodendron tulipifera* **L.:**

Large tree, introduced from North America.

Light wood, indestructible.

Joinery, cabinetry. Temperate zone.

ANNONACEAE

Guatteria ouregou **equals** *G. caribaea* **Urb. (Bois de l'Anglais, Mahot anglais):**

Very large tree, straight trunk, black stringy bark, yields cordage fibers.

Light wood, makes canoes and mats.

Not abundant, from 450 to 680 m elevation.

ROSACEAE

Prunus Dussii **equals** *P. dussii* **Kr. & Urb. (Noyaux de France):**

Sometimes tree, sometimes good-sized.

Hard wood, elastic, much used in wheelwright's work. Temperate zone.

LEGUMINOSEA, CAESALPINIACEAE

Hematoxylum campechianum **equals** *Haematoxylon campechianum* **L. (Bois campêche):**

Usually small tree, rarely large, with craggy trunk.

Hard wood, heavy, elastic, indestructible, having a deep red color and agreeable odor.

Joinery, used for posts and other buried constructions, railroad ties, dyewood.

Abundant in limestone areas of the south to 300 m elevation.

Tamarinus indica **equals** *Tamarindus indica* **L. (Tamarinier):**

Large beautiful tree, many branches, straight trunk.

Hard, dense wood, solid and flexible.

Wheelwright's work, knees of boats.

Native of Asia. Temperate zone.

*The text of this appendix is translated directly from his 1902 text, pages 7–18.

**The name used by Lasaulce 1902.

***Currently valid scientific name based on synonymy.

Hymenoea Courbaril **equals** *Hymenaea courbaril* **L. (Courbaril):**

Large tree reaching 25 m in height and 2 to 3 m in circumference.

One of the best woods, does not warp, a good mahogany color and takes a polish, lasts indefinitely.

Ship timber, carpentry, joinery, etc.

Grows along stream courses in temperate zones. Has become very rare.

LEGUMINOSAE, PAPILLONACEAE

Andira inerunus **equals** *Geoffroea inermis* **W. Wright (Angelin):**

Beautiful tree, medium height, straight trunk reaching 1 m in diameter.

Hard wood, excellent for construction and wheelwright's work.

Abundant. Temperate zone.

Lonchocarpus latifolius **equals** *L. pentaphyllus* **(Poir.) DC. (Savonette jaune):**

Medium height tree, rarely very tall.

Good wheelwright's wood, construction and inlay work.

Abundant. Temperate zone to 450 m elevation.

Lonchocarpus sericeux **equals** *L. sericeus* **(Poir.) HBK (Savonette rivière):**

Average height tree, rarely tall, good wheelwright's wood, construction and inlay.

Temperate zone to 450 m elevation.

Erythrina corallodendron **L. (Bois immortel):**

Small tree, weak wood, spongy, used to make living fences between properties because of its thorns.

Temperate zone to 450 m elevation.

Erythrina indica **equals** *E. variegata* **L. (Bois immortel vrai):**

Large tree, straight trunk, weak spongy wood.

Temperate zone to 350 m elevation.

Dussia martiniciensis **equals** *D. martinicensis* **Kr. & Urb. (Bois gamelle):**

Big tree, straight trunk, and no branches up to a great height.

Weak wood, white, no use for construction.

Rare, 200 to 600 m elevation.

Ormosia doesicarpa **equals** *O. krugii* **Urb. (Angelin palmiste, Angelin bâtard, bois kaklin, grande savonnette, caconnier blanc):**

Large tree, straight trunk, hard wood, used in carpentry.

Abundant, 100 to 600 m elevation.

According to Fournet 1977, 757, not in Martinique today.

LEGUMINOSAE, MIMOSACEAE

Adenanthera pavonina **L. equals** *A. pavonia* **L. (Graines rouges, arbre à graines rouges):**

Large tree giving straight wood, heavy; used in carpentry and paneling. Originally from the East Indies.

Abundant, 0 to 400 m elevation.

Acacia leucocephala **equals** *Leucaena leucocephala* **(Lam.) DeWit (Macatta bourse):**

This tree of small dimensions yields a good wood for inlay work.

Very abundant in temperate zone from 0 to 350 m elevation.

Acacia seleroxylon **equals** *A. scleroxyla* **Tuss. (Tendre à caillou):**

Tree from 12 to 15 m in height and over 0.50 m in diameter.

Sapwood yellow, heart reddish, hard and indestructible.

Used for making stakes, posts, and pile work.

Density: 1.325. Common in the south of the island. 0 to 200 m elevation.

An unidentified acacia in the same region gives such a hard wood that it is known by the name "breakax." It cannot be worked even with the best-tempered tools.

Inga laurina **(Sw.) Willd. (Pois doux):**

Average-sized tree, white wood, not very hard, good for construction.

Density 0.769.

Very abundant up to 650 m elevation.

Inga salutaris **(Pois doux gris): Not identified.**

Large enough tree. Wood hard and clears well, can be used to make clapboard.

Generally abundant from 0 to 500 m elevation.

Inga Martiniciensis **equals** *I. couruscans* **HBK (Pois doux montagne):**

Small tree, marquetry wood.

Here and there from 800 to 1,000 m elevation.

Pithecolobuim unguis-cati **equals** *Pithecellobium unguis-cati* **(L.) Benth. (Diobolle, acacia à bracelets, tendre à cailloux):**

Small tree with spines.

Relatively abundant in dry, rocky places close to the sea. Temperate zone.

LAURACEAE

Nectandra dominicana **(Meissn.) Mez (Laurier gombo):**

Tree 12 to 18 m high and 40 cm in diameter.

Excellent construction wood.

Abundant from 400 to 800 m elevation.

Nectandra antillana **Meissn. (Laurier à cerises):**

Average height, straight. Wood soft, not much value; only useful for interiors.

Abundant from 400 to 800 m elevation.

Nectandra coriacea **(Sw.) Griseb. (Laurier fin):**

Tree from 10 to 17 m high. Excellent construction wood.

Abundant from 400 to 800 m elevation.

Nectandra patens **(Sw.) Griseb. (Laurier Isabelle rouge):**

Average-sized tree. Good construction wood.

Abundant from 400 to 600 m elevation.

Ocotea cernua **Mez (Laurier Isabelle):**

Tree from 18 to 25 m high. Good for woodworking.

Abundant from 400 to 600 m elevation.

Ocotea martinicensis **Mez (Laurier bord de mer):**

Tree from 16 to 22 m high.

Abundant from 200 to 600 m elevation.

Ocotea egersi **equals** *O. eggersiana* **Mez (Laurier noir):**

Tree from 14 to 18 m high. Interior wood yellow, in demand for construction and interior woodwork.

Found at 200 to 600 m elevation.

MALVACEAE

Guazuma ulmifolia **Lam. (Orme pays): [now placed in Sterculiacaea]**

Tree average height, wood soft but good from framing, construction, and cooperage.

Found at 200 to 500 m elevation.

Thespesia populnea **(L.) Soland. ex Correa (Tatalpa):**

Small tree.

Good wood for wheelwright's work and interior woodwork.

Abundant in temperate zone.

Bombax ceiba **equals** *Ceiba pentandra* **(L.) Gaertn. (Fromager): [now placed in Bombacacea]**

Gigantic tree. Wood without value.

Temperate zone.

TILIACEAE

Sloanea massoni **(Sw.) (Châtaignier grandes feuilles): [now placed in Elaeocarpaceae]**

Large tree whose trunk reaches 1 m in diameter. The greenwood is soft; dry it is very hard.

Preferred for interior construction.

Abundant from 400 to 800 m elevation.

Slonea sinematiensis **equals** *Sloanea truncata* **Urb. (Châtaignier petites feuilles):**

Large tree furnishing one of the best construction woods.

Abundant enough from 400 to 800 m elevation.

Slonea, **not identified, equals ?** *Slonea dussii* **Urb. (Châtaignier coco):**

Large, beautiful tree supplying one of the most sought-after woods.

Rare enough from 200 to 500 m elevation.

BIXACEAE

Homalium racemosum **Jacq. (acgnat): [now placed in Flacourtiaceae]**

Average-sized tree from 8 to 16 m high.

Wood used for interior construction.

Density 1.164.

Common from 200 to 400 m elevation.

RUTACEAE

Tobinia punctata **(Bois flambeau noir): [***Pilocarpus racemosus* **Vahl, now very rare?]**

Small tree with branches used for torches that burn easily.

Common in the southern part of the island.

Found at 0 to 200 m elevation.

Zanthoxylon aromaticum **equals** *Zanthoxylum caribaeum* **Lam. (lépiné blanc):**

Average height tree with trunk and branches covered with spines.

Red wood, very hard, not affected by termites.

Excellent for construction in standing water or on land. Excellent furniture wood.

Abundant from 10 to 280 m elevation.

Zanthoxylon Martiniciensis **equals** *Zanthoxylum martinicense* **(Lam.) DC. (lépiné jaune):**

Small tree with the sapwood yellow and the heartwood red and hard.

Used for in-earth construction and for making tool handles of all sorts.

Abundant in the south of the island.

Found at 20 to 400 m elevation.

The last two species are valuable because they grow vigorously in dry areas.

Simaluba officinalis **equals** *Simaruba amara* **Aubl. (Bois blanc): [now placed in Simarubaceae]**

Tree to 20 m high.

Light, whitish wood.

Used for making boards; used for interior wood.

Common from 250 to 700 m elevation.

TEREBINTHACEAE

Bursera balsamifera **equals** *Bursera simaruba* **(L.) Sarg. (Gommier rouge): [now placed in Burseraceae]**

Average-sized tree. Very common.

Used in making canoes, boats, and oars.

Temperate zone.

Mangifera indica **L. (Manguier): [now placed in Anacardiaceae]**

Large tree yielding a wood that is hard as *peupher de France.*

Not used for charcoal.

Common in the whole island from 0 to 400 m elevation.

SAPINDACEAE

Sapiendus saponarica **equals** *Sapindus saponaria* **L. (Savonnette):**

Huge, beautiful tree, trunk very straight and cylindric.

Hard wood, good for wheelwright's work, carpentry, and joinery.

Tree has become rather rare. Found at 0 to 300 m elevation.

Melicocca bijuga **equals** *Melicoccus bijugatus* **Jacq. (Kenettier):**

Large tree, with a straight trunk, twisted at the base. Gives a hard wood, reported used in construction, rare enough.

Found at 0 to 350 m elevation.

MALPIGHIACEAE

Byrosomina spicata **equals** *Byrsonima coriacea* **(Sw.) DC. var.** *spicata* **(Cav.) Ndz. (bois tan):**

Average-sized tree used for drayage vehicles.

Construction and joinery wood.

Fairly common from 200 to 900 m elevation.

Exostemma floribundum **equals** *E. sanctae-lucia* **(Kentish) J. Britton (Quinquina montagne): [now placed in Rubiacea]**

Tree from 8 to 14 m high. Very straight.

Good construction wood. Bark used as a febrifuge.

Abundant from 250 to 800 m elevation.

BIGNONIACEAE

Tecoma pentaphyla **equals** *Tabebuia pallida* **(Lindl.) Miers. (Poirier):**

Large tree, trunk diameter may reach 90 cm; yields one of the best woods of the colony.

It is gray, hard with a fibrous texture; used for making canoes and boats, for construction under water and on soil; used for furniture and by wheelwrights.

Abundant from 0 to 350 m elevation.

Tecoma leuxcoxylon **equals** *Tabebuia leucoxyla* **DC. (Ébénier vert):**

Large tree yielding yellow sapwood and yellow-green heartwood.

A good wood for fine furniture. Almost extinct.

Catalpa Cougimima **equals** *Catalpa longissima* **(Jacq.) Sims. (Chêne noir d'Amérique):**

Tree from 12 to 15 m high, with a wood like that of oak, not attacked by worms.

Used for building ships, for carpentry and wheelwright's work.

Becoming rare. Temperate zone.

VERBENACEAE

Hitex divaricata **equals** *Vitex divaricata* **Sw. (Bois lézard):**

Average height tree, reaches a sizable diameter.

Excellent construction wood, joinery, and wheelwright's work.

Not abundant, from 0 to 500 m elevation.

Citarexylum quadrangulare **equals** *Citharexylum spinosum* **L. (Côtelette):**

Tree from 15 to 20 m high, yielding a wood for construction and joinery.

Found from 300 to 800 m elevation.

BORAGINACEAE

Cordia geraseauthus **equals** *C. alliodora* **(R. & P.) Oken (Bois de Chypre):**

Average-sized tree, yielding a wood for construction and joinery.

Density 0.700

Common. Temperate zone.

STRYACACEAE

Styrax glabnum **equals** *S. glaber* **Sw. (Laurier caraibe):**

Average-sized tree yielding a deep yellow wood, hard and indestructible.

Rare enough, from 250 to 650 m elevation.

Simplocos Martiniciensis **Jacq. (Coca rat): [now placed in Symplocaceae]**

Small tree yielding a wood that can be used for boards and stairs.

Density 0.767

Found from 250 to 650 m elevation.

SAPOTACEAE

Chrysophyllum cainito **L. (Kainitier):**

Average-sized tree yielding a wood used in joinery.

Common in the temperate zone.

Chrysophyllum glabrum **equals** *C. argenteum* **Jacq. (Bois de bouis):**

Small tree from 8 to 12 m high, yielding an excellent construction wood.

Abundant from 200 to 700 m elevation.

Achras sapota **L. (Sapotillier):**

Large tree yielding a reddish, hard, indestructible wood, good for joinery, wheelwright's work, and for any construction in water or on land.

Density: 1.108

Common in the temperate zone.

Mimusops riedleana **equals** *Manilkara bidentata* **A. Chev. (Balata):**

Average-sized tree yielding good construction wood.

From 300 to 600 m elevation.

Found in the south of the island.

Mimusops balata **equals** *Manilkara bidentata* **A. Chev. (Bois balata):**

Very large tree with a straight trunk, yielding one of the best construction woods.

It is hard, elastic, with a reddish brown color. It lasts indefinitely in water and earth.

Abundant up to 700 m elevation.

Guapeba semecarpifolia **equals** *Pouteria semicarpifolia* **Pierre (Bois contrevent):**

Large beautiful tree with a lofty trunk. Wood highly prized for construction. Rare.

Found from 300 m to 800 m elevation.

POLYGONACEAE

Coccoloba uvifera **L. (Raisin bord de mer):**

Medium-sized tree yielding a hard, indestructible wood. Used for construction and for wheelwright's work.

Temperate zone. Found from 0 to 100 m elevation.

Coccoloba pubescens **L. (Raisin grandes feuilles):**

Medium-sized tree yielding a hard, indestructible wood. Used for construction and for wheelwright's work.

Temperate zone. Found from 0 to 300 m elevation.

Coccoloba barbadensis **equals** *C. swartzii* **Meissn. (Bois rouge):**

Elegant tree, large leaves. Hard, red wood, easily worked; carpentry and construction of interiors.

Abundant from 200 to 600 m elevation.

Malpighia urens **L. (Bois capitaine, biere d'amour): [now placed in Malpigheaceae]**

Small tree, hard wood, used for stakes and furniture.

Temperate zone. Found from 0 to 300 m elevation.

MELIACEAE

Trichilia simplicifolia **Spreng. (Bois de l'Anglais):**

Small tree, not more than 5 m high yielding an excellent, hard wood used in construction and in joinery.

Abundant from 300 to 900 m elevation.

Guarera ramiflera **equals** *Gaurea ramiflora* **Vent. (Bois pistolet):**

Small tree yielding a hard wood, highly prized for construction work.

Abundant from 400 to 900 m elevation.

Guarera perottebu **equals** *Guarea perrottetiana* **A. Juss. (Bois coco):**

Average-sized tree yielding a hard wood. Good for construction.

Abundant from 300 to 900 m elevation.

Cedrela odorata **equals** *C. mexicana* **M. J. Roem. (Acajou):**

Large tree with a straight trunk. Wood hard, reddish with lovely shades; highly prized for making furniture. Tree beginning to disappear.

Temperate zone. Found from 0 to 350 m elevation.

From this family, three introduced species are cultivated. These are of great importance and merit large-scale plantings:

1. *Kaya senegalensis* equals *Khaya senegalensis* (Desr.) A. Juss. (Mohogani de Senegal, acajou de Senegal): Very large tree, rare;
2. *Swietenia mahagoni* L. (Acajou de Saint-Domingue): Very large tree, rare;
3. Carapa guyanensis Aubl. (Carapate, bois rouge): Large and beautiful tree, rare.

URTICACEAE

Artocarpus incisa **equals** *A. altilis* **(Park.) Fosberg (Fruit à pain): [now placed in Moraceae]**

Trees of second height class. Indestructible wood, good for joinery.

Abundant from 0 to 600 m elevation.

Artocarpus seminifera **equals** *A. altilis* **(Park.) Fosberg (Châtaignier): [now placed in Moraceae]**

Trees of second height class. Indestructible wood, good for joinery.

Abundant from 0 to 600 m elevation.

COMBRETACEAE

Terminalia catappa **L. (Amandier):**

Tree from 25 to 30 m high, introduced from India. Wood is white or gray brown; good for carpentry and joinery.

Density: 0.751. Temperate zone.

MYRTACEAE

Eugenia octopleura **Krug & Urb. (Goyavier bâtard, grand bois):**

Tree from 10 to 16 m high. Hard, red wood, good for construction.

Abundant from 600 to 1,000 m elevation.

CLUSIACEAE

Calophyllum calaba **L. (Galba):**

Large tree with a trunk reaching to 1 m in diameter. One of the best construction woods.

Found from 0 to 600 elevation.

Mammea americana **(Abricotier):**

Very lofty tree. Red wood, easily worked; good for construction.

Found from 10 to 500 m elevation.

Mammea marcophylla = ? Garcinia humilis **(Vahl) Adams (Abricot bâtard):**

Small tree from 6 to 7 m high. Wood used in carpentry and joinery.

Found from 10 to 400 m elevation.

Tovomita plumieri **Griseb. (Palétuvier grand bois):**

Tree, below average height, yielding an indestructible construction wood.

Abundant from 100 to 500 m elevation.

RUBIACEAE

Genipa americana **L. (Genipa):**

Large tree with very straight trunk. Good wood for joinery and furniture.

Density: 0.730. Common in temperate zone.

Chimarrhis cymosa **Jacq. (Bois de rivière):**

Tree from 12 to 25 m high. Wood used for carpentry and joinery.

Density: 0.890. Abundant from 50 to 600 m elevation in the northern part of the island.

AMENTACEAE

Casuarina equisefolia **equals** *C. equisetifolia* **L. (Filao): [now placed in Casuarinaceae]**

Large tree, originally from the Indian Ocean, giving a whitish wood, heartwood reddish. Very hard, indestructible in the interior of buildings.

Rare. Found from 0 to 600 m elevation

Index

(Plant names in the tables and the appendixes have not been indexed)

Martinique Revisited was composed into type on a Linotron 202 digital phototypesetter in ten point Meridien with two points of spacing between the lines. Meridien was also selected for display. The book was designed by Cameron Poulter, composed by G & S Typesetters, Inc., printed offset by Thomson-Shore Inc., and bound by John H. Dekker & Sons. The paper on which this book is printed bears acid-free characteristics for an effective life of at least three hundred years.

TEXAS A&M UNIVERSITY PRESS : COLLEGE STATION